INTERNATIONAL SERIES OF MONOGRAPHS IN
EARTH SCIENCES
Editor: DEAN EARL INGERSON, Dept. of Geology, University of Texas, Austin 12, Texas

VOLUME 32

ADVANCES IN ORGANIC GEOCHEMISTRY, 1966

ADVANCES IN ORGANIC GEOCHEMISTRY

Proceedings of the Third International Congress

Edited by G. D. HOBSON

and G. C. SPEERS

THE QUEEN'S AWARD TO INDUSTRY 1966

PERGAMON PRESS

OXFORD · LONDON · EDINBURGH · NEW YORK
TORONTO · SYDNEY · PARIS · BRAUNSCHWEIG

Pergamon Press Ltd., Headington Hill Hall, Oxford
4 & 5 Fitzroy Square, London W.1
Pergamon Press (Scotland) Ltd., 2 & 3 Teviot Place, Edinburgh 1
Pergamon Press Inc., Maxwell House, Fairview Park, Elmsford, New York 10523
Pergamon of Canada Ltd., 207 Queen's Quay West, Toronto 1
Pergamon Press (Aust.) Pty. Ltd., 19a Boundary Street, Rushcutters Bay, N.S.W. 2011, Australia
Pergamon Press S.A.R.L., 24 rue des Écoles, Paris 5e
Vieweg & Sohn GmbH, Burgplatz 1, Braunschweig

First edition 1970
Library of Congress Catalog Card No. 64–9310

PRINTED IN HUNGARY
08 012758 4

CONTENTS

PREFACE

THE running of an international congress is a matter of team-work, with individuals and organizations playing different but essential parts in order to ensure its success. This team-work existed for the First International Congress on Organic Geochemistry, in Milan in 1962, for the Second International Congress on Organic Geochemistry, at Rueil-Malmaison (near Paris) in 1964, and was equally in evidence for the Third International Congress, held at Imperial College of Science and Technology, London, S.W.7, on 26–28 September 1966.

The present volume is based on the papers presented in London.

As on the previous two occasions the congress was organized by the European Branch of the Organic Geochemistry Group of the Geochemical Society, the executive committee of the European Branch being: G. D. Hobson (president), E. Eisma (vice president), H. R. von Gaertner, J. Jedwab, G. Long and G. Kulbicki (members).

The executive committee of the European Branch is grateful for financial assistance from the following bodies: British Petroleum Co. Ltd., Burmah Oil Co. Ltd., Esso Petroleum Co. Ltd., Iraq Petroleum Co. Ltd., Mobil Oil Co. Ltd., Shell International Petroleum Co. Ltd., and The Royal Society, London. Its thanks are due to the authors for their most important efforts in preparing and presenting the papers which are here permanently recorded. Dr. D. G. Murchison of the University of Newcastle-upon-Tyne helped extensively with the preliminary and other arrangements. In addition a number of his associates at Newcastle, Mr. J. D. Cranfield and various other personnel at B. P. Research Centre at Sunbury-on-Thames, and Mr. S. E. Coomber and Mrs. A. E. Johns at the Royal School of Mines, London, gave valuable assistance in processing material which was required for the congress. Professor P. C. Sylvester-Bradley and Dr. G. Mueller were so good as to arrange post-congress visits for some participants to see samples of unusual asphaltic matter in the rock collection at Leicester University and also a number of field occurrences in the Carboniferous of Leicestershire and Derbyshire.

The committee also wishes to express its appreciation of Pergamon Press Ltd. having once more undertaken to publish the proceedings.

Finally, it was a pleasure to meet again many of those who had attended the previous congresses, to see new faces, and to become aware of the fact that activities within the realm of organic geochemistry were becoming of interest to workers in fields which are not obviously connected with it.

G. D. HOBSON
G. C. SPEERS

CARBON ISOTOPE STUDY ON METHANE FROM GERMAN COAL DEPOSITS

by

U. COLOMBO,* F. GAZZARRINI,* R. GONFIANTINI,† G. KNEUPER,‡ M. TEICHMÜLLER§ and R. TEICHMÜLLER§

Introduction

It is known that methane (and, in smaller amounts, higher hydrocarbons) is liberated during the coalification process. The amounts of methane developed vary in the different stages of coalification. Calculations on the basis of ultimate analyses of coals of different rank done by Mott (1943), Patteisky (1952, 1964) and Jüntgen and Karweil (1966) lead to the conclusion that greater amounts of methane start to be developed when the stage of bituminous coals with about 30% of volatile matter (dry and ash-free) is reached, whereas the gas developed from lignite and less metamorphosed bituminous coals is mainly carbon dioxide.

Figure 1, from a paper by Patteisky and Teichmüller (1960), shows the trend of the H/C ratio against the O/C ratio during coalification. The hydrogen content decreases sharply when the carbon content reaches 87% (corresponding to about 29% volatile matter). The formation of methane, which is supposed to proceed parallel to the decreasing amount of hydrogen, becomes very important in the stage of coking coals, and continues up to the stage of anthracites, where more than one-third of the total amount of methane is formed (Fig. 2). The total amount of methane liberated during the coalification process of bituminous coals and anthracites is estimated to be of the order of 200 litres per kilogram of coal. Jüntgen and Karweil (1966) have calculated that only a small fraction of the total

* Istituto di Ricerche "G. Donegani", Soc. Montecatini–Edison, Novara (Italy).
† Laboratorio di Geologia Nucleare, Pisa (Italy).
‡ Saarbergwerke Aktiengesellschaft, Saarbrücken (Germany).
§ Geologisches Landesamt Nordrhein – Westfalen, Krefeld (Germany).

methane formed during the coalification process can be retained by the coal.

The portion of the methane which is not retained by the coal migrates away from the coal seam towards the surface. If suitable reservoir space is available in the sediments overlying the coal seams, methane may accumulate in deposits of commercial importance.

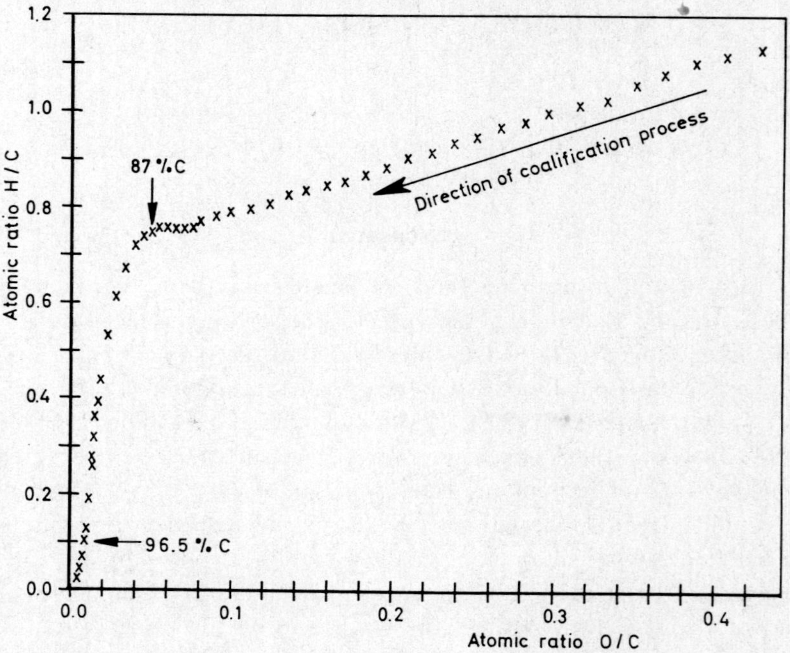

FIGURE 1. – Variation of hydrogen and oxygen contents in coals during the coalification process. Note that decrease of hydrogen is particularly strong in coals in going from 87% C to 96·5% C (after Patteisky and Teichmüller, 1960).

Both chemical and petrographic methods can be used to determine the degree of metamorphism of coals. Among chemical methods, the most common are based on the determination of volatile matter, moisture content and elementary composition of vitrites.

An attempt to use the carbon isotopic composition as an additional tool for studying the coalification process was made by Wickman (1953) on coal samples. He found that metamorphism has no traceable influence on the carbon isotopic composition of coal.

It may seem strange that no isotopic differences are observed in coals of different ranks, although the formation of methane is responsible for the loss of a fraction as high as 14–15‰ of the total carbon during the metamorphic process. The apparent lack of isotopic fractionation contrasts with the fact that the coalification process takes place in a tempera-

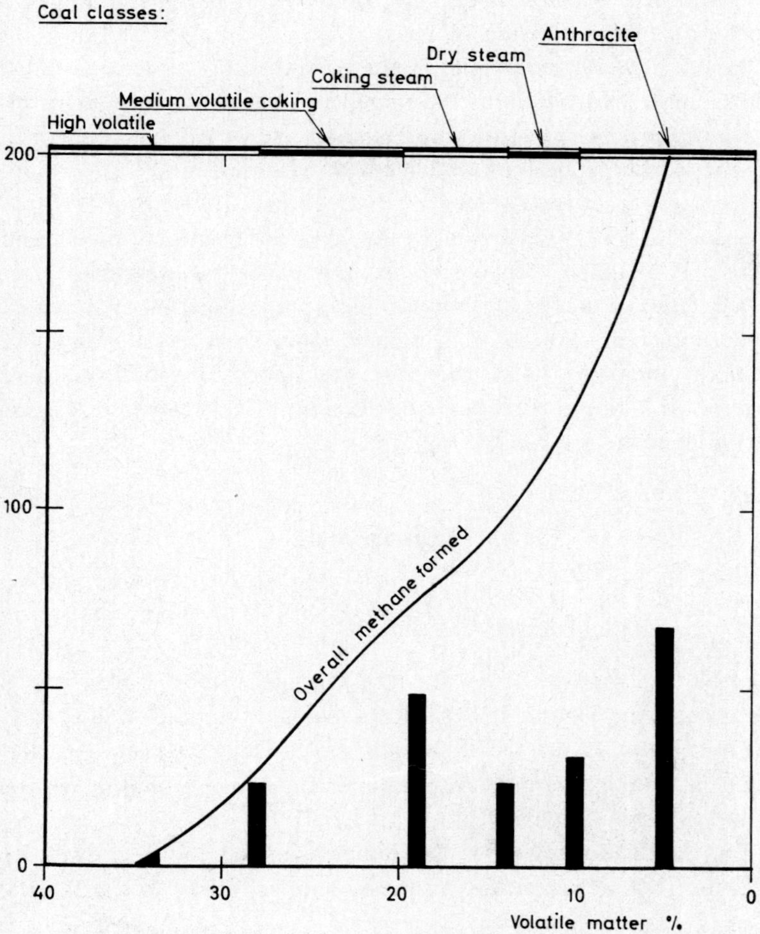

FIGURE 2. – The formation of methane in the various stages of coalification.

ture range at which one would expect isotopic fractionation to occur: bituminous coals and anthracites, in fact are normally coalified at temperatures between 100 and 200° C.

However, it must be noted that only a relatively few molecular groups in the coal structure actually contribute to the generation of methane and of its homologues. If such groups have about the same isotopic composition as the average coal, their total elimination as a result of metamorphism will not lead to any isotopic fractionation in the coal itself, and even their partial elimination will lead to relatively small isotopic variations, which can be masked by the variations of the isotopic composition of the starting material.

It should, however, be possible to detect an isotopic fractionation effect on the volatile hydrocarbons developed in the subsequent stages of the coalification process: the isotopically lighter hydrocarbons should be formed in the earlier stages of coalification, because the C^{12}-C^{12} bonds are ruptured more frequently than the C^{12}-C^{13} bonds (Stevenson *et al.*, 1948; Brodskii *et al.*, 1959). Obviously the greatest fractionation should be observed in methane, because of its low molecular weight.

In any case, a study of the isotopic composition of coal-associated hydrocarbons was thought to be timely, since there are no data of this kind in the literature. It is, therefore, interesting to establish where, in the overall δC^{13} range of methane of different origins, samples associated with coal deposits are placed.

Experimental

1. Description and processing of samples

Over 60 coal samples from Western Germany (mostly from the Saar and Lower Rhine Districts) were made available for this study.

In Figs. 3 and 4, the locations of samples from the Saar District are shown. The samples were taken from different mines, from the Westfalian C, F and Stefanian seams.

In Fig. 5, the locations of the mines sampled in the Lower Rhine District are shown. Practically all these samples belong to the Westfalian A.

Moreover, coal samples from deep boreholes in Western Germany have been studied. The samples ranged from high-volatile bituminous coals to anthracites. The depth at which the coals were taken ranged from 200 to 4000 metres below the ground surface.

In Fig. 6 the treatment of the coal samples is shown. A coal sample, as fresh as possible, was taken and divided in two portions. One portion was used to determine the rank of coalification. For this purpose, the vitrite was isolated and analyzed according to German standard methods.

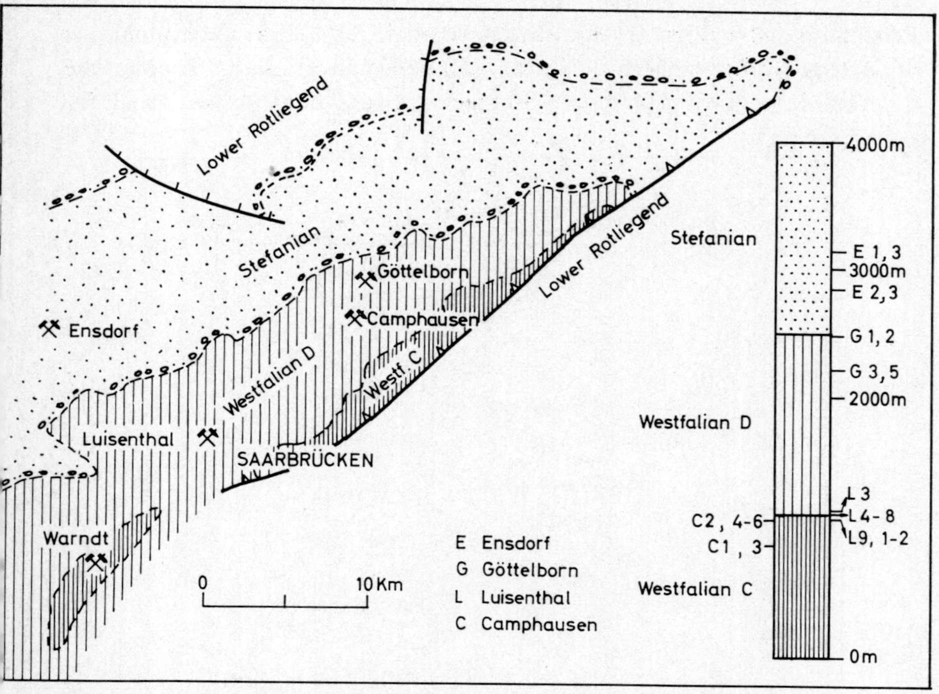

FIGURE 3. – Geological map of the Saar Basin showing the surface of the Carboniferous, the location of the mines, and the stratigraphic position of the different coal samples. The geological profile indicates the thickness of the different coal-bearing strata of the Upper Carboniferous.

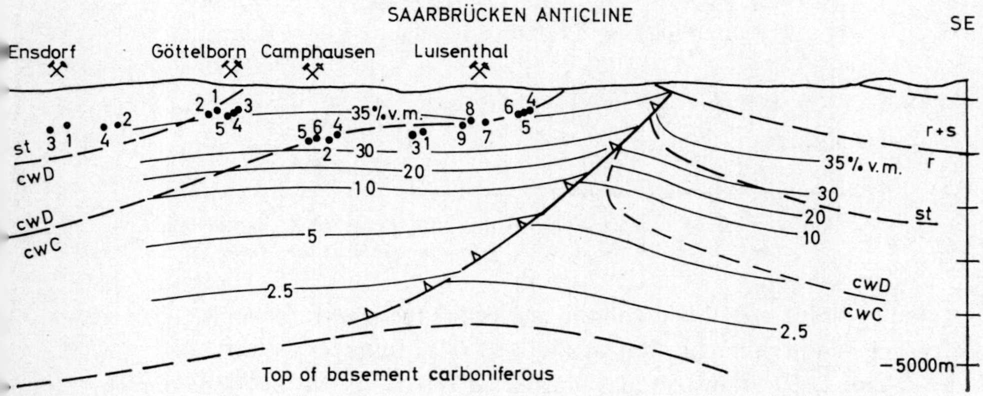

FIGURE 4. – Cross-section through the Saar Carboniferous showing the rank of coal (thin lines = isovols) and the geological location of the different mines and samples (projected on to the section). v.m., volatile matter; cwC, Westfalian C; cwD, Westfalian D; st, Stefanian; s, Lower Triassic; r, Permian.

Proximate and elementary analyses were made, as well as determinations of water-holding capacity, calorific and reflectance values. Among the different rank parameters, hygroscopic moisture content was used for

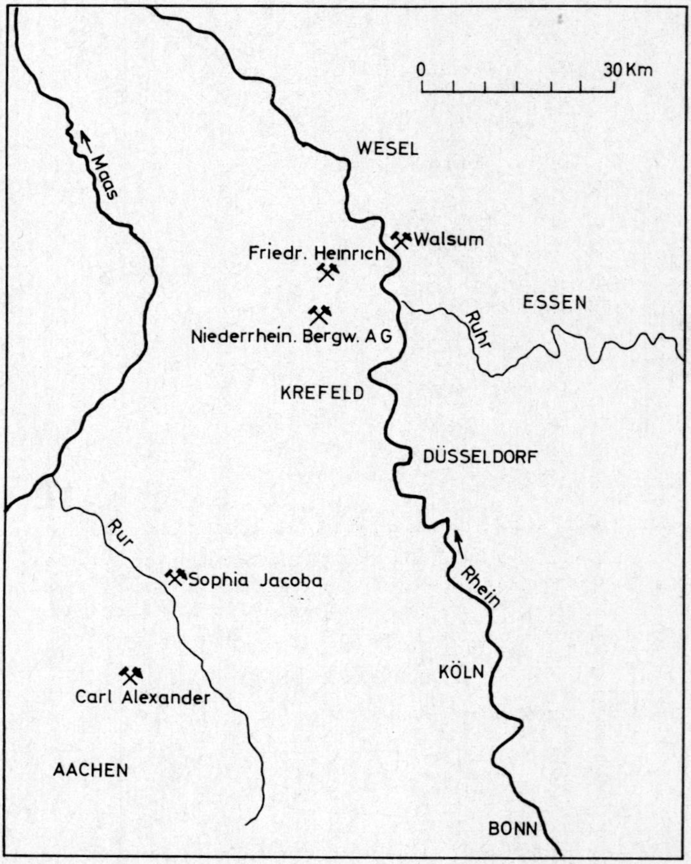

FIGURE 5. – Map of the Lower Rhine and Aachen Districts showing the location of the mines where the samples were taken.

the low-rank coals, and volatile matter for the high-rank coals. The treatment of the other portion of the coal is as follows:

About 100 grams of coal are placed into a special air-tight porcelain container, with several stainless steel balls suitable for a crushing operation. In the laboratory the sample is crushed for a few minutes at room temperature, until particles under 5 mm size are obtained. The gas evolved prior to, and during crushing is then collected (together with the air) directly from the container, using a Toepler pump. The gas collected in this operation is called "free gas". This definition is not scientifically correct, since it is practically impossible to avoid a partial desorption of the sorbed gas. In any case, only a small fraction of the coal hydrocarbons is present in this "free gas". The container is then opened, the

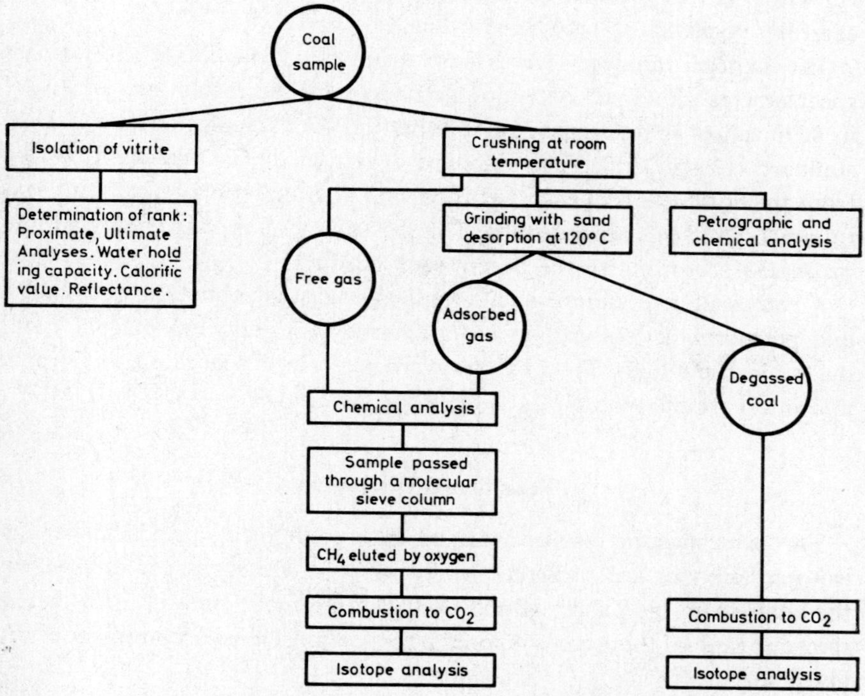

FIGURE 6. – Scheme of sample processing and analysis.

coal sample is divided in two aliquots, one to be used for chemical and petrographic analysis, the other to be processed for collecting the "adsorbed" gas. According to the Gedenk (1963) method, the coal sample (20 grams) is weighed and put back into the same porcelain container, together with 200 grams of quartzitic sand. The air is pumped off and the sample is ground under vacuum for 4 hours. The container is placed in an oven and kept for 30 minutes at 120 °C. The container is then evacuated using a Toepler pump, and the gas desorbed from the sample is collected.

The volumes of both "free" and "adsorbed" gas are measured and the gases are submitted to chromatographic analysis, using a Perkin-Elmer

154 B Vapor Fractometer. In order to determine the carbon isotopic composition of the methane from these gases, the procedure developed by Colombo *et al.* (1965, 1966) was adopted.

Carbon isotopic analysis of several coals of different ranks was also performed. Before running the combustion, such coal samples were treated with HCl to remove the small amounts of carbonates present, and carefully degassed by prolonged evacuation at 120° C.

The isotopic analyses were made using an Atlas M-86, 180° mass spectrometer, equipped with double collector. The results are expressed in δC^{13} values as differences per mil between the samples and the P.D.B. standard (Craig, 1957). The standard deviation of δC^{13} values, resulting from the complete sequence of processing and analytical operations, was found to be of 0·3 per mil units. The gas samples were also analysed and processed according to the described procedure.

A representative aliquot of the crushed coal was submitted to chemical and petrographic analysis, in order to study the influence of coal facies on the isotopic data. The analyses were performed according to German and international methods.

Results and Discussion

The experimental results obtained in the chemical, petrographic and isotopic analyses are presented in Tables 1–4. Petrographic analyses of the coals from the Lower Rhine District were made only in a few cases, since the rank of these coals is so high that exinites have become practically invisible.

A summary of the main chemical, petrographic and isotopic data on coals and related gases is presented in Tables 5 and 6.

The coals exhibit a wide range of metamorphism, as can be seen from the volatile matter and the hygroscopic water contents of the vitrites. The volatile matter ranges from 2·5% in a sample from a West German borehole to 39·8% in coals from the Saar mines.

The carbon isotopic compositions of the degassed coals are all remarkably similar to each other (see Table 1). This confirms the conclusion by Wickman (1953) that metamorphism does not affect the isotopic composition of coals. However, it should be remarked that our coals present isotopic ratios that, when expressed as δ values, are more negative, by about 5 per mil, with respect to the data calculated from Wickman's measurements. Since our data are believed to be quite correct on the basis of the P.D.B. standard, it seems possible that the difference between these results and those found by Wickman is due to a discrepancy between the two standards.

The most important components formed in the coal gases studied are: methane, ethane and higher hydrocarbons, carbon dioxide and nitrogen. The average higher hydrocarbons content of the coal gases is generally rather low compared to that of methane (see Tables 3, 4 and 6), except for a few samples from the Saar district, which exhibit ethane/methane ratios higher than 0·5†. The content of nitrogen in coal gases is widely variable, ranging from 7·66 ml/g to practically 0 (calculated after elimination of the air entrapped by the samples, on the basis of oxygen content). It would be interesting to know how much of the nitrogen found in the samples originated in the coalification process. Gedenk (1963) proposed that, at least in part, the nitrogen calculated as described above is of atmospheric origin, since some of the atmospheric oxygen can disappear by reaction with unsaturated coal components after being adsorbed by the coal. Preliminary experiments in our laboratories seem to confirm Gedenk's finding; nevertheless any correction for the atmospheric contribution to nitrogen is impossible on the basis of oxygen, because coal samples of different rank have different reactivity in respect to oxygen. We have evaluated, in those samples which were associated with large enough volumes of gas, the atmospheric nitrogen content, on the basis of the concentration of argon in the gases. It appears that at least an important portion of the nitrogen actually originated during the coalification process. These findings confirm the observations of Karweil (1955, p. 4) and of Jüntgen and Karweil (1966, p. 258, Fig. 8).

The carbon isotopic composition of the methane exhibits a wide range of δC^{13} (from —70 to —16). Methane from coals taken from the same seam at the top and at the base of each layer presents a similar isotopic composition.

It should be noted that the "free" methane is, as a rule, isotopically lighter than the corresponding "adsorbed" methane: the δC^{13} difference is, on the average, about 5 per mil (see Table 4). This value of 5δ units may roughly be assumed to be the average fractionation factor for the desorption process, the "free" gas being a much smaller volume than the associated "adsorbed" gas. This small isotopic fractionation is attributed to

† It should be noted that the Saar coals commonly are richer in aliphatic compounds (bitumen, resin-wax-component) than Ruhr coals of the same rank (cf. Kröger and Klatt, 1962; Kröger and Hortig, 1966).

TABLE 1. — VITRITE ANALYSIS AND ISOTOPIC COMPOSITION OF COALS

Sample number	Mine	Depth (below sea level) m	Distance from surface of Carboniferous, m	Geological age	Vitrite analysis: Dry and ash free C %	Dry and ash free H %	Dry and ash free v.m. %	Ash free H_2O max %	Ash free H_2O hygr. %	Ash free Combustion heat (Kcal/Kg)	Mean reflectance in oil %	Desorbed coal Ash %	Desorbed coal δC[13] ‰
1. Saar District													
E 1	Ensdorf	356	540	Stefanian	81·29	5·27	38·2	6·4	5·3	7583	0·78	7·4	
E 2	Ensdorf	360	570	Stefanian	81·63	5·47	39·8	4·7	3·8	7801	0·79	7·5	
E 3	Ensdorf	405	605	Stefanian	81·71	5·35	38·7	5·7	5·0	7568	0·78	6·4	
E 4	Ensdorf	410	610	Stefanian	80·98	5·46	37·6	5·0	4·2	7728	0·82	7·9	
G 1	Göttelborn	63	410	Westfalian D	81·14	5·10	34·7	5·1	4·7	7574	0·84	3·7	
G 2	Göttelborn	80	425	Westfalian D	81·63	5·29	34·5	5·1	4·5	7616	0·91	7·6	
G 3	Göttelborn	100	440	Westfalian D	83·07	5·05	34·0	5·1	3·2	7775	0·90	7·1	
G 4	Göttelborn	110	460	Westfalian D	82·65	5·31	35·0	4·9	4·4	7736	0·81	3·3	
G 5	Göttelborn	150	490	Westfalian D	82·92	5·39	34·4	4·9	4·3	7722	0·80	3·5	
L 1	Luisenthal	10	83	Westfalian C	85·66	5·21	33·3	n.d.	2·0	8144	0·97	n.d.	−24·4
L 2	Luisenthal	128	201	Westfalian C	n.d.	n.d.	n.d.	n.d.	n.d.	n.d.	n.d.	n.d.	−24·7
L 3	Luisenthal	150	391	Westfalian C	n.d.	n.d.	n.d.	n.d.	n.d.	n.d.	0·81	13·4	−24·9
L 4	Luisenthal	162	401	Westfalian C	84·33	5·56	37·6	3·1	2·9	8177	0·89	2·9	−25·2
L 5	Luisenthal	165	406	Westfalian C	84·16	5·44	35·8	3·4	3·0	8096	0·90	2·0	−26·0
L 6	Luisenthal	165	405	Westfalian C	84·33	5·62	37·2	3·2	2·1	8174	0·91	3·8	−25·1
L 9	Luisenthal	402	649	Westfalian C	84·54	5·57	34·1	3·1	2·5	8149	0·94	3·0	−25·5
L 10	Luisenthal	714	595	Westfalian C	n.d.	n.d.	n.d.	n.d.	n.d.	n.d.	n.d.	n.d.	−23·6
C 1	Camphausen	527	797	Westfalian C	88·12	5·62	31·6	1·4	1·0	8600	1·19	4·2	
C 2	Camphausen	533	815	Westfalian C	87·14	5·45	32·1	1·7	1·4	8493	1·11	5·8	

C 3	Camphausen	544	814	Westfalian C	87·72	5·46	31·1	1·4	1·1	8602	1·20	3·5	
C 4	Camphausen	585	875	Westfalian C	87·30	5·58	33·6	1·5	1·2	8552	1·08	15·1	
C 5	Camphausen	701	1031	Westfalian C	87·65	5·49	32·0	1·5	0·9	8533	1·14	5·4	
C 6	Camphausen	675	1005	Westfalian C	87·67	5·31	32·4	n.d.	0·7	8542	1·18	7·5	
	2. Lower Rhine and Aachen District												
F 1	Friedrich Heinrich	384	175	Upper Westf. A	89·00	5·02	22·8	1·5	1·1	8603	1·49	2·4	−24·5
F 2	Friedrich Heinrich	521	270	Upper Westf. A	89·64	5·17	22·7	1·4	1·2	8628	1·46	3·3	
F 3	Friedrich Heinrich	530	240	Lower Westf. LA	90·51	4·69	18·2	1·5	1·1	8555	1·75	5·5	−23·6
F 4	Friedrich Heinrich	544	325	Lower Westf. LA	90·82	4·40	15·9	1·5	0·9	8501	1·96	1·8	
F 5	Friedrich Heinrich	580	230	Upper Westf. LA	88·91	5·26	26·4	1·5	1·1	8588	1·30	6·4	
F 6	Friedrich Heinrich	580	230	Upper Westf. LA	88·38	5·12	26·9	1·8	1·3	8508	1·35	7·2	
F 7	Friedrich Heinrich	581	355	Lower WA	90·82	4·40	15·9	1·5	0·9	8501	1·96	3·5	
C.A. 1	Carl Alexander	420	125	Westf. LB	88·56	5·08	24·0	1·7	1·1	8549	1·44	3·4	−23·4
C.A. 2	Carl Alexander	453	353	Upper Westf. A	89·90	4·67	18·5	1·5	1·1	8558	1·73	4·0	
C.A. 3	Carl Alexander	505	455	Upper Westf. A	90·43	4·64	17·0	1·4	1·0	8608	1·79	9·5	
C.A. 5	Carl Alexander	562	285	Upper	89·54	4·93	17·3	1·6	1·0	8622	1·75	1·1	
C.A. 4	Carl Alexander	601	322	Westf. A	89·85	4·88	17·5	1·4	0·8	8602	1·75	10·6	−24·0
N 1	Niederrhein L Bergwerks A.G.	240	40	Lower Westf. A	91·77	4·04	10·2	1·7	1·1	8503	2·27	8·0	−24·1
N 2	Niederrhein L Bergwerks A.G.	330	130	Lower Westf. A	92·14	3·77	7·6	2·7	1·5	8413	2·44	9·5	

TABLE 1 (*cont.*)

Sample number	Mine	Depth (below sea level) m	Distance from surface of Carboniferous, m	Geological age	Vitrite analysis: Dry and ash free C %	Dry and ash free H %	Dry and ash free v.m. %	Ash free H_2O max %	Ash free H_2O hygr. %	Ash free combustion heat (Kcal/Kg)	Mean reflectance in oil %	Desorbed coal Ash %	Desorbed coal δC^{13} ‰
N 3	Niederrhein Bergwerks A.G.	353	133	Lower Westf. A	90·80	4·20	12·6	1·7	1·3	8565	2·10	4·7	
N 4	Niederrhein Bergwerks A.G.	370	170	Lower Westf. A	90·62	4·27	11·9	1·6	1·1	8614	2·05	2·8	
N 5	Niederrhein Bergwerks A.G.	387	170	Lower Westf. A	91·85	3·76	6·7	2·5	1·5	8469	2·60	3·5	
N 6	Niederrhein Bergwerks A.G.	468	250	Lower Westf. A	92·73	3·73	6·1	2·1	1·3	8514	2·71	1·5	−23·7
S 1	Sophia Jacoba	320	90	Upper Westf. A	92·59	3·63	5·2	2·5	1·2	8464	2·90	2·2	
S 2	Sophia Jacoba	390	160	Upper Westf. A	92·51	3·60	4·9	2·4	1·3	8435	2·94	2·7	
				3. West German Borehole									
A 1	Borehole A	2396		Westf. C	92·12	3·95	8·6	n.d.	1·3	n.d.	2·04	4·6	−25·3
A 2	Borehole A	2954		Westf. C	n.d.	n.d.	n.d.	n.d.	n.d.	n.d.	2·83	12·1	−26·0
A 3	Borehole A	3114		Westf. C	94·64	3·13	4·2	2·7	1·0	8506	3·55	43·1	−25·0
A 4	Borehole A	3446		Westf. C	94·03	2·61	3·8	4·6	1·5	8408	4·27	32·8	−25·0
A 5	Borehole A	3882		Westf. C	94·26	2·01	2·5	5·8	1·7	n.d.	4·94	n.d.	−23·8

TABLE 2. – CHEMICAL AND PETROGRAPHIC COMPOSITION OF DESORBED COAL SAMPLES FROM WEST GERMANY

Sample number	Chemical analysis: Proximate: H_2O % a.f.	Ash %	Volatile matter % d.a.f.	Ultimate (dry ash free): C %	H %	O %	N %	S %	Petrographic analysis (vol. %): Vitrinite	Inertinite	Exinite	Micrinite (fine)	Pyrite	Carbonates	Clay	Total minerals
E 1	4·1	7·4	39·6	80·21	5·38	11·19	1·49	1·73	72·6	11·4	12·9	1·1	0·3	1·4	0·3	2·0
E 2	3·4	7·5	39·1	81·18	5·48	10·30	1·46	1·58	77·4	12·1	6·7	0·2	0·8	1·6	1·2	3·6
E 3	4·5	6·4	39·7	80·74	5·13	10·70	1·52	1·91	75·3	11·4	7·7	2·5	0·7	2·4	—	3·1
E 4	3·1	7·9	40·2	79·70	5·50	9·18	1·62	4·00	80·3	6·4	6·1	0·5	4·1	1·4	1·2	6·7
G 1	3·7	3·7	34·7	80·77	4·81	11·95	0·85	1·62	82·4	7·7	6·6	0·8	0·3	1·0	1·2	2·5
G 2	3·5	7·6	41·0	80·29	4·89	11·50	0·91	2·41	78·6	10·6	8·3	—	1·1	1·2	0·2	2·5
G 3	3·0	7·1	41·0	80·88	5·37	11·00	0·94	1·81	72·2	5·0	13·8	1·4	2·9	2·0	2·7	7·6
G 4	3·3	3·3	37·0	81·84	5·17	10·95	0·89	1·15	76·1	7·4	15·3	0·5	—	0·5	0·2	0·7
G 5	3·2	3·5	37·4	82·07	5·31	10·89	0·87	0·86	78·0	7·4	12·5	0·6	—	1·3	0·2	1·5
L 2	1·4	29·3	38·3	82·45	5·40	11·45		0·70	35·6	20·8	18·3	0·7	0·1	0·2	24·3	24.6
L 3	2·0	13·4	44·5	82·95	6·10	8·43	1·04	1·48	61·2	4·7	23·0	1·7	1·4	0·1	7·9	9·4
L 4	2·4	2·9	40·7	84·26	5·78	8·29	0·96	0·71	76·1	4·8	11·7	2·5	1·5	3·3	0·1	4·9
L 5	2·2	2·0	41·1	84·54	5·92	7·76	1·23	0·55	73·8	8·8	14·8	2·0	0·1	1·0	0·1	1·2
L 6	1·7	3·8	44·1	83·99	6·12	8·21	1·10	0·58	69·1	8·2	16·6	3·9	—	0·1	2·1	2·2
L 7	2·3	11·7	38·4	84·70	5·74	7·65	0·97	0·94	80·3	0·8	8·7	1·0	0·5	1·4	7·3	9·2
L 8	2·0	5·1	40·5	84·48	5·94	7·36	1·16	1·06	77·1	4·9	14·9	0·6	0·6	—	1·9	2·5
L 9	2·5	3·0	35·6	85·05	5·55	7·68	0·96	0·76	84·7	3·9	7·5	1·2	0·8	1·6	0·3	2·7

Sample number	Chemical analysis								Petrograpic analysis (vol. %)							
	Proximate			Ultimate (dry ash free)												
	H_2O % a.f.	Ash %	Volatile matter % d.a.f.	C %	H %	O %	N %	S %	Vitri-nite	Inerti-nite	Exi-nite	Micri-nite (fine)	Pyrite	Carbo-nates	Clay	Total min-erals
C 1	0·9	4·2	32·2	87·07	5·55	5·64	1·09	0·65	84·4	5·6	4·8	2·5	—	1·1	1·6	2·7
C 2	1·1	5·8	36·5	87·55	5·78	4·86	1·06	0·75	71·3	4·9	15·6	1·4	0·1	0·4	6·3	6·8
C 3	0·8	3·5	32·9	88·05	5·47	4·88	1·06	0·54	75·5	9·7	8·6	4·5	—	1·5	0·2	1·7
C 4	0·9	15·1	36·9	85·89	5·79	6·49	1·00	0·83	57·2	6·7	11·4	1·0	0·6	8·7	14·4	23·7
C 5	0·6	5·4	34·0	87·93	5·31	4·84	0·97	0·95	73·4	10·5	9·6	3·7	0·3	2·2	0·3	2·8
C 6	0·6	7·5	35·6	86·53	5·31	6·50	0·95	0·71	79·0	2·9	10·8	5·0	—	2·1	0·2	2·3
A 1	1·3	4·6	8·6	92·12	3·95	1·85	1·17	0·91	88·4	10·3	—	—	—	0·3	1·0	1·3
A 2	1·8	12·1	5·4	92·00	2·99	3·15	0·80	1·06	75·5	14·2	—	—	0·5	—	9·8	10·3
A 3	1·1	43·1	7·3	89·51	3·22	6·26	0·07	0·94	53·4	9·1	—	—	Quartz	16·7	20·8	37·5
A 4	1·5	32·8	6·9	92·13	2·97	4·11	0·09	0·88	68·8	11·8	—	—	5.9/0·6		10·9	17·4

TABLE 3. — CHEMICAL AND ISOTOPIC ANALYSIS OF DESORBED GAS SAMPLES FROM COALS OF WESTERN GERMANY

Sample number	Adsorbed gas analysis					
	CH_4/coal (ml/gr)	Hydrocarbons composition vol. %			C_2H_6/CH_4 x 10^3	CH_4 δC^{13} ‰
		CH_4	C_2H_6	C_3^+		
E 1	0·80	80·46	17·25	2·29	214	—28·5
E 2	3·82	81·52	15·80	2·68	194	—33·1
E 3	0·60	46·91	41·70	11·39	889	—26·8
E 4	3·05	78·67	17·50	3·83	222	—33·4
G 1	1·30	99·34	0·66	—	6·7	—47·8
G 2	0·96	96·68	2·40	0·92	25	—44·9
G 3	0·99	96·53	2·11	1·36	22	—56·2
G 4	0·87	99·10	0·90	—	9·4	—46·9
G 5	1·34	97·96	1·73	0·31	18	—41·2
L 1	2·92	99·64	0·36	—	3·6	—43·1
L 2	4·95	99·36	0·64	—	6·5	—44·1
L 3	4·28	98·70	1·30	—	13	—40·2
L 4	0·76	95·00	4·58	0·42	48	—36·5
L 5	0·87	89·22	9·48	1·30	106	—36·5
L 6	2·08	75·06	24·38	0·56	325	—34·8
L 10	4·84	79·11	13·74	7·15	174	—36·1
C 1	2·57	56·97	30·50	12·53	535	—28·3
C 2	0·32	30·12	39·99	30·49	1308	—34·8
C 3	0·65	38·32	33·52	28·16	875	—25·2
C 4	1·26	70·03	24·48	5·49	350	—30·1
C 5	3·24	95·12	3·81	1·07	40	—38·2
C 6	1·18	79·25	10·83	9·92	137	—37·3
F 1	1·48	100·00	—	—	0·0	—55·2
F 2	1·47	98·61	1·39	—	14	—31·2
F 3	2·55	99·60	0·40	—	4·1	—41·7
F 4	4·86	99·27	0·73	—	7·3	—45·6
F 5	1·08	100·00	—	—	0·0	—69·7
F 6	0·36	100·00	—	—	0·0	—63·1
F 7	1·55	97·43	2·57	—	26	—46·1
C.A.1	1·46	99·81	0·19	—	1·9	—70·4
C.A.2	2·19	99·89	0·11	—	1·1	—53·3
C.A.3	1·57	99·86	0·14	—	1·4	—55·1
C.A.4	2·20	99·69	0·31	—	3·1	—43·4
C.A.5	1·61	99·61	0·39	—	3·9	—51·8
N 1	0·01	100·00	—	—	0·0	—49·8
N 2	1·43	99·81	0·19	—	1·9	—44·0
N 3	0·01	100·00	—	—	0·0	—59·7
N 4	1·25	100·00	—	—	0·0	—55·0
N 5	0·82	100·00	—	—	0·0	—60·6
N 6	0·82	100·00	—	—	0·0	—57·6

Sample number	Adsorbed gas analysis					
	CH_4/coal (ml/gr)	Hydrocarbons composition vol. %			C_2H_6/CH_4 x 10^3	CH_4 δC^{13} ‰
		CH_4	C_2H_6	C_3^+		
S 1	1·02	100·00	—	—	0·0	—45·0
S 2	1·94	99·79	0·21	—	2·1	—42·5
A 1	10·34	98·66	1·34	—	14	—16·8
A 2	6·39	99·93	0·07	—	0·7	—25·0
A 3	0·18	94·54	5·46	—	58	—24·4
A 4	1·85	100·00	—	—	0·0	—19·7
A 5	2·62	100·00	—	—	0·0	—23·8

TABLE 4. – HYDROCARBONS CONTENT AND ISOTOPIC ANALYSIS OF THE "FREE" GASES FROM WESTERN GERMANY

Sample number	CH_4/coal (ml/gr)	Chemical analysis (volume %)			δC^{13}‰		δ ads– δ free (*)
		CH_4	C_2H_6	C_3^+	"free"	"adsorbed"	
E 1	0·11	90·48	9·12	0·40	—28·5	—28·5	0
E 2	0·12	88·74	7·73	3·53	—39·0	—33·1	5·9
E 4	0·32	91·16	4·99	3·85	—36·2	—33·4	2·8
G 1	0·05	99·91	0·09	—	—54·3	—47·8	6·5
G 3	0·06	100·00	—	—	—60·4	—56·2	4·2
G 4	0·05	94·63	5·37	—	—48·2	—46·9	1·3
L 1	2·09	99·99	0·01	—	—47·1	—43·1	4·0
L 2	0·75	99·97	0·03	—	—49·8	—44·1	5·7
L 3	0·77	99·85	0·15	—	—45·4	—40·2	5·2
L 4	0·42	98·75	1·19	0·06	—41·6	—36·5	5·1
L 5	0·70	96·15	3·60	0·25	—42·7	—36·5	6·2
L 6	0·11	94·73	5·27	—	—41·3	—34·8	6·5
L 7	0·13	87·27	11·89	0·84	—53·3	—47·0	6·3
L 8	0·14	83·70	14·94	1·36	—51·5	—46·6	4·9
L 10	0·41	98·49	1·39	0·12	—41·7	—36·1	5·6
C 1	1·07	92·41	6·69	0·90	—34·1	—28·3	5·8
C 2	0·10	73·77	21·32	4·91	—44·8	—34·8	10·0
C 3	0·99	92·88	6·28	0·84	—34·0	—25·2	8·8
C 5	0·91	99·81	0·19	—	—47·5	—38·2	9·3
C 6	0·71	99·13	0·78	0·09	—45·6	—37·3	8·3
A 3	0·01	100·00	—	—	—27·2	—24·4	2·8
A 4	0·07	100·00	—	—	—22·6	—19·7	2·9
A 5	0·05	100·00	—	—	—23·3	—23·8	—0·5

* The average of the reported values is 5·11.

TABLE 5. – SOME CHEMICAL, PETROGRAPHIC AND ISOTOPIC DATA ON GERMAN COALS

	Saar District	Lower Rhine District	West German Borehole
Number of coal samples	23	20	5
Depth (m) b. sea l.	From 10 to 701	From 320 to 601	From 2400 to 3880
Volatile matter %	From 31·1 to 39·8	From 4·9 to 26·9	From 2·5 to 8·6
Hygroscopic water %	From 0·7 to 5·3	From 0·8 to 1·5	From 1·0 to 1·7
Vitrinite %	From 35·6 to 84·7	From 61·2 to 83·1	From 53·4 to 88·4
Exinite %	From 4·8 to 23·0	From 0 to 4·4	0
δC^{13} of coal	From —23·6 to —26·4	From —23·4 to —24·5	From —25·0 to —26·0

TABLE 6. – CHEMICAL AND ISOTOPIC DATA OF COAL GASES

		Saar District		Lower Rhine District	West Germany Borehole	
		Free gas	Adsorbed gas	Adsorbed gas	Free gas	Adsorbed gas
Gas volume (ml/g of coal)	Average	0·85	5·37	3·69	0·08	6·20
	Max.	1·62	13·08	11·65	0·13	13·29
	Min.	0·02	0·88	0·67	0·01	2·87
Methane (ml/g of coal)	Average	0·36	1·98	1·49	0·04	4·26
	Max.	1·06	4·95	4·87	0·07	10·34
	Min.	0·01	0·32	0·10	0·01	0·18
Higher Hydrocarbons (ml/g of coal)	Average	0·02	0·39	0·01	0	0·03
	Max.	0·09	1·94	0·04	0	0·14
	Min.	0	0	0	0	0
Nitrogen (ml/g of coal)	Average	0·44	2·79	2·02	0·04	1·91
	Max.	1·50	7·66	6·77	0·06	3·07
	Min.	0	0	0·20	0·01	0·35
δC^{13} of methane (per mil)	Average	—42·1	—37·5	—50·3	—24·4	—22·0
	Max.	—25·2	—25·2	—31·2	—22·6	—16·8
	Min.	—60·4	—56·2	—70·4	—27·2	—25·0

the preferential desorption of the $C^{12}H_4$ from the coals. In theory, even a small fractionation factor may lead to large isotopic enrichments, depending on the extent of the process. Therefore, one might believe that the wide isotopic range of coal methanes was merely due to the more

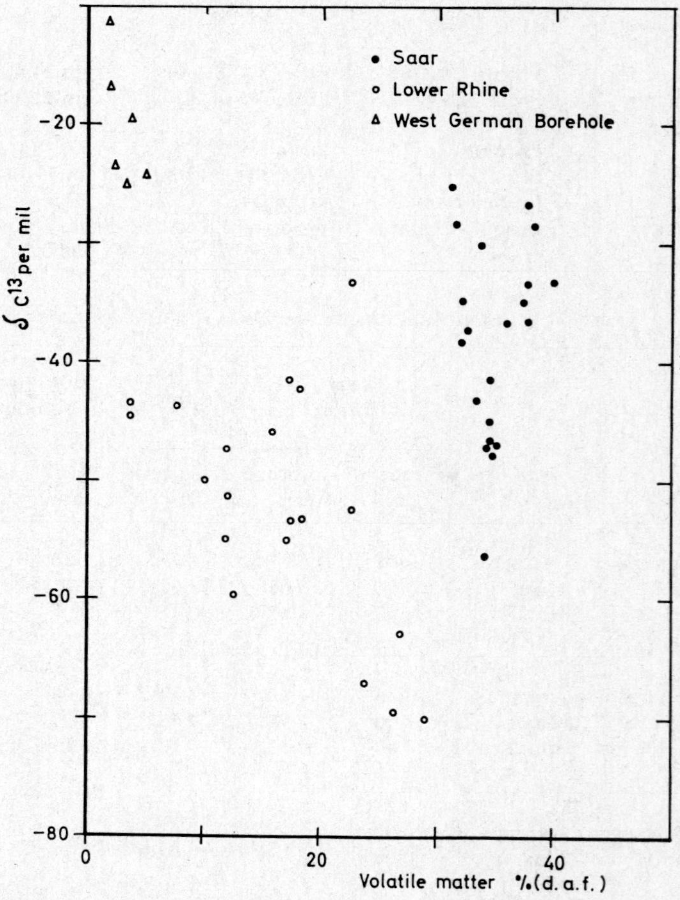

FIGURE 7. – Diagram showing the isotopic composition of coal methanes plotted against the volatile matter content of vitrites.

or less advanced stage of the desorption process within each sample. It should, however, be considered that the desorption rate depends in a great extent on the particle size of coals, as demonstrated by Peters and Warneeke (1933) and Schilling *et al.* (1966). Therefore, because our coal samples consisted of relatively large chunks (several centimetres in diameter), and were analysed shortly after sampling at the mine (no more than 20 days of lag), we believe that desorption after sampling produces only a secondary fractionation effect, and is not responsible for the wide isotopic range now reported for coal methane.

In Fig. 7 the carbon isotopic values of adsorbed methanes are plotted against the volatile matter content of vitrite. The relationship is not satisfactory, particularly for the low-rank coals from the Saar District. If the degree of coalification of the low-rank coals from the Saar is evaluated on the basis of hygroscopic moisture (which seems to be reasonable for coals of very low rank) the relationship seems to improve. In Table 7,

TABLE 7. – MEAN VALUES OF δC^{13} (ADSORBED METHANE), VOLATILE MATTER AND MOISTURE (VITRITE) AND DEPTH OF COAL SAMPLES TAKEN FROM SAAR MINES

Mine	Age	Number of samples	δC^{13}	Volatile matter d.a.f.(%)	Moisture a.f. (%)	Distance from surface of Carboniferous
Ensdorf	Stefanian	4	—30·4	38·6	4·2	581
Göttelborn	Westfalian D	5	—47·4	34·5	3·9	445
Luisenthal	Upper Westf. C	7	—38·8	35·6	2·2	391
Camphausen	Upper Westf. C	6	—32·3	32·1	0·9	890

the average values for the coal samples of the different mines studied in the Saar District are reported. The values for the moisture content refer to the "Damberger Mittelwert" (see Damberger *et al.*, 1964). Low moisture contents seem to correspond to less negative mean values of δC^{13}. But the Ensdorf samples contradict this.

Other relationships between chemical coalification parameters on one hand, and isotopic parameters on the other hand, are very weak too.

According to the data of Tables 2 and 3 no definite relationship between the δC^{13} value of methane and the petrographic and chemical composition of the related coal sample was found.

The lack of a satisfactory relationship between the coal rank or the coal facies on one hand, and the isotopic values of related methanes on the other, is in our opinion to be attributed to interference of physical processes, such as diffusion and effusion of coal gases in the course of migration. The fact that migration may actually play an important role in determining the isotopic composition of the various coal gases, is shown by the good relationship between the δC^{13} of methanes and the ethane/

methane ratios of the Saar coal gases (Fig. 8). In a study of the isotopic composition of Italian natural gases, it has been proposed by Colombo *et al.* (1965, 1966) that migration could cause an isotopic effect on methane, parallel to the fractionation of the different hydrocarbons. An interpretation of carbon isotopic data on coal gases based purely on genetic and

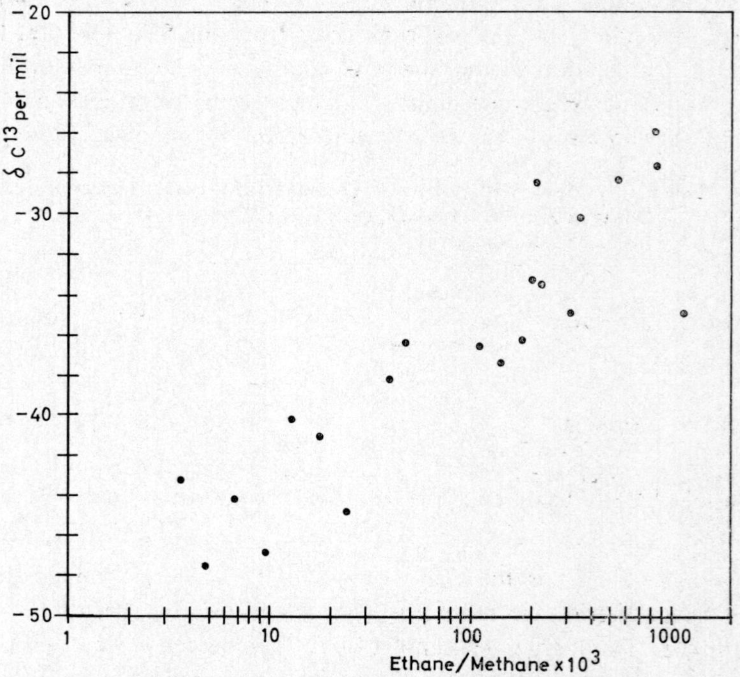

FIGURE 8. — Relationship between the C_2H_6/CH_4 ratio and the δC^{13} of coal methanes in the Saar Basin.

metamorphic factors might, therefore, not be possible. This is particularly true for relatively shallow coals, which are, as a rule, more exposed to desorption and migration effects. Therefore, both genetic and migrational factors should be taken into account in interpreting the chemical and isotopic data of the coal gases in a given area. As an example, the situation of the coal gases from the Saar District, from the surface of the Carboniferous down to 1000 metres of depth is reported schematically in Fig. 9. Down to 400 metres, the gases do not contain appreciable amount of higher hydrocarbons, then the amount increases gradually to reach high values at depths below 550 metres. The same trend is observed in the C^{13} content of the adsorbed methane. As autochthonous higher hydrocarbons will be retained in the coal longer than heavy methane, and as these higher hydrocarbons obviously have disappeared, it must be supposed that the seams in the desorption zone lost their autochthonous gases almost completely before a fractionated, light methane was readsorbed.

The geological history of the Saar coals has been described by Damberger *et al.* (1964). According to these authors the main coalification of the

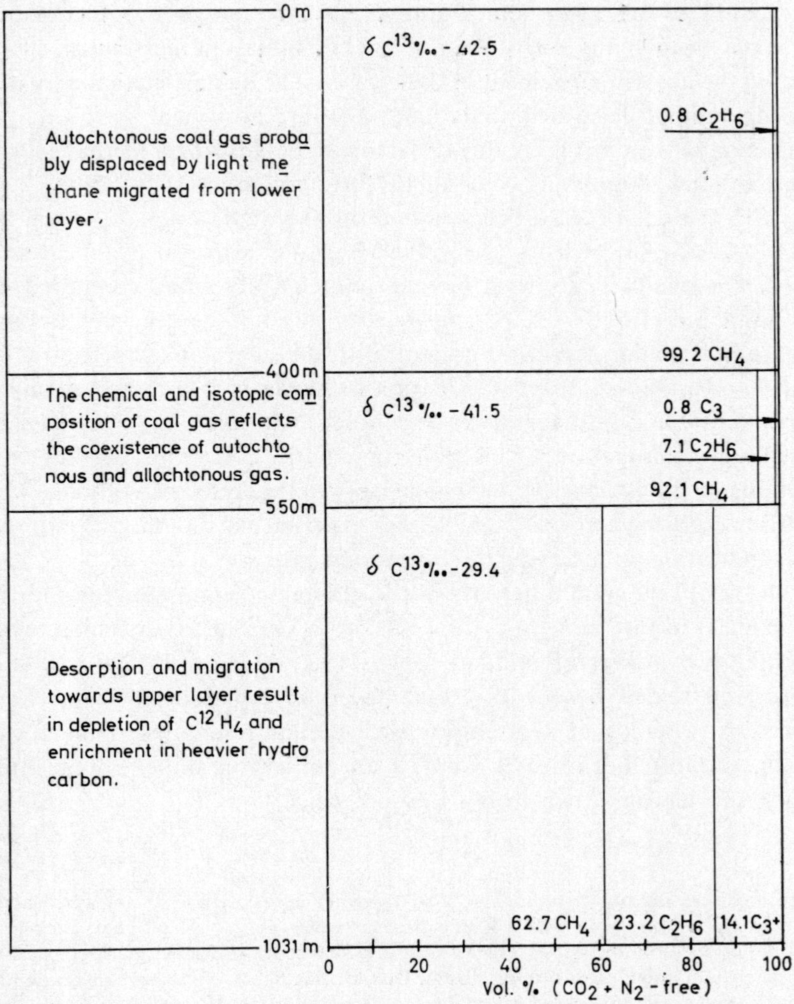

FIGURE 9. – The increase of the amounts of higher hydrocarbons, and $C^{13}H_4$ in adsorbed gas (N_2-, CO_2- free) against the depth below the surface of the Carboniferous in the Saar District.

Saar coals took place in Carboniferous times. After the folding and uplifting of the coal-bearing strata in Lower Permian times, the coals of zone I lay immediately below the surface of the Lower Permian ground surface. At this time desorption must have happened. Ascending coalification gases from deep seams could pass through the wide pore structure of the pressure-released uppermost seams. When later on the Carboniferous was covered by Triassic sediments, the degasified uppermost seams were sealed and could adsorb again allochthonous coalification gases, which ascended from deep-laying coals. An isotopic fractionation must have occurred during the upward movement of these gases. The lightest methanes reached the uppermost degasified coals first and were adsorbed.

A comparison between the C^{13} content of adsorbed methane in the high volatile bituminous coals of the primary zone (δC^{13}—29·4‰) and the C^{13} content of adsorbed methane in the anthracites of borehole A (δC^{13}—22·0‰) shows that the influence of the degree of coalification on the C^{13} content of adsorbed methane is relatively small compared with the influence of diffusion. But this relationship has to be checked by further studies, giving particular attention to the effect of temperature. It is well known that isotopic fractionation decreases with increasing temperatures. In some cases anthracites and other high rank coals may have suffered high temperatures, caused by magmatic heat. This can affect the isotopic composition of methane. Isotopic analyses of methane which is formed during artificial coalification experiments should give more precise indications concerning this relationship.

The relatively great difference of the gas composition between the coals from the Ensdorf mine and the coals from the Göttelborn mine, can be explained by different sealing conditions.* At Ensdorf the coal seams were well sealed by the claystone layers of the Buntsandstein (Lower Triassic), whereas at Göttelborn such sealing is missing. Thus, for instance, at Göttelborn heavy rainfalls are noticeable underground after a much shorter time than in the Ensdorf mine.

* In any case the almost complete lack of higher hydrocarbons in the coal samples from Göttelborn cannot be a result of a particularly low rank of coal, because the amounts of volatile matter and moisture in the coals of Ensdorf are considerably higher than in the coals of Göttelborn. On the other hand, the remarkably high nitrogen content in the adsorbed gas of the coals from Göttelborn and from Ensdorf may be attributed to the low rank of these coals (34 – 38% volatile matter): according to Jüntgen and Karweil (1966, p. 257) nitrogen must be enriched to a considerable degree in low-rank bituminous coals.

The conditions for the Lower Rhine and Aachen Districts are similar to those for the Saar District. The small amounts of higher hydrocarbons in the coals of the Lower Rhine mines down to a depth of at least 455 metres indicate considerable desorption. This has been pointed out by Hedemann (1963). The low $C^{13}H_4$ values correspond to the almost complete absence of higher hydrocarbons in these mines.

The mean values for the coal samples studied from the Lower Rhine and Aachen Districts are:

$$99{\cdot}7\%\ CH_4,\quad 0{\cdot}3\%\ C_2H_6,\quad 0\%\ C_3^+,\quad \delta C^{13}_{CH_4}\ -52{\cdot}0\ ‰$$

in the nitrogen- and carbon dioxide-free adsorbed gas.

According to Hedemann in the Lower Rhine mines the desorption zone reaches down to 700 metres below the surface of the Carboniferous, that is 200 metres deeper than at the Saar. It is interesting that in the Carboniferous of the Lower Rhine District, deep fissures have been found which are filled with Cretaceous sediments (Schaub, 1954). These fissures are 10–20 cm wide and reach as far down as some hundred metres below the surface of the Carboniferous. This shows that even during Cretaceous time the coal-bearing strata were uplifted to a considerable degree, so that open fissure could be formed. The particularly deep reaching degasification of the Carboniferous in the Lower Rhine area might be explained by an especially high and long time of uplift.

A detailed study of the local geology in the different mines led to the conclusion that the base of the desorption zone is not a horizotal plane. Its depth is strongly influenced by the permeability of the rocks. Where porous sediments are present, and where the rocks are strongly folded or broken up by tectonic faults, the desorption zone reaches deeper down than in areas where the layers are predominantly clayey, flat lying, and undisturbed.

Disregarding the pressure release caused by the mining activities, the strata which are today nearest to the surface of the Carboniferous have been released from pressure since their earlier subsidence to greater depths. This pressure release could have caused an opening of the structure of coals and their host rocks, as indicated by the relatively small seismic velocity of such strata.† With decreasing pressure the gases are desorbed from the coal and the gas permeability of the coal increases (cf. Jüntgen and Karweil, 1966, p. 340, and Lidine and Khodot, 1964). Thus pressure

† Even dense granites give a lower seismic velocity when they are uplifted to near the earth's surface.

release and opening of pore structure in shallow zones (500–700 metre below the surface of the Carboniferous) obviously led to a desorption of the autochthonous coalification gases, whereas ascending coal gases from deep coals (which, on account of the higher temperature, still are in the process of coalification) could enter the coal. It must also be taken into account that "coalification" gases from deep bituminous shales may have penetrated the coal seams in the low-pressure upper zones.

The migration of coalification gases will lead to a fractionation of carbon isotopes in a similar way to that which has been described by Colombo *et al.* (1965, 1966) for Italian accumulations of natural gas. The observation that in some cases there are no clear relationships between the isotopic composition of methane and the depth can be explained by the variations in the lithological succession of the productive Carboniferous and the complexity of the tectonic structure. Consequently, the mechanisms, and paths of migration of these gases must have been very complex.

It is probable that at greater depths the relations are different, as may be seen in Fig. 9 and from the few data we have obtained from coals and gases from deep boreholes. Here the supposed relationship between degree of coalification and the isotopic composition of the adsorbed methane in coal samples may, indeed, exist. Therefore, further studies should be concentrated more and more on deep wells and shafts from which cores of coal are available.

The many problems connected with a combined chemical, petrographic and isotopic study of coal gases are, indeed, very complex. Nevertheless, we want to point out that the results here reported, give for the first time the range of the isotopic composition of methane adsorbed in coals of very different ranks. In contrast to the carbon isotopic composition of the coal itself, the δC^{13} value of the adsorbed methane exhibits a very wide range (from -17 to -70). From the view point of coal mining, it may be of interest that, the deep lying primary zone with relatively heavy adsorbed methane corresponds to zones with high contents of free mine gas, as will be reported later.

Acknowledgements

The authors are grateful to Prof. E. Tongiorgi for the interest shown for this work. Useful discussions were held with Dr. H. Damberger, whose contribution is here acknowledged. Thanks are due to G. Ferrero for his valuable assistance in the laboratory work.

The part of the work carried out in the Laboratorio di Geologia Nucleare dell' Università di Pisa was supported by contract C.N.E.N.—C.N.R. No 115/1159/1472.

References

Brodskii, A. M., Kalinenko, R. A. and Lavrovskii, K. P. (1959) On the kinetic isotope effect in cracking. *Int. J. Applied Radiation and Isotopes*, **7**, 118.

Colombo, U., Gazzarrini, F., Sironi, G., Gonfiantini, R., Tongiorgi, E. (1965) Carbon isotope composition of individual hydrocarbons from Italian natural gases. *Nature*, **205**, No. 4978, 1303–1304.

Colombo, U., Gazzarrini, F., Gonfiantini, R., Sironi, G., Tongiorgi, E. (1966) Measurements of C^{13}/C^{12}, isotope ratios on Italian natural gases and their geochemical interpretation. *Advances in Organic Geochemistry*, 1964, Pergamon Press, Oxford, pp. 279–292.

Craig, H. (1957) Isotopic standards for carbon and oxygen and correction factors for mass-spectrometric analysis of carbon dioxide. *Geochim. et Cosmochim. Acta*, **12**, 133–149.

Damberger, H., Kneuper, G., Teichmüller, R. and M. (1964) Das Inkohlungsbild des Saar Karbons. *Glückauf*, **100**, 209–217.

Gedenk, R. (1963) Die Zusammensetzung des Restgases in Kohlen und Nebengestein der Bohrung Münsterland. *Fortschr. Geol. Rheinld. u. Westf.* **11**, 205–238.

Hedemann, H. A. (1963) Zur Frage der Kohlenwasserstoffgase im Oberkarbon. *Erdöl u. Kohle*, **16**, 833–841.

Jüntgen, H. and Karweil, J. (1966) Gasbildung und Gasspeicherung in Steinkohlenflözen, Teil I: Gasbildung – Teil II: Gasspeicherung. *Erdöl u. Kohle*, **19**, 251–258, 339–344.

Karweil, J. (1955) Physikalisch-chemische Probleme der Inkohlung. *Brennstoff-Chemie*, **36**, 1–32.

Kröger, C. and Hortig, H. P. (1966) Zur Berechnung der Inkohlungsgasmengen. *Brennstoff-Chemie*, **47**, 193–224.

Kröger, C. und Klatt, M. (1962) Isolierung und Eigenschaften der Gefügebestandteile von Saarkohlen. *Glückauf*, **98**, 398–406.

Lidine, G. D. und Khodot, V. D. (1964) Les gaz dans les gisements de charbon. *C.R. 5th Congr. Internat. Strat. Géol. Carbonifère*, **1**, 227–238.

Mott, R. A. (1942) The origin and composition of coals. *Fuel in Science and Practice*, **21**, 129–135.

Mott, R. A. (1943) The origin and composition of coals. *Fuel in Science and Practice*, **22**, 20–26.

Patteisky, K. (1952) Das Auftreten des Grubengases und seine Gefahren im Bergbaubetrieb. In: *Grubensicherheit und Grubenausbau*. Bergmännische Fachtagung Leoben 1952. *Montanzeitung*, 1952, pp. 47–56.

Patteisky, K. (1964) *Grubengas- und Schlagwetterkunde*. 135 pages. C. Th. Kartenberg, Herne.

Patteisky, K. and Teichmüller, M. (1960) Inkohlungs-Verlauf, Inkohlungs-Masstäbe und Klassifikation der Kohlen auf Grund von Vitrit-Analysen. *Brennstoff-Chemie*, **41**, 79–84, 97–104, 133–137.

Peters, K. and Warnecke, A. (1933) Physikalische und chemische Untersuchungen über Flözgase. *Glückauf*, **69**, 1181–1190.

Schaub, H. (1954) Kreidesedimente in Spalten des linksrheinischen Steinkohlengebirges. *Geol. Jb.* **69**, 249–254.

Schilling, D. H., Jüntgen, H. and Peters, W. (1966) Die Sorptionskinetik von Methan an Steinkohlen als Grundlage des Ausgasungsprozesses. *Glückauf Forschungsheft* **5**, **27**, 203–214.

Stevenson, D. P., Wagner, C. D., Beck, O. and Otvos, J. W. (1948) Isotope effect in the thermal cracking of propane-1-C^{13}. *J. Chem. Phys.* **16**, 993–994.

Wickman, F. E. (1953) Wird das Häufigkeitsverhältnis der Kohlenstoffisotopen bei der Inkohlung verändert? *Geochim. et Cosmochim. Acta*, **3**, 244–252.

THE SIGNIFICANCE OF CARBON ISOTOPE VARIATIONS IN MARINE SEDIMENTS†

by

JOHN M. HUNT

Woods Hole Oceanographic Institution, Woods Hole, Mass.

Introduction

Several studies have shown that the carbon in terrestrial plants is depleted in C^{13} compared to that in marine plants. This was first reported by Wickman (1952) and Craig (1953). Marine plants, for example as exemplified by Halimeda and Corallina, have per mil deviations (δC^{13}) from the Pee Dee belemnite of around -13 to $-17‰$. Terrestrial plants such as tomato leaves and grass have values around -24 to -26 (the more negative value represents less C^{13} compared to the standard). Park and Epstein (1960) have explained these differences as due to the fact that terrestrial plants utilize more CO_2 from the atmosphere which is leaner in C^{13} than the bicarbonate of sea water utilized by marine plants. They also claimed that there is a kinetic effect causing a slight depletion of C^{13} in plants that obtain their carbon from the atmosphere compared to those that obtain it from sea water. A paper by Silverman and Epstein (1958) showed that the lipid fraction of both terrestrial and marine plants was depleted in C^{13} compared to the total plant, and in this respect it was comparable to petroleum. They suggested that the differences in carbon isotope ratios observed in crude oils were due to different contributions of terrestrial and marine organic matter as the source of carbon for the petroleum. Silverman (1963) showed rather clearly the relationship between plant lipids and crude oils in both the marine and non-marine environment (Fig. 1).

† Contribution No. 1936 of the Woods Hole Oceanographic Institution. This work was supported by National Science Foundation grant GP-4083.

One would expect that the differences in carbon isotope ratios of terrestrial and marine plants would be reflected in the organic matter of the associated sediments. Oana and Deevey (1960) found the δC^{13} of organic matter in lake sediments to range between -28 and $-32‰$. Eckelmann *et al.* (1962) reported the δC^{13} of organic matter in cores from the Pacific and Atlantic Oceans and Mediterranean Sea to be in the range -19 to $-23‰$. Sackett and Thompson (1963) showed a variation from -28 for Mississippi river sediments to -20 for sediments of the upper Gulf of Mexico. They attributed this variation to a mixing of terrestrial and marine derived organic matter, complicated by several other factors.

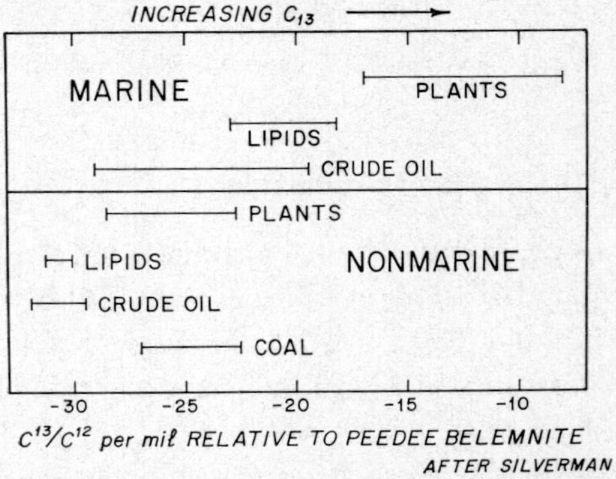

FIGURE 1. — Carbon isotope ranges in natural materials.

Deuser and Degens (1967) have shown that a carbon isotope fractionation of 7 to 9‰ occurs in the hydration of CO_2 to HCO_3^-, but no fractionation occurs when atmospheric CO_2 is dissolved in water. This means that plants growing in river and lake waters of slightly acid pH are utilizing dissolved CO_2 with a δC^{13} of -7, whereas plants growing in sea water with a basic pH are utilizing HCO_3^- with a δC^{13} of 0. This would explain some of the differences reported by Sackett and others between marine and lacustrine organic matter.

The present survey was undertaken to determine if there is a systematic change in the carbon isotope ratios of organic matter of sediments in going from rivers, to bays, and into the open sea along the entire Atlantic Coast of the United States from Maine to Florida. These were compared with organic matter on the Continental Shelf and Slope. This horizontal study of terrestrial and marine variations was partly supplemented with a few cores taken at depth by the Joint Oceanographic Institution Deep Earth Sampling Program.

Analytical Procedure

The samples used in this study were part of the collection obtained in the cooperative program with the U.S. Geological Survey on the study of the Continental Margin. A portion of the dried samples was treated with 0·1 N HCl to remove carbonate carbon. Subsequently the sample was combusted at a temperature of around 900° C, and the generated gases were recycled via an electrically operated Toepler pump in order to convert all the carbon monoxide to carbon dioxide. The CO_2 was passed over hot (500° C) manganese dioxide and copper to remove combinations of sulfur and nitrogen oxides, after the method of Sackett and Thompson (1963). Subsequently the purified CO_2 was analyzed in a Nier—McKinney type mass spectrometer with the results being reported as per mil deviations (δC^{13}) from the ratio of a standard material as follows:

$$\delta C^{13}\,(\text{in}‰) = \frac{(C^{13}/C^{12}\ \text{Sample}—C^{13}/C^{12}\ \text{Standard})}{C^{13}/C^{12}\ \text{Standard}} \times 1000$$

The standard was National Bureau of Standards Tentative Isotope Reference Sample No. 20. All data reported, however, are relative to the Chicago PDB standard. The standard deviation determined from duplicate determinations was 0·15‰ relative to the working standard.

Discussion of Results

The analysis of bottom sediment organic matter in 32 rivers from Maine to Florida shows most of them to have δC^{13} values around −20 at the mouth of the river, compared to −26‰ more than 10 miles upstream. There is, however, considerable variation depending on the extent to which the inland body of water is exposed to the sea. Figure 2 shows some values along the Atlantic Coast from the southern end of Maine to Long Island Sound. Values for rivers in Maine, Massachusetts and Connecticut, vary between −24·2 and −29·2‰ only short distances inland. In contrast, the values in the more open Narragansett Bay are very close to those of marine sediments (−22·2 to −22·7‰). In Nantucket Sound, Cape Cod Bay, and most areas only short distances offshore, the δC^{13} of organic matter is in the range reported for most deep sea sediments, −18 to −22‰. Samples taken in Buzzards Bay and Vineyard Sound adjacent to the Elizabeth Islands have deep sea values of −20·6‰. These results indicate that along the Atlantic Coast in general the terrestrial organic matter is dominant in river sediments only to the mouth of narrow rivers. Out in the more open bays and sound, or on the shelf there is a much greater contribution of marine organic matter resulting in δC^{13} values typical of deep sea sediments. A traverse of samples taken across Nantucket Shoals to the edge of the Continental Rise shows practically no change

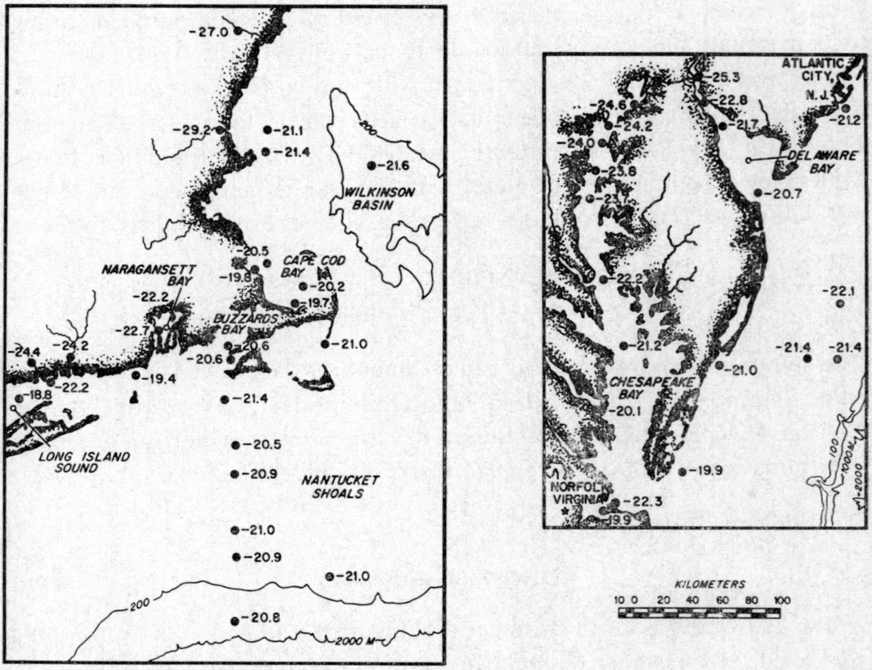

FIGURE 2.

in the δC^{13} of organic matter of surface sediments (range $-20{\cdot}5$ to $-21{\cdot}4$‰). Similar values are also observed in the Wilkinson basin to the north and on the shelf west of this.

Part of the Continental Shelf off the northeastern United States consists of relic sediments that were emergent during the last ice age (about 10,000 years ago), and have not since been covered with sediments. For example, the tusks and molars of mammoths and mastodons have been recovered in several areas, indicating that grazing animals lived in the region that is now sea floor. In two of these areas near Nantucket and on Georges Bank, old peat bogs were found. The peat showed δC^{13} values of $-25{\cdot}9$

and —26·3, respectively. In other words, in 10,000 years this compact organic accumulation did not change from its original terrestrial depletion in δC^{13}. The surrounding organic matter, however, has been altered by the presence of marine organic carbon and possibly reworking by sediment marine organisms, so that there is no evidence of terrestrial derived carbon on the once emergent sediments.

Sackett and Moore (1966) have reported a range from about −9 to −3‰ in the dissolved inorganic carbon of water from the Hudson River going from Palisades, New York, to the Verranzo Bridge near the mouth of the river. They were interested in whether the observed gradients in δC^{13} in rivers could be related to changes in dissolved bicarbonate utilized by the plants. Although no plants were analyzed in our study it was found that the δC^{13} of organic matter in sediments of the Hudson River showed one of the smallest changes of all the rivers studied. Going over the same distance from the Palisades to the Verranzo Bridge the the enrichment in δC^{13} in the organic matter was only from −24·7 to −23·3‰. Values in this area might be altered by the large contribution of organic pollutants from nearby industries.

Figure 2 also shows that the organic matter in sediments of open bays is very close to that of deep sea sediments in δC^{13} values. Values in Delaware Bay range from −20·7 to −22·8‰ almost up to Wilmington, where they finally increase to −25·3‰. The Chesapeake Bay, which is wide, but also over 100 miles long, begins to show the influence of terrestrial carbon only at its upper end. The change is one of the most gradual that was observed anywhere along the Atlantic Coast, from −21·2‰ in the bay to −24·6‰ at its end. Samples that were analyzed 50 miles or more upstream in the James River, York River, and Potomac River varied between −25·0 and −26·1‰. Again samples on the Continental Shelf, even in close proximity to land, showed deep sea values ranging from −19·9 to −22·1‰.

In contrast to the open bays, the enclosed sounds show a definite depletion in C^{13} as shown in Figure 3. All of the river waters pouring into the sea in this area are protected somewhat from the marine influence by a long sand bar running all the way from Cape Henry at the north to Cape Lookout in the south. Although Albemarle Sound is a much wider body of water than Narragansett Bay, the former shows largely terrestrial organic isotope ratio values in the range −24·8 to −27·9‰. This is a 2‰ depletion in C^{13} compared to the Narragansett Bay samples which are not protected by a long sand bar. Again values on the open shelf are comparable to those for deep sea sediments. Figure 3 also shows some samples analyzed along the southern coast in the warm climate of Georgia and Florida. The organic matter upstream in six of the rivers on this map shows values ranging from −25·6 to −28·1‰. The change in δC^{13} on the river south of Savannah is typical of what was first mentioned as being observed in many small rivers along the Atlantic Coast. There is an

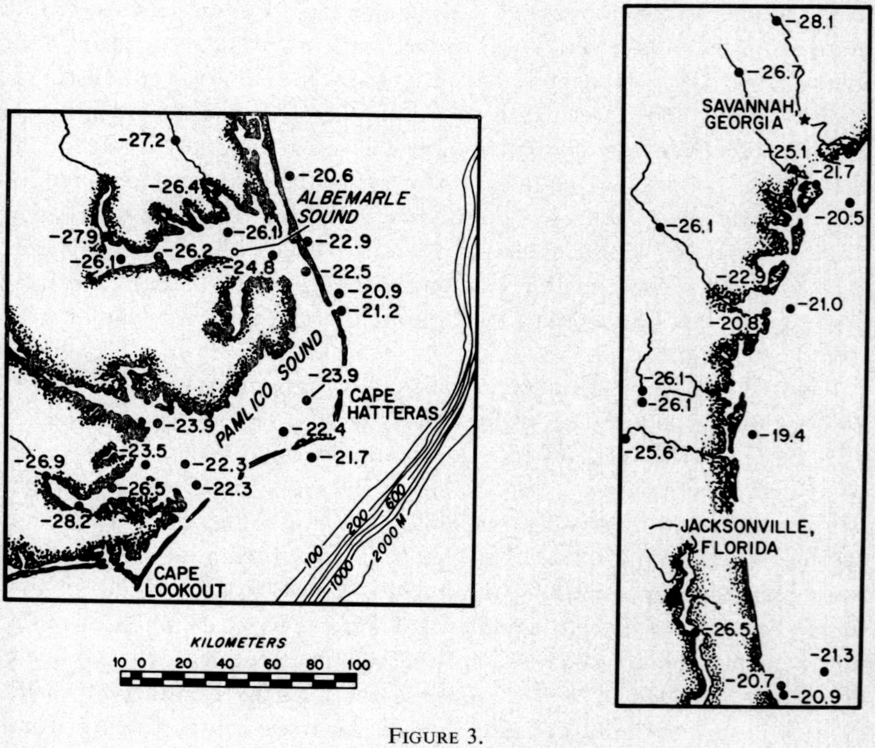

FIGURE 3.

enrichment in C^{13} from −25·1 to 21·7‰ within 10 miles of where the river enters the sea. Samples in the open ocean very near the coast all along this area are in the deep sea sediment range from −19·4 to −22·9‰.

In addition to the surface samples a few analyses were made of the organic matter in four cores obtained by the Joint Oceanographic Institution Deep Earth Sampling Program. The cores were taken in water depths ranging from 75 to 3000 ft. Sediment depths penetrated ranged up to 890 ft. Data on the sampling program have been reported in literature (Joides, 1966). In the present study samples were analyzed in holes

J-2, J-3, J-4 and J-6. Hole J-2 is on the shelf and the others are on the Blake Plateau. In this whole area the δC^{13} of organic matter in surface samples ranges generally between -17 and -22‰ typical of deep sea sediments. In contrast, the samples below 500 ft in all three wells range between $-23{\cdot}5$ and $-25{\cdot}6$‰. Actual values are shown as follows (Table 1). These analyses have a greater error than other data reported in this paper because of the high (40 to 99%) inorganic carbon content. In some cases the inorganic to organic ratio was 1000 to 1, so the usual treatment with HCl left very little carbon.

TABLE 1

Sediment Depth in feet	J-2	J-3	J-4	J-6
Surface	—19·4	—20·3	—18·2	—17·3
100	—	—24·1	—22·3	—22·2
200	—	—20·7	—	—23·8
400	—	—19·3	—	—25·7
450	—	—	—23·1	—
500	—	—24·5	—25·3	—
511	—	—25·1	—24·8	—
700	—24·9	—	—	—
890	—25·6	—	—	—

Several other investigators have reported samples of marine sediments to show a depletion of C^{13} with depth, but in no case were samples analyzed as deep as 500 ft (Eckelmann, 1962; Sackett, 1964). Degens (1967) believes that the depletion in C^{13} with depth is due to the elimination of C^{13} rich fractions of the organic matter, such as proteins and some carbohydrates, with diagenesis. He shows these fractions to have values in the range of -18 and -19‰, respectively. In comparison lignin has values around -24‰. As previously mentioned the lipid fraction is the most depleted in C^{13} with δC^{13} values, generally being more negative than -24‰. Since the lignin and lipids would tend to be preserved in sediments with the gradual elimination of proteins and carbohydrates, this is a reasonable hypothesis for the observed change with depth from about -20 to -25‰.

The variations with depth in hole J-3 may be partly explained by differences in source material. Although no surface samples on the shelf, slope or deep sea in this study were depleted in C^{13} beyond -24 (except peat), there are deep sea surface sediments in other areas depleted in C^{13} to -26 (Sackett *et al.*, 1965).

Conclusions

The isotopic composition of organic matter in 32 rivers along the Atlantic Coast of the United States has been compared with that in sediments of the Atlantic Continental Shelf and Slope. The conclusions from this study were as follows:

1. There is an observed enrichment in δC^{13} of about 6‰ of the organic matter in the sediments of most small rivers going from about 10 miles upstream to the river mouth. This change is about the same as that observed by Deuser and Degens (1967) between dissolved CO_2 and HCO_3^- in water. Since the CO_2 is dominant at the more acid pH of rivers and lakes, and the HCO_3^- at the basic pH of oceans, it appears that the utilization of these two forms of carbon by plants contributes to the differences observed in sediments. The change from river to open sea is fairly abrupt, with values comparable to deep sea marine sediments being found within a few miles offshore.

2. Bays such as the Delaware and Chesapeake have δC^{13} values in their sediments comparable to deep sea sediments as far as 50 miles or more upstream.

3. Sounds enclosed by long sand bars generally show a 1 to 2‰ depletion in C^{13} over deep sea sediments. Rivers feeding into these sounds show terrestrial values in the range -25 to -26‰, even near the river mouth.

4. Relic terrestrial sediments that have not been covered since the land was emergent show δC^{13} values in the organic matter comparable to that of deep sea sediment, except in a compact organic body such as a peat bed. The latter show typical terrestrial values in the range -25 to -26‰.

5. Samples from three holes drilled on the Continental Shelf off Jacksonville, Florida, show δC^{13} values greater than -25‰, at depths greater than 500 ft, in contrast to surface values around -20‰. This depletion in C^{13} is believed to be partly due to diagenesis of the organic matter and partly to other factors such as source material.

References

Craig, H. (1953) The geochemistry of the stable carbon isotopes. *Geochim. et Cosmochim. Acta*, **3**, 53–92.

Degens, E. T. (1967) *Biogeochemistry of Stable Carbon Isotopes*, in press.

Deuser, W. G. and Degens E. T. (1967) Carbon isotope fractionation in the system CO_2 (gas)-CO_2 (aqueous)-HCO_3^- (aqueous). *Nature*, **215**, 1033–1035.

Eckelmann, W. R., Broecker, W. S. Whitlock, D. W. and Allsup, J. R. (1962) Implications of carbon isotopic composition of total organic carbon of some recent sediments and ancient oils. *Bull. Amer. Assoc. Petrol. Geol.* **46,** 699–704.

Joides, (1966) Ocean drilling on the continental margin. Reprinted from *Science*, November 5, 1965, vol. 150.

Oana, S. and Deevey, E. S. (1960) Carbon-13 in lake waters and its possible bearing on paleolimnology. *Amer. J. Sci.* A, **258,** 253–272.

Park, R. and Epstein, S. (1960) Carbon isotope fractionation during photosynthesis. *Geochim. Cosmochim. Acta*, **21,** 110–126.

Sackett, W. M. (1964) The depositional history and isotopic organic carbon composition of marine sediments. *Marine Geology*, **2,** 173–185.

Sackett, W. M. and Thompson, R. R. (1963) Isotopic organic carbon composition of recent continental derived clastic sediments of eastern gulf coast, Gulf of Mexico. *Bull. Amer. Assoc. Petrol. Geol.* **47,** 525–528.

Sackett, W. M., Eckelmann, W. R., Bender, M. L. and Be, A. W . H. (1965) Temperature dependence of carbon isotope composition in marine plankton and sediments. *Science*, **148,** 235–237.

Sackett, W. M. and Moore, W. S. (1967) Isotopic variations of dissolved inorganic carbon, in press.

Silverman, S. R. (1963) *Investigations of Petroleum Origin and Evolution — Mechanisms by Carbon Isotope Studies, Isotopic and Cosmic Chemistry*, pp. 92–101.

Silverman, S. R. and Epstein, S. (1958) Carbon isotopic compositions of petroleums and other sedimentary organic materials. *Bull. Amer. Assoc. Petrol. Geol.* **42,** 998–1012.

Wickman, F. E. (1952) Variations in the relative abundance of the carbon isotopes in plants. *Geochim. et Cosmochim. Acta*, **2,** 243–254.

CARBON ISOTOPE EFFECTS IN METHANE PRODUCTION BY THERMAL CRACKING

by

WILLIAM M. SACKETT, SONGCHAI NAKAPARKSIN and DONALD DALRYMPLE

Department of Chemistry, University of Tulsa
Tulsa, Oklahoma

Introduction

The range in isotopic carbon composition of naturally occurring terrestrial materials has been established through the work of Nier and Gulbransen (1939), West (1945), Wickman (1952), Craig (1953), Silverman and Epstein (1958), etc. The ranges, expressed in terms of

$$\delta C^{13} = \left[\frac{(C^{13}/C^{12})\ \text{Sample}}{(C^{13}/C^{12})\ \text{Standard}} - 1\right] \times 1000$$

where the standard is the Chicago PDB material used by Craig (1953), are given in Fig. 1. It is obvious that the most oxidized carbon, $CaCO_3$, CO_2, dissolved bicarbonate and the most reduced carbon, petroleum and natural gas are relatively enriched in C^{13} and C^{12}, respectively, and that the processes, i.e. photosynthesis, thermal cracking and bacterial decomposition, that convert inorganic carbon to organic products, culminating finally in methane, result in C^{12}-enriched products. This enrichment is presumably due to the slightly greater reactivity of C^{12} relative to C^{13}-containing reacting groups in various compounds.

Not indicated in Fig. 1 are the rather large differences between marine and non-marine deposited calcium carbonates (Clayton and Degens, 1959), and benthonic and planktonic marine plants (Sackett *et al.*, 1965). These and other systematic differences are proving to be useful in determining palaeoenvironments and in characterizing and correlating crude oils and natural gases.

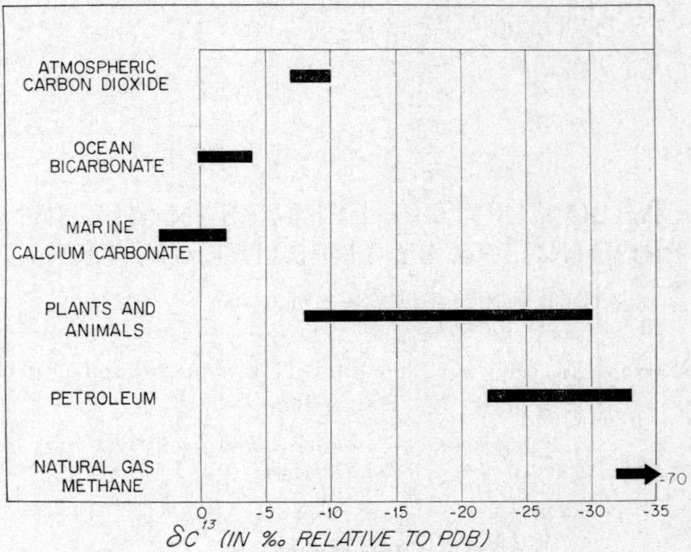

FIGURE 1. — Range of δC^{13} for natural materials.

The purpose of this investigation was to gain a better understanding of the large range of natural methane compositions through a study of the effect of structure, temperature and other factors on the carbon isotope fractionation occurring in the generation of methane by the thermal cracking of organic materials.

Carbon Isotope Fractionation

Isotope fractionation in natural processes is generally attributed to three phenomena: molecular velocity and vapor pressure differences between isotopic species, isotopic exchange equilibria and reaction rate differences due to the substitution of one isotope for another in a reacting chemical compound.

The velocities of different isotopic species of molecules are equal to the inverse square root of the molecular weights. For example for carbon dioxide,

$$\frac{\text{Velocity } (C^{12}O^{16}O^{16})}{\text{Velocity } (C^{13}O^{16}O^{16})} = \sqrt{\frac{45}{44}} = 1{\cdot}011$$

This means that mass 44 CO_2 will collide with a leaf surface 1·1 percent more frequently than mass 45 CO_2, a process used by Park and Epstein

(1960) to explain some of the difference between the isotopic carbon composition of marine and terrestrial plants.

The theoretical basis for isotopic exchange equilibria was put down by H. C. Urey (1935, 1947). Essentially it means that for an isotopic exchange equilibrium process, the isotopes will not distribute themselves randomly, but will tend to disproportionate, leading to equilibrium constants for such reactions as the one below that deviate somewhat from unity. From the observed shifts in spectral line frequencies due to isotope differences in a compound, and using the techniques of statistical mechanics, the ratios of the partition functions of simple molecules may be calculated at various temperatures. From the partition functions, the equilibrium constants for various isotopic exchange equilibria may be calculated. Thus for the isotope exchange reaction,

$$C^{12}O_2 + C^{13}H_4 \rightleftharpoons C^{13}O_2 + C^{12}H_4$$

and from Craig (1953)

	$C^{13}O_2/C^{12}O_2$	$C^{13}H_4/C^{12}H_4$	T, °K
$\frac{Q_2}{Q_1}$	1·2169	1·1369	273·1
	1·1909	1·1220	298·1
	1·1233	1·0821	400
	1·0877	1·0599	500
	1·0656	1·0458	600

it follows that:

$$k = \frac{(C^{13}O_2/C^{12}O_2)}{(C^{13}H_4/C^{12}H_4)} = \frac{1{\cdot}2169}{1{\cdot}1369} = \begin{array}{ll} 1{\cdot}070 & (273{\cdot}1\ °K) \\ 1{\cdot}061 & (298{\cdot}1\ °K) \\ 1{\cdot}038 & (400\ °K) \\ 1{\cdot}026 & (500\ °K) \\ 1{\cdot}019 & (600\ °K) \end{array}$$

By measuring the isotopic composition of methane and carbon dioxide in Yellowstone hot spring gases, Craig (1953) calculated reasonable subsurface temperatures using the above relationships. Zartman *et al.* (1961), using the same procedure, found that the methane and carbon dioxide in glacial till gases were apparently in isotopic equilibrium, since temperatures of 25° C were obtained. Other occurrences of these gases than those mentioned above give anomalous temperatures.

By far the most important factor in producing natural carbon isotope variations is the effect of mass differences on natural reaction rates. The theory for this phenomenon was largely developed by Bigeleisen (1949).

Simply put, for the following hypothetical cracking reaction leading to free radical products,

$$R-C^{12}-C^{12}H_3 \xrightarrow{k} R-C^{12}.+.C^{12}H_3$$

$$R-C^{12}-C^{13}H_3 \xrightarrow{k'} R-C^{12}.+.C^{13}H_3$$

and since $k = Ae^{-E}/RT$ and $A\alpha\sqrt{\dfrac{1}{m}}$, where m (reduced mass)

$$= \frac{M\times M'}{M+M'}; \text{ then } \frac{k}{k'} = \frac{m'}{m}\times \exp -(E-E')/RT.$$

For high temperatures $\dfrac{k}{k'} = \sqrt{\dfrac{m'}{m}}$

For $M = 12$, $M' = 12$; $m = 6$ for $M = 12$, $M' = 13$; $m' = 6{\cdot}24$, $k/k' = 6{\cdot}24/6 = 1{\cdot}02$.

This treatment is far from complete. A discussion of the temperature dependent factor, the effect of symmetry and numerous other aspects is quite complex and beyond the scope of this paper.

Experimentally several investigations have shown isotope effects in the decomposition of carboxylic acids. For example, Ropp *et al.* (1951) found for the dehydration of formic-C^{14} and formic-C^{12} acids,

$$HC^{12}OOH \xrightarrow{k_{12}} H_2O+C^{12}O$$

$$HC^{14}OOH \xrightarrow{k_{14}} H_2O+C^{14}O$$

that k_{14}/k_{12} is 0·889 at 0° C and 0·914 at 25° C. Lindsay *et al.* (1949) found for the decomposition of oxalic acid

$$(C^{12}OOH)_2 \xrightarrow{k_1} C^{12}O_2+C^{12}O+H_2O$$

$$\begin{array}{l} C^{13}OOH \\ | \\ C^{12}OOH \end{array} \xrightarrow{k_2} C^{13}O_2+C^{12}O+H_2O$$

$$\begin{array}{l} C^{12}OOH \\ | \\ C^{13}OOH \end{array} \xrightarrow{k_3} C^{12}O_2+C^{13}O+H_2O$$

that the ratio $k_1/(k_2+k_3)$ is 1·034 at 100° C. Roe and Hellman (1951) determined the intramolecular isotope effect for the decarboxylation of malonic acid. Thus for

$$CH_2(COOH)_2 \xrightarrow{k_1} CH_3OOH + CO_2$$

$$CH_2\begin{cases} C^{14}OOH \\ COOH \end{cases} \begin{matrix} \xrightarrow{k_2} CH_3COOH + C^{14}O_2 \\ \xrightarrow{k_3} CH_3C^{14}OOH + CO_2 \end{matrix}$$

they found that $k_3/k_2 = 1{\cdot}06 \pm 0{\cdot}02$ at $153-154°$ C. For the C^{13} effect Bigeleisen and Friedman (1949) found that $k_3/k_2 = 1{\cdot}02$, at 138° C, predictably about one-half of the C^{14} effect.

Zelinski (1964) found a smaller effect for the decarboxylation of 2-quinoline carboxylic acid, with $k_{12}/k_{14} = 1{\cdot}01$ for temperatures ranging, from 140 to 180° C. These and other published results indicate that there is a 1 to 6 % preferential breaking of the C^{12} relative to C^{13} containing bonds in carboxylic acids.

Relatively little work has been done concerning the fractionation involved in the breaking of carbon-carbon bonds in hydrocarbons. The work that has been reported is questionable, contradictory and incomplete. For example, Stevenson *et al.* (1948) reported an 8% more frequent rupture of the $C^{12}-C^{14}$ bond relative to the $C^{12}-C^{13}$ bond in propane at 500–550° C. Frey *et al.* (1956) report an 8% more frequent rupture for $C^{12}-C^{12}$ relative to $C^{12}-C^{14}$ for propane at 550–600° C. Presumably the effect for the $C^{12}-C^{13}$ bond from their data would be about 4% since the effect is proportional to the mass difference. Brodsky *et al.* (1959) found a 12% preferential breaking of the $C^{12}-C^{12}$ bond relative to the $C^{12}-C^{14}$ bond in ethane at 850°C. This is an enormous effect considering the very high temperatures used in their experiments.

Additional experimental work is needed to clarify the inconsistencies in various reported investigations.

Procedure

Cracking

Hydrocarbons were cracked in pyrex tubes, equipped with break seals, of about 8 ml to 100 ml capacity. The tubes were baked out at 500° C before use to remove trace amounts of adsorbed hydrocarbons and other gases that might have come from glass working. Liquid hydrocarbons were accurately weighed in capillaries. The tubes were cooled to the temperature of liquid nitrogen, evacuated to less than 10 microns, and sealed by fusion of the pyrex glass. The samples were cracked at, temperatures ranging from 350° to 600° C, for times ranging from 2 min to 10 days.

Methane Separation and Combustion

The separation of methane from the cracking products and the combustion of methane to carbon dioxide were performed in a system such as the one shown in Fig. 2. The combustion system consists of a furnace, quartz tube half-filled with copper oxide as a catalyst to convert carbon monoxide to carbon dioxide, one acetone-dry-ice trap and two liquid-nitrogen traps, one of which is used in the separation system. The Toepler pump serves both to separate methane from the cracking products and to cycle the gases through the copper oxide during the combustion.

Referring to Fig. 2, all carbon-containing cracking product gases were trapped at liquid nitrogen temperature (in N_1) except methane which

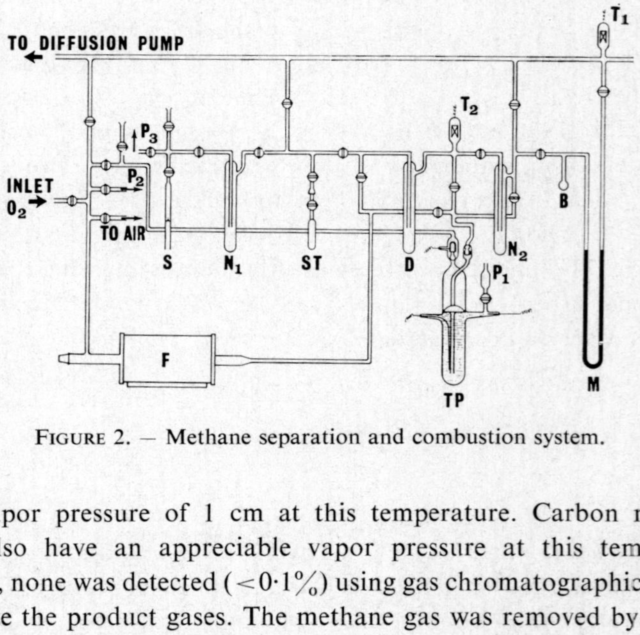

FIGURE 2. – Methane separation and combustion system.

has a vapor pressure of 1 cm at this temperature. Carbon monoxide would also have an appreciable vapor pressure at this temperature. However, none was detected ($<0{\cdot}1\%$) using gas chromatographic methods to analyze the product gases. The methane gas was removed by an automatic Toepler pump (TP) and stored in the combustion tube until the pressure of the system was reduced to 90–100 microns as read on T_2.

The separation was checked by gas chromatography. The ethane/methane ratio was reduced from 1/29 to less than 1/1176 in a sample of natural gas. This is a separation factor greater than 40. To take an extreme case of contamination, consider a separated sample having 99% methane and 1% ethane with δC^{13} values 0 and 12‰, respectively. The following mass and isotope balance calculation may be made:

$$\% \ CH_4 \times (\delta C^{13} CH_4) + \% \ C_2H_4 \times 2(\delta C^{13} C_2H_6) = 100 \times (\delta C^{13} \text{ Total Carbon})$$

$$99 \times (0) + 1 \times 2 \times (+12‰) = 100 \ (\delta C^{13} \text{ Total Carbon})$$

$$\delta C^{13} \text{ Total Carbon} = 0{\cdot}24‰$$

This indicates that 1% of ethane contamination would affect the δC^{13} value of the methane by 0·24‰. This is approximately the standard deviation for replicate determinations of δC^{13}, showing the relative unimportance of ethane contamination.

The methane was converted to carbon dioxide by a procedure similar to that by reported by Craig (1953). This involved combustion of methane over copper oxide at temperatures of 800° – 900° C in the presence of oxygen, and recycling the gases through the copper oxide during the combustion for about 30 min by the use of the automatic Toepler pump (TP). The carbon dioxide was then purified by passing it over the acetone-dry-ice trap (D) for about 15 min to trap water and finally condensed in the liquid-nitrogen trap (N_2) over a period of 15 min.

The volume of carbon dioxide gas was measured in the calibrated manometer (M), and the sample was then condensed into the sample tube (ST) for transfer to the mass spectrometer.

Mass Spectrometric Analysis

The isotopic composition of the carbon dioxide was measured in a Nier-McKinney type mass spectrometer relative to a laboratory standard of powdered charcoal, having a δC^{13} value of −24·8 versus the Chicago PDB standard. This value was determined precisely by comparison with the NBS 21 graphite standard. The δC^{13} values were corrected for 0^{17} contribution, leakage of ball joints in the magnetic values, mass 44 tail under mass 45 peak, and mass 45 background; the overall correction factor was $1{\cdot}080 \pm 0{\cdot}005$.

Results and Discussion

Three aspects were considered in the carbon isotope fractionation occurring during the production of methane from hydrocarbons. They were the effect of chain length of straight-chain hydrocarbons, the effect of temperature in the cracking of normal octadecane, and the fractionation pattern for various types of natural hydrocarbon mixtures. The results of these studies are then used to explain the wide range for the isotopic carbon composition of natural methane occurrences.

Effect of Chain Length

Precise amounts of propane, n-butane, n-heptane and n-octadecane were pyrolyzed at 500° C for varying lengths of time, and the amount and isotopic composition of the methane relative to the parent material were determined as described above. The experimental results and least-squares-determined parameters for straight-line equations of normal and logarithmic functions of δC^{13} are given in Table 1. The numerous octadecane data definitely show a logarithmic dependence, but the propane, butane and heptane data are too few to show how δC^{13} is related to the amount of methane. Knowledge of this dependence is necessary in order to extrapolate these data to zero so that the carbon isotope fractionation occurring during the breaking of a terminal carbon-carbon bond may be determined for the conditions used in the experiments. Although this dependence has not been clearly determined, it is obvious that carbon isotope fractionation in the generation of methane from these four straight-chain hydrocarbons is proportional to the number of carbons in the molecules.

The substitution of a heavy for a light isotope, i.e. C^{13} for C^{12}, lowers the vibrational frequency and the zero point energy of a chemical bond. This means that more energy is required to break a $C^{13}-C^{12}$ relative to a $C^{12}-C^{12}$ bond. It follows that the energy difference between these two types of isotopic bonds is more significant in organic molecules having a low relative to a high carbon-carbon bond dissociation energy, and that for series such as ethane, propane, butane or propane, isobutane, neopentane, with decreasing terminal bond dissociation energies (Table 2) carbon isotope fractionation should increase proportionally.

This supposition may be used to explain the trends indicated in Table 1. Consider the cracking of n-butane yielding free radical products $\cdot CH_3$

and $\cdot CH_2CH_2CH_3$. Three of the numerous reactions that may occur are as follows:

TABLE 1. — CARBON ISOTOPE FRACTIONATION IN THE PRODUCTION OF METHANE FROM STRAIGHT-CHAIN HYDROCARBONS AT 500° C

Parent	δC^{13} CH_4 vs Parent	$\frac{\text{Moles } CH_4}{\text{Moles Carbon in sample of parent}}$	Least squares parameters for normal straight-line equation: $-\delta C^{13} = \delta C^{13}_{CH_4+0} + m \times \frac{\text{Moles } CH_4}{\text{Moles C}}$		$\log \lvert \delta C^{13} \rvert$ * $= \log \lvert \delta C^{13} \rvert_{CH_4=0} + m \times \frac{\text{Moles } CH_4}{\text{Moles C}}$	
			$\delta C^{13}_{CH_4=0}$	m	$\delta C^{13}_{CH_4=0}$	m
Propane C_3H_8	—3·4	0·063	—3·7	+3·9	—3·8	—0·49
	—3·6	0·083				
	—3·0	0·127				
	—3·1	0·187				
n-Butane n-C_4H_{10}	—6·4	0·090	—6·6	+3·6	—9·7	—1·44
	—6·1	0·130				
	—5·9	0·220				
n-Heptane	—7·7	0·070	—8·7	+17·9	—9·6	—1·43
	—4·4	0·218				
	—3·6	0·307				
n-Octadecane n-$C_{18}H_{38}$	—13·1	0·115	—18·2	+38·7	—23·6	—1·82
	—14·4	0·116				
	—13·4	0·137				
	—10·4	0·174				
	—12·4	0·176				
	—8·8	0·224				
	—7·7	0·248				
	—6·7	0·269				
	—5·7	0·311				
	—6·6	0·319				
	—5·7	0·320				
	—6·2	0·331				

* Absolute values are used here and in all subsequent discussions of logarithmic functions of δC^{13}.

(1) Hydrogen abstraction

$$\cdot CH_3 + C_4H_{10} \rightarrow CH_4 + \cdot C_4H_9$$

(2) Thermal decomposition

$$\cdot C_4H_9 \rightarrow CH_3CH_2CH = CH_2 + H$$

(3) Radical interactions

$$\cdot C_4H_9 + \cdot CH_3 \rightarrow CH_3\text{—}CH(CH_3)CH_2CH_3$$

Reaction (2) was chosen to yield an allyl position methyl group and reaction (3) a tertiary carbon. These methyl groups with exceptional low bond dissociation energies (Table 2) should show relatively large carbon isotope fractionations when broken. For a simple hydrocarbon such as ethane, reactions 2 and 3 would not take place to give methyl groups attached to the allyl or to a tertiary carbon. For propane, reaction 2 could not create an allyl methyl group. Thus as larger molecules are cracked, more possibilities exist for the formation of terminal carbon-carbon bonds with relatively small bond dissociation energies and correspondingly larger carbon isotope fractionations.

Temperature Effect

Accurately determined amounts of n-octadecane were pyrolyzed for varying periods of time at 350°, 400°, 500° and 600° C. The temperature range was limited by the unreasonable slowness of the reaction below

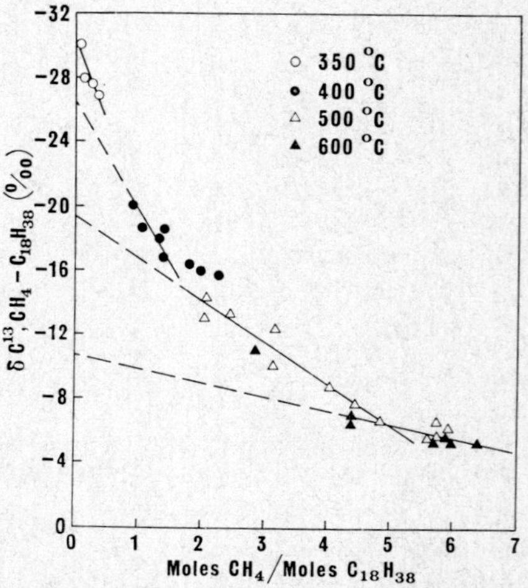

FIGURE 3. – Isotopic composition versus amount of methane for cracking of n-octadecane at various temperatures.

350° C and the softening point of pyrex at about 600° C. As before, the amount and isotopic composition of the methane were determined relative to the octadecane.

When the data for each temperature are plotted separately, it seems reasonable to extrapolate the data to zero to obtain the fractionation for each temperature. When the data are put on one graph as in Fig. 3, it appears that all data points may be described by a smooth exponential curve. This would resolve the problem of apparent breaks in the 400°, 500° and 600° curves.

Least-squares curve-fitting techniques for each set of temperature data give first order semi-log curves described by equations with the form:

TABLE 2. – BOND DISSOCIATION ENERGIES*

Bond	BDE, kcal/mole
CH_3—CH_3	88
C_2H_5—CH_3	85
$(CH_3)_2CH$—CH_3	84
$(CH_3)_3C$—CH_3	80
$C_6H_5CH_2$—CH_3	72
allyl-CH_3	72
CH_3CO—CH_3	82
$HOCH_2$—CH_3	83
$HC{\equiv}C$—CH_3	117
C_6H_5—CH_3	93

* S. W. Benson, Bond Energies *J. Chem. Educ.* **42**, 502–518 (1965).

$$\log |\delta C^{13}|_{CH_4} = K \times \frac{\text{moles } CH_4}{\text{moles carbon}} + \log |\delta C^{13}|_{CH_4} = 0$$

The parameters for the straight-line equation for each temperature are given in Table 3, and the data points and lines are shown in Fig. 4. The intercepts (CH_4=0) of the equations should give the fractionation for each temperature; the slope of the lines is probably related to the number of methyl groups of the type being cleaved from the parent molecule. Considering the mechanism for producing more weakly bonded methyl groups described in the previous section, the ratio of the number of this type of methyl bonding produced to the number being cleaved to form methane would presumably change with temperature, time and other factors. Interpretation of the slope of the lines in terms of mechanism is quite confused and more work is necessary before a better understanding of this question is possible.

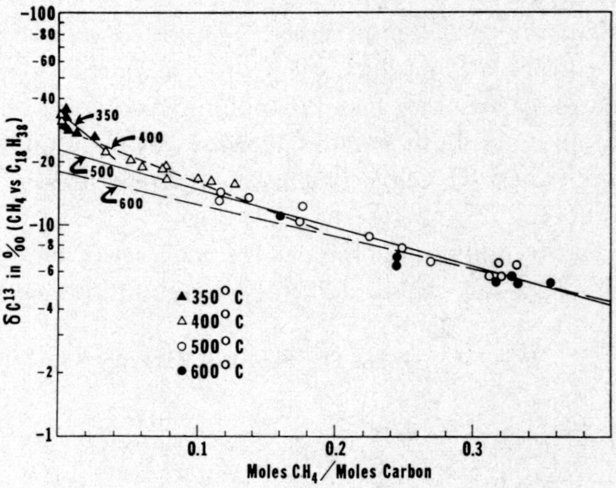

FIGURE 4. – Logarithmic dependence of isotopic composition versus amount of methane for cracking of n-octadecane at various temperatures.

The intercept values of the equations when plotted against temperature fall very nearly on the least-square-determined straight line described by

$$\log |\delta C^{13}|_{CH_4=0} = -0{\cdot}108 \times \frac{\text{Temperature °C}}{100} + 1{\cdot}902.$$

At 0° to 100° C, the temperature range of most natural processes, the fractionation predicted from this equation is 80 to 62‰, respectively. The only work available for comparison purposes is that by Rosenfeld and Silverman (1959). They report a fractionation of 81 and 94‰ at 30° and 23° C, respectively, for the bacterial production of methane during the fermentation of methanol. This process is not too comparable since the bonds being broken are carbon-oxygen rather than carbon-carbon.

Isotope Fractionation for Methane Produced from Naturally Occurring Organic Materials

Samples of a Pennsylvanian crude oil from the Gilcrease sand in Oklahoma, U.S.A., the highly carbonaceous Jurassic Posidonian shale from Germany, and the Pennsylvanian Fleming coal from Kansas, U.S.A., were treated at 500° C as described above for octadecane. The data and

TABLE 3. – TEMPERATURE AND CARBON ISOTOPE FRACTIONATION IN PRODUCTION OF METHANE FROM n-OCTADECANE

Temperature	δC^{13} CH_4, vs. Parent	$\frac{\text{moles } CH_4}{\text{moles C in sample of parent}}$	Least-squares-determined parameters for $\log \lvert \delta C^{13} \rvert = \log \lvert \delta C^{13} \rvert_{CH_4=0} + m \times \frac{\text{moles } CH_4}{\text{moles C}}$: $\delta C^{13}_{CH_4=0}$	m
350° C	—30·0	0·08	—33·8	—5·4
	—28·0	0·13		
	—27·6	0·24		
	—26·9	0·37		
400° C	—22·4	0·59	—29·3	—2·6
	—20·5	0·91		
	—18·7	1·08		
	—18·1	1·33		
	—18·7	1·36		
	—16·8	1·41		
	—16·4	1·80		
	—16·1	1·97		
	—15·8	2·29		
500° C	Same as Table 1		—23·6	—1·8
600° C	—11·1	2·9	18·1	—1·6
	— 7·0	4·4		
	— 6·5	4·4		
	— 5·4	5·7		
	— 5·7	5·9		
	— 5·3	6·0		
	— 5·3	6·4		

the least-squares-determined parameters for the logarithmic dependence of δC^{13} are given in Table 4, together with the n-octadecane parameters. The crude oil and carbonaceous shale show nearly the same pattern as octadecane. The standard deviation for each of these $\delta C^{13}_{CH_4=0}$ values is about 1·5, so that the 2·2 range may be attributed entirely to experimental error. Because of this similarity in behavior, it is presumed that the isotope fractionation associated with the generation of methane from crude oils and carbonaceous shales will show a temperature dependence much like that shown by n-octadecane.

The pronounced difference in the carbon isotope fractionation of methane produced from coal relative to other materials is probably due to a stronger binding of the terminal carbon in coal and/or to appreciable amounts of carbon monoxide (7% in one case) also generated in coal. Finely ground coal samples, having a high surface area and sorption capacity, presumably retain significant amounts of atmospheric oxygen even after the evacuation conditions used of one hour at temperatures up to 300° C. This oxygen may react to give carbon monoxide which is not separated from methane in the procedure used in this study. Thus $\delta C^{13}_{CH_4+CO}$ rather than $\delta C^{13}_{CH_4}$ versus the parent is determined. If the small fractionation of methane from coal is due to stronger bonding in kerogen-like materials rather than CO contamination, then perhaps the isotopic com-

TABLE 4. – CARBON ISOTOPE FRACTIONATION IN THE PRODUCTION OF METHANE FROM VARIOUS ORGANIC MATERIALS AT 500°

Parent	$\delta C^{13}_{CH_4}$ vs. Parent	$\frac{\text{moles } CH_4}{\text{moles carbon}}$ in sample of parent	Least-squares-determined parameters for $\log \mid \delta C^{13} \mid = \log \mid \delta C^{13} \mid_{CH_4=0} + m \times \frac{\text{moles } CH_4}{\text{moles C}}$	
			$\delta C^{13}_{CH_4=0}$	m
Crude oil	—18·7	0·070		
Penn.,	—15·1	0·113	—25·8	—2·1
Oklahoma,	—11·8	0·143		
U.S.A	— 9·8	0·203		
Posidonian	—19·8	0·044		
Shale,	—19·1	0·064	—24·4	—2·0
Jurassic,	—16·6	0·089		
Germany (12% organic carbon)	—15·0	0·110		
n-octadecane	Same as Table 1		—23·6	—1·8
Coal, Penn.,	— 9·9	0·020		
Kansas, U.S.A.	—11·0	0·032		
	— 7·8	0·052	—11·0	—2·1
	— 8·4	0·061		
	— 8·2	0·064		
	— 8·3	0·078		

position of methane formed during cracking may be useful in characterizing the various types of organic material in source rocks.

The isotopic carbon composition of natural methane occurrences has a considerable range. Figure 5 shows reported values for individual producing natural gas wells and other occurrences. In several instances there are several wells producing from the same formation, giving an unrealistic weighting of values for a particular age, i.e. the Pliocene. Some of the isotopically heaviest methanes in hot spring gases from Yellowstone Park, U.S.A., are not shown in Fig. 5. Craig (1953) reports a range of −10·4 to −28·4‰ for these gases, and believes that they have isotopically equilibrated with heavy CO_2 at elevated temperatures. The isotopically light Quaternary gases are in general marsh and glacial till gases produced by bacterial processes. Colombo *et al.* (1964) suggest that the chemical and isotopic data for the many Italian natural gases of Tertiary age, shown in Fig. 5, "are consistent with either of the following alternative

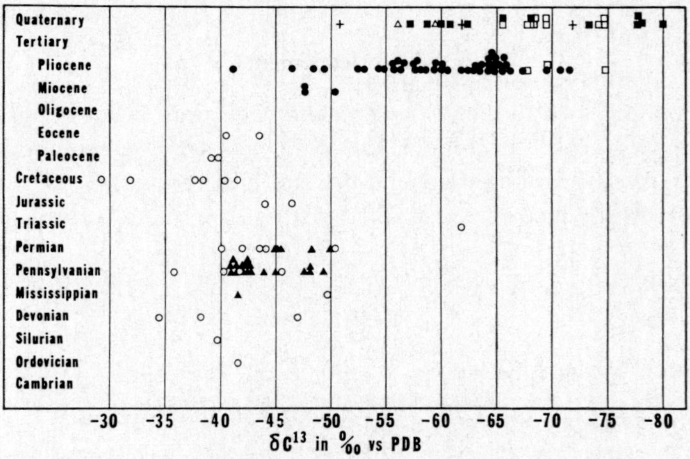

FIGURE 5. — Carbon isotope composition of natural methane occurrences. ■ Oana and Deevy (1960). □ Nakai (1960). ● Colombo *et al.* (1960). ○ Zartman *et al.* (1961). ▲ Newton and Sackett (1966). △ Sacket and Emery (1965) Unpublished data. + Wasserburg *et al.* (1963).

hypotheses; (1) the methane in any given natural gas has an isotopic composition which results from the admixture of an isotopically lighter bacterial methane, and of a heavier methane, formed together with higher hydrocarbons in chemical processes; (2) the observed isotopic distribution is the result of fractionation processes occurring during the migration of the gases."

The work described in this paper indicates that differences in fractionation due to bacterial relative to chemical processes or an isotopic fractionation during migration of methane are not needed to explain the range and evolution of the isotopic composition of natural methanes. It suggests rather, that maturity, temperature of formation and the strength of terminal carbon-carbon bonds in parent organic materials are the primary factors that determine the isotopic composition of all methane occurrences that have not isotopically equilibrated with surrounding carbon-containing components. It then follows that several geological factors become important in determining the composition of old natural gas occurrences. They are the time when the reservoir cap rock was formed, the geothermal gradient in a particular region at the time of formation, and the time of migration. Migration here is not used as a mechanism of isotope fractionation, but as a means of moving the isotopically light gases initially formed in source rocks out into potential reservoirs.

Acknowledgements

The authors wish to thank the Research Corporation for the grant that initiated this work, Dr. John Hunt of Woods Hole Oceanographic Institution for his willing co-operation and NSF grant No. GP4083 and No. GP5179 for financial assistance in this continuing investigation.

References

Benson, S. W. (1965) Bond energies. *J. Chem. Educ.*, **42**, 502–518.

Bigeleisen, J. (1949) The relative reaction velocities of isotopic molecules. *J. Chem. Phys.*, **17**, 675–678.

Bigeleisen, J. and Friedman, L. (1949) C^{13} isotope effect in the decarboxylation of malonic acid. *J. Chem. Phys.*, **17**, 998–999.

Brodsky, A. M., Kalinenko, R. A. and Lavrovsky, K. P. (1959) On the kinetic isotope effect in cracking. *International Journal of Applied Radiation and Isotopes*, **7**, 118–122.

Clayton, R. N. and Degens, E. T. (1959) Use of carbon isotope analyses of carbonates for differentiating fresh-water and marine sediments. *Bull. AAPG*, **43**, 890–897.

Colombo, U., Gazzarrini, F., Gonfiantini, R., Sironi, G. and Tongiorgi, E. (1964) Measurements of C^{13}/C^{13}, isotope ratios on Italian natural gases and their geochemical interpretation. *Advances in Organic Geochemistry* 1964, Pergamon Press, pp. 279–292.

Craig, H. (1953) The geochemistry of the stable carbon isotopes. *Geochim. et Cosmochim. Acta*, **3**, 53–92.

Frey, H. M., Danby, C. J. and Hinshelwood, Sir Cyril (1956) The intramolecular isotope effect in cracking. *International Journal of Applied Radiation and Isotopes*, **7**, 118–122.

Lindsay, J. G., McElcheran, D. E. and Thode, H. G. (1949) The isotope effect in the decomposition of oxalic acid. *J. Chem. Phys.*, **17**, 589.

Nakai, N. (1960) Carbon isotope fractionation of natural gas in Japan. *J. Earth Sciences, Nagoya University*, **8**, 174–80.

Newton, G. R. and Sackett, W. M. (1966) US. Patent No. 3,239,666.

Nier, A. O. and Gulbransen, E. A. (1939) Variations in the relative abundance of the carbon isotopes. *J. Amer. Chem. Soc.*, **61**, 697–698.

Oana, S. and Deevey, E. S., (1960) Carbon 13 in lake waters and its possible bearing on paleolimnology. *Am. J. Sci.*, **258**, 253–272.

Park, R. and Epstein, S. (1960) Carbon isotope fractionation during photosynthesis. *Geochim. et Comochim. Acta*, **21**, 110–126.

Roe, A. and Hellman, M. (1951) Determination of an isotope effect in the decarboxylation of malonic-1-C^{14} acid. *J. Chem. Phys.*, **19**, 660.

Ropp, G. A., Weinberger, A. J. and Neville, O. K. (1951) the determination of the isotope effect and its variation with temperature in the dehydration of formic-C^{14} acid. *J. Chem. Phys.*, **73**, 5573.

Rosenfeld, W. D. and Silverman, S. R. (1959) Carbon isotope fractionation in bacterial production of methane. *Science* **130**, 1658–1659.

Sackett, W. M., Eckelmann, W. R., Bender, M. L. and Be' A. W. H. (1965) Temperature dependence of carbon isotope composition in marine plankton and sediments. *Science*, **148**, 235–237.

Silverman, S. R. and Epstein, S. (1958) Carbon isotopic compositions of petroleums and other sedimentary organic materials. *Bull. Am. Ass. Petr. Geol.*, **42**, 998–1012.

Stevenson, D. P., Wagner, C. D., Beeck, O. and Otros, J. W. (1948) Isotope effect in the thermal cracking of propane-1-C^{13}. *J. Chem. Phys.*, **16**, 993–994.

Urey, H. C. (1947) The thermodynamic properties of isotopic substances. *J. Am. Chem. Soc.*, pp. 562–581.

Urey, H. C., Greiff, L. J. (1935) Isotopic exchange equilibria. *J. Am. Chem. Soc.*, **57**, 321–327.

Wasserburg, G. J., Mazor, E. and Zartman, R. E. (1963) Isotopic and chemical composition of some terrestrial natural gases. *Earth Science and Meteoritics*, North-Holland Publ. Co., Amsterdam, pp. 219–240.

West, S. S. (1945) The relative abundance of the carbon isotopes in petroleum. *Geophysics*, **10**, 406–420.

Wickman, F. (1952), Variations in the relative abundance of the carbon isotopes in plants. *Geoch. et Cosmoch. Acta*, **2**, 243–254.

Zartman, R. E., Wasserburg, G. J. and Reynold, J. H. (1961) Helium, argon and carbon in some natural gases. *J. Geophys. Res.*, **66**, 277–306.

Zelinski, M. (1964) C^{13} and O^{18} isotope effects in the decarboxylation of quinaldic acid. *J. Chem. Phys.*, **41**, 3646.

HYDROGEN ISOTOPIC FRACTIONATION OF WATER PASSING THROUGH TREES

by

R. L. WERSHAW, IRVING FRIEDMAN and S. J. HELLER

U.S. Geological Survey

and

P. A. FRANK

Agricultural Research Service

Introduction

Washburn and Smith (1934) found by density measurements that water in the branches and leaves of a willow tree *(Salix nigra)* had a higher density than river water in the vicinity of the tree. They attributed this increased density to a higher deuterium content in the tree water. Since their work the hydrogen isotopic composition of water in trees has not been investigated by the much more sensitive method of mass spectrometry. However, Gonfiantini, Gratziu and Tongiorgi (1965) have measured by mass spectrometry the O^{18}/O^{16} ratio of water in tree leaves, and they found that the water in the leaves during the daytime was enriched in O^{18} compared to water in leaves from the same trees during the evening, and that water in the external leaves of cabbage and onion had a higher O^{18}/O^{16} ratio than water in the internal leaves. They deduced from these observations that transpiration from leaves increases the O^{18}/O^{16} ratio of the water remaining in the leaves. The increases in deuterium and O^{18} contents of water in leaves undergoing active transpiration would be expected from the results of Craig, Gordon and Horibe (1963) who have shown that liquid water in equilibrium with vapor is enriched in deuterium content by about 7% with respect to the vapor phase, and that the O^{18} content is about 0·9% higher in the liquid than in the vapor.

We have undertaken a detailed study of the hydrogen and oxygen isotopic fractionation taking place in trees, and shall report here some of the results of our hydrogen isotope study. Hopefully this work will permit us to calculate evaporation-transpiration ratios in wooded areas and to understand better the parameters which affect transpiration rates in plants.

Sample Preparation

Water for isotopic analysis was either expressed or distilled from the samples used in this study. The xylem, phloem, and root samples of cottonwood *(Populus sargentii)* and willow *(Salix* sp.*)* trees, all of the cottonwood leaf samples and some of the willow leaf samples were expressed in the apparatus shown in Fig. 1. The expressor was assembled with a sample inside it as shown in the diagram and was squeezed between the two rams

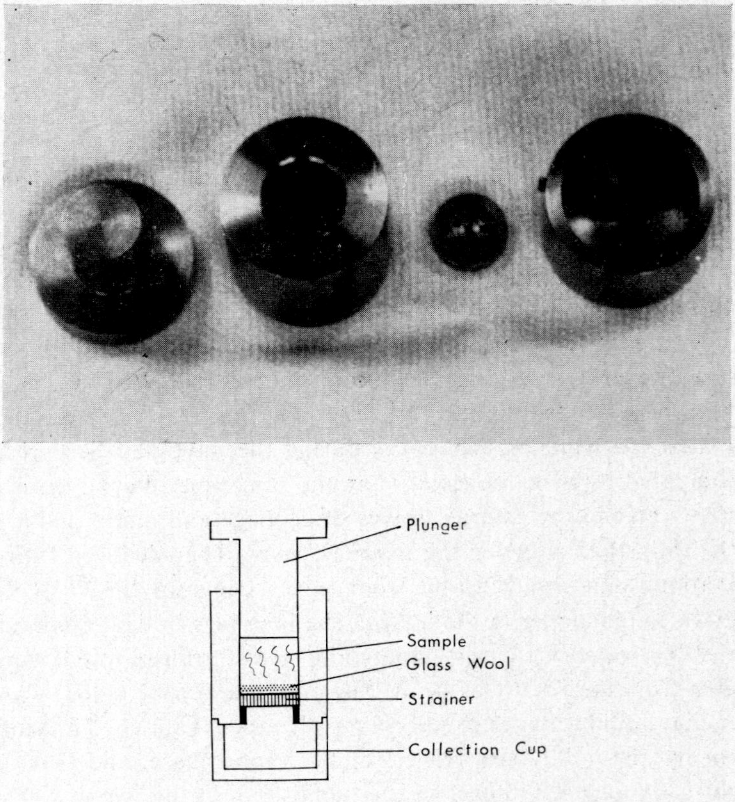

FIGURE 1. – Photograph and drawing of expressor.

of a hydraulic press; the water was collected in the cup on the bottom of the apparatus. Typically samples of 0·5 ml to 2 ml were obtained.

The distillation procedure consisted of placing the sample in a heated tube connected to a cold trap cooled to dry ice temperature. The entire system was evacuated and the tube containing the sample heated to 150° C in order to drive all of the water from the sample into the cold trap.

The water samples were reduced to hydrogen on hot uranium metal and the $\frac{HD}{H_2}$ ratios mass spectrometrically measured as described by Friedman (1953).

Results

Results of the isotopic analyses are reported in this paper as δ_D values where:

$$\delta_D = \frac{\left(\frac{HD}{H_2}\right)\text{Sample} - \left(\frac{HD}{H_2}\right)\text{Standard}}{\left(\frac{HD}{H_2}\right)\text{Standard}} \times 100.$$

The isotopic standard against which all of the samples reported here have been compared is standard mean ocean water (SMOW) which has been defined by Craig (1961).

In order to test the sample preparation procedure a sample of wood was divided into two parts; the water in one of the parts was extracted by distillation overnight and that in the other was extracted by expressing. The δ_D values measured for these two water samples agreed within the limit of error of the mass spectrometer analyses (0·1‰). However, it was found that if the distillation was not carried out to completion, isotopic fractionation of the water took place, as we observed in some of our shorter distillation runs. In all of the samples reported here in which the water was recovered by distillation, the distillation was carried out overnight (at least 16 hours). Because of the long times required for complete recovery of water by distillation we only employed the distillation technique for the soil samples and willow leaf samples where the water could not be recovered by expressing.

The first results which we shall discuss are those of our work on a cottonwood tree growing beside a pond on the eastern end of the Denver Federal Center, Denver, Colorado. 39° 45′ north latitude, 105° west longitude.

Figure 2 is a diagrammatic sketch of the pond and the cottonwood tree growing beside it. A generalized cut-away section of the tree shows the phloem, the living tissue just beneath the bark of the tree through which

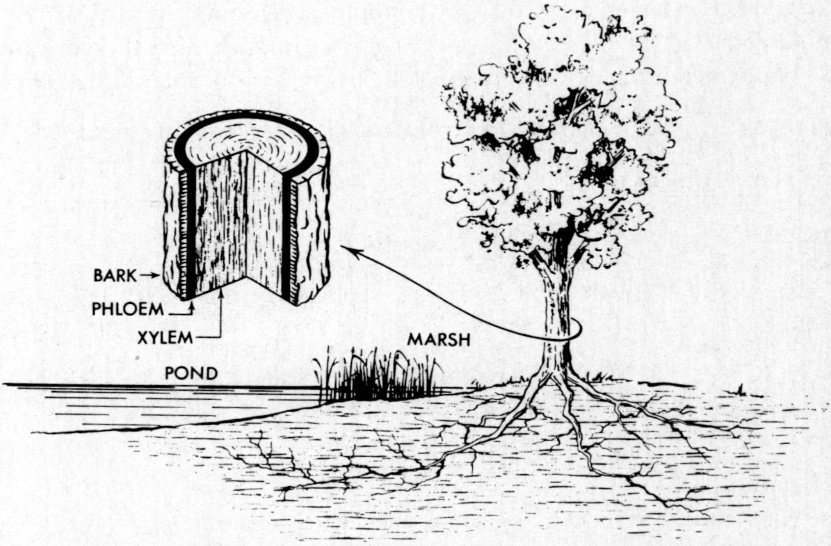

FIGURE 2. — Diagrammatic drawing of cottonwood tree growing beside pond on the Denver Federal Center and generalized cut-away section of the tree.

the nutrients manufactured in the leaves are translocated to various parts of the tree, and the xylem, the part of the tree which carries the water from the roots to the leaves.

The cottonwood tree which was sampled in this work is growing about 1 metre from the edge of the marsh zone of the pond in a sandy clay loam. It is about 10 metres high and 77 cm in circumference at breast height. The pond itself is about 180 metres long and about 120 metres wide at its widest part. Figure 3 is a photograph of the tree that was used in this study; the wounds were made when phloem and xylem samples were taken from it. The water in the pond is derived mainly from ground water seepage, and the storm sewers of the Denver Federal Center. A variety of plants grow around the pond; four different kinds of trees were identified, cottonwood *(Populas sargentii)*, Siberian elm *(Ulmus pumila)*, Russian olive *(Eleagnus angustifolia)*, and willow *(Salix* sp.), and in addition to the trees, milkweed *(Asclepias syriaca)* duckweed *(Lemna Minor)* and rushes *(Scirpus americanus)* grow in the area.

FIGURE 3. — Photograph of the trunk of the cottonwood tree on Denver Federal Center showing wounds made when phloem and xylem samples were taken from the tree.

Several relationships are immediately apparent from Tables 1 and 2 and the corresponding histograms (Fig. 4) of the data obtained from the cottonwood on the Denver Federal Center. (1) In all cases where both water in the trees and water in the pond were analyzed it was found that the water in the trees was heavier (had a higher δ_D value) than the water in the pond. Similarly, the water in the marsh and in the ground is heavier than water in the pond. The relationships of the water in the ground will

FIGURE 4. — Histograms of δ_D values of Denver Federal Center cottonwood *(Populus sargentii)* samples.

be discussed in greater detail below. (2) The water from the phloem generally has an isotopic composition similar to that of the water in xylem; the differences are less than one percent. (3) The water samples from tree leaves are much heavier than any other water samples from the area.

TABLE 1. – δ_D VALUES OF DENVER FEDERAL CENTER COTTONWOOD (*Populus sargentii*) SAMPLES ON DAYS WHEN LEAVES WERE NOT TAKEN

Date	June 13	June 21	June 23	July 18	July 25
	δD	δD	δD	δD	δD
Pond water	—10·8	—10·8	—11·1	—9·5	—9·2
Phloem	—10·5	— 9·8	—10·5	—8·4	—8·2
Xylem	— 9·9	— 9·4	—10·4	—8·7	—8·5
Root				—8·0	—7·5 Sample I
					—7·9 Sample II
Soil					—8·6 Sample I
					—7·9 Sample II
Marsh				—7·5	

The enrichment of deuterium in the water of tree leaves is probably attributable either to isotopic fractionation taking place upon evaporation of water from the leaves (transpiration) or to isotopic fractionation between the carbohydrates being synthesized in the leaves and the water passing through the leaves. It will be shown below that the former process is probably the dominant one.

It will be noted from the data in Table 2 and Fig. 4 that in general the leaves with the heaviest water were taken on clear days and warm days, whereas on the cooler, overcast days the leaf water is lighter. Van Hylckama (1966, and personal communication) and others have pointed out that the amount of transpiration that takes place from a tree is a function of amount of impinging radiation on the tree, the air temperature and the relative humidity, among other things. These data seem to bear this out. However, the morning sample taken before sunrise on August 31, has essentially the same isotopic composition as the sample taken at 1 : 20 in the afternoon on the same day. This may be due to one of the following processes or a combination of them. (1) Transpiration may take place at a high rate on warm nights, or (2) the relatively heavy water in leaves at night may be largely the product of respiration during the night. This would imply that hydrogen in the carbohydrates is as heavy as that

TABLE 2. — δ_D VALUES OF DENVER FEDERAL CENTER COTTONWOOD (*Populus sargentii*) SAMPLES ON DAYS WHEN LEAF SAMPLES WERE TAKEN

Date	July 12	Aug. 12	Aug. 16	Aug. 16	Aug. 25	Aug. 31	Aug. 31	Sept. 9
Time	10 : 10 a.m.†	8 : 00 a.m.	7 : 30 a.m.	3 : 30 p.m.	8 : 00 p.m.	5 : 00 a.m. (Sunrise 5 : 27)	1 : 20 p.m.	12 : 30 p.m.
Cloud cover	Clear	Very overcast	Clear	Overcast	Clear	Clear	Overcast	Partly cloudy
Temperature	27° C	19·5° C	17° C*	Cool	19° C	14·5° C	23° C	30° C
Relative humidity	50%	81%	58%*		37%	71%	47%	22%
	δ_D	δ_D	δ_D	δ_D	δ_D	δ_D	δ_D	δ_D
Pond	—11·1	—9·5	—9·5	—9·5	—9·5	—9·5	—9·5	—9·6 (Measured)
Phloem	—10·6	—8·0			—9·6			—7·1
Xylem	—10·5	—7·7			—9·5			—7·0
Leaves	— 3·1	—5·0	4·2	—5·0	—2·4	—1·6	—1·7	—0·7
Difference leaves — Xylem	+ 7·4	+2·8			+7·1			+6·3
Difference leaves — Pond	+ 8·0	+4·5	+5·3	+4·5	+7·1	+7·9	+7·8	+8·8

* Data from Denver Weather Bureau.

† The times given in this Table are Mountain Standard times and are also local Solar times for the Denver area, because the central meridian (105°) for the Mountain Standard zone passes through Denver.

in the water from which it was synthesized during the light hours. We are going to investigate further the hydrogen isotopic fractionation that takes place between water and the carbohydrates synthesized from it. It may be that high D/H ratios in the water in trees is partially due to a concentration of Protium in the plant tissue, although this does not seem probable. Daum (personal communication) measured transpiration rates of the order of 65 gallons (260 kg) per day for cottonwood trees of similar size to those studied in this paper. We have estimated that the amount of carbohydrate synthesis per day is at least two orders of magnitude less than the amount of water transpired, and therefore even if the enrichment factor for Protium in the carbohydrates were very large we could not entirely account for the high D/H ratios of tree waters by fractionation between the carbohydrates and the water.

TABLE 3. – δ_D VALUES ARVADA COTTONWOOD *(Populus sargentii)**

Date	July 13	July 29
	δ_D	δ_D
Stream	—13·4	—12·8
Phloem	—12·2	—11·7
Xylem	—12·4	—11·6
Root		—10·3
Soil around root		—11·4

* The samples were collected at approximately 4 : 00 P.m. MST.

TABLE 4. – δ_D VALUES OF ARVADA WILLOW *(Salix* sp.*)* SAMPLES*

Date	July 5	July 13	July 24	July 29
	δ_D	δ_D	δ_D	δ_D
Stream	—12·2	—13·4	—12·9	—12·8
Phloem	—11·8	—12·6	—13·0	
Xylem	—11·9	—12·8	—12·8	
Root				—12·3
Soil around root				—12·1
Leaves			— 6·9	— 4·1

* These samples were collected at approximately 4 : 00 P.m. MST.

Data similar to those from the tree on the Federal Center have been obtained from a cottonwood tree and a willow tree about 10 kilometres north of the Federal Center. As indicated by the data in Tables 3 and 4, and Figs. 5 and 6 the water samples from the phloem and xylem are

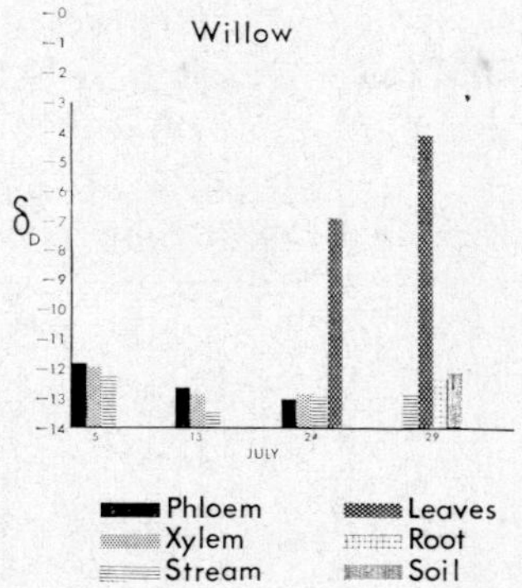

FIGURE 5. — Histogram of δ_D values of Arvada willow *(Salix* sp.*)* samples.

heavier than the ground water, and leaf waters are again much heavier than any of the other water samples.

In order to determine if similar relationships hold for trees growing in more humid areas than Denver (mean annual rainfall 35 cm), samples

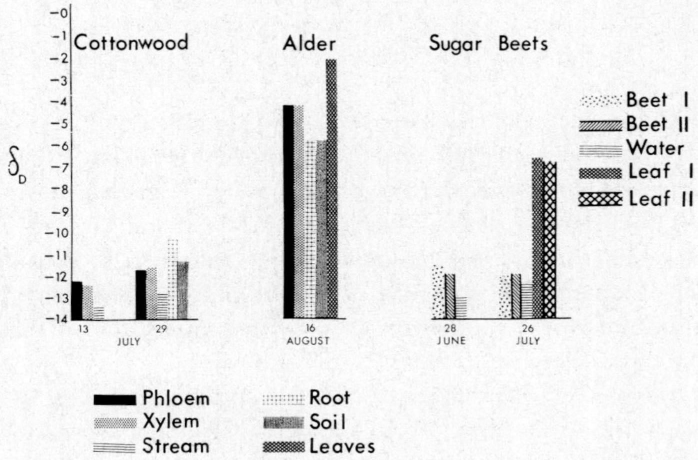

FIGURE 6. — Histograms of δ_D values of Arvada cottonwood *(Populus sargentii)*; Washington alder *(Alnus rubra)* and sugar beet *(Beta vulgaris* L.) samples.

were taken from an alder tree *(Alnus rubra)* growing in the Olympic Penninsula, Washington, where the mean annual rainfall is 360 cm. The data in Table 5 show that the water in the leaves is 3·1% heavier than water in the phloem and xylem of the tree. It will be noted that this difference is less than the average difference for the cottonwood and willow trees in the Denver area, as one would expect it to be because of the higher relative humidity and lower temperature.

TABLE 5. – δ_D VALUES OF WASHINGTON ALDER *(Alnus rubra)* SAMPLES*

	δ_D
Phloem	—5·1
Xylem	—5·1
Root	—5·7
Soil surrounding root sample	—5·7
Leaves	—2·0

Difference between leaves and phloem +3·1

* These samples were taken from a tree growing in the Hoh rain forest, Olympia National Park, Washington, 124° W. Longitude, 47° 50′ N. Latitude at 5 : 45 P.m., August 16, 1966. The estimated relative humidity was 70%, the estimated temperature 18° to 20° C. The mean annual rainfall for the area is 360 cm.

In all of the tree samples discussed above the difference between the δ_D values for the phloem and xylem waters for a given time is small. At first thought one might expect the phloem water migrating down from the leaves to be much heavier than the water moving through xylem, however, Zimmerman (1964) has pointed out that there is considerable migration of water back and forth from the phloem to xylem in trees and this accounts for the similarity in isotopic composition of the water in the two types of tissue.

A few samples of sugar beet plants were obtained from the Agricultural Research Service farm, Fort Collins, Colorado, in order to see if the water in leaves of plants other than trees was also enriched in deuterium, and as shown in Table 6 an enrichment of deuterium of about 5% was found in the leaves. Isotopic measurements have not been made on trees where the transpiration rates are known. However, we plan to do this on some salt cedar trees being grown by Mr. T. E. A. Van Hylckama of the Geological Survey along the Gila River near Phoenix, Arizona. Unfortunately, at the present time we can only make inferences about the relationships that exist between the isotopic composition of the water in the leaves of trees and the amount of transpiration that is taking place from them.

TABLE 6. — δ_D VALUES OF WATERS EXTRACTED FROM SUGAR BEET SAMPLES (*Beta vulgaris* L.)*

Date	June 28	July 26
	δ_D	δ_D
Irrigation water	—12·9	—12·3
Beet Sample I	—11·5	—12·0
Beet Sample II	—11·9	—11·9
Leaf Sample I		— 6·5
Leaf Sample II		— 6·9
Difference (Leaf – Beet) I		+ 5·5
Difference (Leaf – Beet) II		+ 5·0

* These samples were kindly supplied by Dr. Edward E. Schweizer, Crops Research Division, Agricultural Research Service, U.S. Department of Agriculture, Fort Collins, Colorado.

Conclusions

The water in the leaves of a tree is enriched in deuterium with respect to the water in the phloem and the xylem of the tree which, in turn, are enriched in deuterium with respect to the ground water beneath the tree. The enrichment in the leaves is apparently due to isotopic fractionation that takes place during transpiration, and it appears that the amount of fractionation is a function of the rate of transpiration from the tree; the transpiration rate is dependent upon temperature, relative humidity and solar radiation level. Two different processes have probably contributed to the enrichment of deuterium in the water in the phloem and xylem tissue: (1) movement of water from the leaves down into the trunk of the tree, and (2) enrichment of deuterium in the soil water surrounding the tree roots by evaporation. This enrichment of deuterium by evaporation was observed in waters from soil samples taken near the Federal Center cottonwood tree, as indicated by the data in Table 7.

TABLE 7. — δ_D VALUES, SOIL MOISTURE CONTENT AND MOISTURE TENSION OF SOIL DEPTH INCREMENTS*

	δ_D	Soil moisture (% by weight	Moisture tension (bars)
Pond water	—9·6		
Marsh	—7·2		
First inch	—6·1	21·53%	10·0
Second inch	—8·8	25·61%	8·0
Third inch	—8·4	29·59%	6·7

* These samples were taken at 10 : 00 A.m. MST, August 4, 1966, within 1 metre of the cottonwood tree sampled on the Denver Federal Center. The day was overcast, the relative humidity 46% and the temperature 26° C.

The water in the soil in the top 1 inch, where more evaporation is taking place, has a δ_D value of 2% higher than the δ_D values of waters from the lower zones. The data from all of the soil is generally considerably lighter than the residual water in tree leaves. The isotopic composition of the water vapor coming from the soil should therefore be different from that of the water vapor coming from tree leaves, and it should, consequently be possible to calculate, by making a materials balance, the ratio of the amount of water evaporating from soil to the amount of water transpiring from leaves in a given area.

Acknowledgements

The authors would like to acknowledge the assistance of Mr. James Gleason and Mrs. Joy Church in performing the isotopic analyses, and Miss Alberta Anema who helped prepare the samples for isotopic analyses. We wish to thank Mrs. Dorothy Pinckney for collecting the alder tree samples for us and Dr. Edward E. Schweizer of the Agricultural Research Service for providing us with the sugar beet samples. The authors are indebted to Miss Sarah Booker for carefully preparing the manuscript for publication.

References

Craig, H. (1961) Standard for reporting concentrations of deuterium and oxygen-18 in natural waters. *Science*, **133,** 1833–1834.

Friedman, Irving. (1953) Deuterium content of water and other substances. *Geochim. et Cosmochim. Acta*, **4,** 89–103.

Gonfiantini, R., Gratziu, S. and Tongiorgi, E. (1965) Oxygen isotopic composition of water in leaves. In *Isotopes and Radiation in Soil-Plant Nutrition Studies*, Vienna, Int. Atomic Energy Agency, pp. 405–410.

Van Hylckama, T. E. A. (1966) Evaporation from vegetated and fallow soils. *Water Resources Res.* **2,** 99–103.

Washburn, E. W. and Smith, E. R. (1934) The isotopic fractionation of water by physiological processes. *Science*, **79,** 188–189.

Zimmerman, M. H. (1964) Sap movements in trees. *Biorheology*, **2,** 15–27.

GEOCHEMICAL ASPECTS OF THE OCCURRENCE OF PORPHYRINS IN WEST VENEZUELAN MINERAL OILS AND ROCKS

by

J. A. GRANSCH and E. EISMA

Koninklijke/Shell
Exploratie en Produktie Laboratorium
Rijswijk, the Netherlands

Introduction

On the basis of investigations into the presence of indigenous hydrocarbons and residual organic carbon, Philippi (1957) concluded that the West Venezuelan Cretaceous La Luna formation contains outstanding oil source rocks. This conclusion confirmed the opinion of various geologists (Hedberg, 1931; Dufour, 1955) that the La Luna formation has produced the wealth of oil in the West Venezuelan Lake Basin. This type of investigation leads to a classification of the organic matter in sediments but does not directly link the sediments with the oils. The aim of the present investigation is to make such a direct link.

Characteristic Compounds in Crude Oils and Their Source Rocks

In 1936 Krejci-Graf pointed out that crude oils containing certain characteristic compounds should have originated in oil source beds also containing these compounds. We have tried to trace the Lake Basin oils towards the La Luna formation by means of porphyrins.

Van Eggelpoel in 1964 at the Second International Geochemical Congress in Paris reported upon a similar investigation concerning the origin of the Ozouri oil from Gabon.

Porphyrins

The porphyrins occurring in crude oils are generally bound to vanadium and nickel. In sediments they may be bound to iron and sometimes they occur metal-free. Vanadium and nickel porphyrins each show two distinct maxima in the visible part of the absorption spectrum. These maxima differ for the two metal complexes; the main maximum for vanadium complexes is generally at about 5750 A.U., that of the nickel complexes at about 5550 A.U. (Dunning's review on porphyrins in *Organic Geochemistry*, edited by Breger, 1963). Recently Millson, Montgomery and Brown have reported on porphyrin/vanadium complexes having an absorption maximum at about 5900 A.U. (1966).

Porphyrins in West Venezuelan Crude Oils and Sediment Samples

In the absorption spectra of chloroform solutions of West Venezuelan Lake Basin crude oils a maximum at about 5750 A.U. is perceptible above the continuous absorption of the crude oils (Figs. 1 and 2).

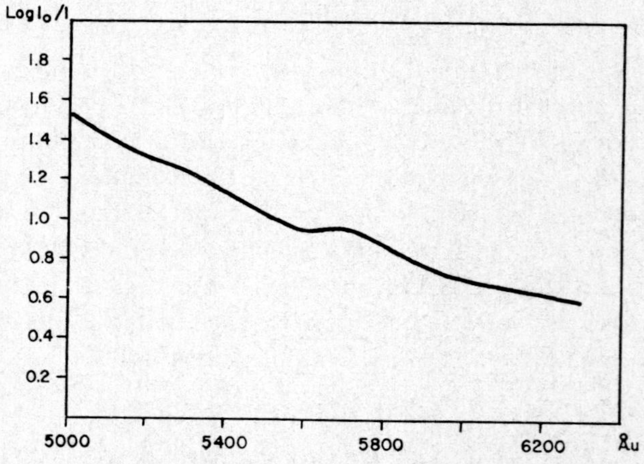

FIGURE 1. – Bachaquero-Crude TMC 5087. % porphyrin as DPEP-V 0·126% conc. = 1·50 g/l. 1 cm cell.

The absorption spectra of chloroform solutions of several Cretaceous sediment samples, particularly those of the La Luna, show this maximum too (Fig. 3).

In a few cases a maximum at about 5550 A.U. has been observed (Fig. 4). The maximum at about 5750 A.U. is ascribed to porphyrin/

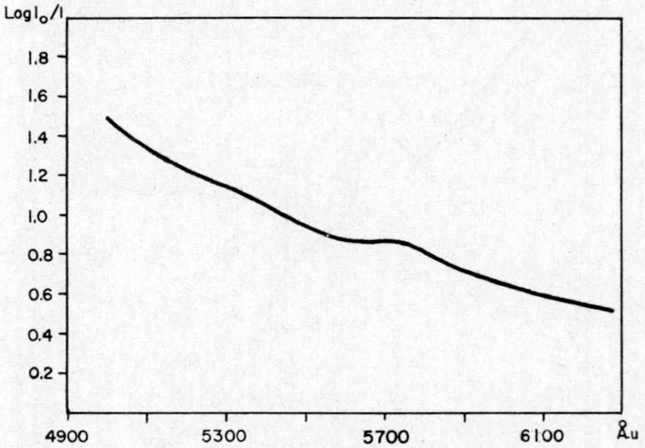

FIGURE 2. – La Paz-Crude TMC 5129% porphyrin as DPEP-V 0·045% conc. = 12 g/l. 0·5 cm cell.

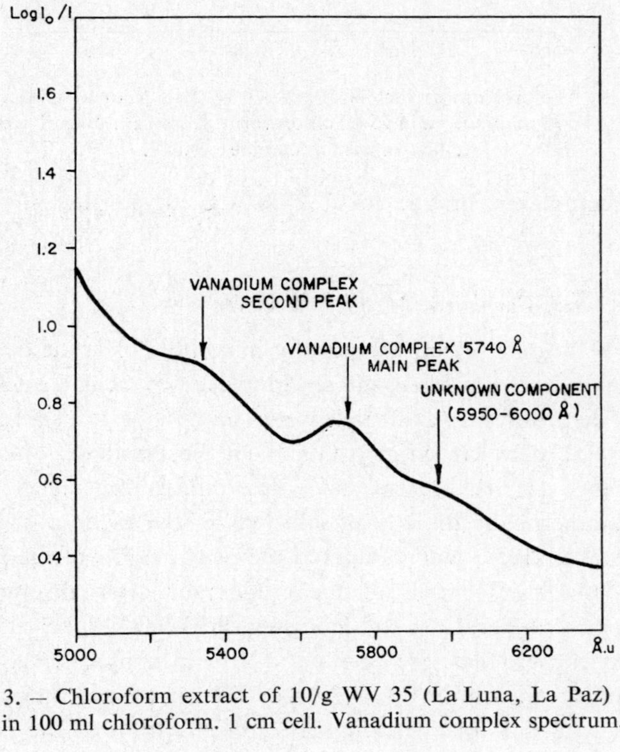

FIGURE 3. – Chloroform extract of 10/g WV 35 (La Luna, La Paz) dissolved in 100 ml chloroform. 1 cm cell. Vanadium complex spectrum.

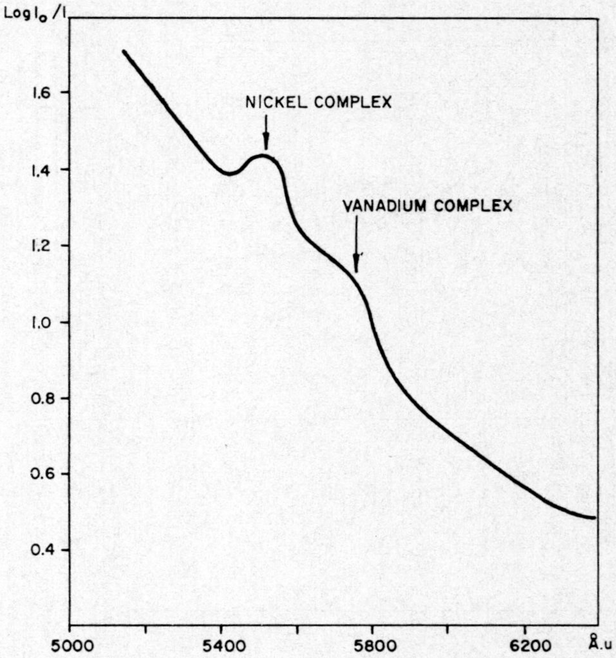

FIGURE 4. — Chloroform extract of 10g VPO 31 (Rio Aponcito Secco) of the Cogollo formation dissolved in 25 ml chloroform. 1 cm cell. Nickel complex as well as vanadium complex visible.

vanadium complexes, that at about 5550 A.U. to porphyrin/nickel complexes.

Determination of Porphyrin/Metal Complexes

Treibs (1934) determined the porphyrin content of crude oils by liberating porphyrins from their metal complexes by acid cleavage with a mixture of hydrobromic acid and acetic acid. This procedure has been further developed by Groennings (1953), by Costantinides (1957) and by Warren Howe (1961). Others have determined the metal complexes spectrophotometrically directly in solutions of the crude oils, e.g. Beach and Shewmaker (1957) who compared the peak heights of the porphyrins with the vanadium content of the crudes; or after chromatographic concentration or separation, e.g. Hodgson, Peake and Baker (1963).

We have chosen the perceptibility of the porphyrin maxima above the continuous absorption of the crude oil or sediment extract themselves as a yardstick for estimating the concentration of the porphyrins.

Crude oils were dissolved in chloroform; sediment samples were crushed and ground and the powders obtained were subjected to extraction with chloroform. The absorption spectra of the chloroform solutions were then measured in the region from about 4500 to 7000 A.U. with a Cary spectrophotometer.

As a standard vanadium complex we used a concentrate of a porphyrin/vanadium complex that at M. Blumer's (1950) recommendation was isolated from a Triassic shale from Serpiano in Switzerland. This standard, which had been concentrated by chromatography and fractional

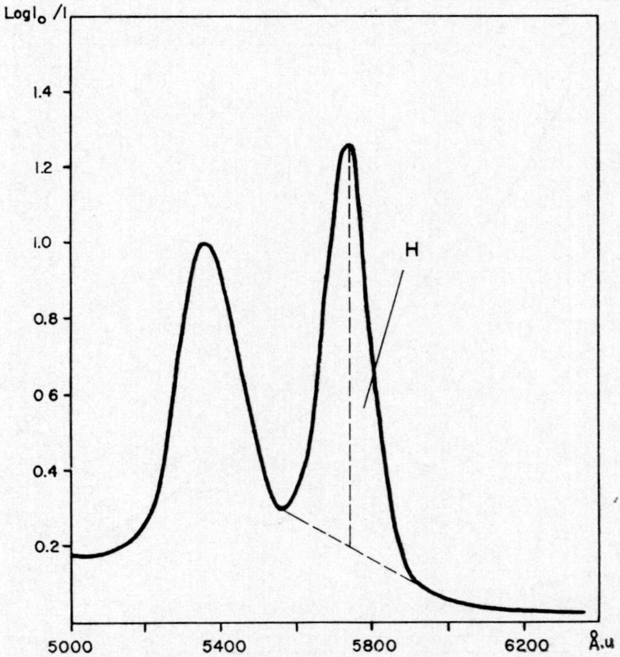

FIGURE 5. — Desoxophyllerythro-etio-porphyrin vanadium complex (DPEP-V) in chloroform, conc. = 38·34 mg/l. 1 cm cell.

precipitation, was purified until the absorption spectrum remained constant (see Fig. 5). The principal constituent of this concentrate is desoxophyllerythro-etio-porphyrin/vanadium complex (DPEP-V). Although we realise that the porphyrins in crude oils are by no means uniform in composition (see for instance Baker, 1966) we have nevertheless assumed that the sediments, crude oils and standard contain only this porphyrin/ vanadium complex.

The tangents to the curves of the porphyrin maxima were used as base lines and the peak to base-line distances H of the standard metal complex (Fig. 5) and h of the crude oil (Fig. 6) represent directly the so-called base-line optical density or base-line absorbance. The base-line absorptivity (1 gram of porphyrins/litre, 1 cm cell) e equals H/g_1, where g_1 is the weight of the porphyrin standard in grams.

The concentration in p.p.m. of the porphyrins in the crude is expressed by $c = h/e \times 10^6 g_2 L$ p.p.m., where g_2 is the concentration of the crude in grams per litre and L represents the optical path length in cm. For the standard used, dissolved in chloroform, $e = 27{\cdot}5$.

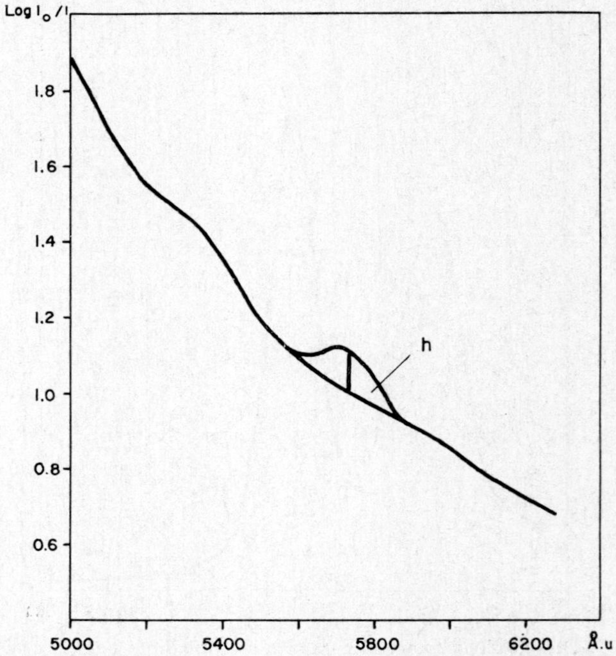

FIGURE 6. — La Paz TMC 5130 crude oil with porphyrin/vanadium complex.

The method was checked by adding increasing quantities of DPEP-V to two crude oils not containing porphyrins, a light one and a heavy one (Fig. 7). The table shows a very satisfactory agreement between the actual and the measured concentrations. The method when applied to a heavy crude (Los Manueles, West Venezuela) appears to be less sensitive in the lower concentrations than when applied to a light crude (Kenali Assam, Sumatra).

ppm DPEP-V added to oil	Base-line opt. dens. at 5740 Å	Calc. ppm of DPEP-V in oil $= \frac{h}{27.5 \times L\ C} \times 10^6$
70	0.392 − 0.340 = 0.052	63
130	0.453 − 0.350 = 0.103	125
270	0.580 − 0.367 = 0.213	258
530	0.855 − 0.420 = 0.435	527
1070	1.357 − 0.495 = 0.862	1045

DETERMINATION OF DPEP-V CONCENTRATIONS IN KENALI ASSAM OIL TO WHICH KNOWN QUANTITIES OF DPEP-V HAVE BEEN ADDED
(30 grams of oil/litre $CHCL_3$, 1 cm cell)

ppm DPEP-V added to oil	Base-line opt. dens. at 5740 Å	Calc. ppm of DPEP-V in oil $= \frac{h}{27.5 \times L\ C} \times 10^6$
0	1.013	0
80	1.017 − 1.017 = 0	0
160	1.023 − 1.015 = 0.008	116
240	1.057 − 1.040 = 0.017	247
400	1.062 − 1.037 = 0.025	364
790	1.090 − 1.043 = 0.047	684
1580	1.215 − 1.111 = 0.104	1513
3170	1.280 − 1.062 = 0.218	3171

DETERMINATION OF DPEP-V CONCENTRATIONS IN LOS MANUELES OIL TO WHICH KNOWN QUANTITIES OF DPEP-V HAVE BEEN ADDED
(2·5 g of oil/litre $CHCL_3$, 1 cm cell).

FIGURE 7. – Calibration of DPEP-V Determination.

			Formation
PLIO-PLEISTOCENE			La Puerta
TERTIARY	MIO-PLIOCENE		Lagunillas
TERTIARY	MIOCENE		La Rosa
TERTIARY	OLIGOCENE		Sierrita, Icotea
TERTIARY	U. EOCENE		Pauji
TERTIARY	UPPER MIDDLE EOCENE		Pauji
TERTIARY	MIDDLE EOCENE		Zamuro
TERTIARY	LOWER MIDDLE TO LOWER EOCENE	Misoa-Trujillo	Pueblo Viejo
TERTIARY	LOWER MIDDLE TO LOWER EOCENE		La concepcion
TERTIARY	PALEOCENE		Guasare
CRETACEOUS	MAESTRICHTIAN		Mito Juan
CRETACEOUS	SENONIAN		Colon
CRETACEOUS	CONIACIAN TURONIAN		La Luna
CRETACEOUS	CENOMANIAN ALBIAN APTIAN BARREMIAN		Cogollo-Tomon
CRETACEOUS	HAUTERIVIAN VALANGINIAN		Cogollo-Tomon
PRE-CRETACEOUS			Basement

FIGURE 8. — Generalized stratigraphic column of Lake Basin in West Venezuela.

Discussion of the results

Figure 8 gives the locations from which the crude oil samples, the core samples and the surface samples are derived.

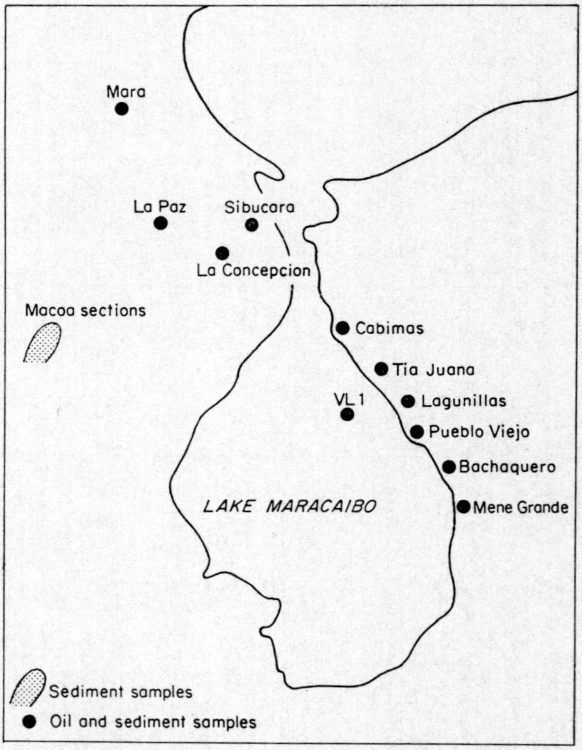

FIGURE 9. — Sample locations in West Venezuela.

Figure 9 gives a generalized scheme of the sequence of the more important geological formations in the Lake Basin (cf. Miller, 1948).

Porphyrin/Vanadium Complexes in Lake Basin crudes

In Fig. 10 the porphyrin content of the crude oils from the Mara Maracaibo province as well as those from the Bolivar Coast have been plotted against the depths of their reservoirs. The La Concepcion crudes and the Sibucara crude from the Mara Maracaibo district and one

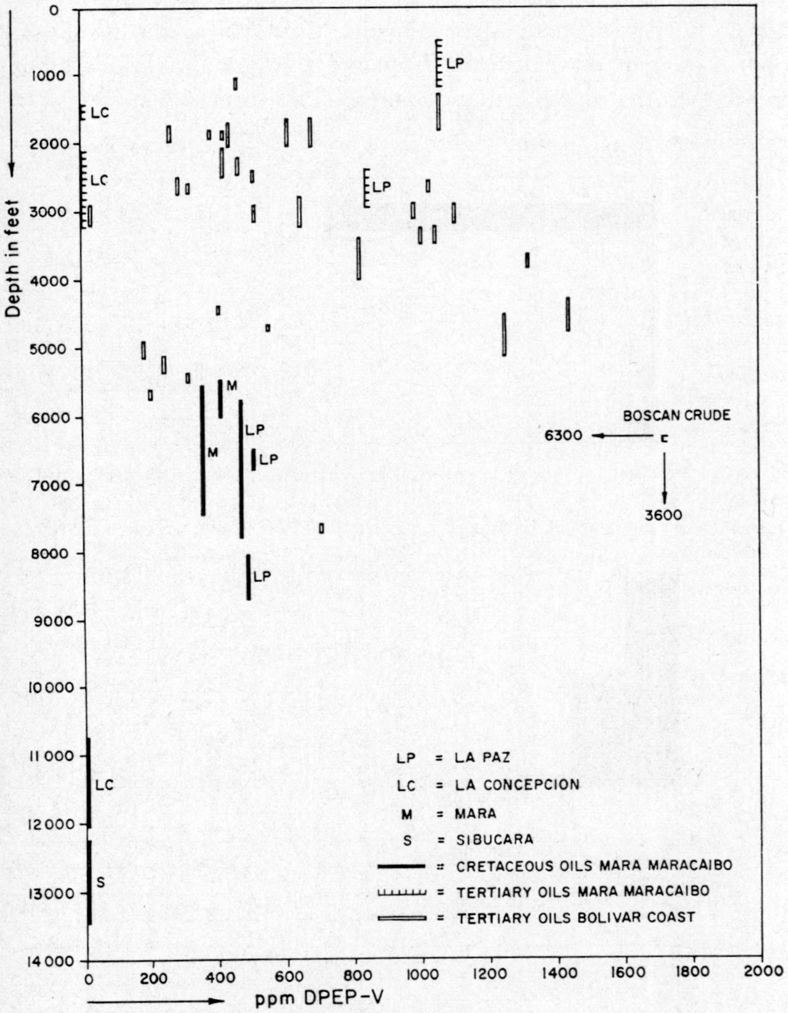

FIGURE 10. — Porphyrin content and reservoir depth of West Venezuelan Lake Basin crude oils.

Pueblo Viejo crude from the Bolivar Coast do not contain porphyrins. The Boscan crude from the Mara Maracaibo district has the highest content, viz. 3600 parts per million. All other crudes contain porphyrin/vanadium complexes, the concentration of which ranges between 200 and 1350 p.p.m.

The porphyrin content of the La Paz crudes tends to decrease with increasing depth, but the La Concepcion crudes and the Bolivar Coast crudes do not show this tendency. In the Mara Maracaibo province the porphyrin content of the crudes from the younger formations tends to be higher than that in the crudes from the older formations.

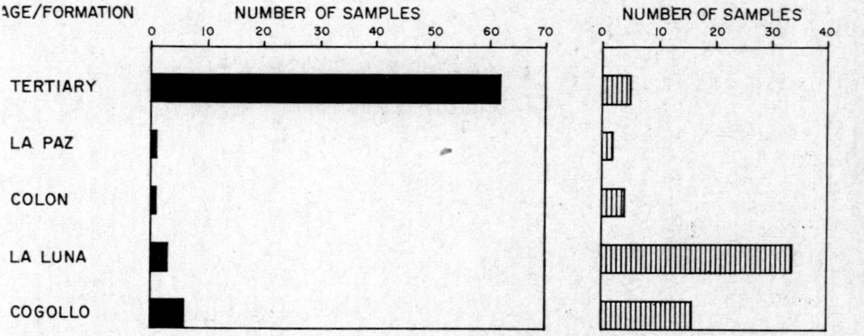

FIGURE 11. — Distribution of porphyrins in West Venezuelan sediment samples.

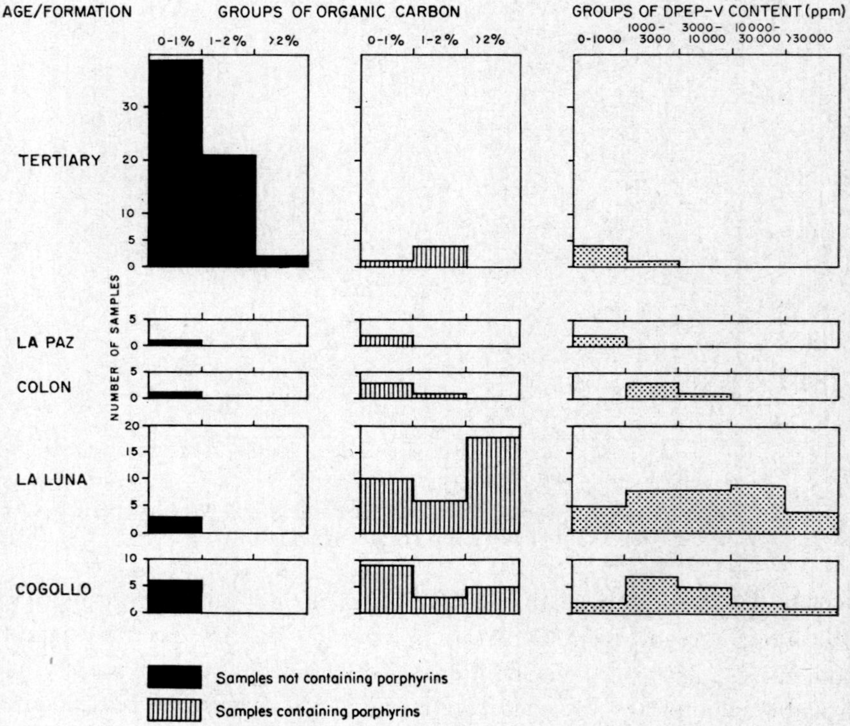

FIGURE 12. — Porphyrin and organic carbon content of West Venezuelan sediment samples.

Porphyrin/Vanadium Complexes in Sediment Samples

Figure 11 shows the number of investigated sediment samples containing or not containing porphyrins. Most of the Cretaceous La Luna and Cogollo samples contain porphyrins, in contrast to the Tertiary samples, the majority of which do not contain them.

Figure 12 indicates the organic carbon and porphyrin content of the various sediment samples analysed.

The Tertiary samples as well as the Cretaceous La Paz shales and Colon limestones generally contain less than 2% organic carbon, whereas several of the Cogollo samples and a large proportion of the La Luna samples contain more than 2% organic carbon.

The porphyrin content attains the highest values in the La Luna and Cogollo samples that have the highest organic carbon content. It should be noted that the porphyrin concentrations in the chloroform extracts are generally much higher than those of the crudes. Figure 13 gives the

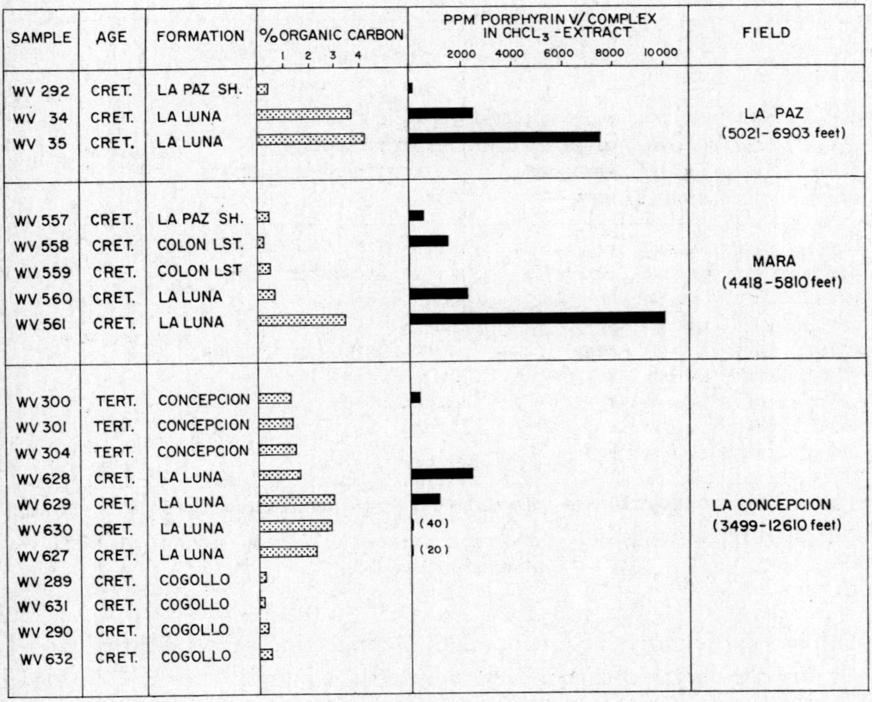

FIGURE 13. — Porphyrins and organic carbon in well sections of Mara Maracaibo area.

organic carbon and porphyrin content of three well sections in the Mara Maracaibo province. All La Luna samples contain porphyrins; in the La Concepcion field two of them have a low porphyrin content (40 and 20 p.p.m.); this will be discussed later.

With the exception of one sample in the Mara field all La Luna samples have a high organic carbon content. As the Mara La Luna sample shows (see Fig. 13), not all the La Luna samples are rich in organic matter. This is also illustrated by Fig. 14 which shows the results of some outcrop

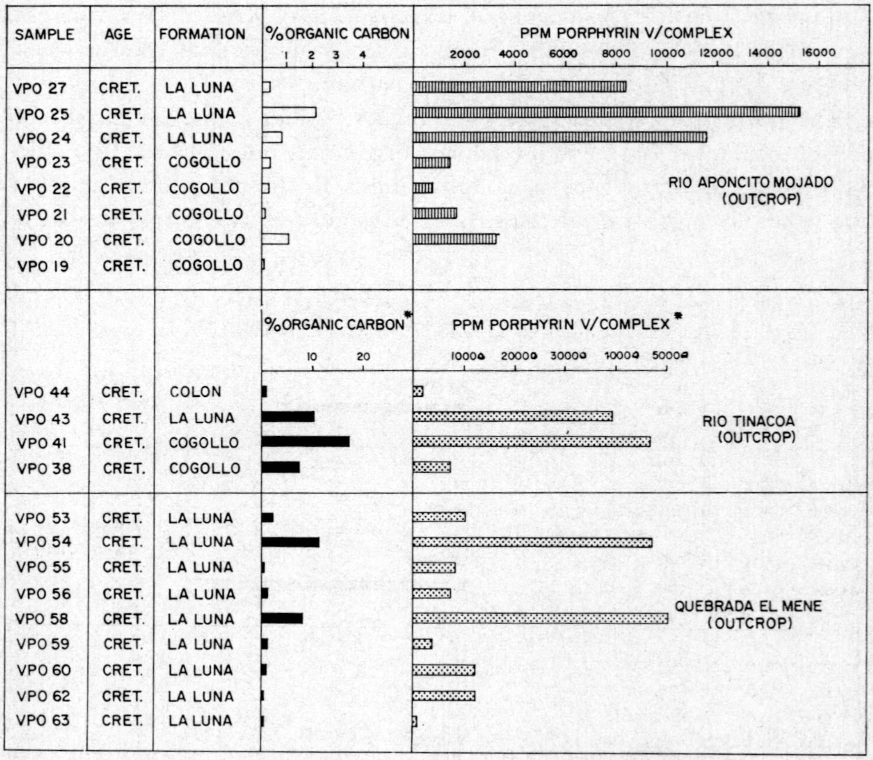

* NOTE THAT ORGANIC CARBON AND PORPHYRIN SCALES ARE DIFFERENT FROM FOREGOING SECTIONS

FIGURE 14. — Porphyrins and organic carbon in outcrop sections of Mara Maracaibo province.

sections: particularly the Quebrada El Mene section shows a large variation in the organic carbon and porphyrin content. The Rio Tinacoa section demonstrates that not only the La Luna but also the Cogollo may have a very high organic carbon content.

Why are porphyrins present in certain rocks and absent in others?

(a) *Environment*

The porphyrins occurring in the Lake Basin are most probably mainly derived from the chlorophylls. These pigments are present in vegetable matter and the remnants of vegetable matter together with inorganic sediment will be deposited. If in a sedimentary basin the amount of organic remains is large, the oxygen in the water is soon consumed, resulting in anaerobic conditions under which chlorophylls may be preserved (Brongersma-Sanders, 1951). When the sediments subside the chlorophylls will be gradually transformed into porphyrins. If vanadium is present in a form accessible to porphyrins, porphyrin/vanadium complexes are formed.

Large parts of the La Luna and some parts of the Cogollo formations have been formed under anaerobic conditions. When these sediments mature hydrocarbons are formed and eventually part of them may start to migrate as crude oil.

Porphyrin/vanadium complexes are soluble in crude oils and it is conceivable that they will migrate with the crude oil.

Migrating oil has to pass several layers before it arrives in a reservoir and on its way some of the oil with the porphyrins may remain in these layers. It is not always possible to indicate whether a rock contains migrated oil or not, but if the percentage of soluble organic matter is high in comparison with organic carbon content we consider it likely that such a rock contains migrated oil.

Most of the Tertiary rocks have a moderately high organic carbon content and the organic matter generally is unfavourable for the generation of hydrocarbons. We think it probable that most of these formations have been deposited under aerobic conditions which are unfavourable for the preservation of chlorophyll. We consider it likely that these depositional circumstances are the reason for the virtual absence of porphyrins in the Tertiary formations.

We will now discuss the possibility that porphyrins present in rocks will be destroyed by high temperature.

(b) *The Influence of Higher Temperature on the Porphyrin Content*

When a sediment subsides, its temperature increases. On account of the increased temperature fossil organic matter changes: a well-known example is expressed by the rule of Hilt: within a given borehole or shaft the fixed carbon content of coals increases as a result of increased temperature.

We have observed that in a few wells La Luna samples do not contain or contain only traces of porphyrins, although the organic carbon, vanadium and fossil content is very similar to that of samples which contain large amounts of porphyrins. Only the C_R/C_T ratio (Gransch and Eisma, 1966) was high. This suggests that porphyrins which may have been present may have been destroyed by thermal reactions. A simple relation between porphyrin content and present depth does not exist. Dr. J. K. A.

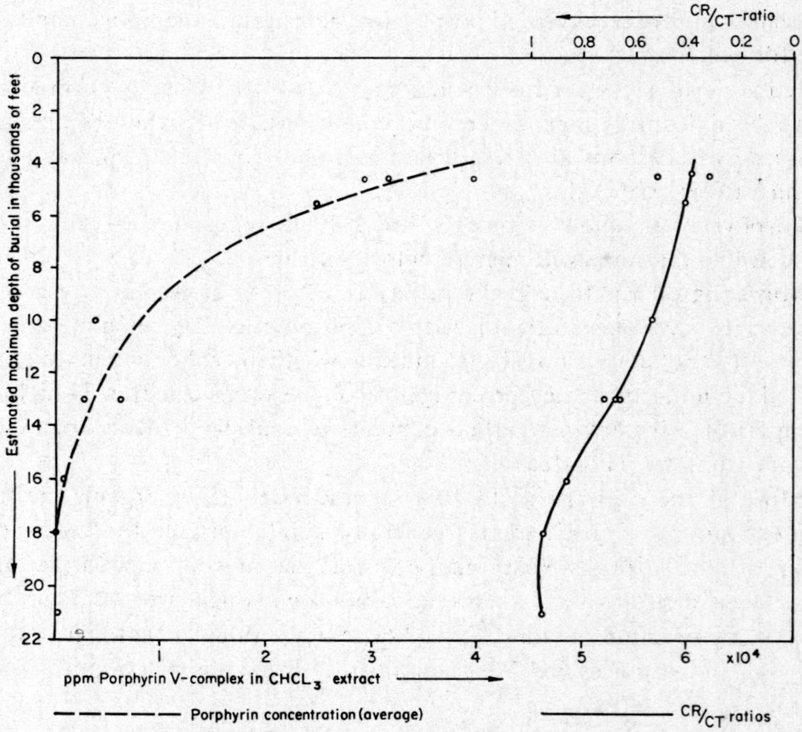

FIGURE 15. — Porphyrins and C_R/C_T ratios of La Luna source rocks against maximum depth of burial.

Habicht of B.I.P.M. supplied us with the maximum depths of burial of the La Luna source beds as estimated from geological evidence. In Fig. 15 we have plotted the average porphyrin content of La Luna source rocks as well as the C_R/C_T ratios against these palaeo-overburdens. The figure strongly suggests that the porphyrin content of the La Luna source rocks decreases with increasing depth of burial. This decrease is accompanied by an increase in the C_R/C_T ratio.

Origin of the Lake Basin crude oils

Mara Maracaibo Area

CRETACEOUS OILS. Figure 16 shows that there is a certain correlation between the concentration of porphyrins in the chloroform extracts from the La Luna core samples from Mara Maracaibo and the porphyrin concentration of the Cretaceous oils.

The average porphyrin content in the various La Luna core samples is much higher than that in the crude oils. Since the porphyrin content in the source rock seems to decrease with increasing depth of burial, the very low porphyrin content of the La Concepcion and Sibucara crudes can be ac-

FIELD	LA LUNA CORE SAMPLES		CRETACEOUS OILS
	Depth, feet	ppm V-complex in $CHCL_3$ extract	ppm V-complex in oils
La Paz	5003-5030	3900	450
Mara	5660-5810	6200	350-400
La Concepcion	10195-11580	840	2*
Sibucara	12757-13088	400	<0.3*

* Porphyrin content determined after chromatographic separation

FIGURE 16. — Comparison of porphyrin content of La Luna core samples with that of Cretaceous crudes in Mara Maracaibo area.

counted for by assuming that these crudes were liberated at a great depth, the La Paz and Mara crudes being liberated at a much smaller depth where the porphyrins were still abundantly present in the source rocks.

TERTIARY LA PAZ CRUDES. Since both the Cretaceous La Paz shales and the La Luna limestones and marls contain porphyrin/vanadium complexes it is difficult to draw a conclusion with regard to the formation from which the Tertiary La Paz crudes have originated (Fig. 13). If, however, the La

Paz shale sample analysed from La Paz is representative for the La Paz shales in this area, it seems more likely that the La Paz crudes originated from the La Luna, which, in contrast to the La Paz shale, is rich in favourable organic matter. Other samples from the La Paz shales from La Paz appear to have C_R/C_T ratios around 0·8, whereas La Luna samples from La Paz have C_R/C_T ratios around 0·5. This indicates that the La Paz shales, with respect to hydrocarbon generation, are less favourable than the La Luna or similar formations.

Bolivar Coast

All the oils occur in Tertiary formations and with the exception of one Pueblo Viejo oil all of them contain porphyrin/vanadium complexes. These oils should therefore be derived from a porphyrin-containing oil source rock.

Samples from several Tertiary formations have been analysed to determine their porphyrin content. Porphyrin/vanadium complexes have been found in three of these formations: the Pueblo Viejo, the Zamuro and the La Concepcion. One out of seven Pueblo Viejo samples, one out of eleven Zamuro samples and two out of thirteen La Concepcion samples contain porphyrin/vanadium complexes.

Several of the Tertiary samples show cracks and fissures and sometimes sand lenses, which means that the possibility of migration cannot be neglected. But even if the porphyrins found in a few of the Tertiary samples are indigenous it is doubtful whether the Tertiary formations showing such an erratic distribution of porphyrins could have contributed at all to oil accumulations that are so extremely rich in porphyrins.

Of the Cretaceous La Luna formation only samples have been analysed from Pueblo Viejo. The La Luna there has a very high C_R/C_T ratio, which is consistent with the great depth of burial. This indicates that the organic matter is carbonized. The La Luna in this area does not contain remnants of benthonic organisms, only of pelagic ones; it has a high organic carbon content and is therefore thought to have been quite capable, before it attained a great depth, of generating crude oil that like the La Luna source rocks elsewhere in the basin contains porphyrins. We therefore consider the bulk of the Bolivar Coast crudes more likely to have originated in the La Luna than in the Tertiary formations, although perhaps a slight contribution to the Tertiary oils from these formations cannot be excluded.

Since most of the Bolivar Coast crudes are rich in porphyrin/vanadium complexes, these crudes should have been liberated from the La Luna source rock when it still contained porphyrins.

Figures 11, 12 and 14 have shown that the organic carbon and porphyrin content of the La Luna varies, indicating that this formation is not everywhere rich in favourable matter. Unlike the Tertiary formations, however, the La Luna contains rich bituminous layers, which should be considered as oil source beds.

The Cogollo formation also contains such bituminous intervals (compare Hedberg, 1951), though perhaps to a lesser extent. Parts of the Cogollo formation, therefore, also contain rich source beds. What has been said, therefore, about the La Luna formation also holds for the Cogollo formation.

General conclusions

It must be emphasized that, in comparison with the vast area and the large number of formations, only a few samples have been analysed. The present evidence in the West Venezuelan basin on the presence of porphyrins in oils and rocks suggests that:

1. The bulk of the oil should have been generated in the La Luna formation, and that
2. In certain areas the oil must have left the source rock before it was at great depth. It does not exclude the possibility, however, that
3. Rocks containing no porphyrins have contributed to a minor extent to the wealth of oil in this basin.

Acknowledgement

The authors wish to than lc the management of Shell Research N. V. for permission to publish this paper.

References

Baker, E. W. (1966) Mass spectrometric characterisation of petroporphyrins. *J. A. C. S.* **88,** 2311.

Beach, L. K. and Shewmaker, J. E. (1957) The nature of vanadium in petroleum. *Ind. Eng. Chem.* **49,** 1157–1164.

Blumer, M. (1950) Porphyrin farbstoffe und Porphyrin-Metall Komplexe in schweizerischen Bitumina. *Helvetica Chimica Acta* **33,** 1627.

Brongersma-Sanders, M. (1951) On conditions favouring the preservation of chlorophyll in marine sediments. *Proc. 3rd W.P.C.*, Section 1, p. 401.

Costantinides, G. and Batti, P. (1957) Compartomento delle porfirine dei bitumen rispetto al l'ossidazione. *La Chimica e l'industriaa*, **39,** 96.

Dufour, J. (1955) Some oil-geological characteristics of Venezuela. *Proc. 4th W.P.C.*, Rome.

Dunning, H. N. (1963) Geochemistry of organic pigments. Chapter 9 of *Organic Geochemistry*, pp. 367–403. Edited by I. Breger, Pergamon Press.

Van Eggelpoel, A. (1964) Comparaison entre les porphyrines extraites de la roche reservoir (argiles silicifiees) d'Ozouri (Gabon) et celles de l'huile. Paper presented at 2nd International Geochemical Congress, Paris.

Gransch, J. A. and Eisma, E. (1966) Characterisation of the insoluble organic matter of sediments by pyrolysis: the C_R/C_T ratio. Paper presented at 3rd International Meeting on Organic Geochemistry, London.

Groennings, S. (1953) Quantitative determination of the porphyrin aggregate in petroleum. *Anal. Chem.* **25,** 938–941.

Hedberg, H. D. (1931) Cretaceous limestone as petroleum source rock in N. W. Venezuela. *A. A. P. G. Bull.* **15,** 229.

Hedberg, H. D. (1951) Discussion of article by C. G. de Juana and L. R. Ponte. Fundamental geological characteristics of the Venezuelan oil basin. *Proc. 3rd W.P.C.*, Section 1.

Hedberg, H. D. (1964) Geologic aspects of origin of petroleum. *A.A.P.G. Bull.* **48,** 1755.

Hodgson, G. W., Peake, E. and Baker, B. L. (1963) The origin of petroleum porphyrins: the position of the Athabasca oil sands. *The K.A. Clark volume*, Edmonton, Canada, pp. 75–100.

Howe, Warren W. (1961) Improved chromatographic analysis of petroleum porphyrin aggregates and quantitative measurement by integral absorption. *Anal. Chem.* **33,** 255.

Krejci-Graf, K. (1936) *Erdöl.* Julius Springer Verlag, Berlin.

Miller, J. B., Edwards, K. L., Wolcott, P. P., Armgard, H. W., Martin, R. and Anderegg, H. (1958) Habitat of oil in the Maracaibo Basin, Venezuela. *Habitat of Oil*, A.A.P.G. p. 601.

Millson, M. F., Montgomery, D. S. and Brown, S. R. (1966) An investigation of the vanadyl porphyrin complexes of the Athabasca oil sands. *Geochim. Cosmochim. Acta*, **30,** 207.

Treibs, A. (1934) Über das Vorkommen von Chlorophyllderivaten in einem Ölschiefer aus der oberen Trias. *Liebigs Annalen der Chemie*, **509,** 103–114.

RESEARCHES CONCERNING PORPHYRINS AND METAL COMPLEXES OF PORPHYRINS FROM THE POSIDONOMYA SHALES OF SWABIA (BRD)

by

WOLFGANG HELLER

Tubingen

Abstract

The Posidonomya shales of Swabia (Lias epsilon – Toarcian) were analysed for products from the reduction of chlorophyll and hemin. For purposes of comparison analyses were also made of material from the bituminous slates (Anis/Ladin) of Monte San Giorgio (Tessin–Switzerland).

Analysis of the reduction products showed free porphyrins as well as metal complexes of porphyrins. The metal complexes were partly physiological substances, but mainly non-physiological products were found. The most remarkable of these were Cu- and V-complexes of porphyrins. In the extracts of the Posidonomya shales and the bituminous slates of Monte San Giorgio no reduction products of chlorophyll were found which still contained Mg. The reason for this is the preponderance of non-physiological metal complexes of porphyrins in the bituminous slates of Monte San Giorgio and the Posidonomya shales, because Mg is easily substituted by V or Cu under relatively mild chemical conditions. Physiological products from the reduction of porphyrins only contain iron. Hemovanadin and hemocyanin were not found.

The porphyrins were extracted as carefully as possible. The Posidonomya shales and the bituminous slates were first pulverized, and then the fine material was extracted with ether in a Soxhlet apparatus. The ether extract was examined for porphyrins by well-known procedures.

The porphyrins were all found in the ether extract. Only residues from these were found in benzol or dioxane extracts. Other means of extraction, e.g. chloroform or cyclohexane gave no better results. Hence ether extracts were preferred. Another reason for this was that all biochemical and clinical chemical analyses for porphyrins use ether extracts. Among the physiological products from the reduction of hemin were 5-amino-laevulinic acid, coproporphyrin and protoporphyrin. No uroporphyrin was found. Incorporation of V and Cu was a secondary process. This incorporation took place with the products from the reduction of hemin and chlorophyll.

PENTACYCLIC TRITERPANES† FROM PETROLEUM AND THEIR SIGNIFICANCE

by

I. R. HILLS and E. V. WHITEHEAD

The British Petroleum Company Limited,
B.P. Research Centre, Chertsey Road,
Sunbury-on-Thames, Middlesex

Introduction

Most geologists generally agree that petroleum, like coal, is derived from the organic remains of plants and, to a lesser extent, the lower animal forms. This belief is supported by the observed optical activity of individual petroleum distillates[1, 2] and also by the enrichment in biological and petroleum products of the stable carbon isotope ^{12}C[3] compared with the isotope ^{13}C. Biochemical processes, especially photosynthesis,[4] enrich the metabolic end-products in ^{12}C relative to the source material.

The increasing use of sophisticated analytical techniques such as gas-liquid chromatography, infra-red, nuclear magnetic resonance and high-resolution mass spectrometry, has led to the isolation and precise identification of molecules with highly significant skeletal features, often referred to as biological markers.

Biological markers should show good chemical stability towards diagenesis, and should not be synthesized in significant proportions by abiogenic processes. Their skeletal features are often specific and vividly related to known biosynthetic sequences.

In terms of petroleum origin the recognition of the exact molecular geometry of individual compounds isolated from natural sources, may be of particular importance. Of secondary importance is a knowledge of the distribution of such markers and their homologously related types.

† The term triterpane has been used to denote the fully saturated analogues of triterpenes.

Biological markers vary widely in structure and include: the porphyrins;[5] n-, 2-, and 3-methyl-alkanes;[6] the isoprenoid alkanes, 2,6,10-trimethylundecane,[7] farnesane,[7, 8, 9] pristane and phytane,[7, 9, 10] and squalane,[11] as well as the isoprenoid acids related to several of these.[12]

The frequent occurrence of optical activity in biochemical systems is too well-known to require elaboration. Certain petroleum distillates[1, 2] and some from coal[13] are optically active and in each case the activity has been associated with saturated pentacyclic hydrocarbons. Several authors [14, 15, 16] have suggested that these polycyclic hydrocarbons may be steranes and triterpanes derived from naturally-occurring steroids and triterpenoids. The occurrence of this hydrocarbon type in shales, some as old as $2{\cdot}7 \times 10^9$ years, has recently been reported.[17]

We have examined petroleum distillates from several crude oils boiling in the range 340°–550° C, taken under reduced pressure so that the temperature of the sample being distilled did not exceed 300° C, particular attention being paid to such distillates from Nigerian crude oil. The results indicate that these distillates contain high concentrations of saturated tetra- and pentacyclic hydrocarbons, believed to be steranes and triterpanes, which reach a maximum concentration in distillates boiling around 475° C.

Experimental

The specific optical rotation of Nigerian distillate fractions plotted in Fig. 1 to boiling point 480° C, increases steadily from $(\alpha)_D^{19} + 1{\cdot}1°$ in distillates b.p. 410° C (at 760 mm corrected) and reaches a maximum $(\alpha)_D^{19} + 6{\cdot}9°$ in the distillate b.p. 478° C; after this, optical activity diminishes progressively until distillates b.p. 500°–550°C show only marginal activity, $(\alpha)_D^{19}$ $+0{\cdot}3°$. A detailed examination of the Nigerian distillate b.p. 340°–480° C was carried out using the standard procedures shown in Fig. 2. Mass spectra of the naphthene plus paraffin (N+P) portion which contained a prominent fragment at m/e 191. showed a high concentration of the hydrocarbon series C_nH_{2n-6} and C_nH_{2n-8}, with carbon number range $C_{27}-C_{31}$.

Initially the (N+P) portion was progressed along two separate lines. After some material had been retained part (68·5 g) was sieved while the remainder (62·4 g) was urea and then thiourea adducted. In the latter case half of the (N+P) portion adducted with urea to give a mixture of normal alkanes, simple isoalkanes and cyclanes. More than half (66% weight) of the adducted hydrocarbon proved to be n-alkane, with carbon numbers in the range $C_{19}-C_{34}$. The remaining 34% of adducted hydrocarbon, from

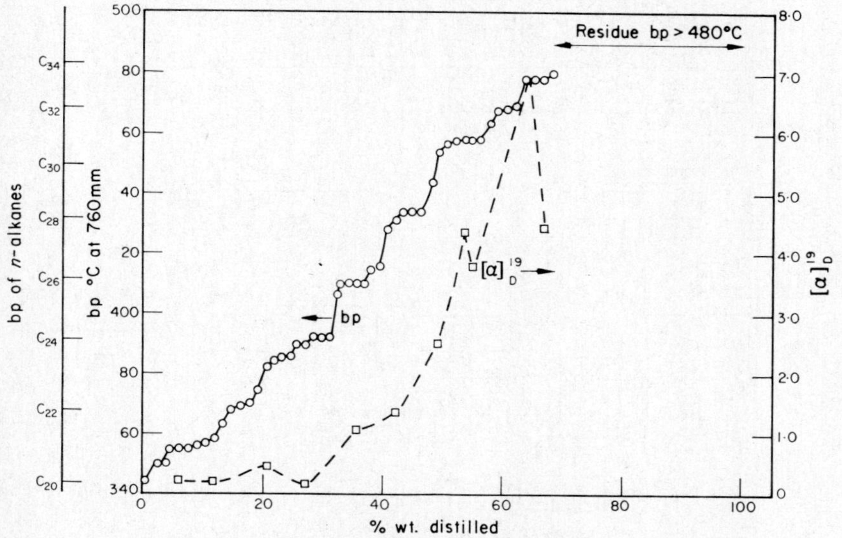

FIGURE 1. — Distillation of Nigerian residue b.p. > 340° C. Spinning band column.

gas-liquid chromatographic evidence, appeared to be mainly 2- and 3-methylalkanes.[18]

Thermal diffusion,[19] carried out in two stages on fractions as indicated in Fig. 2, yielded two similar concentrates with specific optical rotations $(\alpha)_D^{20} + 14{\cdot}5°$; 3·9 g, and $(\alpha)_D^{18{\cdot}5} + 17{\cdot}6°$; 6·4 g, which were combined and after a preliminary separation by liquid-adsorption chromatography over alumina, yielded a resin, 3·7 g, approximately 0·5% weight on crude. A portion of the gas-liquid chromatogram of this resin, the pentacyclic naphthene concentrate, is reproduced in Fig. 3. The prominent peaks are lettered and these letters are used to identify the hydrocarbons which were eventually isolated from this resin. The resin in turn was separated by liquid adsorption chromatography over alumina yielding sixteen fractions. Fractions 1–14 were eluted by purified n-pentane, and fractions 15–16 by purified benzene.

Fraction 3, 0·4072 g, after crystallization from n-pentane, gave a crude crystalline sample of hydrocarbon "F" and a mother liquor. Crystallization to constant melting point from chloroform, yielded 5 mg of hydrocarbon "F" (m.p. 232·5°–234·5° C corrected, purity by capillary gas-liquid chromatography >98% weight). The combined mother liquors were evaporated to leave a residue, which after crystallization from ether, yielded a crude sample of hydrocarbon "D". This in turn was recrystallized from ether/ethanol, chloroform/ether and finally ethyl acetate to yield

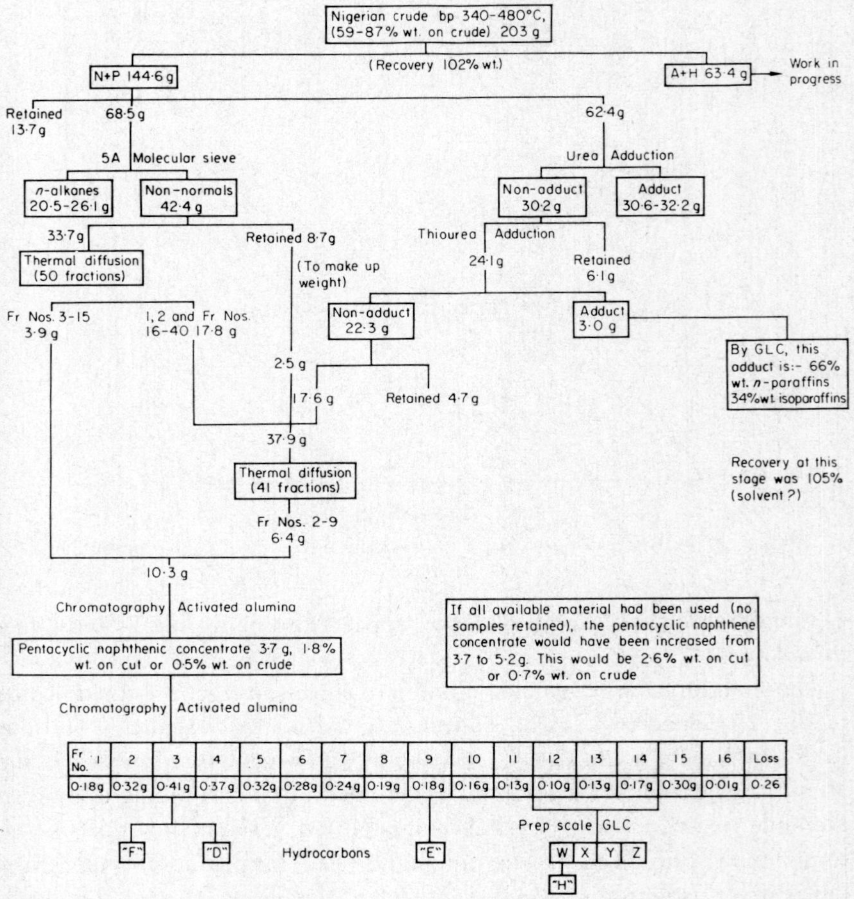

FIGURE 2. — Separation flow diagram.

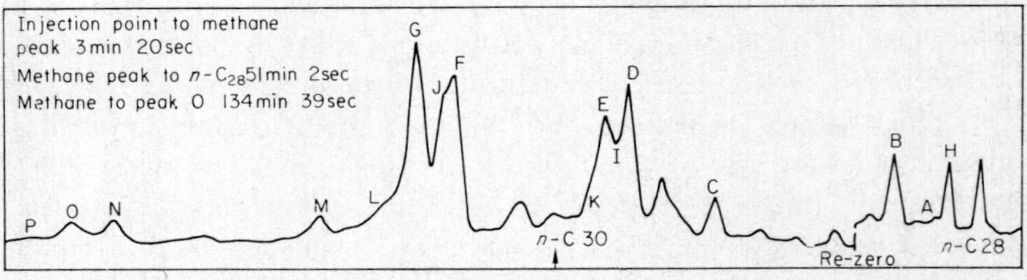

FIGURE 3. — Portion of gas liquid chromatogram of pentacyclic naphthene concentrate with added $n-C_{30}$. 148 ft glass capillary column 0·01 in. I.D. coated SE 30 gum H_2 carrier gas at 252° C. Flame ionization detector.

1 mg of hydrocarbon "D" (purity >95% weight). Hydrocarbon "D" had a corrected melting point of 169°–171° C, the sample then partially solidified and finally remelted at 178°–181° C (solvent of crystallization may be present).

Fraction 9, 0·1790 g, crystallized from benzene and after several recrystallizations from the same solvent gave 2–3 mg of hydrocarbon "E" (m.p. 259·5°–260·5° C corrected, purity >98% weight).

Fractions 12–14 inclusive, 0·4 g, were combined and separated by preparative scale gas-liquid chromatography into four fractions W, X, Y and Z, recovery 70% weight from nine runs.

Fraction W, 0·0374 g, crystallized readily from chloroform giving a crude of hydrocarbon "H". After purification over alumina and recrystallization from chloroform, ethyl acetate and finally cyclohexane it yielded 1·5 mg of hydrocarbon "H" (m.p. 227·5°–230·5° C corrected, purity >98% weight). Fraction Y has a specific optical rotation $(\alpha)_D^{23} + 17°$.

Tests for unsaturation were made with tetranitromethane.[20a, 20b] Lupane, shionane, lanostane, friedelane, zeorinane and our hydrocarbons "D", "E", "F" and "H", all gave a negative reaction. Cycloeucalanane, shionene, friedelene, alnusene and euphene, on the other hand, all gave a positive reaction (yellow coloration). These tests indicate that the hydrocarbons "D", "E", "F" and "H", like the authentic steranes and triterpanes, are fully saturated and do not contain a cyclopropane ring.

The optical activities of hydrocarbons "E" and "F" were determined on approximately 1 mg of sample, using a method developed for organic geochemical work at Sunbury. This method makes use of a model 143 Bendix Electronics Photoelectric Polarimeter, modified to reduce the noise level and to eliminate anomalous colour effects. The sensitivity of the instrument was 0·0001° arc and hence specific rotations could be measured at low concentrations (0·5 mg/ml) using conventional 10 mm low-volume cells. Measurements were made in chloroform solution at 21°C, using the mercury green wavelength of 546 mμ.

The variation of specific rotation with wavelength (in the visible region) was checked by measurements on lupane using a Bendix Electronics Spectropolarimeter type 62 and found to be negligible. Using cycloeucalanane as a standard, little variation of specific rotation with concentration was observed in the range 1–10 mg/ml, but some increase was observed at a concentration of 40 mg/ml.

Analytical data determined on these four solid hydrocarbons "D", "E", "F", and "H" are listed in Table 1.

TABLE 1. – PHYSICAL PROPERTIES OF TRITERPANES

Hydrocarbon and GLC peak	Accurate mass of molecular ion M	Nuclear magnetic resonance data		GLC data (SE 30 gum column)		Optical data	Empirical formula	Melting point °C (corrected, Köfler block)
		Protons in methylene + methine to methyls ±10%	No. of methyl groups per molecule	Relative retention volume n—C_{30} = 1·00	Purity % wt.	$(\alpha)^{21}_{546}$ in chloroform		
D	398·391	—	—	0·93	>95	—	$C_{29}H_{50}$	178·0—181·0
E	412·406	1·1	8	0·95	>98	+21·8°±2·0°	$C_{30}H_{52}$	259·5—260·5
F	412·408	1·1	8	1·10	>98	+22·7°±2·4°	$C_{30}H_{52}$	232·5—234·5
H	370.362	1·5	6	0·60	>98	—	$C_{27}H_{46}$	227·5—230·5
J	—	—	—	1·11	—	—	—	—
Lupane	—	1·2	8	1·11	>98	—15·0°±1·6°	$C_{30}H_{52}$	188·0—189·0
KR 62-148† (gammacerane)	412·409	1·1	8	1·49	>95	+31·9° ±0·4°	$C_{30}H_{52}$	285·0—286·0‡

† Ref. 42.
‡ Uncorrected melting point determined in an open capillary tube.

Discussion

1. *Spectroscopic Data*

High resolution mass spectrometry showed that the hydrocarbons C_nH_{2n-8} at $C_{27}-C_{31}$ were concentrated in the resin obtained by thermal diffusion. Prominent fragments in the mass spectrum at m/e 205, 177, 137, 123 and 109 were now dominated by the intense fragment at m/e 191. Djerassi and his co-workers[21] note the frequency with which this fragment at m/e 191 occurs in the mass spectra of triterpenes. In lupane two different fragments contribute to this peak at m/e 191, making it by far the most intense peak in the upper part of the spectrum. One of these fragments, containing rings A and B of lupane,

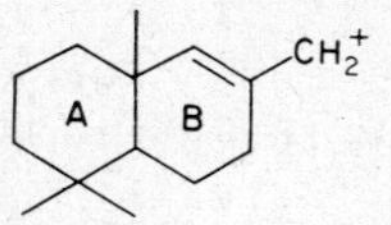

m/e 191

is classed by these authors as the most characteristic fragmentation product of pentacyclic triterpanes.

Mass spectra† of our hydrocarbons "D", "E", "F" and "H" are displayed in Fig. 4 and bear a striking resemblance to two (adiantane and lupane) of the three authentic triterpanes shown in Fig. 5. Each spectrum shows a strong molecular ion M, a significant ion at M-methyl, as well as prominent ions at m/e 191, 137, 123 and 109. Hydrocarbon "D" shows in addition a prominent fragment at m/e 177 which is also the strongest fragment in the spectrum of adiantane. Adiantane, a pentacyclic nortriterpane, $C_{29}H_{50}$, also exhibits prominent fragments at m/e 205, 191, 137, 123 and 109. It does not, however, reinforce any of the peaks in the gas-liquid chromatogram of our resin.

The spectra of friedelane, Fig. 5, and some of its derivatives[21a] show fragments at m/e 149, 217 and 259; these fragments have been associated with steranes.[17] Whilst prominent peaks at m/e 149 and 217 and 218 appear in the spectra of ergostane and cholestane, they are inconsequential in the mass spectrum of lanostane, which yields prominent fragments at m/e 274, 259 and 190. Fragments at m/e 109 and 123 prominent in the

† Only fragments larger than 5% of the base peak are plotted, except in Fig. 9 where, above m/e 235, peaks greater than 2% of the base peak are plotted.

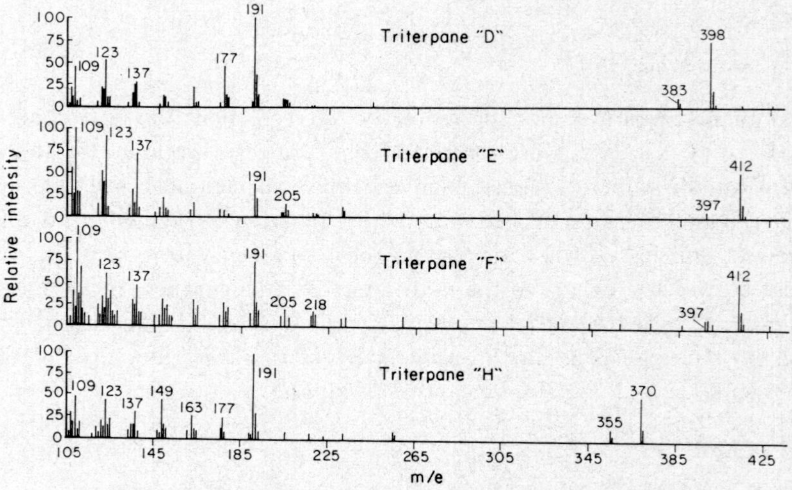

FIGURE 4. — Mass spectra of triterpanes "D", "E", "F" and "H".

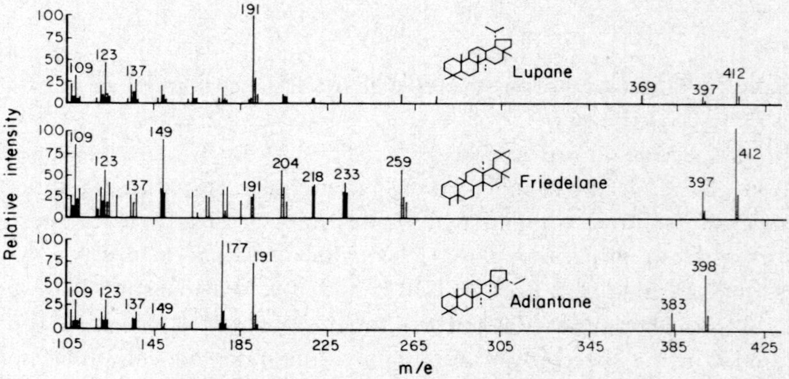

FIGURE 5. — Mass spectra of authentic triterpanes.

mass spectra of triterpanes are also prominent in the mass spectra of some steranes, and an intense fragment at m/e 218 in the mass spectrum of $\triangle^{12}$–oleanene and $\triangle^{12}$ – ursene should be noted.[(21b)]

The mass spectra of *cis*- and *trans*- sclareol oxides:

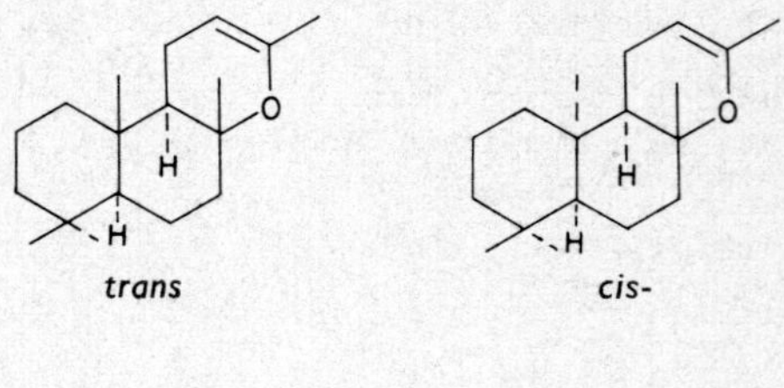

trans *cis-*

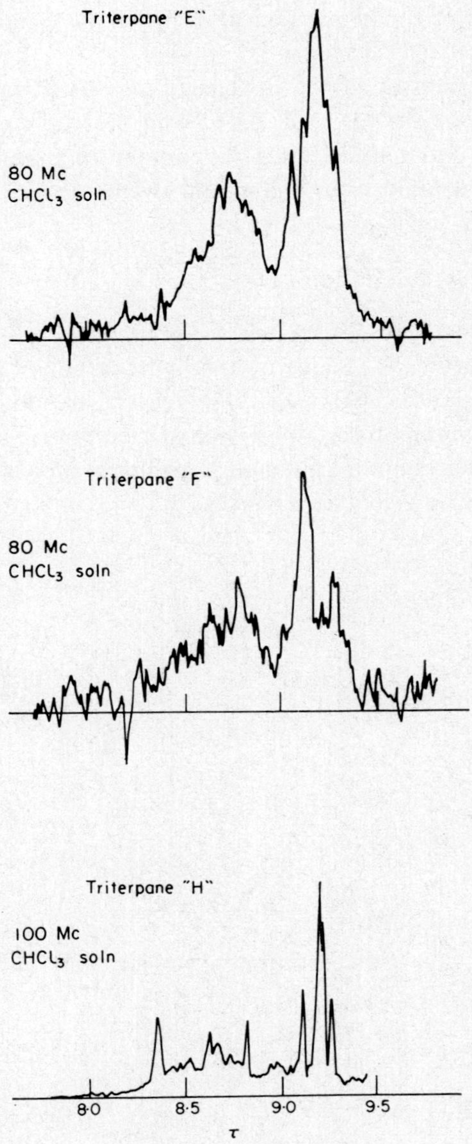

FIGURE 6. — Nuclear magnetic resonance spectra of triterpanes "E", "F", and "H".

also exhibit prominent fragments at m/e 191, 177, 137, 123 and 109. The Russian authors[22] consider that all of these peaks can be related to rings A and B, which of course are typical of the rings A and B in many triterpenes.

The accurate masses of the molecular ions listed in Table 1 indicate that the hydrocarbons "D", "E", "F" and "H" all contain five saturate rings (C_nH_{2n-8}) and that "E" and "F" contain thirty carbon atoms, whilst "D" and "H" contain twenty-nine and twenty-seven, respectively.

2. *Nuclear Magnetic Resonance Data*

Nuclear magnetic resonance spectra, Fig. 6, were obtained from the very small quantities (1–2 mg) of the hydrocarbons "E", "F" and "H" which were available. Although these spectra are not of good quality, they are good enough to establish with fair certainty, that hydrocarbons "E" and "F" each contain eight methyl groups per molecule, whilst hydrocarbon "H" has only six (see Table 1).

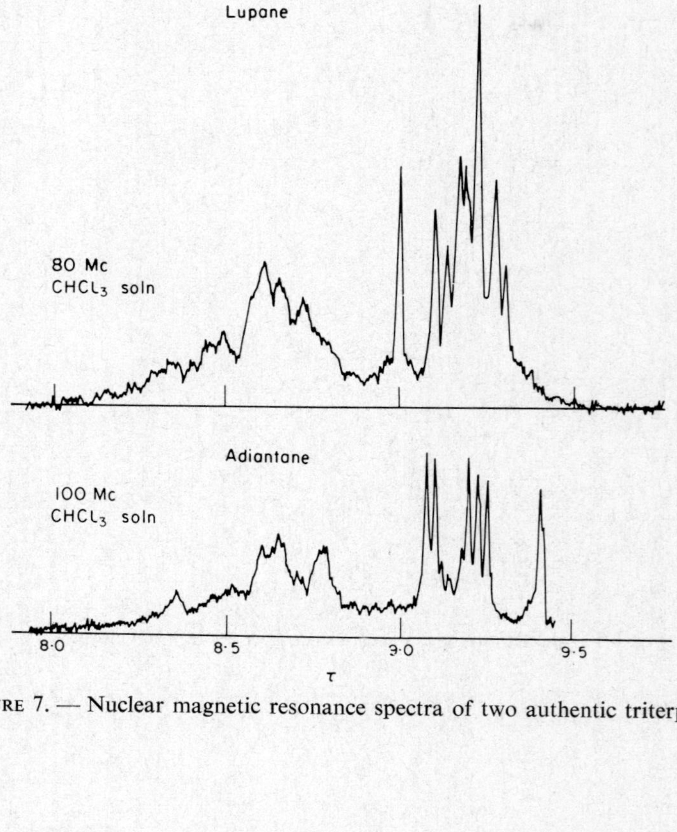

FIGURE 7. — Nuclear magnetic resonance spectra of two authentic triterpanes.

Figure 7 includes the nuclear magnetic resonance spectrum of lupane. Lehn and Ourisson[23] have interpreted the principal features of the lupane spectrum after they had examined the spectra of a large number of lupane derivatives, and Karliner and Djerassi in turn have considered the NMR spectrum of $\triangle^{12}$-oleanene.[21b] Our hydrocarbons have nuclear magnetic resonance spectra which are different from these, but show certain similarities to those of lupane and adiantane (Fig. 7).

3. *Infra-Red Data*

Infra-red spectra of the resin and of the hydrocarbons "E" and "F" isolated from it are shown in Fig. 8, which includes for comparison the spectrum of lupane. Once more the similarities between our hydrocarbons and an authentic triterpane are quite striking. These spectra also indicate that the samples examined are substantially free of hydroxyl (2·75–3·25μ), carbonyl (5·75–6·0μ), and double bonds (6·0–6·3μ).

4. *Gas-Liquid Chromatography*

Authentic lupane and lanostane both reinforce exactly peak J in the gas-liquid chromatogram of the resin (Fig. 3). However, a compound corresponding with peak J has not yet been isolated, and its possible identity with lupane (or lanostane) cannot be proved or disproved at this stage.

All of the authentic steranes and triterpanes that we have examined have retention volumes (Table 2) which fall between n-C_{28} and peak P in the gas-liquid chromatogram of the resin (Fig. 3), but few of these coincide with any of the peaks in our chromatogram: gammacerane (peak O), lupane (peak J) and possibly hopane (peak M), are the exceptions. The presence of a peak at I in Fig. 3 was not immediately obvious but it was discovered between peaks E and D when the gas-liquid chromatograms of the sixteen fractions separated by liquid-adsorption chromatography over alumina, were examined.

The indications from all of these data are that a molecular skeleton based on triterpenoids should be considered for these four hydrocarbons that have been isolated.

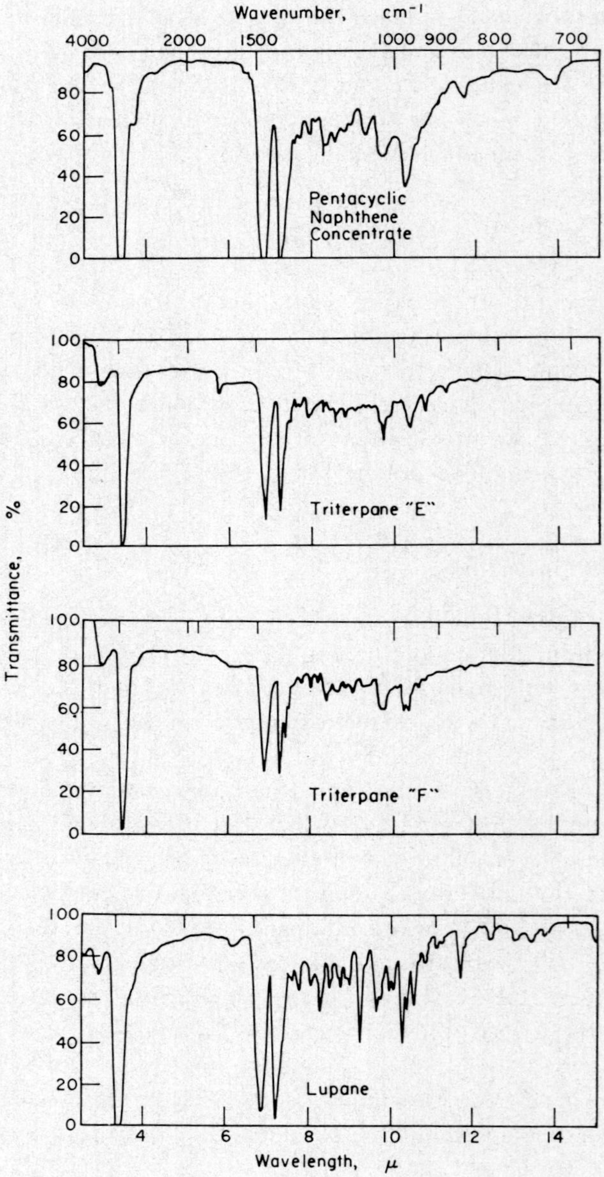

FIGURE 8. — Infra-red spectra.

TABLE 2. — RELATIVE RETENTION VOLUMES OF N-ALKANES, AUTHENTIC STERANES AND TRITERPANES, AND THE PRINCIPAL PEAKS IN THE GAS-LIQUID CHROMATOGRAM OF THE PENTACYCLIC NAPHTHENE CONCENTRATIONS

RRV† $n\text{-}C_{30} = 1{\cdot}0$	GLC Peak	Authentic hydrocarbons	
0·57		$n\text{-}C_{28}$	
0·59			Cholestane
0·60	H		
0·61	A		
0·65	B		
0·76			Ergostane
0·84	C		
0·88			Onocerane III
0·93	D		Onocerane II
0·94	I		
0·95	E		
0·96	K		Fucostane
0·97			Artesane (9 : 19–9–β–Cyclolanostane)
0·99			Onocerane I
1·00		$n\text{-}C_{30}$	Cycloeucalanane
1·03			Adiantane (30-nor-Hopane)
1·10	F		
1·11	J		Lanostane, Lupane
1·14	G		
1·17	L		
1·18			Shionane
1·24	M		
1·25			Zeorinane (Hopane)
1·38			Friedelane (DA-Friedo oleanane)
1·40			Heterolupane (18α–19β-Ursane)
1·45	N		
1·49	O		α-Onocerane (gammacerane, KR 62-148)
1·53	P		

† On a 148′ SE 30 gum capillary column.
Column operating conditions are those noted in Fig. 3.

5. *X-ray Crystallographic Data*

Hydrocarbon "E" is being examined, without the incorporation of a heavy atom, by X-ray crystallography. The unit cell dimensions and density of the monoclinic crystal give a molecular weight of 411, in good agreement with the value determined by mass spectrometry. Inspection of the Patterson syntheses and the Fourier maps suggests a nucleus based on perhydropicene. Four of the five rings, A, B, C and D or B, C, D and

E appear to be six-membered with *trans*- ring fusions and chair configurations. The configuration of the fifth ring is not yet established, nor are the locations of all of the eight methyl groups.

6. *Occurrence of Triterpanes in Crude Oil*

The cumulative evidence leaves little doubt that hydrocarbons "E" and "F" are C_{30} pentacyclic triterpanes, that "D" is probably a nortriterpane and "H" a trisnortriterpane.

The occurrence of hydrocarbons "A" to "O" is not limited to the Nigerian crude oil. Similar resins have been isolated from Kuwaiti, Iranian and Libyan crude oils. These resins yield capillary-column gas-liquid chromatograms with the same peaks A–D, G–H and K–P, but the proportions in each resin are different. The peak identities in each case were established by blending successively, the resin from one crude oil with that from another crude oil and noting the reinforcement of individual peaks in the gas-liquid chromatograms of the blends. Peaks E, F, I and J, prominent in the gas-liquid chromatograms of the Nigerian resin are, however, not seen in the chromatograms of the resins from the other three crude oils. The mass spectrum of each resin, however, shows the prominent fragment ions at m/e 191, 177, 137, 123 and 109, and a high concentration of the hydrocarbon series C_nH_{2n-8} at C_{27}–C_{31}.

7. *Tetracyclic Naphthenes in Petroleum*

Our preliminary examination of a tetracyclic naphthene concentrate, obtained by adsorption chromatography over alumina, of a urea non-adduct revealed the presence of a hydrocarbon series C_nH_{2n-6}, lying in the carbon number range C_{27}–C_{33}. The mass spectrum showed that the tetracyclic napthenes were incompletely separated from pentacyclic naphthenes. The mass spectrum of such a concentrate segregated from a distillate b.p. 340°–480° C from Nigerian crude oil is compared with another segregated from a Libyan distillate b.p. 340°–460° C in Fig. 9. The prominent fragments at m/e 217 and 218, characteristic in the mass spectra of steranes, are not usually prominent in the spectra of pentacyclic triterpanes; the fragment at m/e 149 which is sometimes associated with steranes also appears as a prominent fragment in friedelane (Fig. 5).

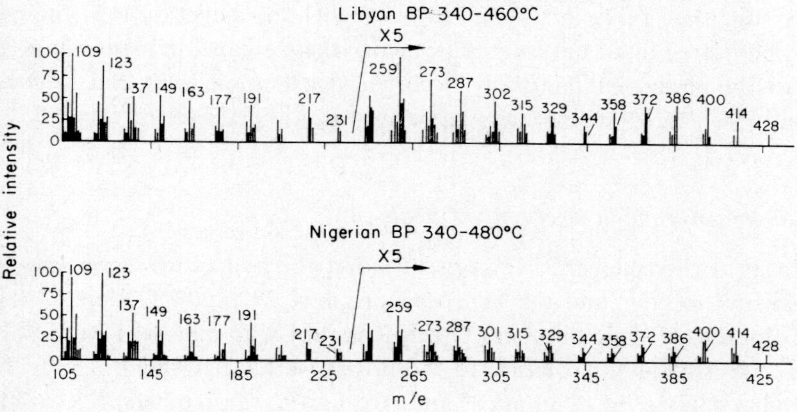

FIGURE 9. — Mass spectra showing hydrocarbon series $C_n H_{2n-6}$, steranes in Nigerian and Libyan crudes.

8. *Tentative Identification of Squalane in Crude Oil*

Our examination of the Libyan distillate, b.p. 340°–460° C, included the preparation of a thiourea adduct from the urea non-adduct of the (N+P) portion. The gas-liquid chromatogram of the hydrocarbons adducting with thiourea contained a prominent peak which was exactly reinforced by authentic squalane. From this observation it is tentatively concluded that this C_{30} isoprenoid alkane is probably present in the waxy distillate range of crude oils.

9. *Alkanes in Crude Oils*

Some Libyan (b.p. 340°–460° C) and Nigerian (b.p. 340°–480° C) waxy distillates have high n-alkane contents, 28% and 23% weight, respectively. The n-alkane distribution of Libyan (Sarir) and Nigerian (Bomu) crude oils[(24)] resemble Uinta Basin crude in having similar CPI values† in approximately the same carbon number range, but differ fıom Uinta Basin[(25)] in having the carbon number maxima of their distributions at C_{15} and C_{17}, respectively. The high concentration of urea-adductable hydrocarbons which appear to be mainly 2- and 3-methylalkanes, in the

† The CPI value is defined as the sum of the concentrations of odd carbon numbered n-alkanes divided by the sum of the concentrations of the even carbon numbered paraffins in the selected carbon number range.

waxy distillate range of Libyan (9% weight) and Nigerian (12% weight) may be related to the natural occurrence of these alkanes in plant waxes[6, 26] or to the equivalent naturally occurring isoalkanoic acid and esters,[27] from which they may perhaps be derived in the presence of clays.[28]

10. *Gammacerane in Green River Shale*

In an earlier paper[29] we suggested that the hydrocarbons isolated by Oakwood *et al.*[1] and the hydrocarbon, m.p. 292°–297° C isolated from a Colorado oil shale bitumen (Green River) by Cummins and Robinson[9] might be similar in structure to those that we have isolated. The hydrocarbon isolated by Cummins and Robinson, designated KR 62-148 (Table 1), was shown subsequently to be gammacerane by high-resolution mass spectrometry and nuclear magnetic resonance.[42] It is related to a primitive ciliated protozoan,[36a] and perhaps to simple plant forms.[42]

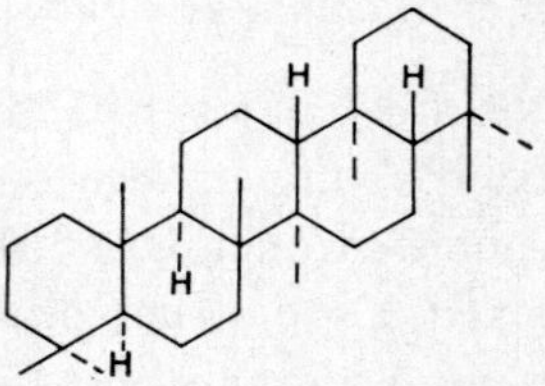

Gammacerane

11. *Other Triterpane Derivatives in Petroleum*

Pentacyclic triterpenoids are known to occur in petroleum. Barton, Overton and Carruthers[30] identified oxyallobetul-2-ene (I) which Carruthers and Cook[31] had isolated earlier from an American crude oil. Later Carruthers and Watkin[32] reported the isolation of 1,2,3,4-tetrahydro-2,2,9-trimethylpicene (II) which they related to triterpenoids.

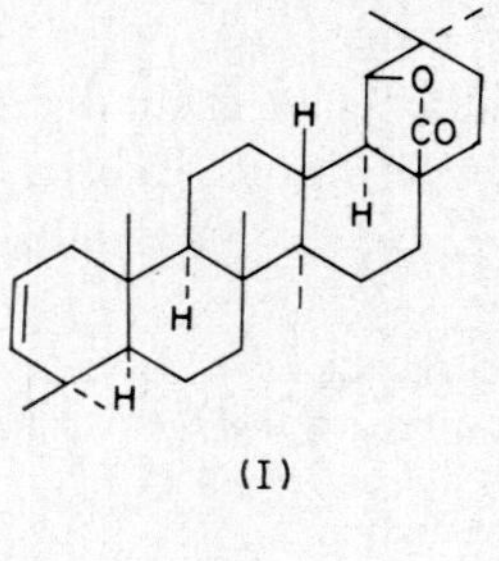

(I)

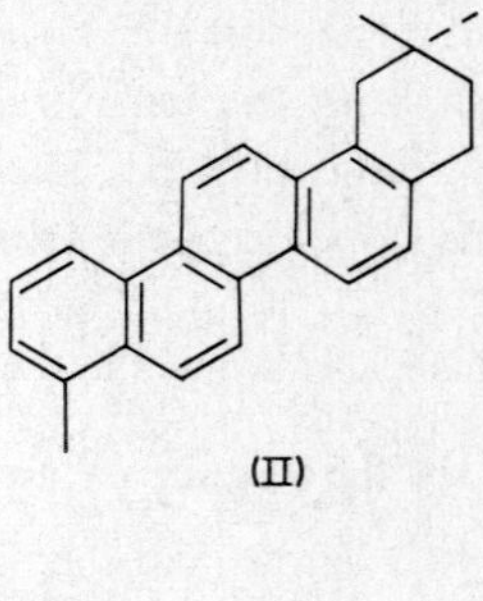

(II)

12. *Triterpenes in Brown Coal*

Šorm and his co-workers[33] have identified a wide range of triterpenoids in brown coal; these include friedelin (III), betulin (IV), oxyallobetulin (V), oxyallobetulone (VI), Δ^2-allobetulen (VII). α-apoallobetulin (VIII).

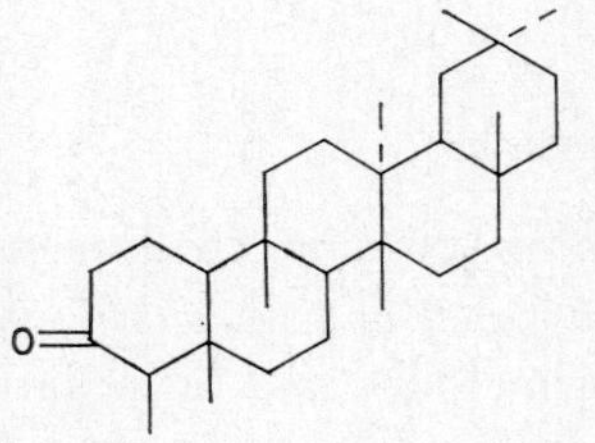

(III)

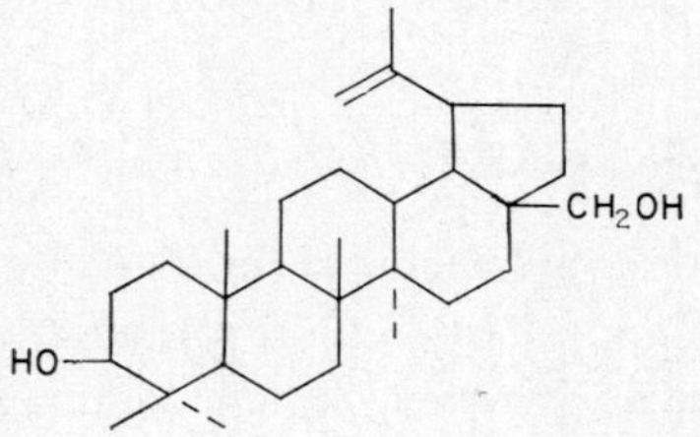

(IV)

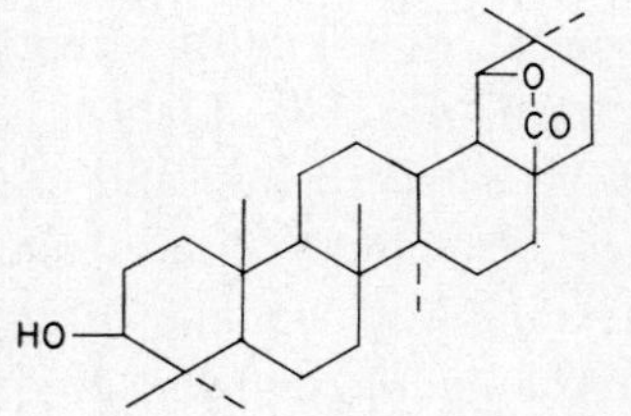

(V)

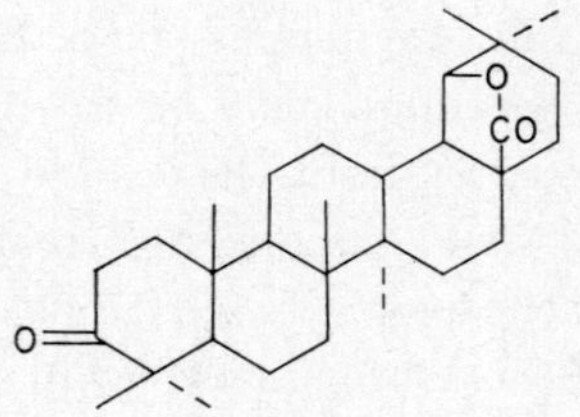

(VI)

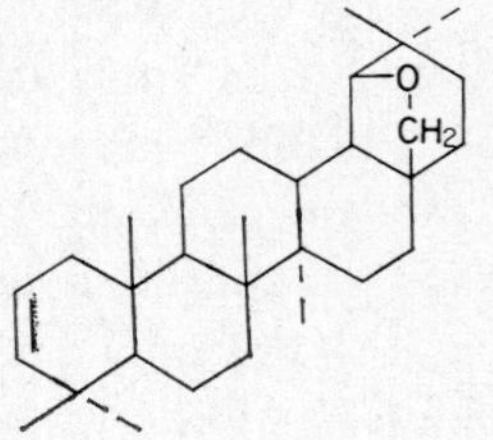

(VII)

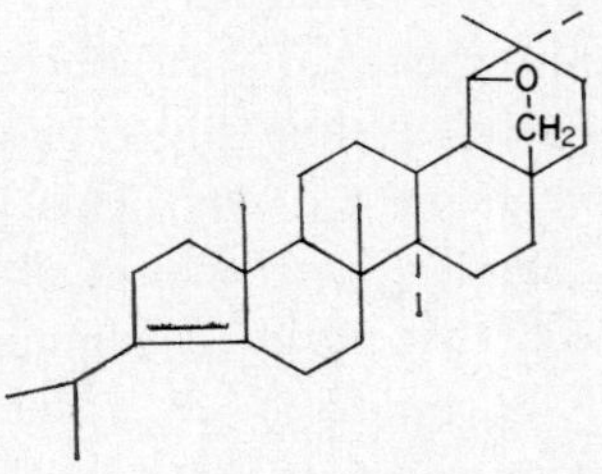

(VIII)

13. *Sterane Derivatives in Petroleum*

Mair and Martinéz-Picó[34] reported the presence of cyclopentanophenanthrene and its 3′-methyl derivative in the American Petroleum Institute's RP6 reference petroleum and related the presence of these two hydrocarbons to the naturally occurring steroids. The sterane concentrate that we have isolated from Nigerian and Libyan crude oils may well be derived from the phytosterols which are abundantly distributed in both aquatic and terrestrial flora from the simple algae to the highly specialized angiosperms. The triterpanes in petroleum are also most probably derived from the naturally occurring triterpenoids which are abundantly but unevenly distributed throughout the plant kingdom.[35a]

14. *Steroids and Terpenes in the Plant Kingdom*

Bacteria are included in the plant kingdom division Thallophyta. They do not appear to contain pentacyclic triterpenes and only rarely steroids. The algae, however, which dominate marine flora and form part of the division Thallophyta, contain an abundance of phytosterols, but apparently few pentacyclic triterpenes.[35b] The remaining two classes, fungi and lichens in the division Thallophyta are sometimes regarded as transition stages in adaption of plant forms to a terrestrial environment. The preferred phytosterols in fungi appear to be based on lanosterol and no pentacyclic triterpenes appear to have been isolated from this class. Lichens, on the other hand, carry both sterols and pentacyclic triterpenes, and examples based on ursane, hopane and rearranged oleanane systems have been reported. Triterpenes based on gammacerane and hopane along with phytosterols have been identified in the plant kingdom division, Bryophyta and Pteridophyta. The Gymnosperms, which dominated the Mesozoic era with the Pteridophytes, contain phytosterols and an abundance of diterpenes; pentacyclic triterpenes, however, have only been found infrequently in this division. The Angiosperms appeared around the Cretaceous period of geological time and rapidly achieved the dominance that they maintain to the present day. This vast division generates phytosterols, simple and complex triterpenes in abundance and considerable variety.[35a]

Sterols occur widely in the animal kingdom but only very rarely are triterpenoids encountered. In general, animals appear to be unable to synthesise triterpenes, perhaps in the absence of chlorophyll, but build instead tetracyclic systems, like cholesterol, derived from lanosterol. The fate of the triterpenes which are in the plant foodstuffs, on which all animals ultimately depend, does not appear to have been elucidated. Tetrahymanol, isolated by Mallory *et al.*[36a] from a protozoan, is one of the few examples of a triterpene isolated from the animal kingdom. These authors in collaboration with Tsuda *et al.*[36b] relate tetrahymanol to gammacerane and note that it is the first pentacyclic triterpenoid alcohol isolated from an animal. Gammacerane is perhaps the simplest pentacyclic triterpane system. Tetrahymanol may possibly be one of the sources of Cummins and Robinson's [9] hydrocarbon isolated from a Green River Oil Shale bitumen.

Our present knowledge of the distribution of triterpenes in the plant and animal kingdom is far from complete; perhaps today it merely scratches the surface. Future work may well reveal new molecular types and clarify and amplify the hazy trends which are presently discernible. These trends, however, suggest that pentacyclic triterpenes are abundantly distributed in terrestrial plants, especially in the more highly specialized Angiosperms, whereas they appear to be rare in the simpler aquatic plant forms, where the division between plant and animal is ill-defined. Sterols, $C_{27}-C_{29}$, on the other hand, appear to be adundantly distributed in both aquatic and terrestrial flora, and their occurrence may be closely associated with the presence of chlorophyll.

If steranes and triterpanes are accepted as biological markers, then the recognition of individual molecules may prove helpful in determining whether a crude oil may have originated from plants. Initially, it may be possible to recognize a marine or terrestrial origin of individual crude oils.

15. *Biological Markers in Geological Time*

The discovery of phytane and pristane in Precambrian sediments[37] and the probable relationship that these hydrocarbons have to chlorophyll and hence to photosynthesis, indicates that the fundamental biosynthetic processes may have changed little in geological time, although the number and complexity of end-products from such processes, have probably increased in time. Bacteria, which are included in the plant division Thallophyta, are probably the simplest and most primitive plant forms known, and again there are reliable reports[38] of their presence in very ancient Precambrian sediments. It has been suggested by ZoBell[39] that bacteria play a considerable part in the transformation of plant remains into fossil fuels. Certain species operate under anaerobic conditions and liberate hydrogen, contributing to the reducing environment normally present in sediments. These conditions are conducive to the saturation of the plant triterpene and steroid skeletons which usually contain at least one double bond. The lipids of which triterpenoids form a part, usually also contain at least one oxygen atom in the form of hydroxyl, ketone, ether, or carboxyl which is often esterified. Bacteriological decarboxylation, before or after hydrolysis, is a possible route to the nor- and trisnortriterpanes reported in this paper. Breger[40] points out that direct bacterial decarboxylation is not known to occur, however, indirect decarboxylation may be possible through β-hydroxy and β-keto acid intermediates which would offer a reduction route to the hydrocarbon. Direct decarboxylation processes, by α-particle bombardment or thermally, should, however, not be overlooked. The actual stages in transition of plant triterpenoids into fossil fuel triterpenoids and the associated aromatics some of which are included in structures I—VIII described earlier, may perhaps be worked out from the studies on brown coal by Šorm and his co-workers.[33]

16. *Optical Activity*

The transition of plant triterpenoids into fossil fuel triterpenoids, involving modification of the functional groups present (e.g. hydroxyl, carbonyl, carboxyl, etc.) will be accompanied by corresponding changes in specific optical rotation. Optical activity will eventually be eliminated when the transition to the fully aromatic form is complete. Transitions involving saturated hydrocarbons (e.g. isomerizations), which would again be accompanied by changes in specific optical rotation, are much more difficult. This would be particularly true for the quaternary asymmetric centres common in the triterpanes.

However, at an asymmetric centre carrying a hydrogen atom, epimizeration can occur through the formation of either carbonium ions or carbanions. The ease of carbanion isomerization will depend on the acidity of the attached hydrogen atom and on the stereochemical stability of the intermediate carbanion that is formed. Isomerization may follow the inversion of the carbanion intermediate, but such changes would certainly alter the identity of an individual triterpane.

17. *Stable Carbon Isotopes*

The evidence reported here leaves many problems unsolved, but indicates that part of Nigerian crude oil may have had its origin in terrestrial plants. The $\delta^{13}C/^{12}C$ value[41] for this Nigerian crude oil relative to the petroleum standard NBS-22 is $+3{\cdot}4\permil$ and for hydrocarbon "F" $+5{\cdot}6\permil$. Both values are well within the range typical of sedimentary organic carbon of marine origin.[3] Further work in this direction is proceeding with a view to resolving this dilemma.

Acknowledgements

Permission to publish this paper has been given by The British Petroleum Company Limited. We thank Professor D. H. R. Barton, Dr. G. Eglington, Dr. T. G. Halsall, Dr. H. Iwamura, Dr. T. J. King, Professor F. B. Mallory, Professor G. Ourisson, Professor T. Takahashi and Professor Y. Tanahashi for samples of steranes and triterpanes which have proved invaluable in this work. We also thank our many colleagues at The Research Centre, Sunbury.

References

1. T. S. Oakwood, D. S. Shriver, H. H. Fall, W. J. McAleer and P. R. Wunz. *Ind. Eng. Chem.* **44,** 2568, 1952.
2. B. J. Mair. *Geochim. Cosmochim. Acta*, **28,** 1303, 1964.
3. S. R. Silverman and S. Epstein. *Bull. Amer. Assoc. Petrol. Geol.* **42,** 998, 1958.
4. P. Echlin. *Science Journal*, **2,** 42, 1966.
5. I. A. Breger (ed.) *Organic Geochemistry*, p. 389. New York, Pergamon Press, 1963.
6a. I. A. Breger (ed.) *Organic Geochemistry*, p. 507, New York. Pergamon Press, 1963.
6b. T. Swain (ed.) *Chemical Plant Taxonomy* p. 187. London, Academic Press, 1963.
6c. J. D. Mold, R. E. Means, R. K. Stevens and J. M. Ruth. *Biochemistry*, **5,** 455, 1966.
7. J. G. Bendoraitis, B. L. Brown and L. S. Hepner, *Anal. Chem.* **34,** 49, 1962.
8a. K. E. Murray, F. E. Huelin and J. B. Davenport. *Nature*, **204,** 80, 1964.
8b. K. Kochloefl, P. Schneider, R. Řeřicha, M. Horák and V. Bažant. *Chem. and Ind.* 692, 1963.
9. J. J. Cummins and W. E. Robinson. *J. Chem. Eng. Data*, **9,** 304, 1964.
10. R. A. Dean and E. V. Whitehead. *Tetrahedron Letters*, 768, 1961.
11. B. O. Lindgren. *Acta Chem. Scand.* **19,** 1317, 1965.
12. J. Cason and D. W. Graham. *Tetrahedron*, **21,** 471, 1965.
13. C. Zahn. S. H. Langer, B. D. Blaustein and I. Wender. *Nature*, **200,** 53, 1963.
14. J. Marcusson. *Chem. Z.* **32,** 377 and 391, 1908.
15a. W. G. Meinschein. *Bull. Amer. Assoc. Petrol. Geol.* **43,** 925, 1955.
15b. W. G. Meinschein. *Geochim. Cosmochim. Acta*, **22,** 58, 1961.
16. M. Louis. *Res. Inst. Français du Petrol*, **19,** 277, 1964.
17. A. L. Burlingame, P. Haug, T. Belsky and M. Calvin. *Proc. Nat. Acad. Sci.* **54,** 1406, 1965.
18. R. A. Dean and E. V. Whitehead. Sixth World Petrol Congr., Frankfurt/Main, 19th-26th June 1963, Section V, Paper 9.
19a. A. Letcher-Jones. *Petroleum Refiner*, **36,** 153, 1957.
19b. W. S. Nathan and D. M. Whitehead. C. R. Thirty-first Congr. Intern. Chem. Indust. Liege, September 1958.
20a. L. F. Fieser and M. Fieser *Advanced Organic Chemistry*, p. 146. New York, Reinhold Publishing Corporation, 1961.
20b. A. Werner, *Ber.* **42,** 4324, 1909.

21a. H. Budzikiewicz, J. M. Wilson and C. Djerassi. *J. Amer. Chem. Soc.* **85,** 3688, 1963.
21b. J. Karliner and C. Djerassi, *J. Org. Chem.* **31,** 1945, 1966.
22. N. S. Wolfson, V. I. Zaretskii, V. L. Sadouskaya, A. V. Semenvosky, W. A. Smit and V. S. Kucherov. *Tetrahedron,* **22,** 603, 1966.
23. J. M. Lehn and G. Ourisson. *Bull. Soc. Chim. France,* 1137, 1962.
24. J. V. Brunnock. *Nature,* **212,** 385, 1966.
25. R. L. Martin, J. C. Winters and J. A. Williams Sixth World Petroleum Congr, Frankfurt/Main, 19th-26th June, 1963, Section V, paper 13.
26a. P. Jarolimek, V. Wollrab, M. Streibl and F. Šorm. *Chem. and Ind.* 237, 1964.
26b. F. Šorm, V. Wollrab, P. Jarolimek and M. Streibl. *Chem. and Ind.* 1833, 1964.
26c. J. D. Waldron, D. S. Gowers, A. C. Chibnall and S. H. Piper. *Biochem. J.* **78,** 435, 1961.
26d. J. B. Harbourne, *Nature,* **206,** 247, 1965.
27. J. D. Mold, R. E. Means and J. M. Ruth. *Phytochemistry,* **5,** 59, 1966.
28. J. Jurg and E. Eisma. *Science,* **144,** 1451, 1964.
29. I. R. Hills and E. V. Whitehead. *Nature,* **209,** 977, 1966.
30. D. H. R. Barton, K. H. Overton and W. Carruthers. *J. Chem. Soc.* 788, 1956.
31. W. Carruthers and J. W. Cook. *J. Chem. Soc.* 2047, 1954.
32. W. Carruthers and D. A. M. Watkin. *Chem. and Ind.* 1433, 1963.
33a. K. Hejno, M. Horák, V. Jarolim, M. Streibl and F. Šorm. *Chem. and Ind.* 1142, 1958.
33b. K. Hejno, M. Horák, V. Jarolim, M. Streibl and F. Šorm. *Coll. Czech. Comm.* **26,** 459, 1961.
33c. K. Hejno, M. Horák, V. Jarolim, M. Streibl and F. Šorm. *Ibid.* **28,** 2318 and 2443, 1963.
33d. K. Hejno, M. Horák, V. Jarolim, M. Streibl and F. Šorm. *Ibid.* **30,** 873, 1965.
34. B. J. Mair and J. L. Martinéz-Picó. *Proc. Amer. Petrol Inst.* **42,** 173, 1962.
35a. P. Boiteau, B. Pasich and A. Rakoto Ratsimamanga, (eds.) *Les Triterpenoids en Physiologie Végétale et Animale,* p. 739. Gauthier-Villars, Paris, 1964.
35b. M. Florkin and H. S. Mason (eds.) *Comparative Biochemistry,* vol. III, chap. 11, p. 614. New York, Academic Press, 1962.
36a. F. B. Mallory, J. T. Gordon and R. L. Conner. *J. Amer. Chem. Soc.* **85,** 1362, 1963.
36b. Y. Tsuda, A. Morimoto, T. Sano, Y. Inubshi, F. B. Mallory, and J. T. Gordon. *Tetrahedron Letters,* 1427, 1965.
37. R. B. Johns, T. Belsky, E. D. McCarthy, A. L. Burlingame, P. Haug, K. H. Schnoes, W. Richter and M. Calvin. Space Sciences Laboratory, University of California, Technical Report, NSG 101–61, Series No. 7, Issue No. 8, 1963.
38. J. W. Schopf, E. S. Barghoorn, M. D. Maser and R. O. Gordon. *Science,* **149,** 1365, 1965.
39a. C. E. ZoBell. *Science,* **102,** 364, 1945.
39b. C. E. ZoBell. *Bact. Revs.* **10,** 1, 1946.
40. I. A. Breger. *J. Amer. Oil Chemists Soc.* **43,** 197, 1966.
41. S. R. Silverman. Private communication.
42. I. R. Hills and E. V. Whitehead, D. E. Anders, and J. J. Cummins and W. E. Robinson. *Chem. Comm.* 752, 1966.

CORRELATION PROBLEMS AMONG CRUDE OILS

by

DIETRICH H. WELTE

Geologisches Institut der Universität Würzburg, Germany

Introduction

The purpose of this paper is to introduce a concept for characterizing individual oils as members of a natural maturity sequence. With this objective, the use of an "internal chemical fingerprint" has been tried for correlation purposes among the petroleums in a closed geological basin.

More details concerning the geological background, the analytical procedures and relevant literature than are given here, can be found in a previous paper (Welte, 1967).

It is well known that crude oils do not originate at the place where they are found. They are generated and released by source rocks and migrate into more porous reservoir rocks. Oil generation and migration is strictly related to the geological development and especially to the subsidence history of a sedimentary basin. During the development of a basin different structures may be alternately opened and closed to various kinds of migrating oils. Thus petroleums from one source bed, released during different subsidence phases, and oils from various other source rocks may be trapped in individual reservoirs.

From this reasoning a twofold problem can be formulated: 1. How can we recognise and distinguish different petroleums released at different subsidence stages from a source bed, and 2. How can we demonstrate whether oils from different reservoirs have a common origin, i.e. a common source bed, or not? The latter is based on the assumption that in source beds there exist homogeneous parts, which have a given type of organic matter. Such a part may be called a source rock unit. In this sense crossing a facies border in the same stratigraphic interval involves passing from one source rock unit to another.

"Characteristics of maturity" and "characteristics of correlation"

Different subsidence stages mean different depth levels and as such a different temperature history. The markers intended to be used to classify oils as members of a natural maturity sequence must therefore be temperature-sensitive and should be indigenous components of source rocks and crude oils. They may be called "characteristics of maturity".

Philippi (1965) has demonstrated in a recent paper that the amount of n-paraffins in potential source beds increases with increasing depth. Simultaneously the relative amount of lower-molecular-weight n-paraffins tends to increase, whereas an odd-predominance among the higher n-paraffins tends to decrease. On heating extracted samples of an oil shale Welte (1965) observed an increasing amount of extractable newly formed n-paraffins with increasing temperature and at the same time a diminishing odd predominance. This finding corroborates Philippi's data.

Therefore, because n-paraffins are found in all petroleums and rock extracts they were selected as characteristics of maturity. For grouping the oils three criteria were used: the absolute amount of n-paraffins, the entire shape of the n-paraffin distribution curve and the degree of an odd-predominance.

For the determination of the degree of relationship among petroleums, internal chemical markers have to be found, that are common compounds in source rocks and crude oils, and that are not affected by migrational effects. They may be called "characteristics of correlation".

It has been suggested, first by Dean and Whitehead (1964) and later also by Welte (1965), that isoprenoid type hydrocarbons could be used for this purpose. They are present in nearly every crude oil and their relative distribution pattern seems to be random. The chemical and physical properties of the isoprenoid-type hydrocarbons are very similar, so that migration processes cannot change the relative amounts of those components very much. Therefore it was decided to use isoprenoid-type hydrocarbons as characteristics of correlation.

Based on the foregoing considerations it seemed desirable to investigate the petroleums of a small closed basin, in order to test the above outlined concept. The area considered here is the Gifhorn trough of Northwestern Germany.

Oil samples from the Gifhorn trough

The Gifhorn trough is part of the larger North-west German basin (Fig. 1).

The geology of this area has been described, mainly by Hedemann (1956), Roll (1956), Hecht (1959) and Philipp *et al.* (1964). It has a length of about 100 km and a width of 50 km. The NNE-SSW striking axis of the trough moved from the Triassic to the Cretaceous some 10–20 km

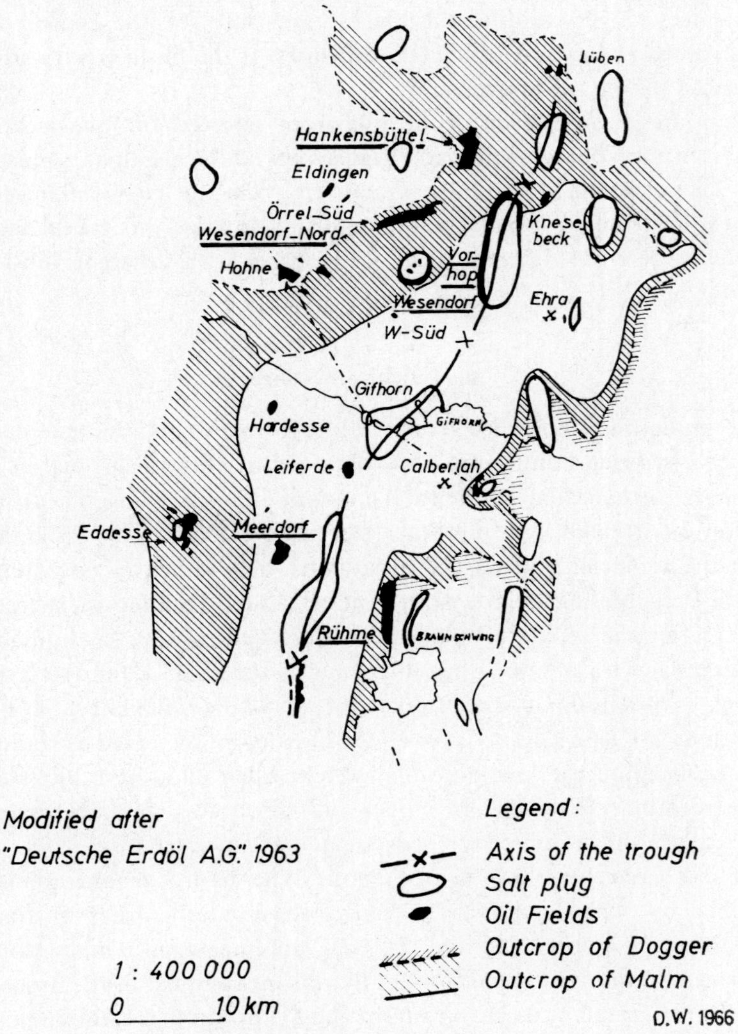

FIGURE 1. — Map of the Gifhorn trough.

from west to east. During the Triassic mainly non-marine shales and sands, and some marine carbonates and evaporites, were deposited. At the beginning of the Jurassic predominantly marine shale sedimentation occurred. The Liassic and Dogger especially consist of very thick shale sequences that are interrupted by sandstones only in the Lias-alpha, Dogger-beta and Dogger-epsilon (Cornbrash).

Another important feature of the Gifhorn trough is the occurrence of salt plugs. They originate in the underlying Zechstein and penetrate the sedimentary cover at different places, especially in the centre of the trough along the axis. The structural history of the basin is very strongly influenced by the salt movement.

From this basin eleven crude oils were selected for analysis. They were from the above-mentioned Dogger-beta and Lias-alpha sandstones, with the exception of two oils, which were from the Neocomian (Lower Cretaceous) from the top of a salt plug (Table 1). Care was taken, that the samples were from different parts of the basin with different subsidence histories.

Analytical procedure

The crude oils were freed from water by centrifuging and thereafter a saturated hydrocarbon fraction was obtained by elution chromatography. For the determination of the total amount of n-paraffins the saturated fraction was treated with molecular sieves, type 5A (van der Wiel, 1965). From the molecular sieves the n-paraffins can be nearly quantitatively desorbed by boiling them with n-pentane and n-heptane in succession.

Thereafter the separated n-paraffins were analysed by gas-liquid chromatography, with a flame-ionization unit as detector. From the chromatograms a n-paraffin distribution curve could be calculated. The pure n-paraffins n-C_{12}, n-C_{15}, n-C_{16}, n-C_{18}, n-C_{24} and n-C_{32} served as standards.

In this method the low-molecular-weight n-paraffins were lost because of evaporation. However, this does not affect the significance of the results, since all the oils have been treated identically.

For the concentration of isoprenoid type hydrocarbons a method described by Schlenk (1950, 1951) has been used. In order to diminish the number of components for the GLC-analyses a urea adduction was performed on the saturated fraction. By this means most of the n-paraffins and such iso-paraffins as have just one methyl group in a suitable position can be eliminated (Schlenk, 1949). The rest of the fraction consisting

TABLE 1. — CHARACTERISTICS OF MATURITY AND CORRELATION OF PETROLEUMS FROM THE GIFHORN TROUGH

Übersichtschema

Entwicklungsreihe und Verwandtschaftsgrad von Ölen des Gifhorner Troges.

I	II	III		IV
Bohrung Horizont u. Trägertiefe in m	mittl. Tiefenlage des mutmaßlich. Muttergesteins in m	Entwicklungsreihe: Reifemerkmale		Verwandtschaftsgrad: Korrelationsmerkmale
		n. Paraffine Gew. %	n. Paraffinverteilungskurve	
We 1018 Neokom 395,3 – 434,4	– 600 (Salzstockdach!)	1		
We N 8 Dogger Beta 1387,4 – 1397,0	– 1000 Dogger Beta	2		
We 1069 Neokom 330,28 – 382,4	– 600 (Salzstockdach!) Rhät (?), Lias (?)	6		
Rü 11 Dogger Beta 694 – 709	– 1700 Dogger Beta	9		
We 45 Dogger Beta 978,75 – 1027,0	– 2000 Dogger Beta	10		
Ha M 10 Dogger Beta 1588,0 – 1502,25	2000 – 2500 Dogger Beta	8		
We 57 Lias Alpha 1855,4 – 1902,28	– 2000 Lias Alpha	11		
Vo 52 Dogger Beta 1783,0 – 1790,7	– 2500 Dogger Beta	9		
Me 30 Dogger Beta 1668 – 1674	– 2600 Dogger Beta	13		
Me 22 Dogger Beta 1697 – 1705	– 2600 Dogger Beta	15		
Me 21 Dogger Beta 2002 – 2006,5 2026 – 2036,5	– 3100 Dogger Beta	27		

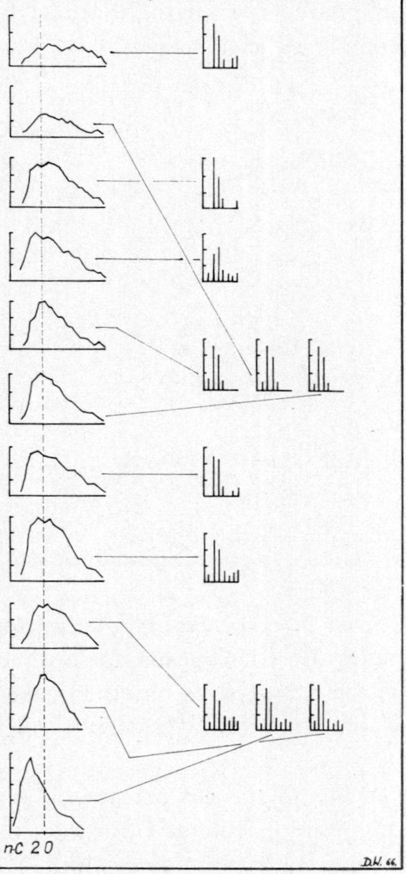

I: Well, horizon, and reservoir depth in metres.
II: Depth of probable source rock when oil was released.
III: Natural maturity sequence, characteristics of maturity.
IV: Degree of relationship among petroleums, characteristics of correlation.

The petroleums are ranked according to increasing maturity.

especially of multi-branched iso-paraffins and cyclic components, was subjected to a double thio-urea adduction. By the thio-urea adduction molecules which show a regular spacing of methyl substitutions, such as the isoprenoid type hydrocarbons, are preferentially concentrated. This concentrate was analysed by means of GLC, using a 2-metre packed column (Apiezon L), a 3-metre packed column (Silicon high vacuum grease) and a 50-metre capillary column (Apiezon L). Commercial purified pristane (2,6,10,14-tetra methylpentadecane) and the saturated degradation products of phytol, that had been heated in the presence of clay were employed as standards.

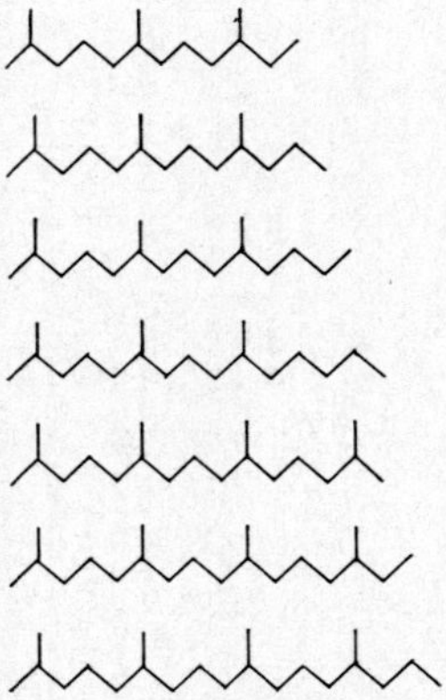

	Name:	B.P.	n - PAR
A FARNESANE	2,6,10, - TRIMETHYLDODECANE	254°	n C 14
B	2,6,10, - TRIMETHYLTRIDECANE	270°	n C15
C	2,6,10, - TRIMETHYLTETRADECANE		
D	2,6,10, - TRIMETHYLPENTADECANE	287°	n C 16
E PRISTANE	2,6,10,14,-TETRAMETHYLPENTADECANE	302°	n C 17
F PHYTANE	2,6,10,14,-TETRAMETHYLHEXADECANE	316°	n C 18
G	2,6,10,14,-TETRAMETHYLHEPTADECANE	330°	n C 19

FIGURE 2. — Isoprenoid-type iso-paraffins tentatively identified in petroleums.

Since so far the identity of the isoprenoids has not been established completely by other means, for instance, by IR-analyses or mass spectrometry, the components observed have been named with letters. Figure 2 shows the isoprenoids tentatively identified by GLC and the attached letters.

By means of the gas chromatograms the relative amounts of the isoprenoids were calculated from their peak areas. The peaks were cut out and their weights were determined. To get a basis for comparison, peak E (pristane) was taken arbitrarily as 100, and all other peaks were related to it. The resulting "fingerprint" diagrams are given in Fig. 3.

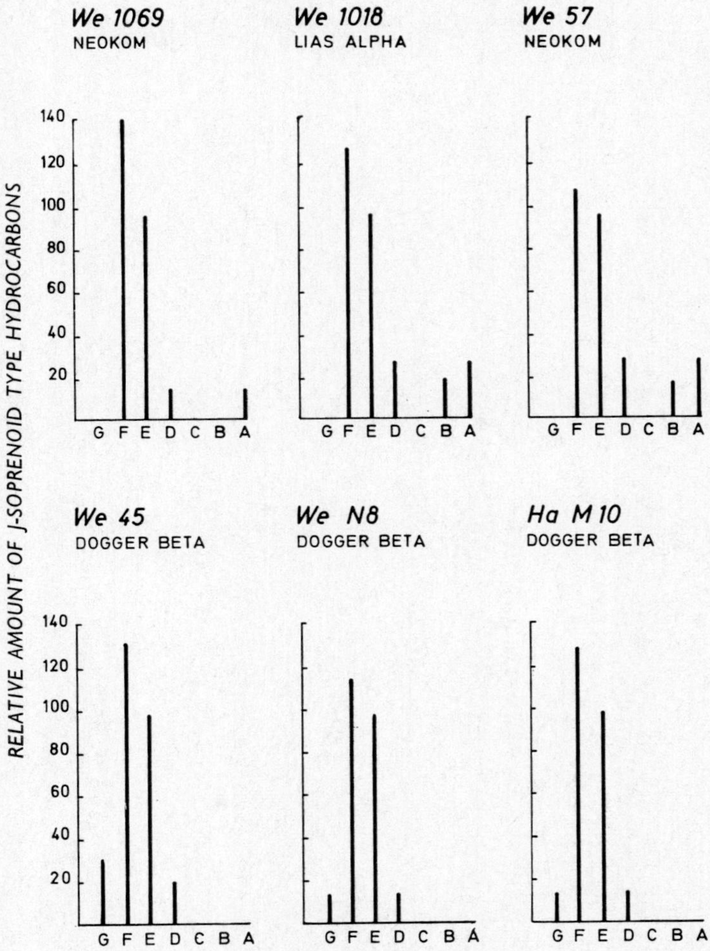

FIGURE 3.— Characteristics of correlation based on relative distribution of isoprenoid-type iso-paraffins in petroleums.

Discussion of the analytical results

As a consequence of the above considerations the crude oils investigated have been grouped in a relative sequence according to increasing amounts of n-paraffins, the rise and shift of a maximum in the n-paraffin distribution curve toward shorter chain lengths, and according to the disappearance of a slight odd-predominance among the higher n-paraffins. Figure 4 shows three typical n-paraffin distribution curves (We 1018,

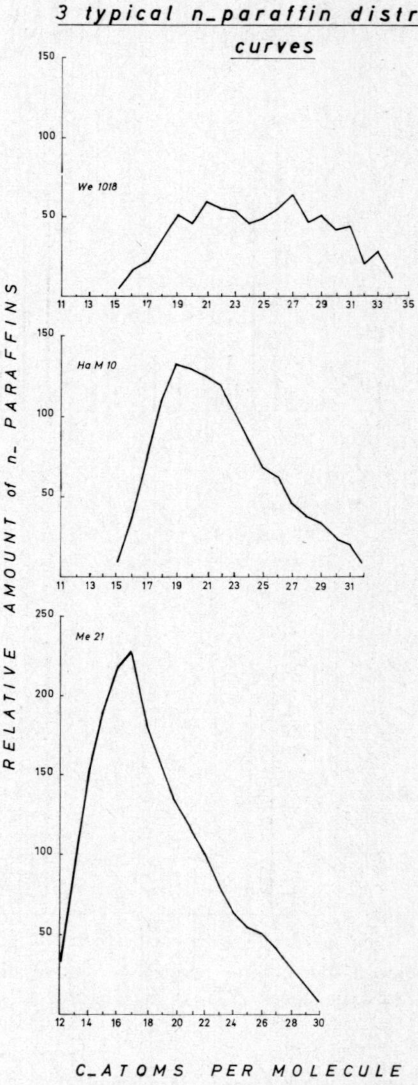

FIGURE 4. — Characteristics of maturity based on relative distribution of n-paraffins in petroleums.

Ha M 10, Me 21) from the beginning, the middle and the end of this sequence.

In this way a natural evolutionary series was established. It starts with relatively immature oils, low in n-paraffins and with a relic of an odd-predominance, and at the end there are more mature oils with a high amount of n-paraffins, a strong maximum in the low-molecular-weight region and a smooth distribution curve. This evolutionary sequence is shown in Table 1. It is to be emphazised that this is a relative sequence that should not be directly compared with petroleums from other basins.

In the Gifhorn trough the oils from the Wesendorf salt plug and from Wesendorf-Nord represent the relatively immature beginning of the evolutionary series, in the middle there follow the oils from Rühme, down flank from the Wesendorf salt plug, Hankensbüttel and Vorhop, and at the end there are the mature oils from Meerdorf.

The different members of this evolutionary series may originate from one source rock unit, having been released during different time intervals at different depth levels, especially if they are taken from stratigraphically comparable reservoir sands (Dogger-beta) and neighbouring oil fields. Therefore, an attempt was made to detect a possible relationship between the crude oils with the help of the isoprenoid-type hydrocarbons, Figure 3 shows six examples of different and similar distribution patterns for isoprenoids with 15–21 carbon atoms. A compilation of those "fingerprints" of all the crude oils is given in Table 1.

From the figures and the table it is obvious that the dominant components in all the oils are peaks F (phytane) and E (pristane). Both of them are two or three times more abundant than the other components. With the exception of Rü 11, F (phytane) is more abundant than E (pristane). After F and E the most frequent component is D (2,6,10-trimethyl-pentadecane), followed by G, A, B, and finally C. Phytane and pristane range up to 0·5 – 0·7% of the total crude. If a component is not shown on a diagram, this means that its content is below 0·05%. In Fig. 3, the "fingerprints" of the only three oils that do not come out of the Dogger-beta sandstones are shown. The oils, We 1069 (Neocomian) and We 1018 (Neocomian), are produced from the top of the Wesendorf salt plug; the oil We 57 (Lias-alpha) comes from down flank.

The two oils from the top of the salt plug are similar in their distribution pattern as regards the components F, E, D and A; however, the component B is present in one oil only. On this basis the oils are dissimilar. The same holds true for the oil from down the flank, We 57, that exhibits completely different ratios among the components F, E, D and A from the oils from the top of the plug.

Another oil from the east side flank of the Wesendorf salt plug is produced out of the Dogger-beta sandstone. This sample, We 45 (Fig. 3), is distinguished from the other Wesendorf oils by a relatively high amount of G and the absence of the components C, B, A. However, there is a certain similarity between this oil and another one (We N 8) that is found a few kilometres farther north and that is also produced from the Dogger-beta sandstone. The two distribution patterns of We 45 and We N 8 are not identical, but they are much more similar than the previous ones. Another Dogger-beta oil (Ha M 10) from the Hankensbüttel oil field, still farther north, shows a very strong similarity to We N 8.

The isopre- noid distribution patterns of the other oils are shown in Table 1. The oils from the Vorhop salt plug in the north (Vo 52) and from the Rühme field (Rü 11) in the south are not similar to each other or to any other oil. The three samples of the Meerdorf field which is also situated in the south, show a high degree of similarity to each other, but are different from the other oils.

Characteristics of maturity and correlation, and their geological interpretation

The basic concept for this investigation was the hypothesis that according to the available analytical data on rock extracts from various depth ranges (e.g. Philippi, 1965), and according to our knowledge about petroleum, there should be an evolutionary sequence among the crude oils of a given basin that parallels the evolution of the organic matter in source rocks.

With the help of the homologous series of n-paraffins it was possible to establish such a sequence. However, it has also to be shown whether this sequence is really a product of the subsidence history of a particular source rock area within the basin. Beyond that meaningful relationships between crude oils on the basis of chemical "fingerprints" can be established only if there is also sufficent geological information.

In other words, the geochemical results can only be interpreted in the right manner if the relevant geological facts are also taken into consideration. Among the geological information about the Gifhorn trough the paper by Philipp *et al.* (1964) in particular served as basis for interpreting the geochemical results.

A detailed analysis of the uplift and subsidence of the areas under consideration has been made by the latter authors. With a vast amount of geological information and especially with the help of measurements on shale densities (Roll, 1956) and investigations on quartz diagenesis in reservoir sands (Füchtbauer, 1961) so-called "diagrams of subsidence" have been constructed. In this way the depth position of the most probable source beds and the approximate time of migration could be estimated (Philipp *et al.*, 1964). More details about this are given in another paper (Welte, 1967).

In what follows the different oils are discussed. For this purpose it may be very useful to give a general impression of the present structure of the Gifhorn trough. Figure 5 shows a structural map, contoured on the top of the Dogger-beta sandstones (Philipp *et al.*, 1964).

At the beginning of the maturity sequence there are oils from the top of the Wesendorf salt plug, We 1018 and We 1069. In spite of the fact that they both come from a Lower Cretaceous sandstone and from wells that are under 700 metres apart, they do not show similar characteristics of correlation. Furthermore, although both are relatively immature, We 1069 has a higher rank of maturity than We 1018. It is highly improbable that the oils could have migrated to the top of the salt plug from down flank out of the Lias-alpha or Dogger-beta reservoirs. This is because those oils from the flank, We 57 and We 45, are different from the oils at the top in respect of their maturity and correlation index (Table 1). In addition, geological reasoning (Roll, 1956) would also exclude migration from the flank to the top.

Therefore, it may be concluded, that the oils from the top of the salt plug are either of very local origin or perhaps from the basinward dipping closely-neighbouring, Cretaceous sediments. Since the Cretaceous reservoir sands and the covering Tertiary are separated by an unconformity caused by erosion and subsequent transgression (Roll, 1951) it is impossible that oil accumulation occurred before the beginning of the Tertiary. Thus the oil accumulations on the top of the Wesendorf salt plug and the others considered in this publication, developed at about the same time, that is during a period from the end of the Cretaceous to the beginning of the Tertiary (Philipp *et al.*, 1964). If we compare the two samples, We 1018 and We 1069, with the others, it is evident that these two immature petroleums must have originated in source rocks with relatively little overburden. This conclusion is in complete accordance with the observation of Roll (1956), that the shales on top of the salt plug as well as those adjacent to it, have never been burried deeper than about 500–600 metres, and these are the only conceivable source beds.

The discrepancy in maturity between We 1018 and We 1069 may be explained as a consequence of the mobile, disturbed surface of the salt

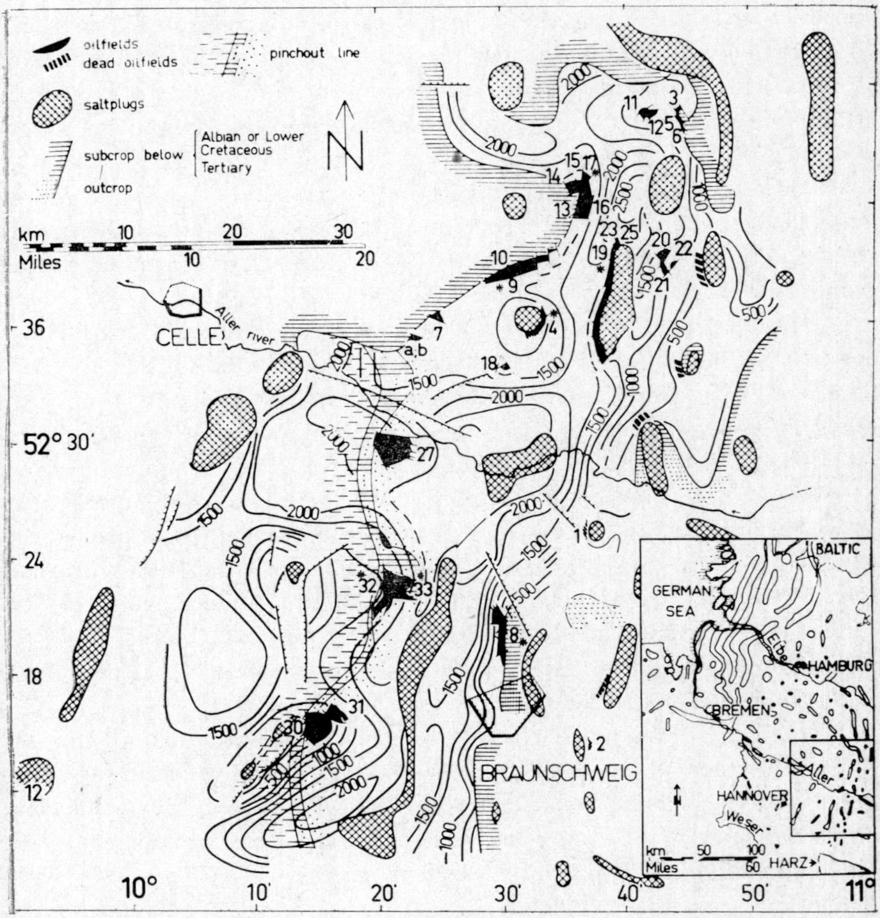

FIGURE 5. — Structure map of Gifhorn Trough, contoured on top of the Dogger-beta sandstones. Numbers with an asterisk mark the oil fields from which a sample has been analysed. (This map is a reproduction from the paper by Philipp *et al.*, 1964.)

plug. Possibly We 1018 is an immature oil, the reservoir of which was cut off from its drainage area very early, whereas the We 1069 oil could be diluted further by the supply of more mature oil.

The oil sample, We N 8, is also grouped among those at the beginning of the maturity sequence, being a low-grade-maturity oil. Since its position in the maturity sequence is between We 1018 and We 1069 it would be expected, that the corresponding source rock had a depth position of about 500–600 metres at the time when the oil was released.

The overburden of the reservoir at the beginning of the oil accumulation has been determined by Philipp *et al.* (1964) to be about 500 metres. However, further subsidence and thus compaction of the neighbouring potential source beds continued after the Pre-Albian uplift, and lasted until the end of the Tertiary. In this time interval the reservoir and its drainage area had been buried to a depth of 1400 metres.

A increase in maturity is, in the first place, a consequence of higher temperature. Since the heat conductivity of salt is three to four times higher than that of shale and since salt plugs are known to have a higher temperature then their surrounding it is easily understandable that an oil which originated on the top of a salt plug (We 1069) is more mature than an oil of comparable or even greater depth from an undisturbed normal stratigraphic column (We N 8).

The characteristics of correlation demonstrate a relationship between We N 8 and Ha M 10. This also makes sense in connection with the geology, since both oils are produced from the coarse, highly permeable upper part of the Dogger-beta sandstone. The probable source rocks are the enveloping shales of this sandstone (see also Welte, 1967).

However, Ha M 10 is much more mature than We N 8. This is because the drainage area of the Hankensbüttel oil has been burried to a depth of about 2500 metres in the centre of the trough (Fig. 5).

The similarity in the distribution of isoprenoid hydrocarbons (characteristics of correlation) and the difference for the n-paraffins (characteristics of maturity) between Ha M 10 and We N 8 may demonstrate a certain independence between the two parameters. It is to be expected, and has already been demonstrated (Robinson *et al.*, 1965), that the composition in isoprenoids may change with depth and temperature.

However, it is felt that the distribution pattern of the isoprenoid hydrocarbons is influenced much more by the primary source material, whereas the homologues series of n-paraffins is strongly influenced by the action of temperature, especially because of the ease of generating new n-paraffins of low molecular weight. There is some parallelism between the above reasoning and the ideas of Gedenk (1966).

The characteristics of correlation also show that there is a certain similarity between We N 8, Ha M 10 and the Dogger-beta oil, We 45, from the east flank of the Wesendorf salt plug. In contrast to the other two the last oil is not produced from the highly permeable upper part of the Dogger-beta sandstone, but from the less permeable lower part. The trap is located at about 1000 metres where the sandstone meets the flank of the salt plug. The Dogger-beta sandstone dips toward the basin and in the centre of it, east and north-east of the salt plug, reaches a depth of about 2000 metres (Fig. 5). It is, therefore, understandable why We 45 is found among the middle group of the maturity sequence, close to Ha M 10.

A possible relationship between We N 8 and Ha M 10 on the one hand and We 45 on the other, that may indeed be deduced from the isoprenoid fingerprints, can neither be supported nor refuted on geological grounds.

Sample We 57 from the southern flank of the Wesendorf salt plug cannot be related to any other oil. This is in accordance with the geological information, since the oil is produced from the Lias-alpha sandstone, the enveloping shales of which also possess source rock properties. Its maturity is comparable to We 45 and Ha M 10. According to density measurements those Lias-alpha shales have never been burried deeper than 2000 metres (Roll, 1956). This in turn is the depth position to be expected, when comparing We 57 with the stage of maturity of We 45 and Ha M 10.

The last oil out of the northern part of the Gifhorn trough is Vo 52 from the north-western flank of the Vorhop salt plug. Like the We N 8 oil and Ha M 10 it is produced from the upper part of the Dogger-beta sandstone, from a shoestring sand. However, there is no similarity in the distribution of isoprenoids between Vo 52 and the latter oils. The Dogger-beta dips toward the north into a deeper part of the basin. About 2 km from the salt plug the Dogger-beta reaches a depth of 2500 metres (Fig. 5). As regards maturity, the oil is somewhat more advanced than We 57.

From the southern part of the Gifhorn trough four oils have been investigated, Rü 11, Me 30, Me 22 and Me 21. The sample Rü 11 from the Rühme oilfield on the eastern rim of the basin does not show any relation to any other oil. It is produced from the westerly dipping Dogger-beta sandstone, that forms a trap against the transgressive Cretaceous. According to its relative position in the maturity sequence, and according to experience with the other oils, especially with We N 8 (We 1069 cannot be used for comparison, because of the different temperature situation around a salt plug), it may be deduced that its source area had been burried to 1600–1700 metres. The findings of Philipp *et al.* (1964) confirm this deduction (see also Fig. 5).

The end of the maturity sequence is formed by the oils from the Meerdorf field, about 10 km, West of Rühme. The similarity of these three oils, Me 30, Me 22 and Me 21, on the basis of their isoprenoid hydrocarbons, is in agreement with the geological situation. The three different oils come from three different tectonic blocks. However, these three blocks already existed before the accumulation of oil in the Dogger-beta reservoir (Philipp *et al.*, 1964). This is also the reason why Me 21 is more mature than the other two oils. It is produced from the deepest block.

The high maturity of the Meerdorf oils relative to the others is a result of the very strong subsidence especially in this area. On the diagram of subsidence it can be shown that the Meerdorf area was buried during the Cretaceous to 2500–3000 metres. However, at the end of the Cretaceous and during the Tertiary there was again uplift and erosion (Philipp *et al.*, 1964), so that at present the producing sands are at depths of 1700 and 2000 metres, respectively.

Conclusions

In the beginning a twofold question about the recognition of maturity stages and about the possibility of demonstrating a relationship between crude oils was set down. In order to answer these two questions it was necessary to find chemical parameters good enough for differentiating crude oils and suitable for reflecting the effects of differences in geological history.

The homologous series of the n-paraffins has been used for identifying maturity stages in petroleums. It can be shown that the maturity of an oil is mainly dependent on the temperature history of its source rock. That means that deeply buried source beds give off more mature oils than source rocks in a shallow position. The influence of the present depth of the reservoir rock seems to be negligible.

For correlation purposes the so-called isoprenoid-type hydrocarbons have been used. The similarity of their relative distribution patterns, as in the case of fingerprints, may give a hint as to a possible relationship between certain oils. However, in this connection it has to be demonstrated, whether such a relationship is also geologically justified. With the knowledge of the distribution of some key components in petroleums and of the stratigraphy and tectonic evolution of the Gifhorn trough the oils can be classified into a natural series expressing their developmental history.

The attempt to detect relationships between different crude oils with the help of isoprenoid-type hydrocarbons has, in principle, also been successful. However, it has to be borne in mind that more data are necessary, using other key components, in order to test and refine the correlation principle demonstrated here.

Acknowledgments

The Deutsche Forschungsgemeinschaft, Bad Godesberg, made this project possible by generous financial support.

The gas chromatographic analyses were performed with instruments made available by Gewerkschaft Elwerath, Hannover. On this account, and for his great interest, the author wants to thank Dr. A. Roll of the latter company. Furthermore, the writer acknowledges gratefully many stimulating discussions and much valuable advice from Dr. W. Philipp, also of Gewerkschaft Elwerath, Hannover.

The writer is grateful to the following oil companies, which made sample material available: Gewerkschaft Elwerath, Hannover; Deutsche Erdöl A. G., Hamburg; Preussag, Hannover; Mobil Oil A. G., Celle.

For their help in the laboratory the author wants to express his thanks to Mrs. U. Grütze, Miss M. Hub and Mr. Dipl. Geol. G. Ebhardt.

References

Dean, R. A. and Whitehead, E. V. (1964) The composition of high boiling petroleum distillates and residues. *Proc. 6th Wld. Petrol. Congr.*, Sect. V, Paper 9. Hamburg.

Füchtbauer, H. (1961) Zur Quarzneubildung in Erdöllagerstätten. *Erdöl u. Kohle, Erdgas, Petrochem.* **14,** 169–173.

Gedenk, R. (1966) Zur geochemischen Beurteilung und Genese von Kohlenwasserstoffansammlungen. *Erdöl u. Kohle, Erdgas, Petrochem.* **19,** 720–727.

Hecht, F. (1959) Migration, Tektonik und Erdöl-Lagerstätten im Gifhorner Trog (Nordwestdeutschland). *Erdöl u. Kohle,* **12,** 303–315.

Hedemann, H. A. (1956) Präalbische Tektonik am Nordwestrand des Gifhorner Troges. Geotektonisches Symposium (Hans-Stille-Festschrift), 213–227. Stuttgart.

Philipp, W., Drong, M. J., Füchtbauer, H., Haddenhorst, H. G., and Jankowsky, W. (1964) The history of migration in the Gifhorn Trough (NW-Germany), *Proc. 6th Wld. Petrol. Congr.*, Sect. I, Paper 19. Hamburg.

Philippi, G. T. (1965) On the depth, time and mechanism of petroleum generation. *Geochim. et Cosmochim. Acta,* **29,** 1021–1049.

Robinson, W. E., Cummins, J. J. and Dinneen, G. U. (1965) Changes in Green River oil-shale paraffins with depth. *Geochim. et Cosmochim. Acta,* **29,** 249–258.

Roll, A. (1951) Jurassic Troughs and Oil Deposits in Northwest Germany. *Proc. 3rd Wld. Petrol. Congr.*, Sect. I, Leiden.

Roll, A. (1956) Zur Strukturgeschichte der Salzstöcke von Wesendorf und Hohenhorn. Geotektonisches Symposium (Hans-Stille-Festschrift), 228–245. Stuttgart.

Schlenk, W. jr. (1949) Die Harnstoffaddition der aliphatischen Verbindungen. *Liebigs. Ann. Chem.* **565,** 204–240.

Schlenk, W. jr. (1950) Einschlußverbindungen des Thioharnstoffs. *Experientia,* **6,** 292 (Kurzmitteilung).

Schlenk, W. jr. (1951) Die Thioharnstoff-Addition organischer Verbindungen. *Liebigs. Ann. Chem.* **573,** 142–162.

Welte, D. H. (1965a) Kohlenwasserstoffgenese in Sedimentgesteinen: Untersuchungen über den thermischen Abbau von Kerogen unter besonderer Berücksichtigung der n-Paraffinbildung. *Geol. Rdschau.* **55,** 131–144.

Welte, D. H. (1965b) Relation between petroleum and source rock. *Bull. Amer. Assoc. Petrol. Geol.* **49,** 2246–2268.

Welte, D. H. (1967) Zur Entwicklungsgeschichte von Erdölen auf Grund geochemisch-geologischer Untersuchungen. *Erdöl u. Kohle, Erdgas, Petrochem.*, **20,** 65–77.

Wiel, van der, A. (1965) Molekularsiebe und Gas-Flüssigkeits-Chromatographie als Hilfsmittel zur Bestimmung der Konstitution von n-Paraffin, *Erdöl u. Kohle, Erdgas, Petrochem.* **18,** 632–636.

ÉTUDE SUR LES TRANSFORMATIONS PHYSIQUES ET CHIMIQUES DES PÉTROLES EN LIAISON AVEC LES CONDITIONS GÉOLOGIQUES

par

R. S. Byramjee

Compagnie Française des Pétroles

et M. A. Bestougeff

Compagnie Française de Raffinage

I. Introduction

L'étude de la transformation des pétroles est un des principaux problèmes de la géochimie de ces caustobiolites. Il est très important d'établir le rapport entre l'effet immédiat des différents facteurs géologiques et géochimiques et les propriétés physico-chimiques des huiles brutes.

Les transformations physiques et chimiques que subit le pétrole au cours de son histoire sont généralement admises; cependant les points de vue sur le sens et sur le mécanisme de ces transformations divergent considérablement.

Dans un certain nombre de travaux, les auteurs acceptent ou refusent la possibilité de la transformation des pétroles bruts dits "paraffiniques" en ceux dits "asphaltiques" ou "naphténiques" et vice-versa, plutôt par sentiments personnels que selon les preuves chimiques sûres.

Les termes mêmes pour désigner cette transformation (maturation, métamorphisme[10, 13, 14]) sont assez mal définis et sont empruntés à d'autres domaines. Nous préférons utiliser dans ce travail le terme "transformation" en distinguant :

1° *la transformation physique* provoquée par les phénomènes physiques tels que l'évaporation, la ségrégation, la filtration, l'adsorption, etc.
2° *la transformation chimique* due à l'action des facteurs internes ou externes, tels que les multiples changements de la structure des molécules, de leur taille, de leurs fonctions chimiques, etc.

La composition chimique du pétrole est donc la combinaison de ces multiples actions physico-chimiques qui se manifestent selon les conditions géologiques variables au cours de toute l'évolution des pétroles bruts.

Du point de vue géologique, les facteurs que nous examinons habituellement sont les suivants:

a) *Conditions géologiques actuelles*

— Température
— Pression
— Profondeur
— Âge
— Lithologie { réservoir / couverture }
— Type de piège
— Hydrodynamique

b) *Histoire géologique*

— Roches-mères
— Migrations (compaction, tectonique, hydrodynamique)
— Enfouissement maximum
— Époques de mise en place, etc.

Le problème ainsi présenté est très vaste et extrêmement complexe. C'est pourquoi le but de ce travail se limite à souligner les points importants de la transformation.

En laissant de côté la question de l'origine du pétrole et les détails de sa migration, on commence l'examen à partir d'un pétrole brut déjà formé pouvant contenir différents groupes de constituants (gazeux, liquides et solides à la température ambiante) et pouvant se déplacer plus ou moins loin à travers des roches.

Les études assez poussées sur la composition chimique d'un nombre assez important, de pétroles bruts d'une origine géologique connue effectuées ces dernières années[3, 7] *ont démontré la possibilité de grouper les pétroles bruts des différents gisements appartenant à une région, en plusieurs types chimiques.*

Les pétroles bruts appartenant au même type chimique peuvent cependant se distinguer à certains égards.

Ces différences sont liées habituellement aux conditions *géologiques particulières des gisements;* ainsi l'étude de la composition chimique des

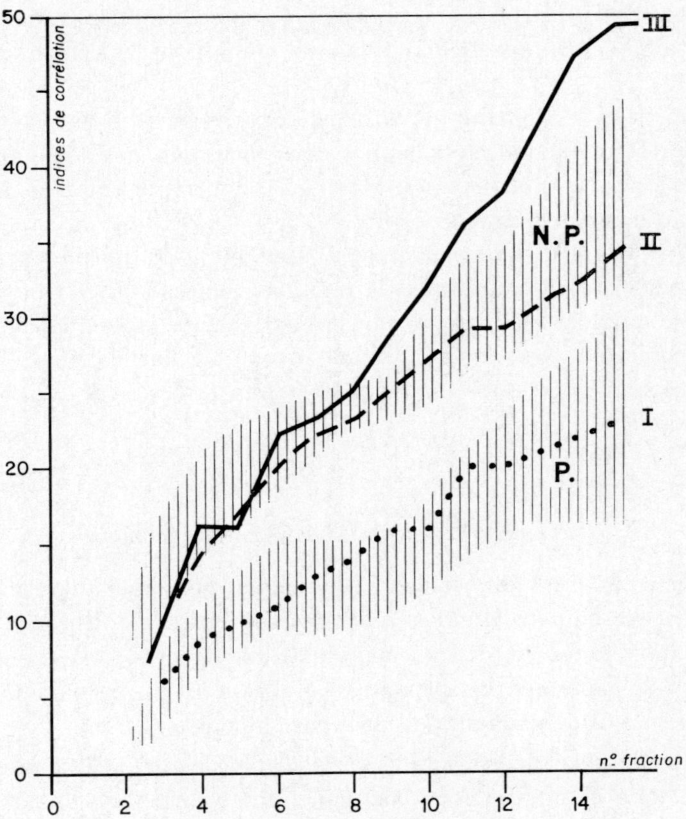

FIGURE 1.

pétroles débouche sur le problème de corrélation entre les caractéristiques physico-chimiques des pétroles et les facteurs géologiques d'environnements locaux.

Jones et Smith[(7)], par exemple, à la suite de l'analyse de 310 échantillons de bruts dans un Bassin Permien ont établi 5 principaux types (et de 15 à 20 sous-types). Il a été possible en outre dans certains cas de rattacher ces types de pétroles aux conditions géologiques connues ("basinal" or "shelf environment").

L'analyse d'une série de pétroles du Nord Sahara effectuée en collaboration avec Vasse et Semat (C.F.R.) nous a permis également d'établir 3 types de pétroles liés, d'une part à la répartition stratigraphique et, d'autre part avec les conditions tectoniques (fig. 1).

Le type I correspond aux pétroles paraffiniques de la classification de H. M. Smith. Ces pétroles se rencontrent le plus souvent dans des réservoirs stratigraphiquement plus jeunes que ceux renfermant les pétroles de types II et III.

Le type II est "naphténo-paraffinique" et correspond à des pétroles plus riches en constituants naphténiques et aromatiques que les pétroles du premier type. Le brut de Hassi-Messaoud (non représenté sur la figure) entre dans cette deuxième catégorie.

Le type III se distingue du type II par une quantité plus grande de constituants aromatiques dans les fractions lourdes. On y trouve également une quantité plus forte de constituants hétéro-atomiques. Ces différences coïncident avec des conditions tectoniques particulières: la structure géologique correspondante comporte des failles et des discordances importantes.

II. Généralités sur la transformation phisique

Les différents "environnements" géologiques provoquent la transformation physique du pétrole qui se répercute sur leur composition. Il est très difficile dans beaucoup de cas concrets de distinguer le changement de la composition chimique du pétrole dû à l'action des facteurs purement physiques de celui résultant des réactions chimiques.

L'indice le plus sûr de l'action seulement physique est la conservation des structures chimiques des constituants du pétrole. *On peut admettre en première approximation que la présence ou l'absence dans un pétrole brut donné des constituants d'une certaine structure ainsi que la teneur relative et absolue de ces structures constitue le reflet de son histoire chimique.*

Il est nécessaire cependant d'examiner en détail les possibilités pratiques d'application de ce critère dans les études géochimiques des pétroles.

a) *Évaporation et ségrégation des fractions légères*

L'évaporation change considérablement la composition fractionnée des pétroles, mais ne modifie pas considérablement sa composition chimique. On a constaté souvent que les compositions chimiques du condensat des gisements de gaz et de l'essence des différents gisements de pétroles appartenant à la même formation géologique sont très voisines.

Ainsi au Nord Sahara, la comparaison du condensat de Hassi R'Mel et des essences des indices de pétrole de la même région, provenant de la même formation (Trias) montre que ces produits appartiennent au même type chimique.

En particulier ils figurent dans le même groupe "paraffinique" de la classification de Smith. Or, du point de vue géologique, on a toutes les raisons de penser que leurs faciès d'origine sont les mêmes.

Si l'évaporation s'accompagne d'autres phénomènes physiques, les corrélations peuvent devenir très difficiles.

Sokolov, dans son récent ouvrage[14], souligne la nécessité de tenir compte dans tout examen de la transformation du pétrole, non seulement des processus d'accumulation mais également des phénomènes de dispersion qui vont de pair. Chaque gisement de pétrole et de gaz s'accroît ou diminue suivant les vitesses réciproques de deux phénomènes (entrées et sorties). Les processus géologiques conduisant à la destruction des pièges favorisent la perte du pétrole et du gaz. Les gisements de gaz subissent un rapide dégazage. Les gisements de pétrole s'appauvrissent en perdant les fractions légères par évaporation : le produit restant dans la couche se transforme en bitume (voir chapitre *Oxydation).*

b) *Ségrégation des fractions lourdes*

Il s'agit surtout de la séparation ou de la sédimentation des constituants dont la solubilité dans le pétrole est relativement faible, à basse température. La formation des dépôts contenant principalement les hydrocarbures solides (paraffin wax) et les constituants asphalténiques au cours de l'extraction du pétrole est bien connue. Le même phénomène provoqué par les mêmes facteurs physiques (brusque dégazage, changement de la température accompagné de filtration) conduit à la ségrégation partielle des constituants solides dans les roches.

Nous avons étudié en détail un pétrole brut de Libye provenant de la Cyrénaïque du Sud et dans le Crétacé Inférieur à environ 3000 m de profondeur.

Ce pétrole brut contient 14,3% d'asphaltènes et 17% de paraffines solides (paraffin wax) et 0,16% de soufre. Sa composition chimique globale est approximativement la suivante :

Groupes de constituants	% par rapport au pétrole brut		
Paraffines normales	18,5		
Paraffines ramifiées	13,8		
Total paraffines		32,3	
Naphtènes	32,4	32,4	
Total saturés			64,7
Aromatiques	3,8		
Naphtèno-aromatiques	7,2		
Total aromatiques		11,0	11,0
Hétérocycliques	10,0		
Asphaltènes	14,3		
Total		24,3	24,3
GRAND TOTAL	100,0	100,0	100,0

La détermination de la courbe de distribution des paraffines normales montre deux maxima (fig. 2) : l'un vers C_{12}–C_{13} et l'autre vers C_{21}–C_{23}. Cette distribution bimodale est exceptionnelle.

En effet, jusqu'à présent on n'a observé que deux types de courbes : les unes présentant un seul maximum vers C_5–C_9 et les autres, un seul maximum, vers les termes supérieurs.

On peut donc penser que la distribution bimodale observée peut s'expliquer par un phénomène de ségrégation, ou éventuellement par un mélange (voir III, d).

Il semble que la ségrégation des fractions lourdes des pétroles soit un phénomène beaucoup plus répandu dans la nature que l'on ne le croit généralement.

En particulier, on peut admettre que les variations assez sensibles de la teneur en paraffines solides (paraffin wax) qu'on rencontre dans les gisements voisins appartenant à une même formation géologique (par exemple au Sahara), sont dues plutôt aux phénomènes physiques de ségrégation qu'à une transformation chimique hypothétique.

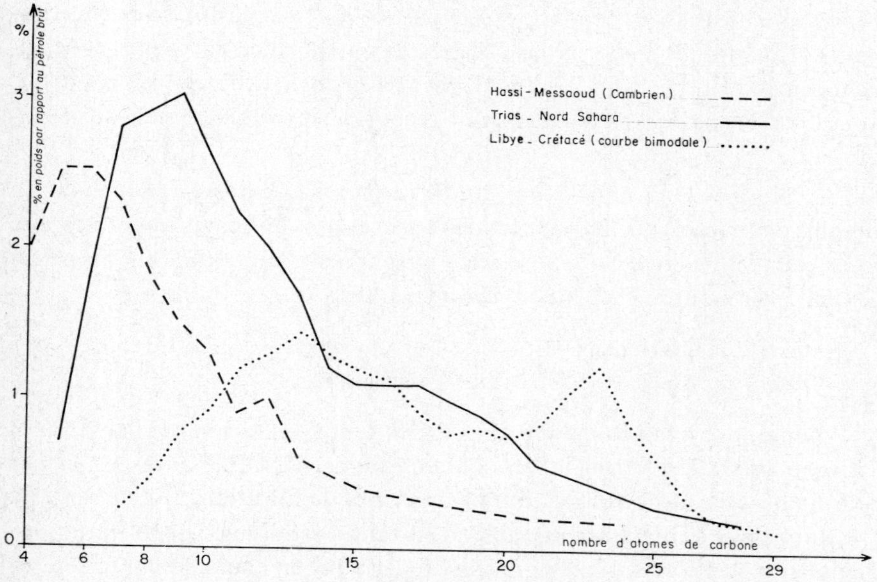

FIGURE 2.

c) *Filtration et adsorption*

Ces phénomènes interviennent essentiellement au cours de la migration du pétrole et du gaz.

On examine ces deux phénomènes ensemble étant donné leur liaison étroite. En effet, la filtration du pétrole à travers les roches s'accompagne toujours d'une adsorption plus ou moins grande suivant l'état et la nature des roches. En outre, l'adsorption ne peut pas se manifester sans mouvement de la matière, c'est-à-dire filtration à travers des couches. Il n'existe aucun doute que ces phénomènes peuvent modifier considérablement la composition du pétrole. Ce qui est moins sûr c'est la détermination exacte du degré de changement de la composition fractionnée et chimique, provoqué par ces facteurs dans les différents cas particuliers.

Les essais pour reproduire ces changements effectués jusqu'à maintenant au laboratoire(16) donnent relativement peu de renseignements directement transposables dans les conditions de gisement. On peut croire que l'adsorption sélective provoque un effet plus fort sur la composition chimique des pétroles que toutes les autres actions physiques. Par exemple, en comparant la composition chimique des pétroles de différents gisements, Jones et Smith concluent que le passage des pétroles bruts à travers les couches absorbantes conduit, d'une part à la perte sélective des constituants aromatiques légers et des composés soufrés, et, d'autre part à l'enrichissement en fractions légères d'essence et en hydrocarbures naphténiques[7]. Ceci permet en particulier de reconstituer certains trajets de migration.

Au Nord Sahara nous avons étudié plusieurs gisements comprenant au moins deux réservoirs séparés par des couches plus ou moins imperméables. Les pétroles des réservoirs inférieurs et supérieurs appartiennent au même type, mais montrent les différences suivantes :

—réservoir inférieur : plus riche en aromatiques
—réservoir supérieur : plus riche en naphtènes

Du point de vue géologique nous avons toutes les raisons de penser qu'il y a eu migration verticale des hydrocarbures. Les différences observées résultent donc, à notre avis, des phénomènes de filtration et d'adsorption.

Rappelons qu'un autre exemple de filtration est donné par les variations de la composition isotopique des hydrocarbures en cours de migration, en particulier pour les isotopes du carbone. Les études de Colombo[5] et de Silvermann[12] ont montré qu'au cours de la migration les hydrocarbures légers s'enrichissent en isotope C_{12}.

d) *Mélanges dans les gisements*

La figure 3 représente les indices de corrélation de quatre pétroles du Sahara. On y retrouve les types I (pétrole "A") et II (pétroles "B" et "C") dont nous avons précédemment parlé. Par contre, la courbe correspondant au pétrole X est très différente. On pourrait croire qu'elle définit un type nouveau : ce serait une erreur.

En effet, il faut d'abord remarquer que la courbe X recoupe les différentes zones de la classification de Smith. Ensuite, en se reportant au contexte géologique, on constate que le pétrole X provient à coup sûr d'au moins deux sources différentes. Au contraire, le pétrole B, géographiquement voisin de X, ne provient que d'une seule source.

Ainsi le brut X est un mélange et on peut croire que, d'une façon générale, *une courbe de C.I. qui recoupe les zones de Smith correspond très probablement à un mélange.*

Un autre exemple de mélange est donné par Jones et Smith[7] à propos des pétroles de Simpson et Ellenburger. Les critères utilisés dans ce cas

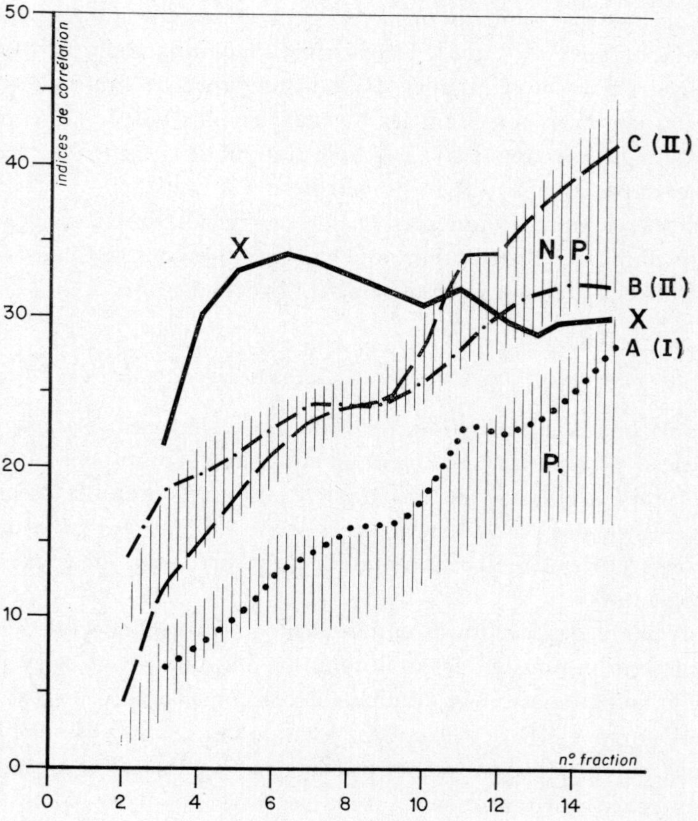

NB. La courbe X correspond à un mélange.

FIGURE 3.

pour mettre en évidence plusieurs sources sont les mêmes que ceux que nous venons d'indiquer, mais ils sont moins marqués.

III. Généralités sur la transformation chimique

Les transformations chimiques du pétrole peuvent être classées en deux catégories :

1°) celles provoquées par les forces internes : interactions des constituants du pétrole, réarrangements intermoléculaires, etc.

et

2°) celles résultant de l'action des agents externes principalement du soufre et de l'oxygène.

Nous acceptons comme point de départ les vues développées par Andreev et Dobryansky[1, 6] que la transformation chimique du pétrole est un phénomène universel et régulier, lié à la tendance de toute substance à former des systèmes possédant les niveaux les plus bas de l'énergie libre dans les conditions données. Ce critère s'applique à toutes les réactions chimiques entre les constituants du pétrole.

Parmi ces réactions, il faut citer surtout la dégradation, la condensation, l'isomérisation, la redistribution de l'hydrogène et des radicaux (substituants) dans les cycles et éventuellement l'hydrogénation.

Limites du changement des structures chimiques au cours de l'évolution

Il semble que la modification des structures chimiques des constituants du pétrole se produit très lentement au cours de l'évolution et n'aboutisse que rarement à la transformation complète des motifs primitifs (le squelette carbone). Cette idée est confirmée par la présence dans le pétrole de nombreux corps dont la structure chimique est voisine sinon identique à celle des "progenitors".

La formation des structures entièrement nouvelles s'effectue à la suite de la dégradation poussée des molécules les plus complexes et éventuellement de la recombinaison de différents débris ainsi formés.

L'étude comparative des réactions chimiques et des conditions physico-chimiques de l'évolution des pétroles permet de faire les constatations suivantes en ce qui concerne les modifications possibles des structures chimiques.

1°) *La chaîne droite* (n-paraffines, chaînes latérales droites) *en se dégradant donne* principalement *le même type de structure*. Il peut se produire, en présence des catalyseurs, une certaine isomérisation avec la formation de chaînes surtout monoramifiées[3 bis].

2°) *La chaîne ramifiée* (iso-paraffines, chaînes latérales) *en se dégradant donne la même structure ramifiée*, mais peut fournir également, suivant la constitution du produit de départ, une structure linéaire plus courte.

L'isomérisation et la cyclisation (formation des cycles saturés) sont possibles en présence des catalyseurs naturels appropriés. Mais ces dernières réactions jouent probablement un rôle secondaire dans la transformation chimique du pétrole.

3°) *Les cycles saturés se comportent très différemment suivant leur structure* et la présence d'autres groupes dans les molécules.

Les cycles à 6 chaînons peuvent assez facilement se déshydrogéner en participant activement à la redistribution extra et intermoléculaire de l'hydrogène. Ces réactions s'effectuent encore plus facilement dans le cas des structures polycycliques.

L'isomérisation des cycles à 6 chaînons en cycles à 5 chaînons s'effectue assez facilement. Par contre, la réaction inverse est moins probable.

La dégradation des molécules cycliques saturées dotées de chaînes latérales commence par le détachement des chaînes longues. Les chaînes courtes sont plus stables et le groupe méthyl attaché au cycle possède le maximum de résistance.

La réaction de décyclisation n'est pas à exclure, mais la rupture des noyaux demande des conditions spécifiques pour rendre possible l'hydrogénolyse dans l'environnement géologique (voir § hydrogénation).

Il faut noter, en outre, que pour les structures polycycliques saturées, la cyclisation est toujours graduelle et la modification du squelette carboné est très progressive. Par exemple, les naphtènes tricycliques deviennent bicycliques avant de se transformer en monocycliques et finalement en structure aliphatique.

4°) Les structures cycliques aromatiques se dégradent suivant un schéma analogue à celui des cycles saturés. La différence principale réside dans la *stabilité extrêmement grande du cycle aromatique*. Une fois formé, ce système cyclique résiste pratiquement à toute épreuve. Les hydrocarbures aromatiques mono et bicycliques peuvent cependant intervenir dans les multiples réactions de redistribution des radicaux (substituants) comme c'est représenté sur la figure 4.

L'identification de ces structures isomères et la détermination de leur proportions réciproques présente un intérêt considérable pour l'étude géochimique des pétroles.

Jones et Smith insistent beaucoup sur l'importance du "profil aromatique" (redistribution des hydrocarbures aromatiques dans les essences) pour la caractérisation des types de pétroles. À cette occasion on peut noter le travail très intéressant de Burova et Khodjaev[(4)] qui ont analysé les essences provenant des pétroles bruts d'Ousbekistan extraits de deux formations différentes (Paléocène et Crétacé). Ils ont trouvé, en accord avec d'autres auteurs, que la teneur en hydrocarbures aromatiques est en général plus forte dans les essences du Crétacé, c'est-à-dire provenant des pétroles plus anciens. Le rapport benzène/toluène dans les mêmes essences s'élève à 1/5 ou 1/7 contre 1/3 ou 1/4 pour celles du Paléocène.

Le rapport quantitatif des hydrocarbures aromatiques bisubstitués pour ces deux groupes est réciproquement: meta $>$ para $>$ ortho et méta $>$

ortho-para. Parmi les aromatiques en C_9, le 1, 2, 4 trimethyl-benzène prédomine nettement, cependant que la somme des 1, 3, 5 et 1, 2, 3 tri-methylbenzènes est égale à la teneur de l'isomère 1, 2, 4 dans les essences du Crétacé et très inférieure pour celles du Paléocène.

Une essence provenant du Jurassique analysée en détail également ne contient que des traces de benzène et peu de toluène : l'éthylbenzène est

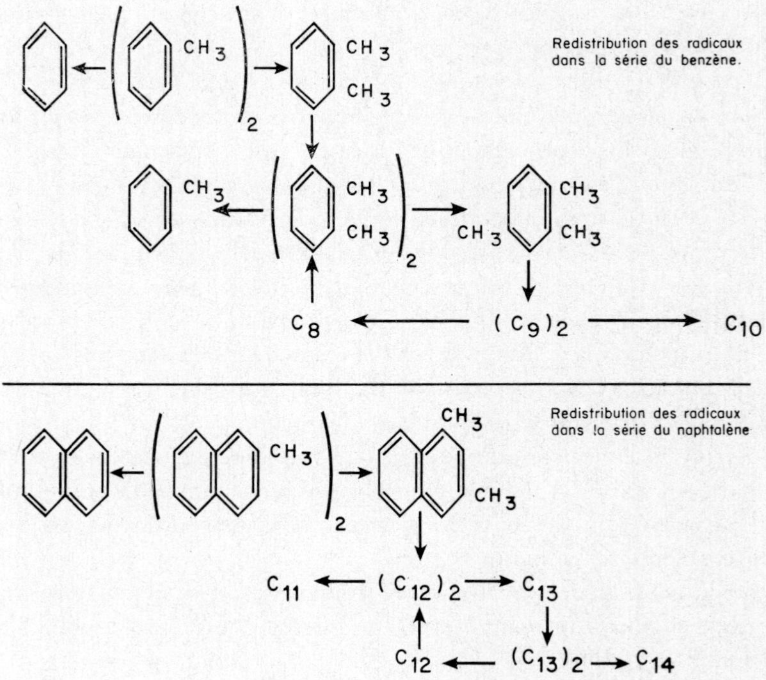

FIGURE 4.

le constituant prédominant, tandis que les proportions réciproques des aromatiques bisubstituées en C_8 meta, ortho et para sont très voisines.

5°) Les structures aromatiques *polycycliques* ainsi que celles hétéro-cycliques se comportent dans les réactions de dégradation (départ des chaînes, redistribution des radicaux, etc.) comme tous les systèmes aroma-tiques. Ils ont en outre une *forte tendance à la condensation*. Les composés formés à la suite de ces dernières réactions conservent sensiblement la structure de base des molécules constituantes qui ne perdent que leur individualité.

6°) Les structures *cycliques* "*mixtes*" (naphténo-aromatiques), qui contiennent simultanément dans leurs molécules les cycles aromatiques et saturés, *sont les plus réactives* et les plus mobiles.

Elles participent probablement d'une façon massive dans la transformation chimique des pétroles sous réserve néanmoins des limitations des réactions énumérées précédemment pour les différents systèmes cycliques.

Ce bref aperçu des possibilités de transformation chimique des différentes structures *montre que la transformation chimique des pétroles ne s'effectue pas uniformément pour tous ses constituants, mais différemment pour chaque type de structure et d'une façon graduelle avec conservation au moins en partie de la structure de base.*

Dégradation

La dégradation est la principale réaction que subissent les molécules du pétrole en fonction du temps et de la profondeur d'enfouissement. Elle conduit à la formation des quantités de plus en plus grandes d'essences et de gaz.

Jones et Smith[7] constatent qu'il y a en général un accroissement de la teneur en essence avec l'âge et la profondeur d'enfouissement.

On peut citer de nombreux travaux rapportant cette constatation faite à l'occasion d'études géochimiques dans différentes régions.

Cependant la transformation chimique du pétrole due à la dégradation est très complexe. Une étude approfondie à la fois de la composition chimique du pétrole, des gaz et éventuellement des matières organiques des roches, ainsi que la connaissance de l'environnement géologique, sont nécessaires pour pouvoir comprendre et exploiter les données analytiques dans la conduite de l'exploration pétrolière. On n'a pas encore trouvé une méthode universelle satisfaisante. L'étude géochimique se contente encore de rechercher les "indices", les "marqueurs", etc.

Voici l'énumération des principaux indices proposés et utilisés par différents auteurs :

1. Le "profil naphténique" analogue au "profil aromatique" déjà mentionné, proposé et exploité par Jones et Smith (% naphtènes dans les fractions de distillation B.O.M.).

2. Le rapport $\frac{\text{\% d'hydrocarbures paraffiniques}}{\text{\% d'hydrocarbures naphténiques}}$ pour différentes fractions (Kartzev[8]).

3. Le coefficient de "paraffinicité" des hydrocarbures naphténiques

Cp/Cn, où Cp = % d'atomes de carbone dans les chaînes et Cn = % d'atomes de carbone engagé dans les cycles (Sergienko[11]).

4. Le rapport $\dfrac{\%\ \text{d'hydrocarbures monoaromatiques}}{\%\ \text{d'hydrocarbures biaromatiques}}$ (Kurbsky[9])

D'après Kurbsky l'augmentation de ce rapport dans les fractions légères correspond au taux de dégradation du pétrole.

Si la teneur en résines augmente ou diminue en même temps, on doit admettre que la dégradation est accompagnée par les réactions d'oxydation ou le phénomène d'adsorption.

5. Le déplacement des maxima sur la courbe de distribution des n-paraffines en fonction du poids moléculaire représente, à notre avis, un indice important. Le déplacement de ce maximum vers les poids moléculaires faibles correspond dans une certaine mesure au taux de dégradation et est caractéristique pour les pétroles bruts provenant des roches anciennes (fig. 2).

Des courbes analogues peuvent être établies pour les différentes séries d'aromatiques dans le cas où les analyses détaillées sont disponibles. Jones et Smith indiquent ce déplacement du maximum de la teneur en aromatiques (prise ensemble) pour différents types de pétroles. Ceci s'observe en particulier pour les pétroles de types I et II du Nord-Sahara. L'évolution avancée des pétroles du type II conduit à une augmentation de la teneur en aromatiques dans les fractions légères (cf. par exemple la composition du brut paléozoïque de Hassi-Messaoud).

6. Teneurs des différents termes de la série des isoprenoïdes.

Condensation

Les réactions de condensation (et polycondensation) jouent un rôle très important dans la transformation chimique des pétroles.

Deux facteurs au moins sont nécessaires pour que ces réactions aient lieu effectivement, à savoir:

1°) la présence des constituants polycycliques (hétéro-naphténo-aromatiques) en quantité notable, et

2°) la présence des agents actifs, surtout le *soufre et l'oxygène* (voir § sulfuration, oxydation).

En effet, les pétroles bruts soufrés contiennent un taux élevé de résines et d'asphaltènes, tandis que les pétroles bruts "anciens" et d'autres caustobiolites se trouvant à l'abri de l'action du soufre et de l'oxygène sont pauvres en ces constituants.

Cependant les réactions de condensation ont lieu tout de même dans ce dernier cas mais à une échelle assez réduite. En l'absence de structures hétérocycliques convenables, il ne se produit qu'une condensation des structures polyaromatiques disponibles.

Ainsi la transformation chimique globale d'un pétrole brut léger, type dit "paraffinique", dont la majeure partie des constituants est saturée et assez volatile en une huile asphaltique lourde, est chimiquement *irréalisable.* Cela est juste non seulement au point de vue des réactions de condensation des hydrocarbures mais également (comme on le verra plus loin) par les voies de sulfuration et d'oxydation.

Redistribution de l'hydrogène et des radicaux (substituants)

L'absence parmi les constituants du pétrole des constituants oléfiniques d'une part, et l'existence des réactions de dégradation qui doivent obligatoirement conduire à leur formation (sans apport extérieur de l'hydrogène) d'autre part, ont posé le problème de la redistribution de l'hydrogène (hydrogénation interne).

Le fait de la redistribution de l'hydrogène à la suite de la transformation chimique du pétrole est indéniable et peut être facilement mesuré par le dosage de l'hydrogène.

Cependant, le mécanisme de cette redistribution qui englobe une série de réactions très complexes dans les conditions géologiques est inconnu.

Il est très souhaitable de transposer nos connaissances purement chimiques dans ce domaine sur le plan géochimique, à l'aide des expériences directes. C'est la seule voie qui permettra de sortir des généralités et des suppositions.

La possibilité de l'hydrogénation des constituants du pétrole par l'hydrogène gazeux venant de l'extérieur n'est généralement pas admise, étant donné que les gaz des gisements ne contiennent pas ou contiennent très peu de ce constituant.

Sokolov a réalisé les analyses plus précises de gaz et fait le calcul de la quantité globale de l'hydrogène pouvant se former dans les roches profondes par l'action sur l'eau des éléments radio-actifs des oxydes de métaux et d'autres substances minérales. Ce calcul montre que la quantité d'hydrogène qui peut imprégner les roches par migration verticale assez facile pour ce gaz est largement suffisante pour hydrogéner tout le pétrole[(14)].

Il reste cependant à démontrer que cette réaction d'hydrogénation se réalise effectivement dans les conditions géologiques.

La redistribution des radicaux que nous avons déjà examinée s'effectue probablement sur une échelle assez grande et assez souvent dans les conditions géologiques. Sans cette hypothèse, il est impossible d'expliquer la diversité des isomères et leur répartition variable dans les pétroles bruts de différents types, tels que ceux de l'Ousbekistan[4].

Action des agents externes

a) SULFURATION. La sulfuration modifie considérablement la composition chimique des pétroles et en partie la structure chimique de leurs constituants.

Il se forme des types particuliers de pétroles bruts soufrés. Dans la classification chimique des pétroles bruts on est obligé de les traiter à part (classification d'après les indices de corrélation de H. M. Smith).

On admet que la plus grande partie des composés soufrés du pétrole résulte de l'action du soufre sur les constituants de ce dernier. Il est connu également que le soufre réagit assez facilement à des températures dépassant 150–170° avec différents hydrocarbures, dont les paraffines.

Nous avons effectué un certain nombre d'essais en vue d'étudier cette action sur différentes fractions du pétrole et de déterminer la répartition des composés soufrés avant et après ce traitement.

Les distillats et les résidus provenant des pétroles bruts à faible teneur en soufre ont été chauffés avec 5 % d'une poudre fine de soufre à 160–180° pendant 5 heures dans un ballon muni d'un agitateur mécanique et d'un réfrigérant à reflux.

Le soufre qui n'a pas réagi et l'hydrogène sulfuré ont été éliminés en agitant énergiquement (agitateur à secousses) le produit de réaction avec le mercure pendant 5 heures à la température ambiante. Le produit liquide, après avoir été séparé par extraction au Soxhlet du mercure et des sulfures de mercure formés, est chromatographié sur colonne.

Les résultats de ces essais, dont deux exemples figurent sur les tableaux 1 et 2, permettent de tirer les conclusions suivantes:

1°) *L'action du soufre dans les mélanges d'hydrocarbures de différentes structures est très sélective.*

Les hydrocarbures aromatiques se combinent en premier lieu avec le soufre ; les monoaromatiques sont attaqués plus faiblement que les constituants contenant plusieurs cycles aromatiques. Les hydrocar-

TABLEAU 1.

TRAITEMENT AVEC LE SOUFRE D'UN DISTILLAT HUILEUX DU PÉTROLE BRUT HASSI-MESSAOUD (P.M. 350 env.)

FRACTIONS		DISTILLAT AVANT TRAITEMENT CHROMATOGRAPHIE SUR ALUMINE ACTIVÉE à 250°					DISTILLAT APRÈS TRAITEMENT CHROMATOGRAPHIE SUR ALUMINE ACTIVÉE à 250°				
				% de Soufre					% de Soufre		
N°	Nature	% en poids	n_D^{20}	Total dans la fraction	Thioph. *	Par rapport à la charge	% en poids	n_D^{20}	Total dans la fraction	Thioph. *	Par rapport à la charge
Charge	-	100,0	1,4997	0,39	88,5	100,0	100,0	1,4985	1,23	77,2	100,0
1	Saturés	38,2	1,4609	0,05	-	4,9	40,9	1,4623	0,05	-	1,7
2	Saturés	21,7	1,4687	0,05	-	2,6	17,8	1,4677	0,05	-	0,8
3	Monoaromatiques	13,3	1,4955	0,07	-	2,3	7,2	1,4808	0,18	-	1,0
4	Mono et bi-aromatiques	4,1	1,5148	0,09	-	1,0	9,4	1,5253	2,32	86,2	17,7
5	Polyaromatiques	18,8	1,6021	1,49	100 env.	71,1	19,6	1,5965	3,87	81,4	61,8
6	Polyaromatiques	3,9	1,6336	1,84	91,3	18,1	5,1	1,6523	4,14	65,2	17,0

* - Le soufre "Thiophénique" est calculé par rapport au soufre total dans la fraction -

TABLEAU 2.

TRAITEMENT AVEC LE SOUFRE D'UN RÉSIDU DÉSASPHALTÉ ** DU PÉTROLE BRUT HASSI-MESSAOUD

FRACTIONS		MALTÈNES AVANT TRAITEMENT CHROMATOGRAPHIE SUR SILICAGEL DAVIDSON 70					MALTÈNES APRÈS TRAITEMENT ** CHROMATOGRAPHIE SUR SILICAGEL DAVIDSON 70				
				% de soufre					% de soufre		
N°	Nature	% en poids	n_D^{20}	Total dans la fraction	Thioph.*	Par rapport à la charge	% en poids	n_D^{20}	Total dans la fraction	Thioph. *	Par rapport à la charge
Charge	-	100,0	1,571	0,46	66,3	100,0	100,0	1,565	1,02	76,0	100,0
1	Saturés	31,6	1,4851	0,05	-	3,5	27,6	1,4843	0,05	-	1,4
2	Monoaromatiques	10,1	1,5053	0,17	-	3,7	8,4	1,4984	0,18	-	1,5
3	Biaromatiques	11,4	1,5300	0,37	-	9,1	12,2	1,5193	0,54	-	6,5
4	Biaromatiques	1,5	1,5464	0,56	100 env.	1,7	44,5	1,553	1,69	72,8	73,8
5	Polyaromatiques	40,1	1,604	0,78	61,5	67,7	1,4	1,606	1,55	76,8	2,0
6	Hétéroatomiques (résines)	5,3	1,628	1,25	88,0	14,3	5,9	-	2,56	64,5	14,8
-	Asphaltènes	0,0	-	-	-	-	8,0***	Fusion 205°	3,63	17,9	-

* - Le soufre Thiophénique est calculé par rapport au soufre total dans la fraction. -

** - Asphaltènes précipités par l'heptane normal. -

*** - Le produit après traitement avec le soufre se compose de 92 % de maltènes *(dont la chromatographie est donnée dans ce tableau)* et de 8 % d'asphaltènes formés pendant le traitement. -

bures saturés et en particulier les naphtènes sont à peine touchés dans le cas de distillat.

2°) *La réactivité des constituants du pétrole augmente sensiblement en fonction de l'accroissement de leur poids moléculaire.*

On observe la formation de quantité notable d'asphaltènes aux dépens surtout des constituants polyaromatiques et hétéro-atomiques ("résine") dans le cas de traitement des fractions extraites des résidus.

3°) La répartition des composés soufrés dans les produits traités en laboratoire est voisine de celle rencontrée dans les fractions analogues des pétroles soufrés naturels.

4°) Au cours de l'attaque par le soufre d'un pétrole donné, *il se forme des composés du type sulfures en proportion plus grande par rapport au produit initial que celle des "thiophéniques".*

Ainsi le taux de "sulfures" pour le distillat après le traitement augmente de 12,8 à 22,7 % et le taux de sulfures dans les composés soufrés nouvellement formés s'élève à 27,4 %.

	Distillat non traité		Distillat traité			
			Total — % S		% S nouvellement formé	
	% dans le produit	% relatif	% dans le produit	% relatif	% dans le produit	% relatif
Soufre total dont :	0,39	100	1,23	100	0,84	100
— les sulfures	0,05	12,8	0,28	22,7	0,23	27,4
— les thiophènes	0,34	87,2	0,95	77,3	0,61	72,6

Les observations que nous venons de mentionner donnent des indications plausibles sur les causes de la répartition particulière et de l'importance variable des différentes séries de composés soufrés dans les pétroles.

Le très important problème sur l'origine et le mécanisme d'action du soufre ou de ses dérivés sur les pétroles dans les conditions géologiques étudiées dans une série de travaux dépasse, vu son ampleur, le cadre de notre communication.

Néanmoins quelle que soit l'origine du soufre * qui agit sur le pétrole, il est indéniable que son action pouvant intervenir à des époques différentes

* D'après les connaissances actuelles, la plupart du temps le soufre est lié aux évaporites et provient de la réduction des sulfates.

de son évolution modifie profondément la composition chimique et donne naissance à la formation des asphaltènes et des résines en grande quantité, à partir des structures polycycliques déjà existantes dans l'huile brute. En ajoutant ces résultats aux données géologiques, on peut en tirer des renseignements sur les trajets de migration et l'époque de mise en place du pétrole.

b) OXYDATION. La transformation des constituants du pétrole due aux réactions d'oxydation est évidente. Les difficultés résident dans la détermination :

1°) de l'échelle et du mécanisme de ces réactions dans les conditions géologiques et
2°) des sources de l'oxygène

On admet qu'en général le milieu des roches, surtout aux grandes profondeurs, n'est pas oxydant†. Pour rendre possible l'oxydation, il doit exister des voies de pénétration pour les agents d'oxydation. Outre l'oxygène atmosphérique qui ne peut intervenir qu'à une profondeur assez limitée, on imagine toutes sortes d'hypothèses avec la participation parfois des bactéries que l'on amène à l'endroit et au temps voulu pour le besoin de la cause.

Il est bien établi que l'oxydation du pétrole coïncide le plus souvent avec des zones où l'huile s'est trouvée proche de la surface, soit directement, soit par l'intermédiaire de failles, ou bien au voisinage des grandes discordances qui correspondent à des périodes anciennes d'érosion.

Quant au mécanisme et aux résultats de l'oxydation, ces questions nous intéressent plus particulièrement dans le cadre de notre travail. L'examen de la littérature ainsi que nos données expérimentales, non encore publiées, permettent de faire les constatations suivantes qui sont dans une certaine mesure analogues aux résultats trouvés pour l'action du soufre.

1°) *L'action de l'oxygène sur un mélange de constituants du pétrole de structures différentes est très sélective.*

Les composés polaires et polycycliques sont oxydés de préférence. Les hydrocarbures monoaromatiques sont attaqués plus faiblement que les constituants polyaromatiques. Les hydrocarbures saturés ne s'oxydent que très faiblement surtout dans le cas des distillats.

† Il resterait à étudier l'intéraction des "couches-rouges" avec les hydrocarbures.

2°) Les produits oxydés provenant de réactions sont d'autant plus instables que la température de réaction est plus élevée. Une série de réactions secondaires avec dégagement de H_2O et CO_2 accompagne ainsi l'oxydation proprement dite. Parmi ces réactions secondaires, il faut noter surtout la condensation des produits oxydés et non oxydés grandement facilitée par la présence d'atomes d'oxygène et de soufre dans les molécules de constituants qui subissent la réaction.

TABLEAU 3.

INFLUENCE DE LA STRUCTURE DU PRODUIT DE DÉPART SUR LES TENEURS ET LES PROPRIÉTÉS DES PRODUITS OXYDÉS

ANALYSE DES PRODUITS AVANT TRAITEMENT									APRÉS TRAITEMENT, TENEURS EN ASPHALTÉNES ET RESINES LOURDES				
Produit de départ (numéro échant.)	Poids Moléculaire	n_D^{20}	% S total	% S relatif sulfures	R_T	R_A	% C_R	% insoluble heptane	% insoluble heptane froid	(1) % soluble heptane chaud (1)+(2) = 100 %	(2) % de (1) insoluble heptane chaud = (3) + (4)	(3) % de (1) soluble éther sulfurique = (2) - (4)	(4) % de (1) insoluble éther sulfurique = (2) - (3)
1	570	1,5623	7,5	40 %	5,4	2,0	49	0	7,4	72,2	27,8	-	-
2	650 - 700	1,5872	5,3	2 %	6,0	4,2	55	0	13,8	21,4	76,9	15,4	61,5
3	700	1,60 (env.)	5,0	-	7,0 (env.)	5,0 (env.)	60 (env.)	0	66,7	11,5	88,5	14,6	73,9

LÉGENDE

Nombre total des cycles dans la molécule ——— R_T
Nombre des cycles aromatiques ——— R_A
Nombre d'atomes de carbone engagés dans les cycles ——— C_R

La réaction de condensation assez poussée conduit à la formation des constituants hétérocycliques de poids moléculaire élevés et finalement aux asphaltènes.

Les constituants saturés du pétrole ne donnent pas directement des asphaltènes. On peut les obtenir assez difficilement par l'action prolongée de l'oxygène. Dans ce cas il se forme d'abord les polycycles aromatiques et hétéroatomiques (avec participation de l'oxygène) qui permettent de passer ensuite aux molécules du type asphalténiques. Cependant, ces asphaltènes diffèrent sensiblement de ceux extraits du pétrole.

La condensation des hydrocarbures aromatiques polycycliques seule (ou des molécules cycliques saturées) ne conduit pas non plus à des produits semblables aux asphaltènes naturels.

Pour montrer l'influence de la structure sur la possibilité de formation des composés du type résines et asphaltènes, nous avons oxydé dans les mêmes conditions ("soufflage" à l'air à 260° pendant 6 heures), trois fractions étroites provenant des résidus de distillation du pétrole brut soufré du Moyen-Orient (voir tableau 3).

Les rendements en poids des asphaltènes (de 7,4 à 66,7 %) et les répartitions entre les asphaltènes et les "résines lourdes" (de 11,5 à 72,2 %) sont complètement différents compte tenu de la nature des produits de départ.

Aucune oxydation, non seulement dans les conditions géologiques mais même au laboratoire à la température assez élevée de 260°, *n'est capable de transformer un pétrole brut léger constitué principalement par des hydrocarbures (saturés) paraffiniques et naphténiques en huiles lourdes asphaltiques.*

Ces considérations peuvent éclairer d'un jour nouveau le problème de la ceinture asphaltique des bassins.

c) ACTION BIOLOGIQUE. La composition chimique des pétroles peut être profondément et sélectivement modifiée par l'action des bactéries spécifiques. Ce domaine a attiré, un moment donné, une attention considérable des chercheurs et les premiers résultats des nombreuses études ont soulevé de grands espoirs d'explication universelle. Actuellement la situation revient à l'état plus normal grâce à deux faits capitaux:

1°) L'action bactérienne, en modifiant la composition chimique, ne peut pas modifier la structure des molécules et produire par exemple à partir des naphtènes monocycliques, des naphtènes tricycliques.

2°) Il est pratiquement impossible de faire intervenir et de prouver l'existence des bactéries dans tous les endroits où la transformation chimique du pétrole peut avoir lieu.

Il semble ainsi que l'action biologique dans la transformation chimique du pétrole est restreinte et s'applique seulement à certains cas spécifiques.

Conclusions

1. Les travaux en géochimie n'en sont qu'à leur début et il reste beaucoup à faire pour améliorer nos connaissances.

Aujourd'hui, évitant de nous lancer dans des théories générales, nous avons voulu présenter une analyse des mécanismes fondamentaux et des transformations qui sont possibles dans la nature.

2. L'étude de la composition chimique des pétroles permet d'établir des parentés entre les différents bruts à l'échelle d'un bassin grâce à la connaissance des transformations physico-chimiques possibles ou impossibles en fonction de l'histoire géologique du bassin.

3. La corrélation entre la composition chimique des pétroles et conditions géologiques n'est actuellement possible que dans certains cas seulement. Le progrès dans ce domaine est conditionné par le perfectionnement des méthodes analytiques et le développement des expériences directes géochimiques (réalisation des transformations des pétroles dans les conditions géologiques).

4. L'étude de la *structure* chimique des constituants du pétrole permet de différencier l'action des facteurs physiques d'une part, et chimiques d'autre part, intervenant dans la transformation des huiles brutes.

5. L'étude fragmentaire de la composition chimique des pétroles et la détermination des "indices" et des "marqueurs" sont insuffisantes pour utiliser efficacement ces données dans la pratique de l'exploration.

La connaissance poussée de la composition chimique des gaz et des pétroles, de la structure de leurs constituants, complétée éventuellement par données chimiques sur les matières organiques des roches environnantes, est nécessaire pour relier tout pétrole à son (ou ses) faciès d'origine et déterminer le degré d'évolution qu'il a subi.

6. La transformation chimique des pétroles ne s'effectue pas uniformément pour tous ses constituants, mais différemment pour chaque type de structure et d'une façon graduelle avec conservation au moins en partie de la structure de base.

Bibliographie

1. P. F. Andreev, A. J. Bogomolov, A. F. Dobryansky et A. A. Kartzev. *Transformation du pétrole brut dans la nature*. Gostoptekhisdat, Leningrad, 1958.
2. M. A. Bestougeff, Étude sur les "paraffines" du pétrole. In *Advances in Organic Geochemistry*, Edit. Hobson et Louis. vol. 24, 197–211. Pergamon Press, 1964.

3. R. Byramjee, Semat, Vasse et M. Bestougeff. Étude géochimique comparée des pétroles bruts sahariens. Rapports internes C.F.P., 1965–1966.
4. E. G. Burova et G. Khodjaev. Résultats d'études de la composition chimique des essences des pétroles d'Ousbekistan. *Neftechimia*, Edit. Sergienko, Acad. Scienc. S.S.R. de Turkmenie, 135–144, 1963.
5. U. F. Colombo, F. Gazzarini et G. Sironi. Carbon isotope composition of individual hydrocarbons from Italian natural gases. *Nature*, **205**, 27, 1965.
6. A. F. Dobryansky, *La chimie du pétrole*. Gostoptekhisdat, Leningrad, 1961.
7. T. S. Jones et H. M. Smith, Oil composition and stratigraphy in Permian Basin. In *Fluids in Subsurface Environments*, Edit. Young and J. E. Galley. Amer. Ass. Petrol. Geol., Mémoire No. 4, pp. 101–204, 1965.
8. A. A. Kartzev, *Hydrogéologie des gisements de pétrole et de gaz*. Gostoptekhisdat, Moscou, 1963.
9. G. P. Kurbsky, *Neftechimia*, pp. 289–298. Gostoptekhisdat, Moscou, 1963.
10. M. Louis. Geochemical transformation of petroleum. Institut Français du Pétrole, Réf. I.F.P. No. 8060, 1963.
11. S. R. Sergienko, *Les constituants de poids moléculaire élevé du pétrole*. Edit. Khimia, Moscou, 1964.
12. S. R. Silverman. In *Isotopic and Cosmic Chemistry*, Edit. Craig and Miller, Amsterdam, 1964.
13. S. R. Silverman. Migration and segregation of oil and gas. In *Fluids in Subsurface Environments*, pp. 53–65, Amer. Ass. Petrol. Geol., Mémoire No. 4, 1965.
14. V. A. Sokolov, *Processus de la formation et de la migration de l'huile et du gaz*. Edit. Nedra, Moscou, 1965.
15. H. G. Thode et J. Monster. Sulfur-isotope geochemistry of petroleum. In *Fluids in Subsurface Environments*, pp. 367–377, Amer. Ass. Petrol. Geol., Mémoire No. 4, 1965.
16. A. Van Eggelpoel. Étude sur l'évolution des huiles au cours des migrations, etc. *Revue de l'Institut Français du Pétrole*, **14**, 1595–1614, 1959, et **15**, 69–92, 1960.
17. V. C. Vyshemirsky, *Les conditions géologiques du métamorphisme des charbons et des pétroles*. Edit. de l'Université de Saratov, U.R.S.S., 1963.

INFLUENCE OF TEMPERATURE, PRESSURE, AND A CLAY MINERAL ON THE EVOLUTION OF THE CHEMICAL STRUCTURE OF AN AROMATIC FRACTION OF A CRUDE OIL

by

Y. Califet and J. L. Oudin

Institut Français du Pétrole

Abstract

Investigations have been made on the evolution of an aromatic fraction of a crude oil as a function of temperature (180° C), of pressure (300 kg/cm^2) the presence or absence of a clay mineral (sodium montmorillonite), and of a 3% salt solution, during 21 days in an inert atmosphere (helium).

Light paraffinic hydrocarbons were formed, as well as heavy products (resins—asphaltenes) in small amounts. The presence of montmorillonite increased the rate of the formation of heavy products, and water reduced it.

INFLUENCE DU MODE D'ÉCHANTILLONNAGE SUR LA COMPOSITION CHIMIQUE DES FRACTIONS LÉGÈRES D'UNE HUILE BRUTE

par

M. POULET et J. ROUCACHE[1]

Introduction

Pour pouvoir comparer valablement entre eux des bruts différents, il est nécessaire de connaître:

— le degré de précision des analyses effectuées,

— l'influence du mode de prélèvement sur la composition de l'échantillon étudié.

Nous avons profité de l'étude géochimique, entreprise sur le gisement d'Hassi Messaoud (Algérie) pour étudier ces divers points. Dans cet exposé, nous présenterons brièvement le schéma analytique utilisé ; nous étudierons successivement ensuite:

— les erreurs analytiques liées aux techniques de mesure (écarts sur des analyses effectuées en double),

— les erreurs liées à la reproductibilité de l'échantillonnage (comparaison de deux analyses effectuées sur des échantillons prélevés à 6 mois d'intervalle dans le même puits),

— l'influence du mode de prélèvement sur la composition des fractions légères.

Nous comparons les résultats obtenus sur le même brut prélevé en monophasique à haute pression, à pression atmosphérique et à une pression de 30 kg/1 cm^2.

De nombreux ingénieurs ont participé à ce travail:

— M. Rouge et M. Vivier, ingénieurs CFP(A)[2] qui se sont occupés des

(1) Ingénieurs de recherche à l'Institut Français du Pétrole (I.F.P) 92, RUE IL-MALMAISON-FRANCE.

(2) CFP(A) = Compagnie Française des Pétroles (Algérie).

prélèvements effectués dans la partie nord du gisement d'Hassi Messaoud (concession CFP(A)),

— M. Raynal, ingénieur IFP qui a réalisé le dégazage des échantillons prélevés sous pression et de l'analyse de la fraction gazeuse,

— Mme Guichard et M. Boulet, ingénieurs IFP qui ont réalisé, sous la direction de M. Buzon, l'analyse des individus en C_7 C_8 C_9 et l'analyse des familles de saturés,

— M. Cornu et M. Ulrich du Centre d'Etudes Nucléaires de Grenoble qui ont effectué l'analyse des aromatiques.

La préparation des échantillons et la synthèse des données a été effectuée à l'IFP (Division Géologie) par Mme Roucache et M. Poulet.

Nous remercions également les directions de la CFP et de la CFP(A) qui nous ont autorisés à faire mention des résultats obtenus sur des bruts d'Hassi Messaoud.

1. Schema analytique d'un brut de C_1 à C_{30}

Dans le cas ou le brut est prélevé à une pression supérieure à la pression atmosphérique, il est dégazé et l'on détermine par chromatographie en phase gazeuse la teneur en méthane, éthane... jusqu'au C_7+. Puis une charge de 2 litres de brut dégazé est distillée de la façon suivante:

1ère coupe	0–60° C	coupe 0	pression atmosph.
2ème coupe	60–210° C	coupe 1	
3ème coupe	210–255° C	coupe 2	
4ème coupe	255–290° C	coupe 3	
5ème coupe	290–336° C	coupe 4	sous vide de 10 à 1 mm de Hg.
6ème coupe	336–375° C	coupe 5	
7ème coupe	375–410° C	coupe 6	
8ème coupe	410–470° C	coupe 7	

La coupe 1 est fractionnée en 5 coupes

60– 80° C	coupe 1–1
80–105° C	coupe 1–2
105–135° C	coupe 1–3
135–160° C	coupe 1–4
160–210° C	coupe 1–5

Étude des fractions légères

Ayant analysé lors du dégazage sur le liquide et sur le gaz les hydrocarbures jusqu'aux C_7, nous déterminons sur les coupes de distillat 1–1 et 1–2 les individus en C_7 par chromatographie en phase gazeuse.

On analyse par spectrométrie de masse les monocycloparaffines en C_8 de la coupe 1–3.

On détermine par spectrométrie de masse la proportion d'Alkylcyclopentanes et d'Alkylcyclohexanes dans la coupe 1–4 (coupe C_9).

On dose par chromatographie en phase gazeuse les alkylbenzènes contenus dans les coupes 1–2, 1–3, 1–4.

Étude des fractions de point d'ébullition > 210° C (détermination des familles d'hydrocarbures saturés)

Sur les coupes 2 et 3 (coupes C_{11}–C_{14}; C_{14}–C_{16}), on détermine directement dans la coupe de distillat par spectrométrie de masse (spectro MS_2) les familles suivantes :

paraffines	C_nH_{2n+2}
cycloalcanes non condensés	C_nH_{2n}
cycloalcanes condensés	C_nH_{2n-2}
alkylbenzènes	C_nH_{2n-6}
indanes + tétralines	C_nH_{2n-8}
alkylnaphtalènes	C_nH_{2n-12}
hydrocarbures	C_nH_{2n-14}
hydrocarbures	C_nH_{2n-16}
benzothiophènes	

Sur les coupes 4–5–6–7 (coupes C_{16}–C_{19}, C_{19}–C_{22}, C_{22}–C_{25}, C_{25}–C_{30}) on procède à une séparation par chromatographie liquide des deux familles: hydrocarbures saturés, hydrocarbures aromatiques.

Les hydrocarbures saturés sont analysés par spectrométrie de masse (MS_2), on obtient la composition en volume de :

paraffines
cycloparaffines non condensées
cycloparaffines condensées (2 à 6 noyaux)

Les hydrocarbures aromatiques sont analysés par spectrométrie de masse (spectromètre Atlas, type CH_4). On obtient pour les coupes 1–5 à 6 la répartition en poids des hydrocarbures aromatiques de formule générale C_nH_{2n-p} en fonction de n et p.

2. Échantillonnage représentatif du brut du gisement

2.1. *Erreurs analytiques liées aux techniques de mesure* (cf. planche 1)

Nous envisageons successivement :

— *la reproductibilité de la distillation analytique* (graphique 1)

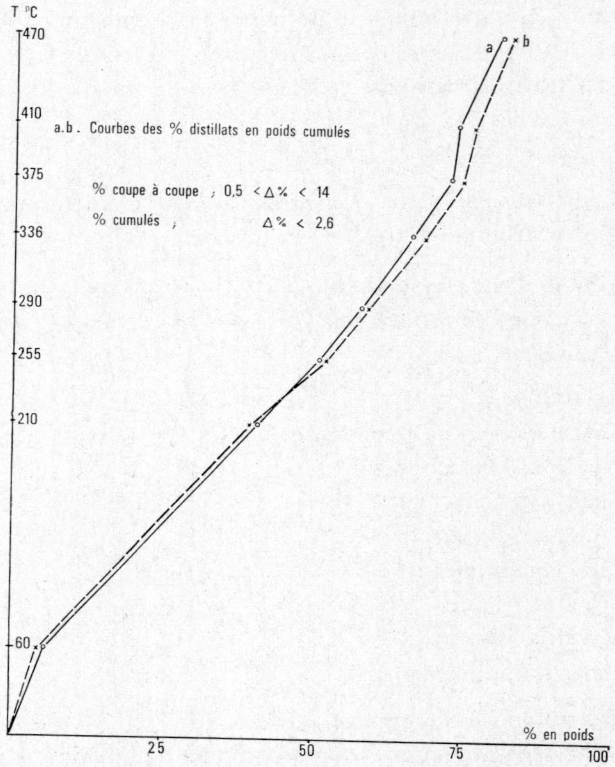

Graphique 1. — Distillation analytique d'un même brut PA.

Comparés coupe à coupe, les écarts sont assez importants ; par contre, les bilans sont bien reproductibles ($\Delta = 2{,}6\ \%$).

— *la reproductibilité des analyses par spectrométrie de masse des hydrocarbures saturés* (spectromètre de masse MS_2) *des hydrocarbures aromatiques* (spectromètre de masse Atlas) (cf. graphiques 2, 3, 4, 5)

La reproductibilité est excellente ; qu'il s'agisse de résultats sur échantillons provenant de la même préparation par chromatographie liquide,

6

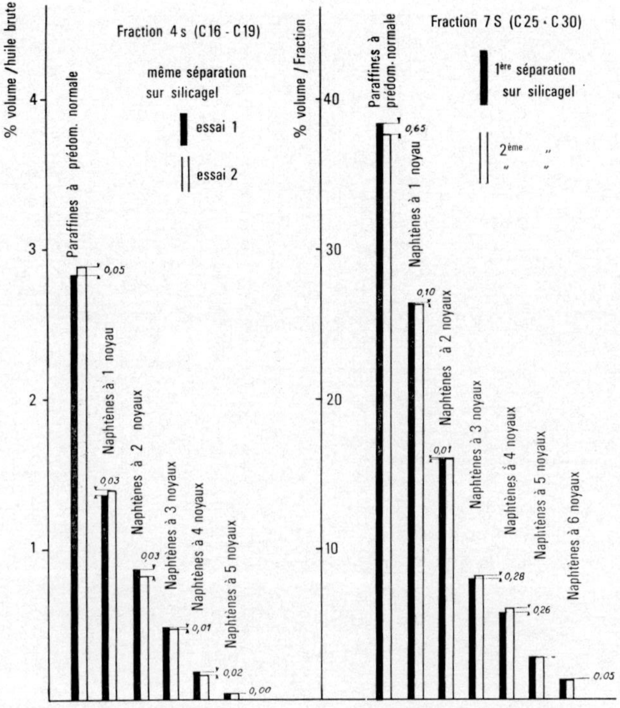

GRAPHIQUE 2. — Analyse des hydrocarbures saturés par spectrometrie de masse (MS_2).

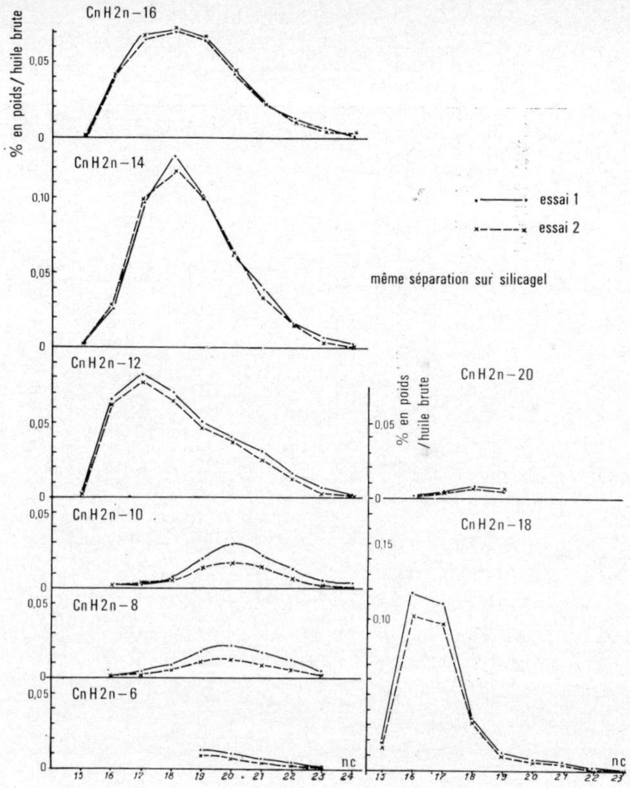

GRAPHIQUE 3. — Analyse des hydrocarbures aromatiques par SM (Atlas CH_4), coupe 5 ($C_{19}-C_{22}$).

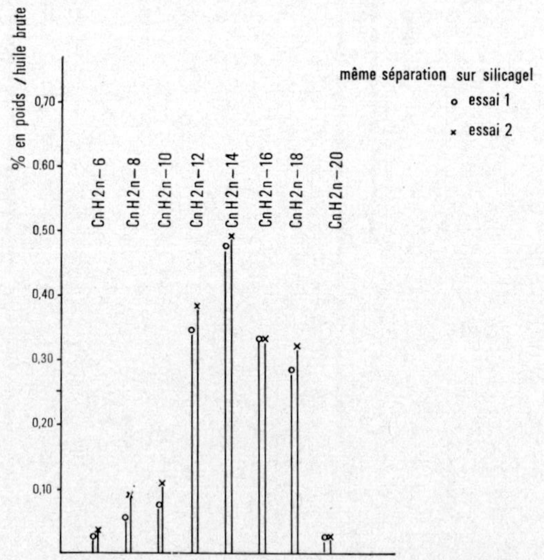

GRAPHIQUE 4. — Analyse par spectographie de masse (Atlas CH_4) Bilan des familles aromatiques dans une coupe 5 ($C_{19}-C_{22}$).

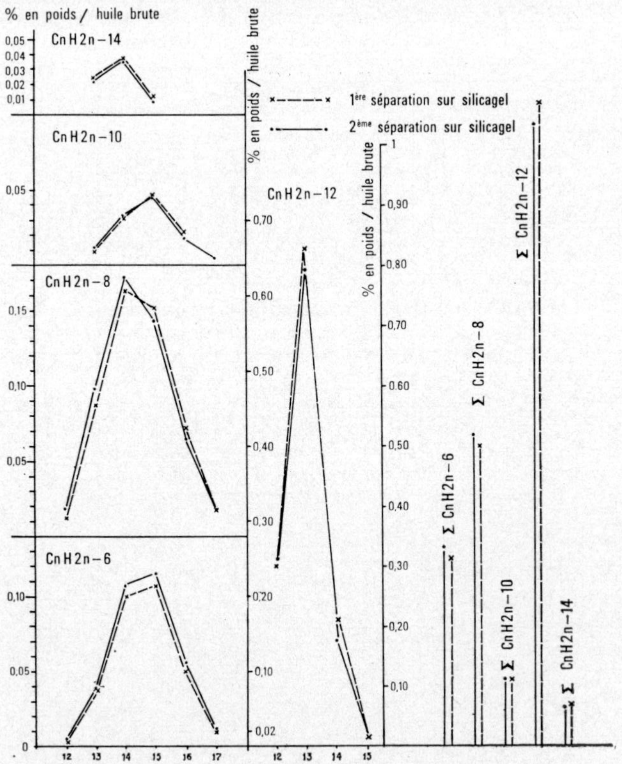

GRAPHIQUE 5. — Analyse des hydrocarbures par SM (Atlas CH_4), coupe 3 ($C_{14}-C_{16}$).

ou qu'il s'agisse de résultats sur échantillons provenant de deux préparations différentes par chromatographie liquide (les différences sont comprises entre 1 et 3 %).

— *la reproductibilité des analyses par chromatographie en phase gazeuse*

Dans l'ensemble, là aussi nous atteignons une bonne reproductibilité, surtout pour les valeurs supérieures à 1 %, qu'il s'agisse du dosage des

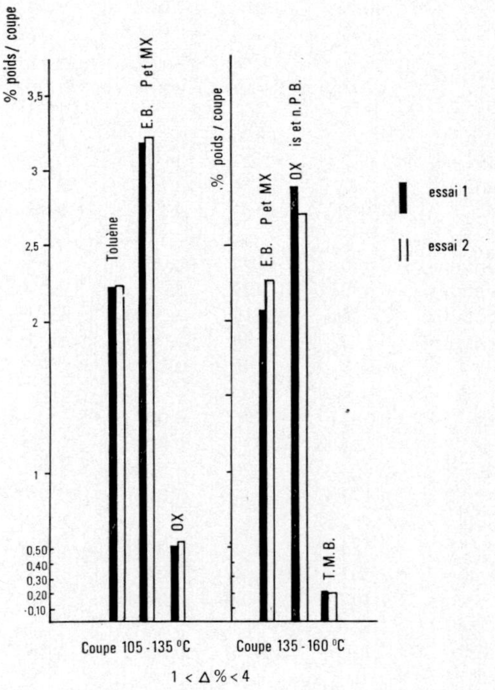

GRAPHIQUE 6. — Alkyl benzènes: reproductibilité analytique sur une même coupe.

individus des coupes 60–80° C, 80–105° C ou qu'il s'agisse du dosage des alkylbenzènes dans les coupes 105–135° C, 135–160° C (différences comprises entre 1 et 4 %,* pour les valeurs supérieures à 1 %) (cf. tableau 1, graphique 6).

* On notera quelques valeurs pour lesquelles l'écart est plus important; par exemple le 2,3 DMp; les isomères du DMCP; il s'agit de composés pour lesquels les pics sont très mal résolus et les surfaces sont difficiles à intégrer correctement.

TABLEAU 1. — REPRODUCTIBILITÉ DE L'ANALYSE PAR CHROMATOGRAPHIE EN PHASE GAZEUSE DES INDIVIDUS DES COUPES 60–80; 80–105° C

Individus	brut A					
	coupe 60–80° C			coupe 80–105° C		
	Essai 1	Essai 2	△	Essai 1	Essai 2	△
Propane	0,043	0,059	0,016	0,048	0,096	0,048
Iso-butane	0,018	0,019	0,001	0,042	0,066	0,024
N-butane	0,224	0,240	0,016	0,424	0,523	0,099
Iso-pentane	1,076	1,090	0,014	0,374	0,414	0,040
N-pentane	4,248	4,278	0,030	0,675	0,720	0,045
2.2 Diméthylbutane	0,090	0,090	0,	0,	0,	—
2. Méthylpentane	18,082	18,123	0,041	0,339	0,334	0,005
3. Méthylpentane	11,357	11,510	0,153	0,333	0,369	0,036
N-hexane	39,131	39,310	0,179	4,463	4,572	0,109
2.2.Diméthylpentane	0,190	0,194	0,004	0,158	0,160	0,002
2.4.Diméthylpentane	0,493	0,497	0,004	0,690	0,735	0,045
3.3.Diméthylpentane	0,006	0,010	0,004	0,136	0,150	0,014
2.3.Diméthylpentane	0,173	0,189	0,016	2,779	2,581	0,198
2.2.3 Triméthyl-butane	0,017	0,006	0,011	0,007	0,013	0,006
2. Méthylhexane	0,781	0,776	0,005	9,030	8,592	0,438
3. Méthylhexane	0,635	0,633	0,002	12,600	12,181	0,419
3. Ethylpentane	0,000	0,000	0,000	1,213	1,221	0,008
N-heptane	0,177	0,171	0,006	25,967	25,497	0,470
Cyclopentane	0,384	0,453	0,069	0,016	0,034	0,018
Méthylcyclopentane	10,263	10,205	0,058	2,772	2,712	0,060
Cyclohexane	3,797	3,779	0,018	5,805	5,673	0,132
1.1. Diméthylcyclo-pentane	0,150	0,175	0,025	1,537	2,005	0,468
1. tr 3	0,160	0,148	0,012	3,064	2,920	0,144
1. cis 3	0,255	0,158	0,097	2,349	2,057	0,292
1. tr 2	0,238	0,204	0,034	5,834	5,695	0,139
1. cis 2	0,	0,	—	0,534	0,549	0,015
Ethylcyclopentane	0,	0,		0,920	0,834	0,086
Méthylcyclohexane	0,029	0,028	0,001	11,191	10,621	0,570
Benzène	7,986	7,658	0,328	1,575	1,634	0,059
Toluène	0,	0,	—	3,917	3,619	0,298
Cts lourds (différence à 100)	0,	0,	—	1,208	3,421	

2.2. *Reproductibilité de l'échantillonnage* (cf. planche 1)

Deux échantillons ont été prélevés à six mois d'intervalle sur le même puits.

On a chaque fois prélevé :

— un échantillon en monophasique (bouteille Colman de 1 litre) sur lequel a été effectué l'analyse jusqu'à la fraction 2 (jusqu'aux C_{10})

— un échantillon à pression atmosphérique sur lequel on a effectué l'analyse complémentaire (C_{10+}), la quantité de liquide provenant du dégazage de la bouteille haute pression étant insuffisante pour une analyse complète.

Pour pouvoir faire un bilan complet du brut, on a posé : (% C_{10+}) brut dégazé = (% C_{10+}) prélevé à pression atmosphérique, hypothèse dont on vérifiera le bien-fondé plus loin.

QUALITÉ DE L'ÉCHANTILLONNAGE MONOPHASIQUE. Comparaison des résultats analytiques obtenus sur les deux prélèvements monophasiques du brut «B» réalisés à six mois d'intervalle: Ecarts entre B_1 et B_2 prélevés respectivement le 4.4.1965 et le 29.10.65. (Cf. planche 2.)

— pour les hydrocarbures légers (méthane, éthane... hexane) les écarts sont compris entre 2 et 10 % (cf. graphique 7).

— les deux distillations donnent des résultats très voisins. Pour les coupes 80–210° C les écarts sont compris entre 1 et 3,5 %. Pour les coupes 210–375° C les écarts sont compris entre 0,1 et 5,2 %. Seules les coupes > (375–410° C) montrent un écart important supérieur à 30 %; en effet, il faudrait distiller cette coupe sous un vide poussé, de façon à ne pas trop chauffer la charge, dans nos conditions opératoires, il y a certainement un début de cracking au delà de cette température, c'est pourquoi nous limitons notre étude aux hydrocarbures en C_{25} (410° C) (cf. graphique 8).

— Comparaison des résultats coupe à coupe. L'analyse par spectrométrie de masse (MS_2) des familles d'hydrocarbures dans les coupes 105–135°C, 135–160° C, 160–210° C, donne des résultats bien reproductibles, les écarts sont compris entre 0,6 et 4 % (cf. graphique 9).

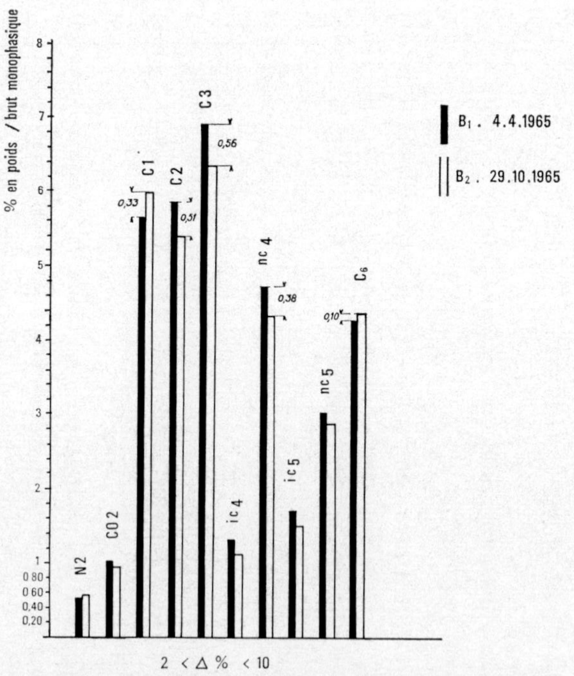

GRAPHIQUE 7. — Comparaison des résultats obtenus entre deux échantillons d'un même puits prélevés à 6 mois d'intervalle. Brut prélev. monophasique Analyse des hydrocarbures légers par chrom. en phase gazeuse.

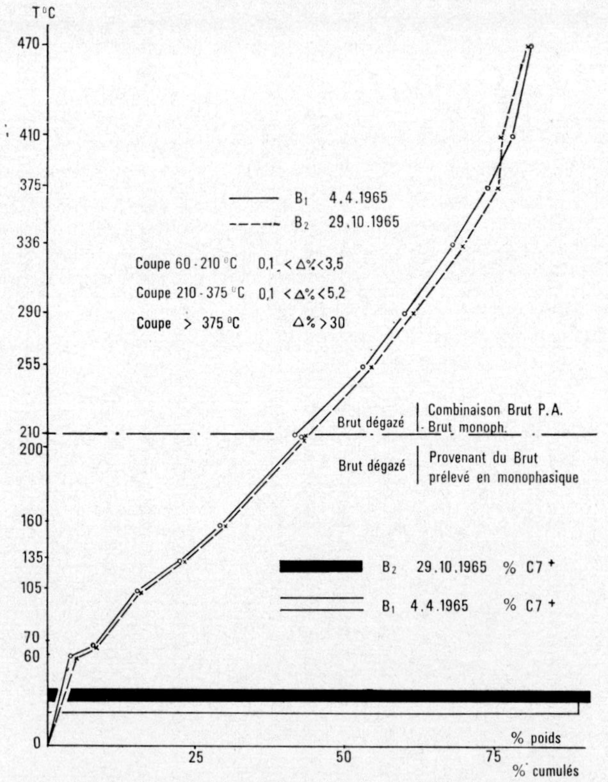

GRAPHIQUE 8. — Comparaison des résultats obtenus entre deux échantillons d'un même puits prélevés à 6 mois d'intervalle. Comparaisons des deux distillations.

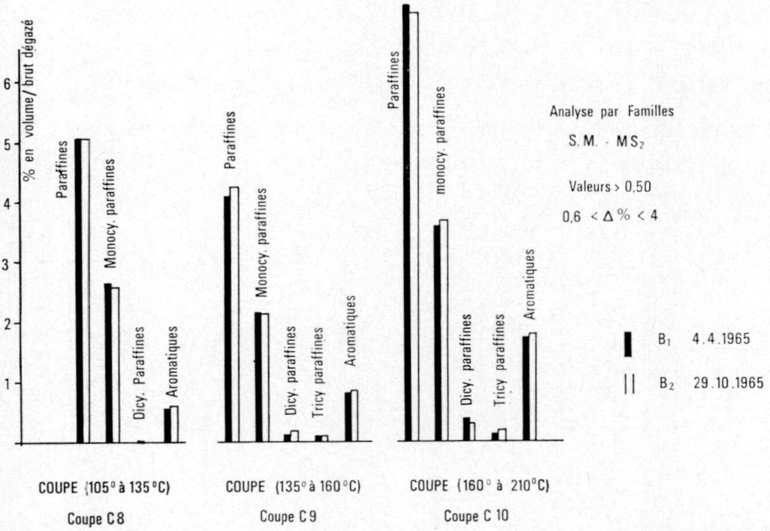

GRAPHIQUE 9. — Comparaison des résultats obtenus entre deux échantillons d'un même puits prélevés à 6 mois d'intervalle.

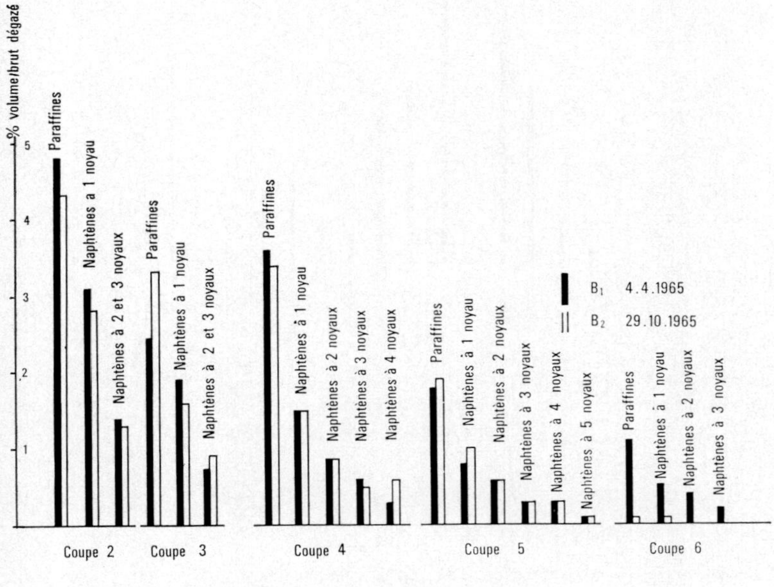

GRAPHIQUE 10. — Comparaison des résultats obtenus entre deux échantillons d'un même puits prélevés à 6 mois d'intervalle. Comparaisons coupe à coupe des familles d'H.C. saturés SM – MS_2.

L'analyse des familles d'hydrocarbures saturés dans les coupes (2, 3, 4, 5, 6) 210° C à 410° C, montre une moins bonne reproductibilité, coupe à coupe, surtout pour la coupe 6 (375–410° C) (cf. graphique 10).

— Analyse des individus en C_7 par chromatographie en phase gazeuse (cf. graphique 11). On remarque une bonne reproductibilité des résultats.

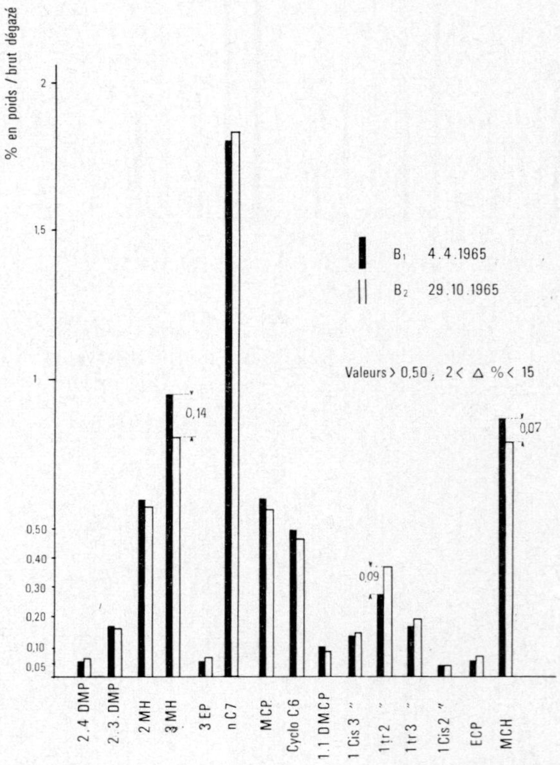

GRAPHIQUE 11. — Comparaison des résultats obtenus entre deux échantillons d'un même puits prélevés à 6 mois d'intervalle Coupe (60° à 150° C) Individus en C_7 (Chr. ph. gaz.) Brut prél. monoph.

— Analyse des monocycloparaffines en C_8 par spectrométrie de masse (MS_2) (cf. graphique 12).

— Analyse des alkylbenzènes $\leqslant C_9$ par chromatographie en phase gazeuse (cf. graphique 13). Pour ces deux types d'analyses, les écarts sont compris entre 7 et 20 % ; mais il s'agit de la détermination de petites valeurs $\ll 1$ %, ce qui explique cette mauvaise reproductibilité des résultats.

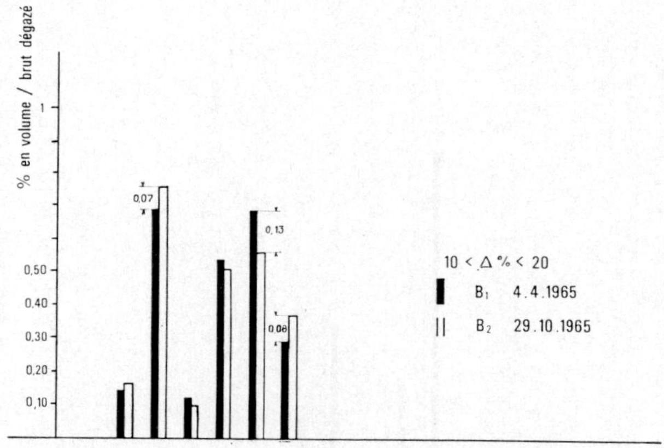

GRAPHIQUE 12. — Comparaison des resultats obtenus entre deux echantillons d'un même puits prélevés à 6 mois d'intervalle Coupe (105° à 135° C) Monocycloparaffines en C_8 SM – MS_2.

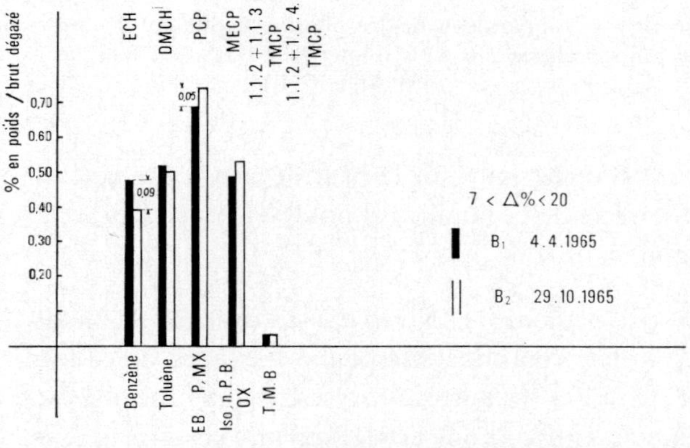

GRAPHIQUE 13. — Comparaison des résultats obtenus entre deux échantillons d'un même puits prélevés à 6 mois d'intervalle Coupe (60° à 160° C) Alkylbenzène < C9 (Chr. ph. gaz.).

— Analyse des familles d'hydrocarbures aromatiques par spectrométrie de masse (Atlas CH_4) (cf. graphique 14).

Pour les familles $C_n\ H_{2n-8,\ 2n-10,\ 2n-12,\ 2n-14,\ 2n-16,\ 2n-18}$, la reproductibilité est très bonne, elle l'est moins pour la famille $C_n\ H_{2n-6}$, car le bilan des $C_n\ H_{2n-6}$ provient de résultats analytiques obtenus à partir de techniques différentes (chromatographie en phase gazeuse, spectrométrie de masse) ; de plus l'analyse du benzène, du toluène et de quelques alkylben-

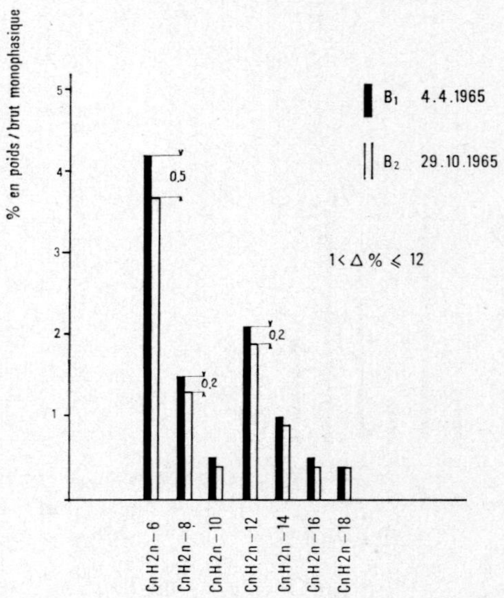

GRAPHIQUE 14. — Comparaison des résultats obtenus entre deux échantillons d'un même puits prélevés à 6 mois d'intervalle Bilan des familles aromatiques (SM Atlas CH_4).

zènes en n = 8 et 9 est faite sur le brut dégazé provenant du brut monophasique, et le reste de la famille est analysé sur le brut prélevé à pression atmosphérique.

CONCLUSION (cf. planche 1). Nous constatons que la distillation analytique n'est pas bien reproductible coupe à coupe; de ce fait, comme elle est à la base de toutes les autres analyses, chaque fois que l'on comparera les bruts coupe à coupe, nous enregistrerons des écarts assez importants pour un même échantillon. Mais chaque fois qu'il s'agira de résultats portant sur l'ensemble du brut, nous constaterons que la reproductibilité de l'échantillonnage est très satisfaisante, en effet dans ce cas (analyse des individus en C_7), (analyse des familles aromatiques), les écarts entre B_1 et B_2 ne sortent guère de la fourchette des erreurs analytiques dues aux différentes techniques de mesures.

2.3. *Variations de compositions chimiques liées au mode de prélèvement*

Dans le but d'étudier l'influence du mode de prélèvement, on a effectué sur trois puits de la concession CFP(A) d'Hassi Messaoud un échantillonnage complet : en monophasique en tête de puits ou au fond *«échantillon C»*; à 30 kg au circuit d'essai *«échantillon G»*; à pression atmosphérique *«échantillon P.A.»*.

Pour chacun des puits, on a réalisé sur les trois prélèvements (C – G – P.A.) une analyse des hydrocarbures $\leqslant C_{10}$, seulement pour les prélèvements (G et P.A.) une analyse des hydrocarbures $> C_{10}$, car la quantité prélevée en monophasique était insuffisante.

2.3.1. INFLUENCE DU MODE DE PRÉLÈVEMENT SUR LA COMPOSITION DE LA FRACTION DE BRUT INFÉRIEURE À C_{10}. — *Influence sur la teneur en hydrocarbures* $\leqslant C_7$. On constate un déficit important pour l'échantillon (P.A.) et un gain notable pour l'échantillon (G), ce qui entraîne dans le premier cas une concentration des C_7^+ et, dans le second cas, une dilution des C_7^+ (cf. tableau 2).

TABLEAU 2. — INFLUENCE DU MODE DE PRÉLÈVEMENT SUR LA TENEUR EN HYDROCARBURES $< C_7$ DANS LE BRUT RAMÈNE À PRESSION ATMOSPHÉRIQUE À ENVIRON 15°C

	Brut à pression atmosphérique en tête du puits	*Prélèvement Gerzat*		*Prélèvement Monophasique*	
		Brut analysé	Brut dégazé	Brut analysé	Brut dégazé
A					
% Légers $< C_7$	6,88	11,33	13,76	7,90	9,61
% C_7^+	93,12	88,67	86,24	92,10	90,39
		K† = 0,972		k† = 0,981	
B					
% Légers $< C_7$	1,50	11,83	13,43	7,98	10,54
% C_7^+	98,50	88,17	86,57	92,02	89,46
		K = 0,982		k = 0,972	
C					
% Légers $< C_7$	4,45	12,32	14,01	8,52	8,90
% C_7^+	95,55	87,68	85,99	91,48	91,10
		K = 0,980		k = 0,996	

† C_7^+ analysé × (K ou k) = C_7^+ dégazé —
Facteur K ou k engendré par les pertes survenant lors des manipulations du brut dégazé après le dégazage.

Influence sur la répartition des hydrocarbures en C_6 et C_7 (C_6 de point d'ébullition > n C_6).

Échantillon (P.A.). On constate pour chaque individu un déficit qui varie d'un brut à l'autre, d'un individu à l'autre, et ces variations sont assez «anarchiques», aucune extrapolation n'est possible (cf. graphiques 15 et 16).

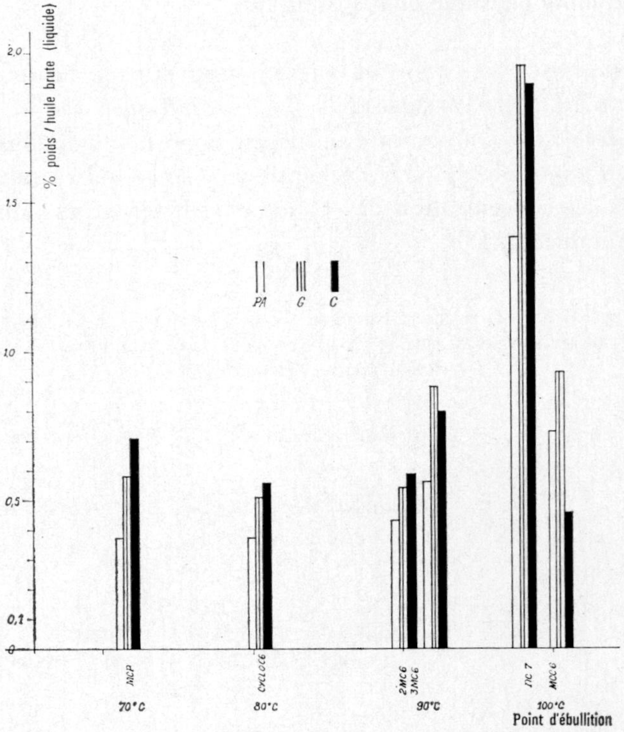

GRAPHIQUE 15. — Brut C. Influence du mode de prélèvement sur les C_7.

Échantillon (G). On remarque que seuls, les individus à point d'ébullition les plus élevés sont en quantité plus importante que dans l'échantillon C., ces individus n'ont pas subi de pertes et se trouvent certainement enrichis par effet de concentration (cf. graphiques 15 et 16).

Influence sur certains individus en C_8 (coupe 105–135° C).

Echantillon (P.A.). Comme précédemment, on enregistre généralement un déficit, variable pour chaque individu (cf. graphique 17).

Echantillon (G). Les résultats sont également en majeure partie déficitaires (cf. graphique 17).

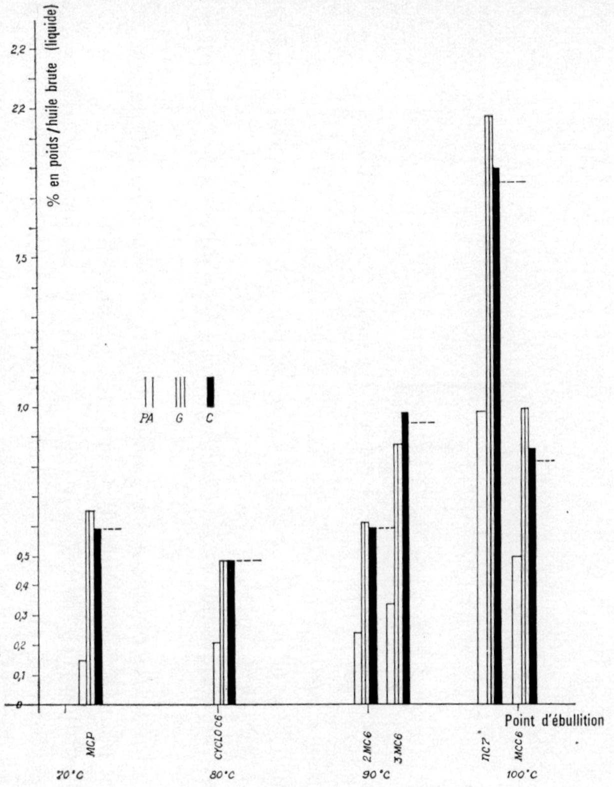

GRAPHIQUE 16. — Brut B. Influence du mode de prélèvement sur les C_7.

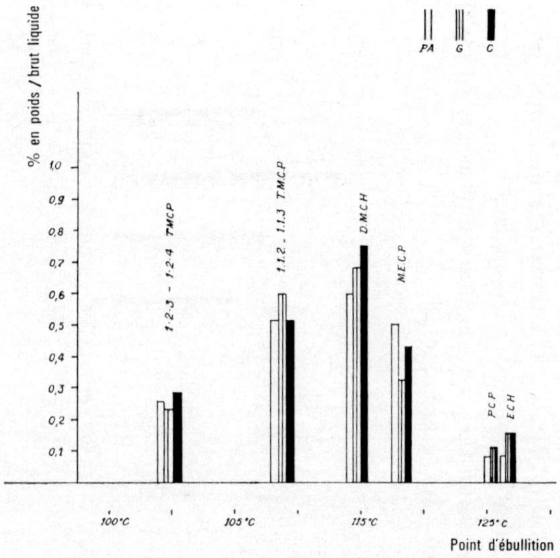

GRAPHIQUE 17. — Brut C. Influence du mode de prélèvement sur les monocycloparaffines en C_8.

Influence sur les aromatiques: benzène et alkylbenzènes ≤ C_9 *dans les coupes* (60–80. . ., 80–105, 105–135, 135–160° C)

Échantillon (P.A.). On remarque un déficit pour tous les individus et surtout très marqué pour le benzène.

Échantillon (G). Les résultats sont assez voisins des chiffres de l'échantillon (C). (cf. graphique 18).

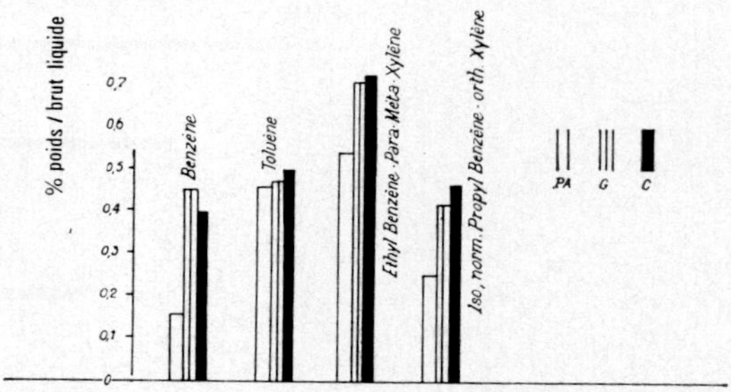

GRAPHIQUE 18. — Sur les alkyl benzènes. Brut C.

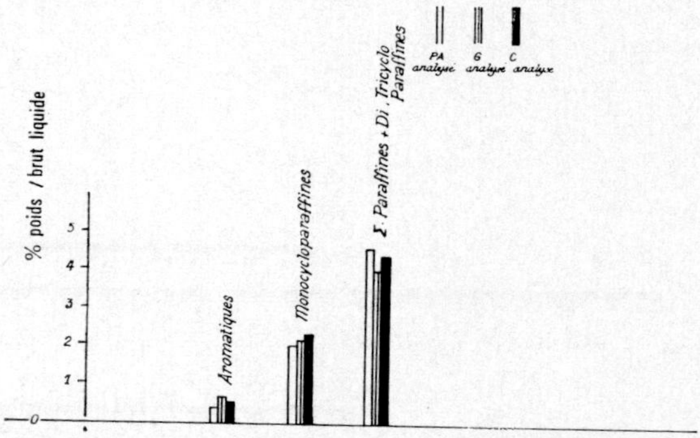

GRAPHIQUE 19. — Sur la coupe 1.3. Coupe 105–135° C brut C.

Influence sur la coupe 105–135° C (C_8–C_9).

Échantillon (P.A.). On constate un léger déficit des monocycloparaffines (8 % environ).

Les aromatiques ont un pourcentage très voisin de celui de l'échantillon (C).

La somme des pourcentages (paraffines di- et tricycloparaffines) est très voisine pour les échantillons P.A. et C. (cf. graphique 19).

Échantillon (G). On constate pour la teneur en aromatiques très peu de différence avec l'échantillon C. Par contre il y a un léger déficit pour les monocycloparaffines et la somme (paraffines di- et tricycloparaffines). (cf. graphique 19).

Influence sur la coupe 135–160°C (C_9–C_{10}) (cf. graphique 20).

Les échantillons P.A., G., C., ont à peu près les mêmes pourcentages d'hydrocarbures aromatiques et d'hydrocarbures saturés, ce qui suppose encore une très légère perte de ces hydrocarbures sur les échantillons P.A. et G.; en effet, si les C_9^+ étaient une constante, quel que soit le mode de

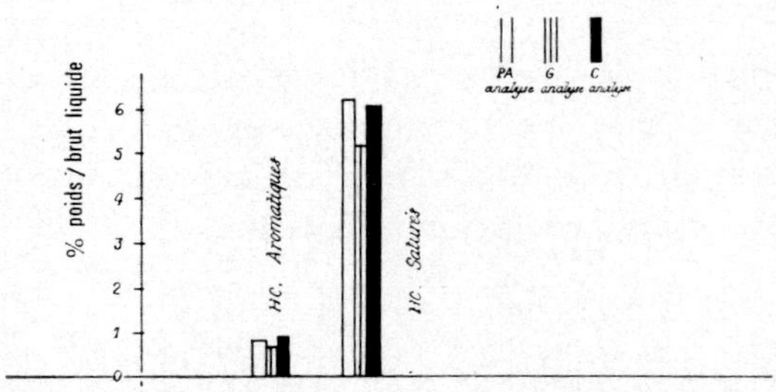

GRAPHIQUE 20. — Sur la coupe 1.4. Coupe 135–160° C brut C.

prélèvement l'échantillon (P.A.), pour lequel C_7^+ P.A. > C_7^+ C., devrait par effet de concentration avoir des chiffres plus forts que ceux de l'échantillon (C.,); et l'échantillon (G.) pour lequel C_7^+ G. < C_7^+ C., devrait par effet de dilution avoir des chiffres légèrement inférieurs à ceux de l'échantillon (C).

2.3.2. CONCLUSION

Echantillon (P.A.). Nous avons constaté dans la coupe des C_7 une perte sur tous les individus, en quantité variable d'un brut à l'autre :

— pour les hydrocarbures en C_8, le déficit concerne certains individus, pas forcément les mêmes, suivant les bruts étudiés, et il s'ensuit un enrichissement d'autres individus, par effet de concentration ;
— pour les alkylbenzènes, le benzène est l'individu le plus touché, certainement à cause de son point d'ébullition assez bas.

Échantillon (G). Ce mode de prélèvement est douteux, car le brut (G.) est plus léger que le brut (C.) dégazé ; en effet, dans le cas des bruts de Messaoud étudiés, la densité à 20° C d'un brut (G.) dégazé est de 0,793, alors que celle du même brut (C.) est de 0,807. Cet excès de légers perturbe en particulier la composition des coupes C_7, C_8 (monocycloparaffines).

Par contre, tant pour l'échantillon (P.A.) que l'échantillon (G.) les perturbations dues au mode de prélèvement ne sont que légères sur la coupe 135–160 (C_9, C_{10}).

De ce fait, nous considèrerons la coupe 160–210° C comme étant bien représentative du brut, quel que soit le mode de prélèvement, et nous posons :

$$\% \, C_{10}^{+} \, \text{P.A.} \times k' = \% \, C_{10}^{+} \, \text{C.}$$
$$\% \, C_{10}^{+} \, \text{G.} \times K = \% \, C_{10}^{+} \, \text{C.}$$

où $k' \leqslant 1$ $\begin{cases} \text{correction de l'augmentation de la concentration des } C_{10}^{+} \text{ P.A.} \\ \text{due à la perte des fractions légères} \end{cases}$

où $K \geqslant 1$ $\begin{cases} \text{correction de la diminution de la concentration des } C_{10}^{+} \text{ G.} \\ \text{due à l'excès de légers dans le brut.} \end{cases}$

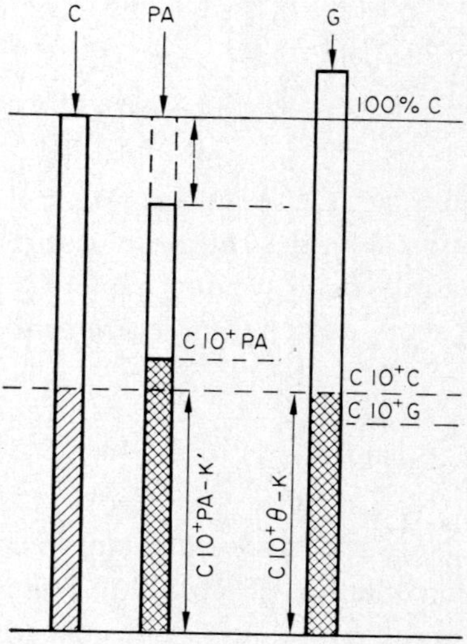

Nous avons appliqué cette correction et vérifié que l'influence du mode de prélèvement ne se fait plus ressentir au delà des hydrocarbures en C_{10}.

Influence sur les familles aromatiques $C_n H_{2n-p}$.

n varie de 6 à 24 pour p = 6, 8, 10, 12, 14, 16, 18 (cf. graphiques 21 à 27) ;

dans lesquels nous reporterons pour chaque famille

1 — le pourcentage en poids de $(C_n H_{2n-p})$ P.A. (brut P.A. analysé)

2 — le pourcentage en poids de $(C_n H_{2n-p})$ G. (brut Gerzat analysé)

3 — le pourcentage en poids de $(C_n H_{2n-p})$ P.A. $\times$ k′ (brut dégazé)

4 — le pourcentage en poids de $(C_n H_{2n-p})$ G$\times$K (brut Gerzat corrigé)

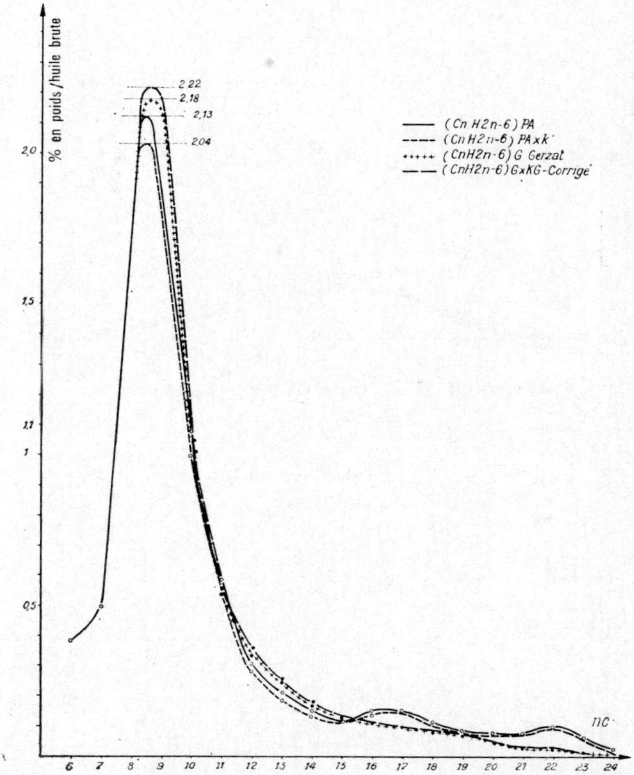

GRAPHIQUE 21. — Sur les $C_n\ H_{2n-6}$ brut C.

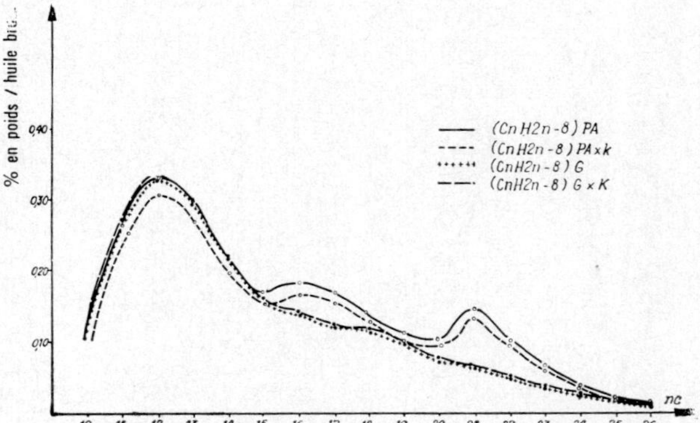

GRAPHIQUE 22. — Sur les $C_n\ H_{2n-8}$ brut C.

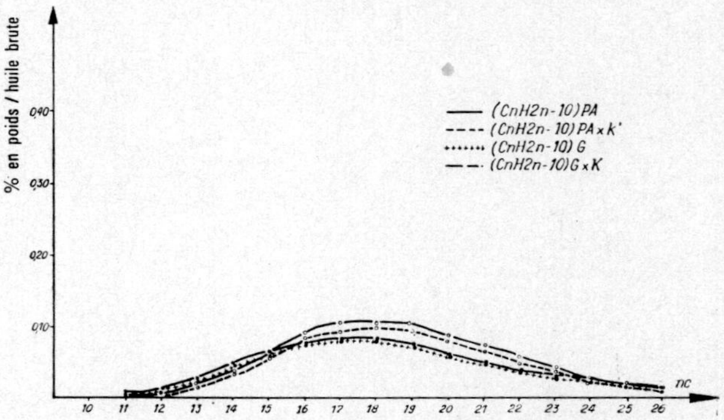

GRAPHIQUE 23. — Sur les C_n H_{2n-10} brut C.

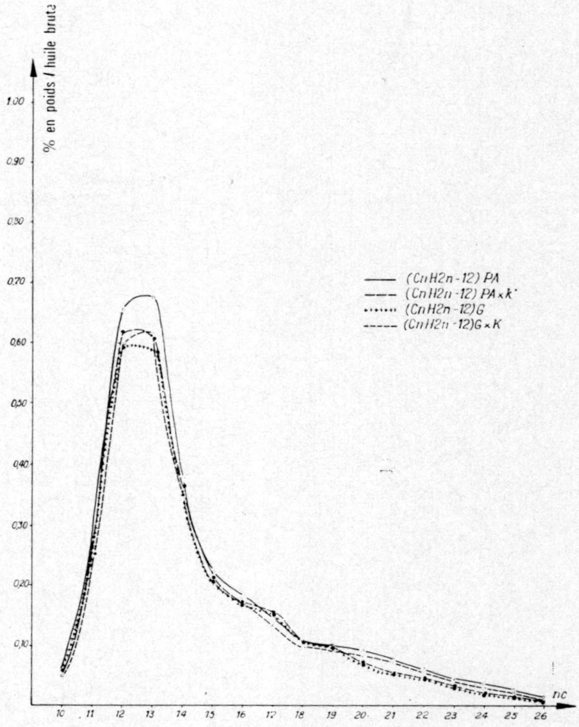

GRAPHIQUE 24. — Sur les C_n H_{2n-12} brut C.

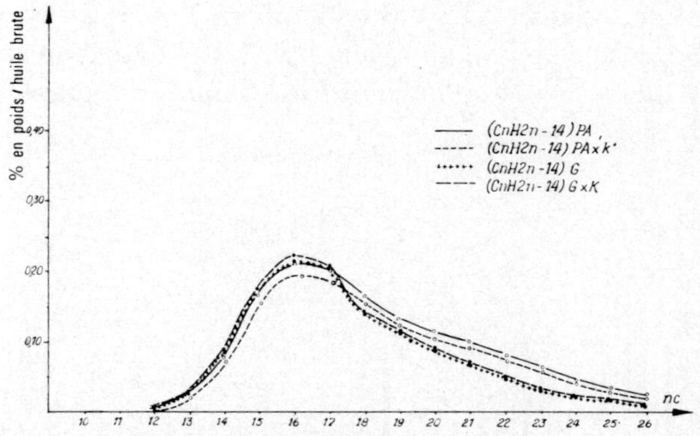

GRAPHIQUE 25. — Sur les $C_n H_{2n-14}$ brut C.

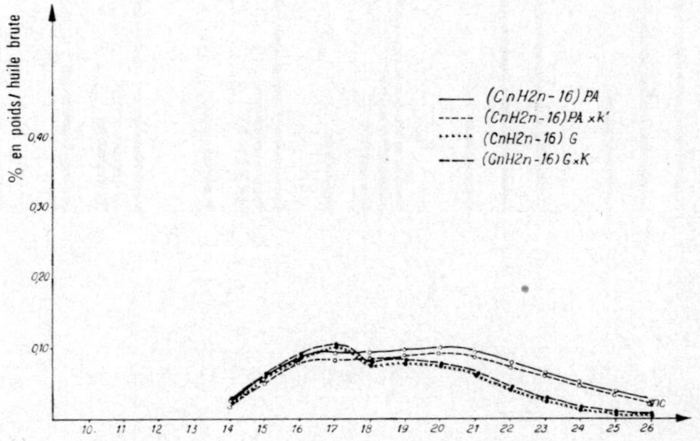

GRAPHIQUE 26. — Sur les $C_n H_{2n-16}$ brut C.

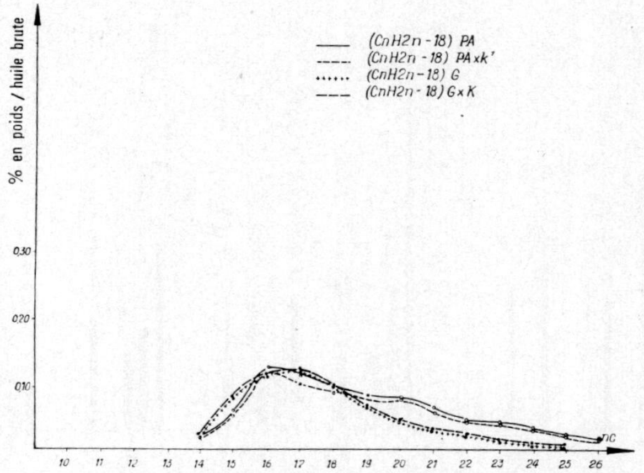

GRAPHIQUE 27. — Sur les $C_n H_{2n-18}$ brut C.

En 3 et 4, les pourcentages des échantillons P.A. et G. analysés ramenés aux chiffres que l'on aurait obtenus pour l'analyse d'un échantillon C.

Influence sur les familles d'hydrocarbures saturés (cf. graphiques 28–29).

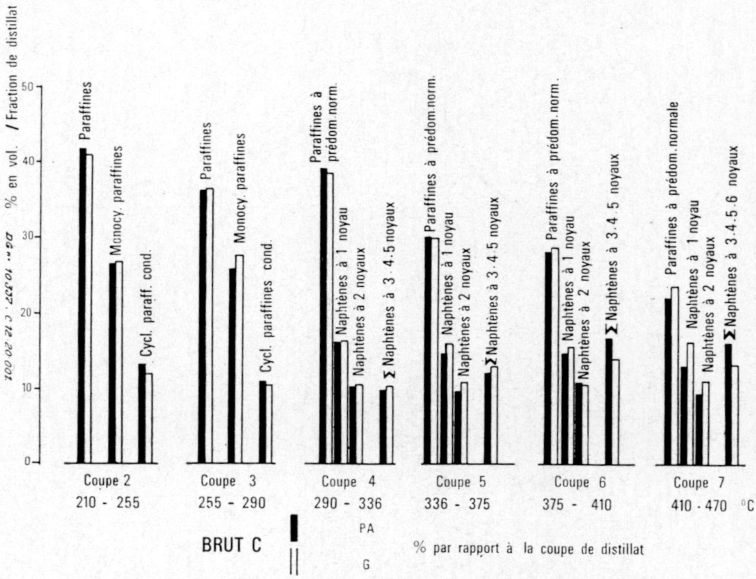

Graphique 28. — Répartition des H.C. saturés.

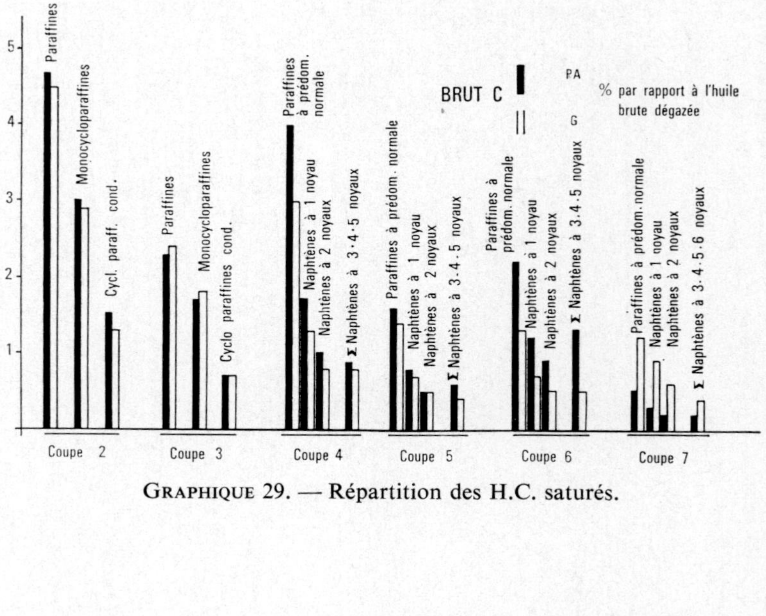

Graphique 29. — Répartition des H.C. saturés.

Lorsque l'on considère les teneurs coupe à coupe, par rapport au brut total (cf. graphique 29) on remarque pour les coupes 6 et 7 une différence importante entre l'échantillon G et l'échantillon P.A. Ces différences n'existent pas lorsque l'on considère les pourcentages de chaque famille par rapport à la coupe de distillat (cf. graphique 28). Comme nous l'avons mentionné précédemment, la mauvaise reproductibilité de la distillation est cause de ces écarts.

Cette remarque mise à part, nous obtenons pour la fraction C_{10}^{+} des résultats voisins quel que soit le mode de prélèvement.

3. Conséquences pratiques pour l'étude des bruts

Dans le cas des bruts étudiés dont le GOR est voisin de 200, on a constaté que les prélèvements à la pression atmosphérique et à 30 kg/cm^2 ne représentent le brut in situ qu'au delà des C_9. Le dégazage qui affecte le brut introduit des variations de composition quantitative supérieures aux erreurs analytiques dans la fraction liquide C_6 à C_9 ; ces variations en outre sont particulières à chaque puits échantillonné. Seul le prélèvement monophasique en tête de puits et au fond permet une analyse valable des constituants liquides légers. La comparaison des deux analyses effectuées sur le même brut prélevé à 6 mois d'intervalle montre que le compromis prélèvement monophasique (étude des légers) + prélèvement à pression atmosphérique (étude complémentaire) représente bien le brut.

En conclusion, si l'on veut comparer entre eux, des bruts de gisements, où le prélèvement en monophasique n'est pas possible (techniquement ou économiquement) l'échantillonnage à pression atmosphérique suffit de préférence aux prélèvements «Gerzat» plus compliqués et qui ne donnent pas de résultats plus exacts. Dans ce cas on comparera les bruts entre eux :

— par la présence ou l'absence d'individus en C_6, C_7, C_8, C_9.

— par les pourcentages des différentes familles d'hydrocarbures qui composent la coupe C_{10}^{+}.

GEOCHEMICAL PROSPECTING FOR PETROLEUM

by

P. Leplat

Service Géochimie, Labofina, Brussels

Abstract

Examination of the rare publications on geochemical prospecting for oil shows that most workers have used only one method. It is believed that only by the use of a combination of techniques can valid conclusions be drawn.

A preliminary study showed that there are means for making numerous analyses under standard conditions in a routine manner on a large number of samples.

In the first stages of the laboratory work the aim was to develop simple methods. Amongst these are thin-layer microchromatography and gaseous phase pyro-chromatography.

Using pyrograms of rocks, the disadvantages of analyses made in the dynamic regime in a Pye chromatograph were discussed in relation to analyses made in a flame-ionization chromatograph in the static regime.

Typical micro-chromatograms and typical pyrograms of petroleum fractions and of oil shows were presented.

Qualitative interpretations and particularly quantitative interpretations of some pyrograms of rocks made it possible to demonstrate the value of pyrochromatography in the gaseous phase in both surface and subsurface prospecting for oil.

In a second study samples were taken with an auger at depths of about 2 m on one of the Company's concessions. The results obtained by the different analytical techniques permit the discussion of the validity of this approach in locating favourable zones for the accumulation of oil at depth.

APPLICATION OF DATA PROCESSING METHODS TO GEOCHEMICAL PROSPECTING FOR PETROLEUM

by

R. E. GERARD

Geoservices, Paris, France

Introduction

An important method of geochemical prospecting for oil and gas is the study of anomalous patterns of hydrocarbon gases concentration in soil or subsoil formations. Usually, geochemical anomaly maps are made by contouring on the individual sample values, as given by their chemical analysis.

This technique has the following disadvantages:

The contouring rules are never clearly stated and vary from one interpreter to the next, thus allowing plenty of room for wishful thinking.

There is a rapid variation in sample value from one sample location to the next. This makes sampling look too sparse and, therefore, contouring difficult. This situation is complicated further when sampling is irregular due to terrain conditions.

Different types of lithological formations have different gas retention properties, as well as different responses to the chemical treatment necessary to retrieve the adsorbed gases. It is then not possible to compare zones of different facies.

To obviate these difficulties, we have elaborated a procedure which reduces sample values to a common datum and specifies rules for mapping eventual anomalies.

Model

This procedure is based upon the following assumptions:

The measured value of a geochemical indicator (here, the soil gas content) is a random variable depending on sample location and several other parameters which may or may not be random variables. We will call these parameters the characteristic parameters.

For fairly homogeneous conditions over a surveyed area, and when there is no petroleum accumulation at depth, the geochemical indicator is normally distributed (actually, it is a truncated normal distribution).

For fairly homogeneous conditions, but when the sampled area overlies petroleum-bearing formations, the geochemical indicator distribution is perturbed. The set of sample values resolves into a set of normally distributed values or background, and an anomalous set.

When conditions are not homogeneous over the surveyed area, the set *S* of all the measured values of a geochemical indicator is constituted of subsets *S*1, *S*2, *S*3, etc., each of them being characterized by a particular normal distribution of the geochemical indicator and by a particular combination of characteristic parameters. In other words, there is a certain background for certain conditions such as type of geological outcrop, depth of sampling, etc.

These assumptions have been experimentally tested by surveying areas overlying dry and productive structures. These experiments showed that the hypothesis of normal background distribution cannot be reasonnably rejected.

Procedure

When conditions are not homogeneous, the plot of the sample values against their cumulative relative frequency, on probability paper, can be approximated by a polygonal line (Fig. 1). The samples can be grouped into several sets, each of them being characterized by a certain range of carbonate content and a certain sampling depth. For each set, if the sample carbonate content range and the sampling depth range have been well chosen, the plot of sample values against their cumulative relative frequency is a straight line on probability paper (Fig. 2). We found that carbonate content and sampling depth were essential in discriminating between sets of samples by studying regressions of the sample values on these parameters (Figs. 3 and 4). In some instances, it is necessary to introduce other parameters such as lithological description, grain size, etc. A simplified procedure is then as follows:

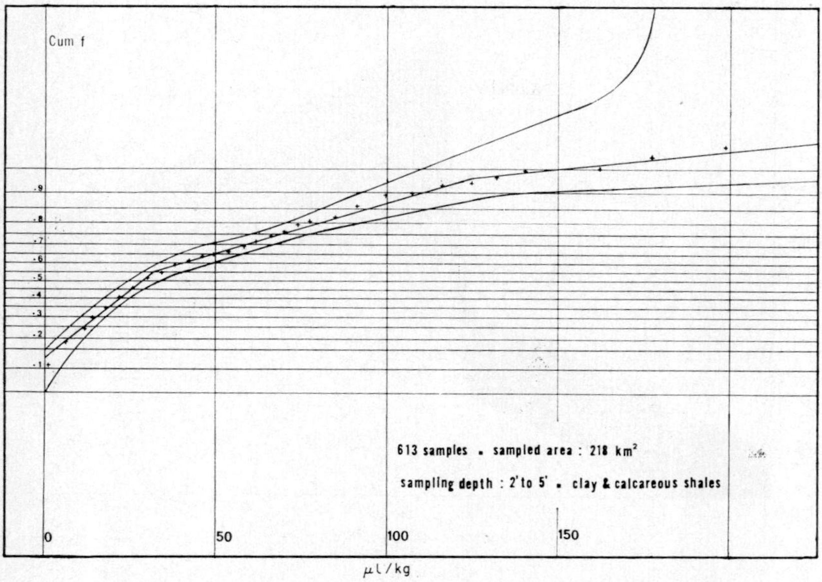

FIGURE 1

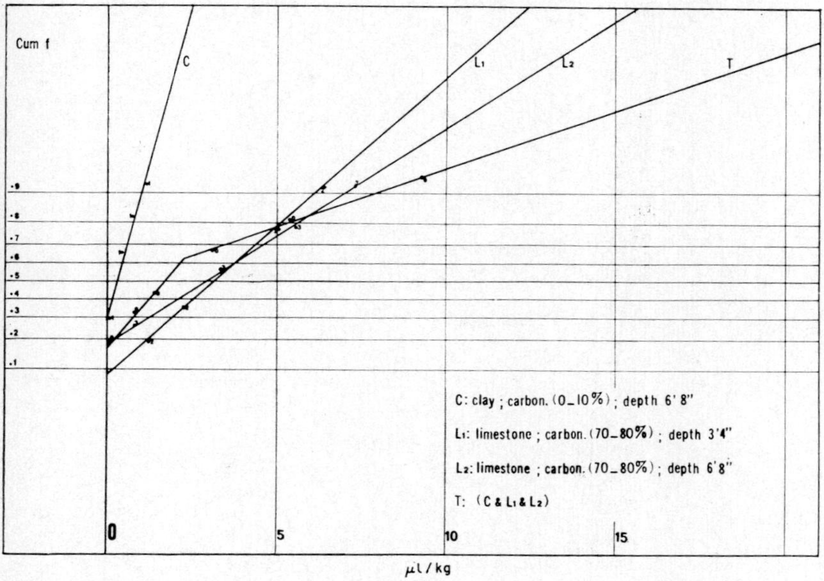

FIGURE 2

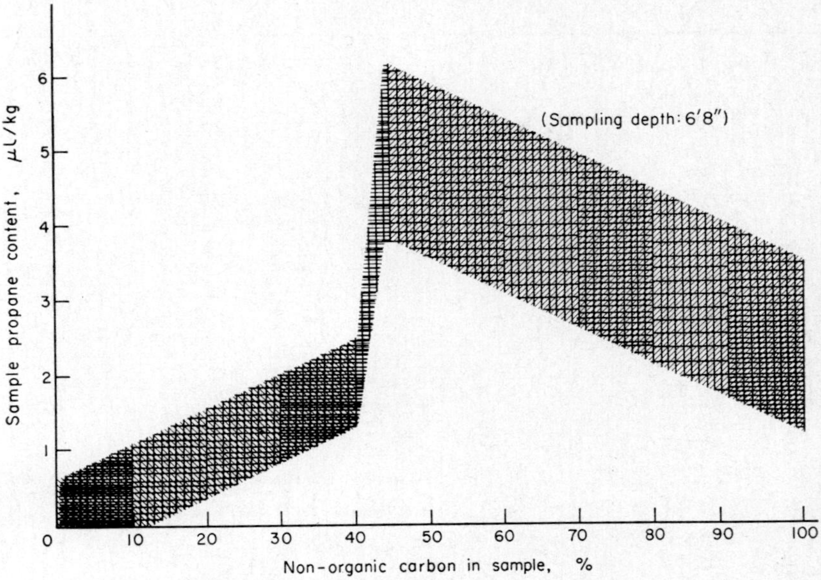

FIGURE 3

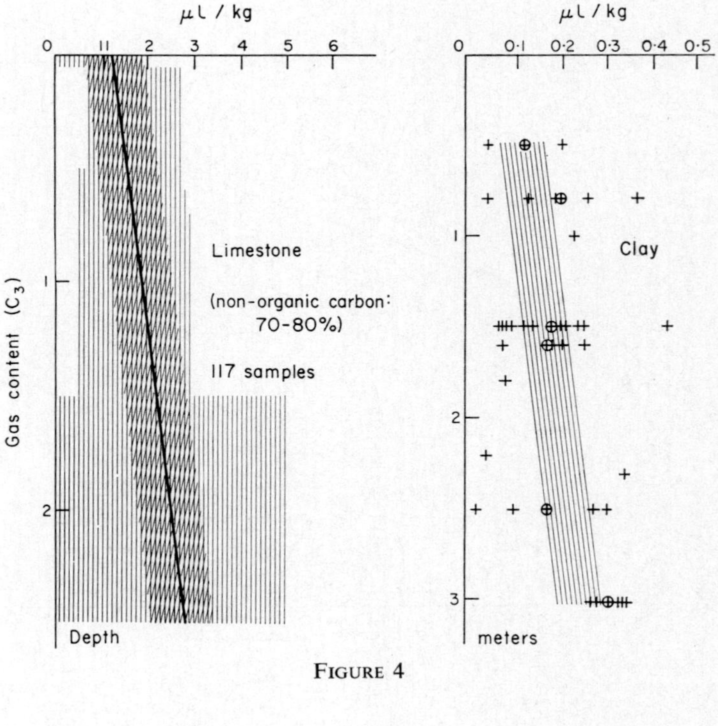

FIGURE 4

Plot the sample values versus their cumulative relative frequency on probability paper.

Find the best polygonal line that passes by the plotted points.

The first segment to the left of this polygonal line defines on the sample value axis a certain range. Consider the group of samples whose values belong to that range.

Find the characteristic parameter that belongs to the majority of samples in this group.

Plot the values of all the samples presenting this character against their cumulative relative frequency on probability paper.

This procedure is repeated as necessary until the set S of samples is separated into subsets of normally distributed samples.

For each subset, the background average is estimated by the median m of the sample value distribution. If s is the standard deviation from the median, we define, at the four sigmas level, the lower limit of the anomaly as the median plus four times the standard deviation. Each sample belonging to a particular subset and whose value x is larger than this limit, will be affected a weight defined by the formula (Fig. 5):

$$T = (x-m)/s)-4$$

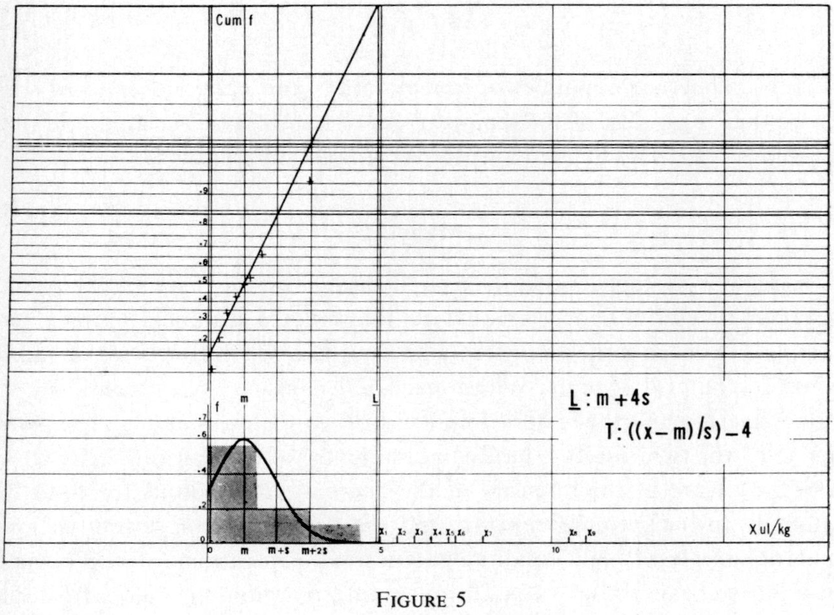

FIGURE 5

T is the Student's t statistic relative to a sample value x less four, so, for the lower limit of the anomaly T is null. The T– values are computed for each subset using the corresponding median and standard deviation

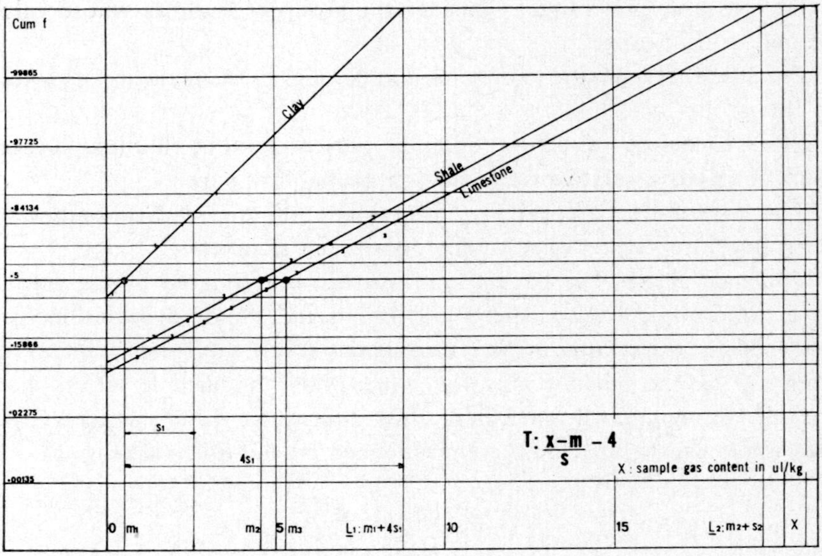

FIGURE 6

(Fig. 6). They are nondimensionnal numbers and can be compared with each other regardless of the sample facies or depth. The T–values are then plotted on the map at their respective sample locations.

Mapping

A regular isometric grid is then overlaid on the map. The spacing between two grid nodes is equal to the average distance between two samples. On each node is centred a unit surface which is a circle of radius equal to the grid spacing. Within each unit surface, we compute the T–values average and the relative frequency of anomalous samples. The product w of the two results is plotted at the centre of the unit surface (Fig. 7). w is a measure of the intensity of the anomaly. It accounts for both the anomaly to background contrast and the probability of getting an anomalous sample within a small neighbourhood. The anomaly map is made by contouring on the w– values using an interpolation technique. It should

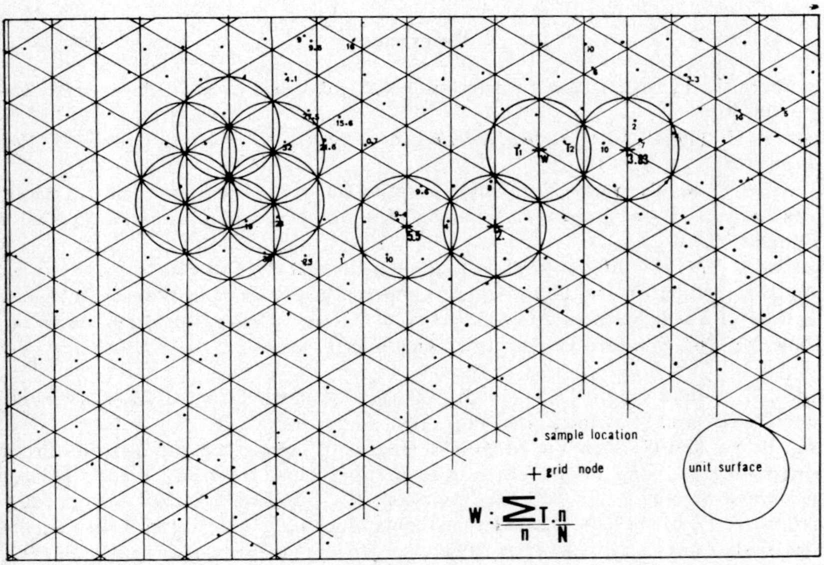

FIGURE 7

be noted that the resolution is limited by the grid spacing, and therefore by the sampling density.

If the area to be mapped is large, there is a possibility of regional variation or trend for the background. To account for this variation, the surveyed area is divided in smaller intermediate areas with overlapping. Each surface is treated independently. With a sampling density of three to five samples per square kilometer, experience shows that a ten by ten kilometers square is a good choice for this intermediate surface.

Conclusions

The data interpretation method which has been described is a diagnostic technique whose purpose is to bring out anomalous patterns when mapping a geochemical indicator. It is based on a model that sheds no light on the mechanisms for the origin of geochemical anomalies. It is therefore imperative, in a further step of interpretation, to correlate the geochemical map with pertinent data obtained from geophysical and/or geological surveys.

References

Agterberg, F. P. (1965) Statistical techniques. *Geol. Surv. Canada*, Paper No. 65–2, pp. 58–60.

Cramer, H. (1963) *Mathematical Method of Statistics*. Princeton University Press, Princeton.

Mezencev, O. K., Ratnikov, V. M. and Skorospelkin, S. A. (1964) Traitement statistique des données des recherches géochimiques. Traduit de *Razvedka Okht. Nedr. S.S.S.R.*, **30,** No. 10, 10–15.

Miesch, A. T. (1964) Effects of sampling and analytical error in geochemical prospecting, Proceedings of the Third Annual Conference sponsored by Stanford University School of Earth Sciences and University of Arizona College of Mines, June 24 to June 29, 1963, Stanford University Publications, *Geological Sciences*, **9,** No. 1, 156–70.

Miller, R. L. and Kahn, S. S. (1965) *Statistical Analysis in the Geological Sciences*. John Wiley and Sons Inc., New York.

Yufa, B. Ya. and Gurvich Yu. M. (1964) The use of the median and quartiles in estimating normal and anomalous values of a geochemical field. Translated from. *Geokhimiya*, No. 8, pp. 817–24; *Geochemistry International*, No. 4, pp. 801–7

Zakharova, E. M. (1964) Méthodes mathématiques dans l'exploitation des résultats des analyses des masses lors des recherches géochimiques. Traduit de *Vest. Moskov. Univ., Geol.*, No. 6, pp. 28–35.

PORE FLUID IN SHALES AND ITS GEOCHEMICAL SIGNIFICANCE

by

G. LONG, S. NEGLIA and E. RUBINO

Agip Direzione Mineraria, Servizio Geochimico, S. Donato Milanese, Italy

Introduction

Both in hydrodynamic studies and in those trying to explain the concentration process for water in the subsurface the hypothesis has been used that shale formations are flushed by a flow of water, but little attention has been given to the determination of the composition of the interstitial water saturating the shales. If a shale is considered as a system formed of a solid phase, the dry shale, and of a liquid phase, the interstitial water, the study of the composition of such interstitial water should give useful indications as to the type of water that was migrating through the formation at the moment when the sample was taken.

The present paper aims to show that in fact, at least in those cases dealt with by the authors, there is a relationship between the composition of the interstitial water of a shale and that of the aquifers close to it.

Before presenting the experimental results, it seems useful to discuss briefly the factors affecting both the porosity of a shale and the composition of its interstitial water during geological time.

Shale porosity

In a recent paper Rowell (1965) has shown that the porosity of a shale can be of two kinds: intercrystalline, due to the pores between the clay crystals, and intracrystalline, due to the water which coats the external surface of the crystals and the internal surface of the plates of the expandable clay minerals.

The intercrystalline porosity depends directly upon the degree of orientation and packing of the clay crystals: the porosity will be higher in a disordered clay, and lower in a well oriented one. It must be noted that the orientation of the crystals is favoured by low salinity of the interstitial water during both the sedimentation and the compaction phases. For a NaCl salinity greater than 0·4 m/l (Rowell, 1965) the attraction between clay crystals, resulting from edge-face bonds, prevails over the repelling forces between the crystals faces, consequently the clay minerals settle down in a disorderly way with formation of aggregates. Under these conditions the initial intercrystalline porosity would be high and would tend to persist in the early stages of compaction until the pressure causes the particles to orientate (Engelhardt–Gaida, 1963).

The intracrystalline porosity depends upon the surface area of the clay and the thickness of the water film coating it. This thickness depends upon the presence on the surface of exchangeable cations which tend to hydrate and to form an electric double layer.

The electric double layer is made of a layer of ions adsorbed on the clay surface (Stern layer), and of ions free to move into the diffuse double layer (Gouy layer). The distribution of the cations between the two layers depends on the salinity, on the base exchange capacity, on the valence and on the particular properties of the cations (Kruyt, 1952). It must be noted that the higher the valence, the stronger will be the tendency for the cations to stay in the Stern layer.

The Stern layer has a maximum thickness of 5 Å, equal to two layers of water, while the diffuse layer can be much thicker. Van Olphen (1954) has studied the attractive and repelling forces which act on the clay crystals, distinguishing the long range ones with a field of action greater than 5 Å — comparatively weak and responsible of the expansion of the diffuse double layer — and the short range ones with a field of action lower than 5 Å, responsible of the hydration of the Stern layer. It is interesting to note that the energy of the electric double layer, given as a function of its thickness, goes through a minimum at about 5 Å (Van Olphen, 1954) and that in clays saturated with a water, the salinity of which is greater than 0·4 m/l, the diffuse double layer is destroyed, while only the Stern layer with a thickness of 5 Å (Norrish, 1954) remains. This value corresponds to the presence between the opposite surface of two crystals or of two plates of clay, in the case of expandable clay minerals, of four layers of water with a unit thickness of about 2·5 Å.

In compacted sediments, which are the subject of petroleum geology, only short range forces between crystals are present because the inter-planar separation is lower than 10 Å. This figure for a montmorillonitic clay, with a surface area of 800 m^2/g, corresponds to a porosity value of about 52%.

The gradual expulsion of the four layers of the interplanar water, at constant temperature, takes place at different pressure values. On the basis of thermodynamic considerations, Van Olphen (1963) has calculated that at 25°C the fourth, the third, the second and the first water layer would be removed at pressures of 200, 1300, 2500 and 5400 atm, respectively. At 50° C the last two layers of water would be removed at lower pressures, namely at 660 and 4450 atm. These theoretical values agree satisfactorily with the experimental tests. Engelhardt–Gaida (1963) have shown that montmorillonite compacted at 800 kg/cm^2 has a porosity value of 33%, and when compacted at 3000 kg/cm^2, a porosity value of 20%. These values correspond to a montmorillonite with two and one water layer, respectively.

The Engelhardt–Gaida data appear to confirm that, at least in shales compacted to porosity values of the order of 33%, intercrystalline porosity would no longer be present. This finding would suggest either that the clay crystals are well oriented or that they are bent or broken in order to mould onto the underlying crystals, thus eliminating the voids between them. These possibilities appear to be confirmed, in the case of kaolinite, by electron micrographs taken on samples prepared by the replica method (Bates and Comer, 1955).

Another problem is that of the presence of chloride anions in the shales. When intercrystalline porosity still exists, the chlorides can be dissolved in the water which fills the voids. Many considerations, however, induce us to admit that chlorine anions can be accommodated in the diffuse double layer and also in the Stern layer. In fact working with the Engelhardt–Gaida data concerning the salinity of the internal clay solution at various stages of compaction, Table 1 and the curves shown in Fig. 1 were obtained.

All reported analytical data are referred to 1 cc of dry sample of sodium montmorillonite (about 2·6 g). The samples 1, 2 and 3 were saturated with 1·007, 0·576, 0·157 N solutions of sodium chloride, respectively.

An interpretation of the straight part of the curves shown in Fig. 1 can be given by supposing that the sodium chloride concentration is different in the two inter- and intracrystalline solutions, and that during the early stages of compaction only the intercrystalline solution comes out.

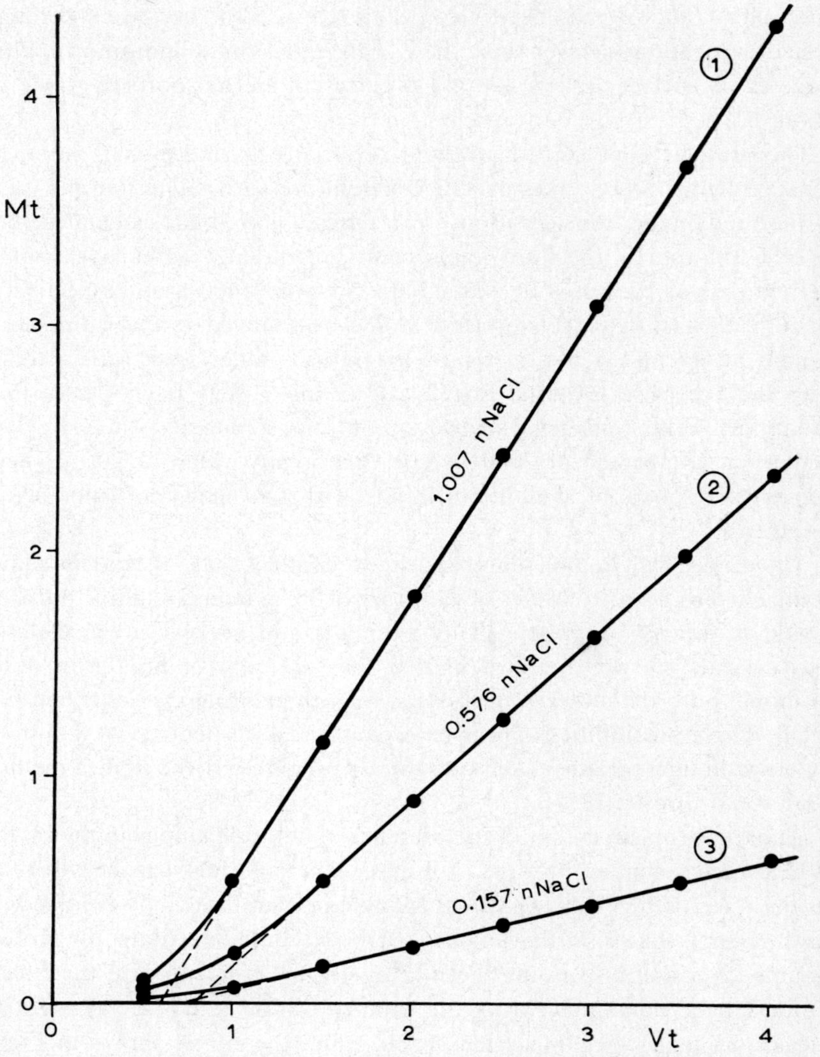

FIGURE 1. — NaCl content change of the pore solution versus void volume in compressed montmorillonite clay (from Engelhardt–Gaida, 1963, experimental values, fig. 12, p. 928).

Simple material balance gives:

$$M_t = M + \overline{M} = mV + \overline{m}\overline{V} \tag{1}$$

and

$$V_t = V + \overline{V} \tag{1'}$$

Combining equation (1′) with equation (1) yields:

$$M_t = -(m-\overline{m})\overline{V}+mV_t \qquad (1'')$$

where

$V_t, V, \overline{V}$ = total volume voids, intercrystalline volume, intracrystalline volume, respectively (cc);

$M_t, M, \overline{M}$ = meq Cl^- present in the volumes V_t, V and $\overline{V}$, respectively;

$m_t, m, \overline{m}$ = M_t/V_t, M/V and $\overline{M}/\overline{V}$, respectively, are total salinity, intercrystalline salinity, intracrystalline salinity (eq/1).

Equation (1″), as applied to the straight part of the curves of Fig. 1, reduces to:

$$M_t = a+mV_t \qquad (1''')$$

where:

$$a = (m-\overline{m})\overline{V} = \text{constant}$$

and the intercrystalline solution concentration is:

$$dM_t/dV_t = m \qquad (1'^{V})$$

TABLE 1. – NaCl CONTENT OF THE PORE SOLUTION VERSUS VOID VOLUME (FROM ENGELHARDT–GAIDA, 1963, EXPERIMENTAL VALUES, FIG. 12, p. 928)

V_t	1		2		3	
	m_t	M_t	m_t	M_t	m_t	M_t
0·5	0·18	0·09	0·08	0·04	0·03	0·01
1·0	0·54	0·54	0·22	0·22	0·07	0·07
1·5	0·77	1·15	0·36	0·54	0·10	0·16
2·0	0·90	1·80	0·45	0·89	0·12	0·24
2·5	0·97	2·43	0·50	1·25	0·14	0·34
3·0	1·02	3·08	0·54	1·63	0·15	0·43
3·5	1·06	3·70	0·57	1·99	0·15	0·54
4·0	1·08	4·32	0·58	2·34	0·16	0·64

The exponential part of the curves of Fig. 1 can be interpreted by supposing that it is related to the compaction stage of the intracrystalline voids. In this case the equation becomes:

$$M_t = \overline{m}\overline{V} \qquad (1^{V})$$

and the intercrystalline solution concentration, equal to the total concentration of the pore solution of clay is:

$$m_t = \overline{m} = M_t/\overline{V} \tag{1^{V'}}$$

Figure 1 shows the results of some experiments with sodium montmorillonite at pressure up to 800 atm. In the highest attainable pressure range (800–3200 atm) the direction of change is reversed (Engelhardt-Gaida, 1963, p. 929, Fig. 14). If the porosity is reduced from 30% to 20%, the concentration of the pore solution begins to rise.

An interpretation of this experimental finding can be given by supposing that the chloride ions, which are still present in the two last layers of the clay hydration water, come out with difficulty when the second water layer is removed, with the consequence that the salinity of the interstitial water, contrary to expectation, increases instead of diminishing.

The anions, in fact, can be accommodated in the Stern layer. Grahame (1947), developing the double layer Stern theory, has suggested that the anions, having an ionic radius smaller than that of cations (the anions, with the exception of fluorine, are partially dehydrated), are placed in a plane even closer to the clay surface, "inner Helmholtz plane", while the cations are in a second outer plane, "outer Helmholtz plane". Grahame has supplied evidence that in the Stern layer only the anions are really chemisorbed with loss of a part of their hydration shells, whereas the cations remain hydrated and are only attracted by electrostatic forces (Kruyt, 1952).

Donnan equilibrium

Whenever a homogeneous, compacted clay, deprived of intercrystalline porosity, is immersed in water, two solution phases are formed: the external solution phase and the internal solution phase. The latter is strongly influenced by the fixed charges on the clay surface.

The equilibrium distribution of water and ions between the two solution follows the Donnan (1924) equation and theoretically it should be possible to calculate the composition of one of them, i.e. the external one, when the composition of the other is known.

In the case in which a clay is immersed in a solution containing Na^+, K^+, Ca^{++}, Mg^{++} and Cl^- ions (which is very frequent in nature), the equilibrium conditions are given by equations (2), (3), (4), (5) and (6):

$$\eta_{Na^+} = \bar{\eta}_{Na^+} \tag{2}$$
$$\eta_{K^+} = \bar{\eta}_{K^+} \tag{3}$$
$$\eta_{Ca^{++}} = \bar{\eta}_{Ca^{++}} \tag{4}$$
$$\eta_{Mg^{++}} = \bar{\eta}_{Mg^{++}} \tag{5}$$
$$\eta_{Cl^-} = \bar{\eta}_{Cl^-} \tag{6}$$

where η = electrochemical potential of the external solution,
$\bar{\eta}$ = electrochemical potential of the internal solution.

The electrochemical potential is related to the chemical potential by the equation:

$$\eta_i = \mu_i + ez_i V \tag{7}$$

where μ_i = chemical potential of ion i,
e = unit charge of the ion,
z_i = valence of ion i,
V = electric potential of the solution.

The chemical potential in turn is related to the ionic concentration by the equation:

$$\mu_i = \mu_{0i} + RT \ln y_i m_i \tag{8}$$

where μ_{0i} = chemical potential of ion i taken at the standard state,
y_i = molar activity coefficient of ion i,
m_i = molar concentration of ion i.

The activity coefficients of the single ions are related to the mean activity coefficient of the salt by the equation:

$$y\pm = (y_+^{n+} \cdot y_-^{n-})^{1/n} \tag{9}$$

where $y\pm$ = mean molar activity coefficient of the salt,
$n+$ = number of positive ions coming from the dissociation of one molecule of salt,
$n-$ = number of negative ions coming from the dissociation of one molecule of salt
$n = n_+ + n_-$

Combining equations (2), (3), (4), (5) and (6) with equations (7), (8) and (9) gives:

$$\bar{m}_{Na^+} \cdot \bar{m}_{Cl^-} \cdot (\bar{y}_{NaCl})^2 = m_{Na^+} \cdot m_{Cl^-} \cdot (y_{NaCl})^2 \tag{10}$$
$$\bar{m}_{K^+} \cdot \bar{m}_{Cl^-} \cdot (\bar{y}_{KCl})^2 = m_{K^+} \cdot m_{Cl^-} \cdot (y_{KCl})^2 \tag{11}$$
$$\bar{m}_{Ca^{++}} \cdot (\bar{m}_{Cl^-})^2 \cdot (\bar{y}_{CaCl_2})^3 = m_{Ca^{++}} \cdot (m_{Cl^-})^2 \cdot (y_{CaCl_2})^3 \tag{12}$$
$$\bar{m}_{Mg^{++}} \cdot (\bar{m}_{Cl^-})^2 \cdot (\bar{y}_{MgCl_2})^3 = m_{Mg^{++}} \cdot (m_{Cl^-})^2 \cdot (y_{MgCl_2})^3 \tag{13}$$

The mean activity coefficients of the salts NaCl, KCl, $CaCl_2$ and $MgCl_2$ have been experimentally determined in water solution by Harned and Owen (1958) as a function of the concentration of the salts dissolved in water. The mean activity coefficients of the same salts dissolved in the internal solution of compacted clays are not known: the only paper on this subject, limited to sodium chloride and to the field of high porosity values in which intercrystalline porosity is still present, is that by Bernstein (1960). Nevertheless from the Bernstein data it would appear that in the field of high salinity (> 3 m/l), the activity coefficients of the two solutions tend to be equal.

Therefore, as first approximation, if we suppose that the activity coefficients of the salts in the two solutions are equal and that the concentration of the ions in the internal solution are known, equations (10), (11), (12) and (13) form a system of four equations in five unknowns. A fifth equation is provided by the condition of electroneutrality for the external solution:

$$m_{Na^+} + m_{K^+} + 2m_{Ca^{++}} + 2m_{Mg^{++}} = m_{Cl^-} \tag{14}$$

The solution of the system versus the unknown m_{Cl^-} is given by the equation:

$$(m_{Cl^-})^3 - \overline{m}_{Cl^-} \cdot (\overline{m}_{Na^+} + \overline{m}_{K^+}) \cdot m_{Cl^-} - (\overline{m}_{Cl^-})^2 \cdot 2(\overline{m}_{Ca^{++}} + \overline{m}_{Mg^{++}}) = 0 \tag{15}$$

Once m_{Cl^-} is known, the concentration of the monovalent ions (m_M) and bivalent (m_B) of the external solution can easily be calculated from equations (10), (11), (12), (13), neglecting both external and internal activity coefficients.

$$m_{Na^+} + m_{K^+} = m_M = \frac{(\overline{m}_{Na^+} + \overline{m}_{K^+})\overline{m}_{Cl^-}}{m_{Cl^-}} \tag{16}$$

$$m_{Ca^{++}} + m_{Mg^{++}} = m_B = \frac{(\overline{m}_{Ca^{++}} + \overline{m}_{Mg^{++}})(\overline{m}_{Cl^-})^2}{(m_{Cl^-})^2} \tag{17}$$

Filtration of saline solution through shales

The hypothesis that natural shales behave as semipermeable membranes, i.e. permeable to water and not to the salts dissolved in them, has been suggested by De Sitter (1947) and taken up by other authors (see White, 1965).

The filtering properties of a compacted clay have been proved in laboratory tests by McKelvey and Milne (1962).

The source of the motive energy for the subsurface waters is believed to be the pressure exerted on the fluids by the load of the sediments and by tectonic thrusts, bearing in mind that in porous strata extending to the outcrops, the hydrostatic pressure of meteoric waters could be effective (see Moulenes, 1964).

On the basis of the considerations set out before, the flow of the water inside the shales can be roughly depicted as streamlines of water which follow the outlines of the clay mineral crystals or plates, and having a variable composition according to whether the water flow goes through the double layer spaces or through the intercrystalline pores outside the field of action of the forces acting on the clay surface. Between the water which fills every single intercrystalline pore and the water which coats the clay surface, Donnan equilibrium will exist, with the consequence that the anion concentration in the pore water will be higher than in the double layer. On the whole, the composition of the interstitial water in a shale will then be the sum of the contributions given by the water both in each single pore (intercrystalline porosity) and on each clay surface (intracrystalline porosity).

This system of flow can be summarized as follows:

1. The water that migrates from the aquifer into the intercrystalline voids of the shales maintains its composition.

2. The exchange position sites on the clay surface are preferentially saturated with cations according to the composition of the water in the intercrystalline pores (Kruyt, 1952).

3. If the intercrystalline pores, because of compaction, do not communicate any longer or vanish, the water that migrates from the aquifer into the shale bed will flow only through the double layer thus modifying its composition. A Donnan equilibrium will then be established between the whole of the intracrystalline water of the shale bed and the water of the aquifer. As a consequence, the water in the aquifer will tend to concentrate, when water flow goes on during geological time.

4. In the course of filtration through a shale, a saline solution is modified both in respect of concentration of the salts and in respect of the composition of the salts. The bivalent cations particularly will be more strongly adsorbed on the clay surfaces and will be less mobile. On the other hand, the monovalent cations will be more mobile with the consequence that the solution which comes out of the shale will be enriched in them. The anions too, show differences in mobility which can be related

to preferential adsorption on the clay surface. In particular the bicarbonate ion seems to be more mobile than the chloride ion.

These considerations arise from the assumption that the hydrated ions can go freely through the narrow interstices inside a compacted shale, whose width is in the range of 2,5–10 Å. On the other hand if the hydrated ionic radii, as given in the literature (Grim, 1953, p. 148; Klotz, 1950, p. 331) are taken into consideration, no ion should be able to go through interstices with a width lower than 6 Å. In our opinion such figures, inter alia very varied, can not be used because they have been obtained under conditions of temperature, pressure and concentration widely different from those prevailing in nature.

The filtering properties of a shale explain the distribution of the brines as a function of the depth of burial in the subsurface reported by Chebotarev (1955) and thoroughly examined by White (1965).

Membrane-concentrated waters particularly rich in calcium chloride are characteristic of deep aquifers covered by shale formations; membrane-filtered waters, characterized by the presence of sodium bicarbonate, are found in the aquifers above strata of compacted shale. The concentrated waters having the chloride ion neutralized by a cation of a weak base, calcium, are lightly acid and will tend to dissolve calcite; the dilute waters, having the sodium ion partly neutralized by a weak anion, bicarbonate, will show an alkaline pH and will favour the dissolution of silica and the precipitation of calcite.

Experimental results

Analytical procedures

The determination of the pore solution composition in a shale is made by means of the following measurements:

1. *Apparent density*. The apparent density is determined on a core sample weighing 50–100 g. The bulk volume measurement is carried out by using a water pycnometer after water-proofing the sample by spraying on it a thin coat of nitrocellulose paint. The weight of the sample is obtained after drying in an oven at 110° C.

2. *Real density*. The real density is determined on a sample which has been ground until it goes through an 80-mesh sieve. The measurement is carried out on 15-g sample, dried at 110° C, by the hydrostatic balance method, using xylene as a liquid.

3. *Chlorides.* They are extracted in a Soxhlet apparatus from a finely ground sample (2–50 g), using an 80% methanol solution. The time of extraction is 6 hours. The chloride determination on the extracted liquid is carried out by the Mohr method.

4. *Base exchange capacity (BEC) and exchanged ions.* The base exchange capacity is determined on a 1-g sample, according to the Mackenzie (1951) procedure. In the centrifuged liquid, containing the exchange ions, the sodium and potassium concentrations are determined by the atomic adsorption spectrophotometer (Perkin–Elmer model 303). The same liquid has been used many times, without success, for the determination of calcium and magnesium in exchange positions: the treatment with ammonium acetate dissolves the calcium and magnesium carbonates, almost always present in the shales, thus making it impossible to measure the calcium and magnesium in exchange positions. The content of these ions in exchange positions is then calculated as a difference, using the equation of the electric balance of the internal solution:

$$\text{BEC}+(\text{Cl}^-) = (\text{Na}^+)+(\text{K}^+)+(\text{Ca}^{++})+(\text{Mg}^{++}) \quad (1)$$

where all concentrations are expressed in meq/100 g of dried sample.

5. *Clay mineral composition.* The clay mineral analysis has been made to be compared with the base exchange capacity of the sample. The procedure followed has already been described in a previous paper (Long, Neglia and Favretto, 1962).

6. *Surface area.* The surface area of the samples has been determined by the Diamond and Kinter (1958) method, using 1·0 g of sample. As a conversion factor of the glycerol retention data to values of surface area, we have chosen the ratio between the theoretical surface area for montmorillonite of 800 m^2/g (see Dyal and Hendriks, 1950) and the glycerol retention value of 19·4%, a mean value of different montmorillonite determinations (see Kinter and Diamond, 1958, Table 2, pp. 323). Thus a glycerol retention of 1% would be equivalent to a monolayer area of 40·5 m^2/g.

The precision of all the determinations listed above, apart from the clay mineral analysis, is quite satisfactory. The scatter from the mean of six determinations is lower than ±3%.

A systematic error could be involved in the bulk volume determination since this is carried out at room conditions instead of at underground conditions. However it is not likely that a lowering of pressure can modify the volume of a compacted shale, since it is a plastic body and once compacted keeps its volume. Also volume variations due to temperature changes are usually negligible because the cubical expansion coefficient of the silicates is very low, in the range of $0{\cdot}3 \times 10^{-4}/1°$ C.

Sampling and results

The determination of the composition of the pore solution has been made on 19 samples (from the Collesano formation of Lower Miocene age) taken in a Sicilian gas-producing well and on 30 shale samples of Triassic age from a North African oil-producing well. For simplicity the two wells will be referred to as "well G" (gas) and "well O" (oil), respectively.

The results of all analytical determinations are shown in Tables 2 and 3 together with the depth at which the samples are taken. The same data,

TABLE 2. — WELL G–ANALYTICAL DATA

Sample N°	Depth m	Cl^- meq/100g	Vv ml/100g	BEC meq/100g	Na^+ meq/100g	K^+ meq/100g	Carbonate (as $CaCO_3$)%	Surface area m^2/g	Clay minerals composition		
									M.L.	K.	I.
147	2404·7	0·06	0·6	14·8	10·2	2·8	3·4	105·3	40	15	45
148	2454·2	0·03	0·9	15·4	11·4	2·8	2·0	97·2	40	20	40
149	2454·3	0·02	0·7	20·8	14·5	4·0	0·4	109·3	30	25	45
150	2455·0	0·02	0·7	18·8	12·5	4·1	2·6	109·3	30	30	40
151	2457·0	0·03	0·7	25·2	17·9	5·1	0·4	125·5	35	25	40
152	2457·4	0·02	1·2	23·3	15·9	4·9	5·6	141·7	35	20	45
154	2457·6	0·02	0·8	21·8	15·8	4·6	4·7	145·8	40	15	45
156	2461·7	0·02	0·6	20·5	14·4	3·6	3·1	133·6	35	30	35
157	2461·8	0·02	1·0	21·3	14·0	4·4	3·5	129·6	35	25	40
159	2462·6	0·02	0·4	14·1	9·5	2·4	3·9	101·2	40	25	35
160	2552·8	0·04	0·5	12·4	8·1	1·6	7·5	72·9			
161	2554·5	0·01	0·2	7·4	4·8	1·0	11·8	44·5			
162	2556·7	0·01	0·8	13·0	8·5	1·5	5·3	81·0			
163	2556·9	0·01	0·8	14·6	9·7	1·9	5·2	93·1			
165	3010·0	0·09	1·0	13·5	9·5	1·1	6·5	93·1			
166	3010·2	0·08	1·0	15·2	10·3	1·5	5·4	101·2			
167	3011·2	0·08	0·9	14·4	10·0	1·3	6·4	101·2			
168	3011·4	0·06	0·8	15·3	10·4	1·5	7·7	101·2			
169	3011·7	0·06	0·8	13·7	10·0	1·5	6·5	93·1			

M.L., Mixed Layers; K, Kaolinite; I, Illite.

TABLE 3. — WELL O–ANALYTICAL DATA

Sample N°	Depth m	Cl- meq/100g	Vv ml/100g	BEC meq/100g	Na^+ meq/100g	K^+ meq/100g	Carbonate (as $CaCO_3$)%	Surface area m^2/g	Clay minerals composition %		
									M.L.	K.	I.
51	2403·5	17·5	4·1	11·6	15·5	6·0	18·5	113·4	10	—	90
52	2403·9	19·3	4·5	13·2	16·7	6·7	13·1	113·4	20	—	80
53	2404·1	18·7	4·4	12·2	16·7	5·7	21·5	105·3	—	—	100
54	2404·5	21·9	5·3	13·5	18·9	6·4	6·2	117·4	25	—	75
55	2404·9	17·7	3·8	11·4	15·3	5·9	19·2	91·3	15	—	85
56	2405·2	17·3	3·9	12·0	15·0	5·9	18·4	109·3	20	—	80
57	2405·5	16·6	4·2	12·5	16·5	6·3	17·4	109·3	20	—	80
58	2405·8	19·3	4·1	12·8	17·0	5·9	13·9	105·3	20	—	80
59	2406·0	22·0	4·5	16·2	19·5	7·5	8·0	129·6	—	—	100
60	2406·1	19·9	4·7	14·2	17·4	6·1	4·6	133·6	30	—	70
61	2406·7	16·3	4·0	13·7	17·2	6·1	9·4	125·5	20	—	80
62	2407·4	13·2	4·0	15·6	17·5	6·5	4·0	125·5	30	—	70
63	2407·8	16·9	4·3	15·9	18·0	6·7	7·7	129·6	25	—	75
64	2408·5	13·3	3·4	12·5	15·4	5·2	0·6	105·3	25	5	70
65	2408·6	15·4	3·5	12·9	15·4	5·3	2·3	121·5	20	5	75
66	2419.3	19.4	4.5	15.5	19.4	6.6	0.6	133.6	20	5	75
67	2421·0	11·8	2·8	12·0	13·3	5·0	0·0	129·6	20	20	60
68	2421·1	14·9	3·4	13·2	15·5	5·0	0·0	109·3	20	10	70
70	2423·3	13·3	3·0	11·1	12·4	4·3	1·4	81·0	20	5	75
71	2436·4	20·8	4·9	14·0	19·8	6·8	1·4	121·5	15	—	85
72	2437·0	17·4	4·2	11·4	16·4	5·8	13·9	101·2	15	10	75
73	2437·5	19·6	4·3	13·2	17·9	5·7	7·5	117·4	15	—	85
74	2464·3	19·8	5·6	15·8	18·3	6·7	13·3	141·7	30	—	70
75	2465·8	21·6	5·7	18·2	22·1	6·8	0·9	149·8	35	—	65
76	2466·5	20·1	5·1	19·1	22·1	7·7	0·3	162·0	40	15	45
77	2466·8	21·8	4·8	18·8	22·7	7·4	0·6	174·1	30	10	60
78	2467·5	19·2	5·3	18·4	22·5	6·2	0·3	182·2	45	—	55
79	2468·3	20·7	5·0	21·2	21·2	6·8	0·0	190·3	35	5	60
80	2469·7	22·9	5·1	18·9	22·7	6·9	0·3	174·1	40	—	60
81	2471·8	15·7	3·8	13·4	16·6	4·5	0·3	129·6	25	25	50

M.L., Mixed Layers; K, Kaolinite; I, Illite.

in order to follow qualitatively the reciprocal trend of the different parameters, are drafted as a log in Figs. 2 and 3.

Examining these data, the following can be observed:

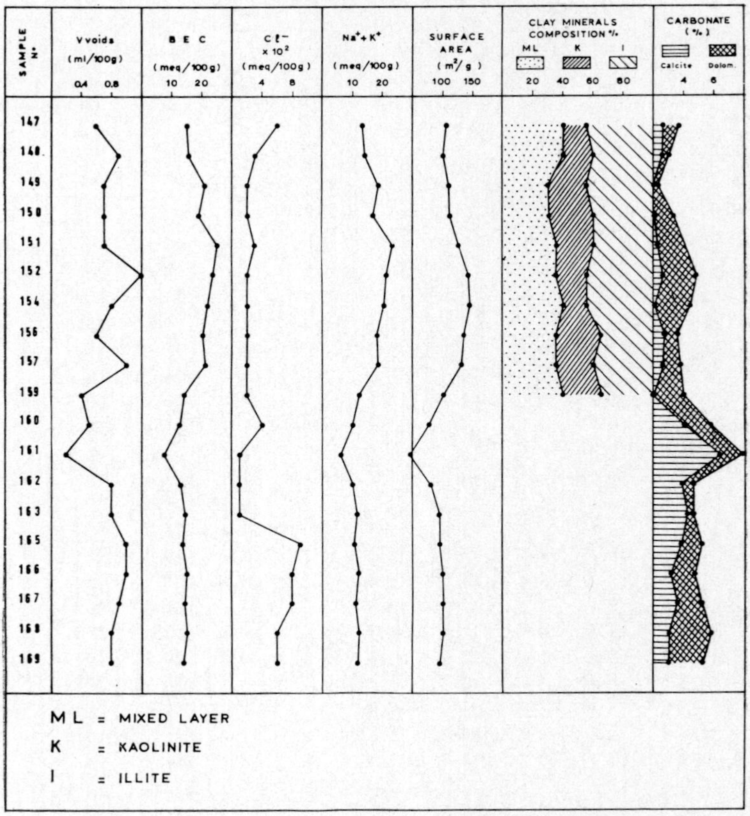

FIGURE 2. — Well G analytical data. Vertical variations.

WELL G (Fig. 2)

1. The quantity of chloride in relation to the BEC values is very low.
2. The sum of the monovalent cations, $Na^+ + K^+$, closely follows the BEC parameter and has almost equal values.
3. The base exchange capacity cannot be correlated either with the chloride content or with the void volume.
4. The clay minerals associated with these sediments are mainly composed of illite, kaolinite and illite-montmorillonite mixed-layers of ordered type in the samples taken to the depth of 2462·6 m, and of disordered type for the samples taken at greater depth. Owing to the difficulty of analyzing the disordered mixed-layers, the mineralogical composition has been reported only for the samples taken to the depth of 2462·6 m.

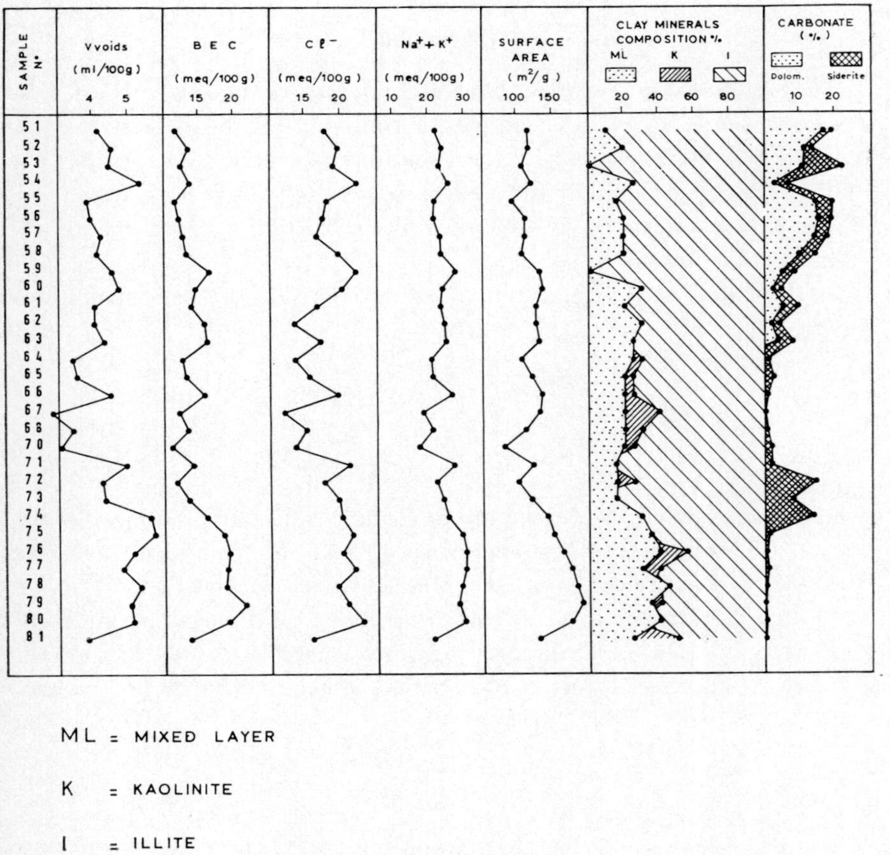

FIGURE 3. — Well O analytical data. Vertical variations.

WELL O (Fig. 3)

1. The chloride content is very high, in the range of 10–20 meq/100 g.
2. The sum $Na^+ + K^+$ is of the same magnitude as the chloride.
3. The void volume is high and agrees with both the BEC and the chloride content values.
4. The clay minerals are almost completely illite-montomorillonite mixed-layers.

Discussion

The relationship between base exchange capacity and surface area for both well O and well G samples, is shown in Fig. 4. In the two cases a linear relation expressed by the equations $BEC = 0{\cdot}11\ S$ and $BEC = 0{\cdot}16\ S$ exists for well O and well G, respectively.

The above linear correlation may be explained, from a theoretical point of view, by assuming:

1. Both *BEC* and *S* values of a shale are the sum of the contributions of the single clay minerals:

$$BEC = \frac{ax}{100} + \frac{by}{100} + \frac{cz}{100} \tag{1}$$

$$S = \frac{a'x}{100} + \frac{b'y}{100} + \frac{c'z}{100} \tag{2}$$

where a, b, c, and a', b', c' are the *BEC* and *S* values of montmorillonite, illite and kaolinite, respectively. (For simplicity only the more common clay minerals are considered).

x, y, z are the percentages of the above mentioned clay minerals. The percentage of the inert material is not considered, since it does not contribute either to *BEC* or to *S* values.

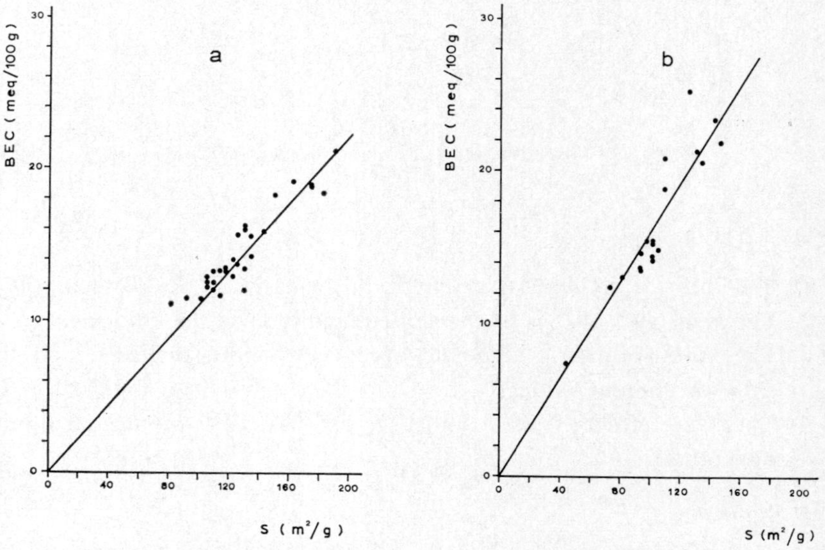

FIGURE 4. — Relationship of *BEC* with surface area. (a) Samples from well O. (b) Samples from well G.

2. The nature of the clay minerals brought into a basin should remain constant during the time of deposition of a formation, or at least the part of it concerning the samples under examination; furthermore the eventual successive diagenetic processes, according to Weaver (1958), do not modify the crystal skeleton but may only collapse the montmorillonite plates, giving rise to illite-montmorillonite mixed layers. Thus it is assumed that the *a*, *b*, *c* and *a′*, *b′*, *c′* values of the clay minerals do not undergo marked changes with geological time.

3. The percentage ratios of the single clay minerals in the samples considered, remain constant.

From these assumptions we obtain:

$$\frac{BEC}{S} = \frac{ax+by+cz}{a'x+b'y+c'z} = \frac{a+\dfrac{yb}{x}+\dfrac{zc}{x}}{a'+\dfrac{yb'}{x}+\dfrac{zc'}{x}} = \text{constant} \qquad (3)$$

The data on the mineralogical composition of well O and well G samples (Table 2.) indicate that the ratios y/x and z/x are practically constant. (At well O the kaolinite content is small and therefore negligible.)

The *BEC* (base exchange capacity) and *Vv* (void volume) shown in the correlation diagram (Fig. 5), also indicates the existence of a roughly linear relation between the two parameters. In the case of well G the

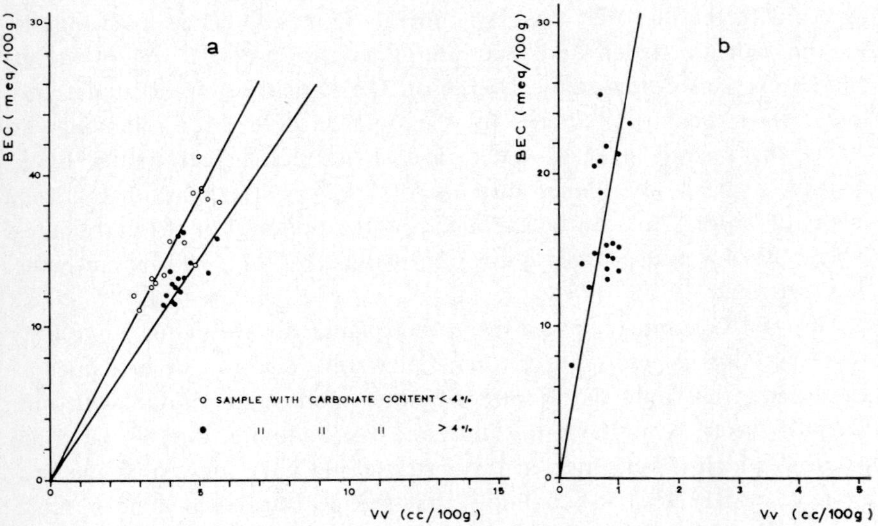

FIGURE 5. — Relationship of *BEC* with *Vv*. (a) Samples from well O. (b) Samples from well G.

equation of the straight line, passing through the origin, is $BEC = 22{\cdot}0\ Vv$. In the case of well O two straight lines, both passing through the origin, can be drawn: one, $BEC = 3{\cdot}75\ Vv$, for the samples with a carbonate content $< 4\%$, and the second, $BEC = 2{\cdot}95\ Vv$, for the samples with a carbonate content $> 4\%$.

The experimental results shown in Fig. 5, may be easily explained if we assume that in all the sets of samples analyzed the intercrystalline porosity is nil, and that the interplanar distance is constant. The void volume will then be equal to the volume of the film of water coating the surface area of the clay. Under this assumption, important conclusions can be drawn concerning the calculation of the interplanar distance in a shale and the reliability of the Donnan equation in a shale-water system.

The interplanar distance can be easily calculated from the equation:

$$Vv = \frac{S \times h}{200} \tag{4}$$

where Vv = void volume (ml/100),
S = surface area (m^2/g),
h = interplanar distance (Å).

For well O, samples with a carbonate content $< 4\%$, the mean value of h is 5·9 Å, corresponding approximately to two layers of water. This agrees with the fact that the clay minerals in well O shales exist mainly in the calcium form and that the geostatic pressure value, about 550 kg/cm^2 according to the Dickinson (1951) method of calculation, is lower than the one predicted by Van Olphen (1963) for removing, at 50° C, the second layer of water. The shales rich in carbonates have, instead, a mean interplanar distance of 7·5 Å, corresponding to three layers of water. This can be explained by the presence of the carbonates which, having much larger grain dimensions, would partly balance the geostatic load.

The well G samples have a mean interplanar distance equal to 1·4 Å, corresponding to less than one layer of water. This can be explained if we assume that only the montmorillonite fraction is hydrated with one layer of water, while both the illite and the kaolinite, due to the compaction, which is very high in this case, should have lost even their last layer of water. This would imply that the pressures which have acted on well G sediments were much higher than those which acted on well O sediments and that, since the depth of the sediments is practically the same, these high pressures were of tectonic origin. The schistosity of the well G shales and the high degree of silica cementation of the sandstones close to the shales seem to confirm this hypothesis.

The void volume-chlorides correlation is also very good in well O (Fig. 6). This can be explained by the fact that the internal salinity of a shale – the external salinity being constant —depends upon the base exchange capacity of the shale and its degree of compaction. Since these

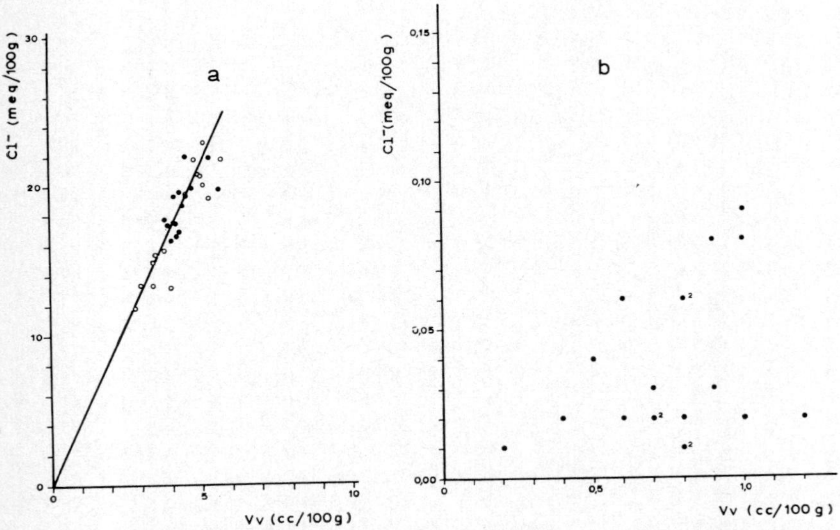

FIGURE 6. — Relationship of chloride content (Cl^-) with volume void (*Vv*). (a) Samples from well O. (b) Samples from well G.

parameters in the samples are practically constant, the internal salinity also remains constant. Furthemore because both the chloride content and the void volume of the samples are high, they are less subject to analytical error.

In well G on the contrary, the relation between the chloride content and the void volume is less good, probably because we are dealing with small values of both parameters. Since well G shales as well as those of well O do not have intercrystalline porosity they should contain the chloride ion inside the electric double layer. It should be possible in this case to calculate, starting from the pore solution composition, that of the near-by aquifer in thermodynamic equilibrium with it.

In Tables 4 and 5 are reported all the data necessary to solve the third degree equation and the calculated values of the chloride, monovalent

TABLE 4. — WELL O — DATA FOR THE CALCULATION OF THE EXTERNAL SOLUTION SALINITY

Sample N°.	A eq/l.	$\overline{m}_{Cl}$ eq/l.	$\overline{m}_M$ eq/l.	$2\overline{m}_B$ eq/l.	$\overline{m}_M \cdot \overline{m}_{Cl}$	$(\overline{m}_{Cl})^2$	$(\overline{m}_{Cl})^2 \cdot 2\overline{m}_B$	$m_{Cl} = x$; $x^3 - (\overline{m}_M \cdot \overline{m}_{Cl})x - (\overline{m}_{Cl})^2 2\overline{m}_B = 0$	m_{Cl}	m_M	$2m_B$	$\frac{m_M}{m_{Cl}} \times 100$	$\frac{2m_B}{m_{Cl}} \times 100$
51	2·8	4·3	5·2	1·8	22·4	18·5	33·3	$x^3 - 22·4x - 33·3 = 0$	5·3	4·2	1·1	79·2	20·8
52	2·9	4·3	5·2	2·0	22·4	18·5	37·0	$x^3 - 22·4x - 37·0 = 0$	5·4	4·1	1·3	75·9	24·1
53	2·8	4·2	5·1	1·9	21·4	17·6	33·4	$x^3 - 21·4x - 33·4 = 0$	5·3	4·0	1·3	75·5	24·5
54	2·5	4·1	4·8	1·9	19·7	16·8	31·9	$x^3 - 19·7x - 31·9 = 0$	5·1	3·9	1·2	76·5	23·5
55	3·0	4·6	5·6	2·1	25·8	21·2	44·5	$x^3 - 25·8x - 44·5 = 0$	5·8	4·4	1·4	75·9	24·1
56	3·1	4·4	5·3	2·1	23·3	19·4	40·7	$x^3 - 23·3x - 40·7 = 0$	5·5	4·2	1·3	76·4	23·6
57	3·0	3·9	5·4	1·5	21·1	15·2	22·6	$x^3 - 21·1x - 22·6 = 0$	5·1	4·1	1·0	80·4	19·6
58	3·1	4·7	5·6	2·2	26·3	22·1	48·6	$x^3 - 26·3x - 48·6 = 0$	5·9	4·4	1·5	74·6	25·4
59	3·6	4·9	6·0	2·5	29·4	24·0	60·0	$x^3 - 29·4x - 60·0 = 0$	6·2	4·7	1·5	75·8	24·2
60	3·0	4·2	5·0	2·2	21·0	17·6	38·6	$x^3 - 21·0x - 38·6 = 0$	5·2	4·0	1·2	76·9	23·1
61	3·4	4·1	5·8	1·7	23·8	16·8	28·6	$x^3 - 23·8x - 28·6 = 0$	5·4	4·4	1·0	81·5	18·5
62	3·9	3·3	6·0	1·2	19·8	10·9	13·1	$x^3 - 19·8x - 13·1 = 0$	4·7	4·2	0·5	89·4	10·6
63	3·7	3·9	5·7	1·9	22·2	15·2	28·9	$x^3 - 22·2x - 28·9 = 0$	5·3	4·1	1·1	77·3	22·7
64	3·7	3·9	6·0	1·5	23·4	15·2	22·8	$x^3 - 23·4x - 22·8 = 0$	5·3	4·4	0·9	83·0	17·0
65	3·7	4·4	5·9	2·2	26·0	19·4	42·7	$x^3 - 26·0x - 42·7 = 0$	5·8	4·5	1·3	77·6	22·4
66	3·4	4·3	5·8	2·0	24·9	18·5	37·0	$x^3 - 24·9x - 37·0 = 0$	5·6	4·4	1·2	78·6	21·4
67	4·3	4·2	6·5	2·0	27·3	17·6	35·2	$x^3 - 27·3x - 35·0 = 0$	5·8	4·7	1·1	81·0	19·0
68	3·9	4·4	6·0	2·2	26·4	19·4	42·7	$x^3 - 26·4x - 42·7 = 0$	5·8	4·5	1·3	77·6	22·4
70	3·7	4·4	5·6	2·6	24·6	19·4	50·4	$x^3 - 26·6x - 50·4 = 0$	5·8	4·2	1·6	72·4	27·6
71	2·8	4·2	5·4	1·7	22·7	17·6	29·9	$x^3 - 22·7x - 29·9 = 0$	5·3	4·3	1·0	81·1	18·9
72	2·7	4·1	5·3	1·6	21·7	16·8	26·9	$x^3 - 21·7x - 26·9 = 0$	5·2	4·2	1·0	80·8	19·2
73	3·1	4·5	5·5	2·1	24·7	20·2	42·4	$x^3 - 24·7x - 42·4 = 0$	5·7	4·3	1·4	75·4	24·6
74	2·8	3·5	4·5	1·9	15·7	12·2	23·2	$x^3 - 15·7x - 23·2 = 0$	4·6	3·4	1·2	73·9	26·1
75	3·2	3·8	5·1	1·9	19·4	14·4	27·4	$x^3 - 19·4x - 27·4 = 0$	5·0	3·9	1·1	78·0	22·0
76	3·7	3·9	5·8	1·8	22·6	15·2	27·4	$x^3 - 22·6x - 27·4 = 0$	5·3	4·3	1·0	81·1	18·9
77	3·9	4·5	6·3	2·2	28·3	20·2	44·4	$x^3 - 28·3x - 44·4 = 0$	6·0	4·7	1·3	78·3	21·7
78	3·5	3·6	5·4	1·7	19·4	13·0	22·1	$x^3 - 19·4x - 22·1 = 0$	4·9	3·9	1·0	79·6	20·4
79	4·2	4·1	5·6	2·8	23·0	16·8	47·0	$x^3 - 23·0x - 47·0 = 0$	5·6	4·1	1·5	73·2	26·8
80	3·7	4·5	5·8	2·4	26·1	20·2	48·5	$x^3 - 26·1x - 48·5 = 0$	5·9	4·4	1·5	74·6	25·4
81	3·5	4·1	5·5	2·1	22·5	16·8	35·3	$x^3 - 22·5x - 35·3 = 0$	5·4	4·2	1·2	77·8	22·2
									5·4	4·2	1·2	78·0	22·0

and bivalent cation contents of the external solution. Mention should be made that letter A indicates the molar concentration of the shale anion, i.e. the *BEC* value divided by the void volume.

In order to obtain a representative analysis of the water saturating the aquifer, the arithmetic mean of the concentrations of chloride, mono-

TABLE 5. — WELL G – DATA FOR THE CALCULATION OF THE EXTERNAL SOLUTION SALINITY

Sample N°.	A eq/l.	$\overline{m}_{Cl}$ eq/l.	$\overline{m}_M$ eq/l.	$2\overline{m}_B$ eq/l.	$\overline{m}_M \cdot \overline{m}_{Cl}$	$(\overline{m}_{Cl})^2$	$(\overline{m}_{Cl})^2 \cdot 2\overline{m}_B$	$m_{Cl} = x$ $x^3-(\overline{m}_M \cdot \overline{m}_{Cl})x-(\overline{m}_{Cl})^2 2\overline{m})_B = 0$	m_{Cl} eq/l.	m_M eq/l.	$2m_B$ eq/l.	$\frac{m_M}{m_{Cl}} \times 100$	$\frac{2m_B}{m_{Cl}} \times 100$
147	24·7	0·10	21·7	3·1	2·2	0·01	0·0	$x^3-2{\cdot}2x = 0$	1·5	1·5	0·0	100	0
148	17·1	0·03	15·7	1·4	0·5	0·00	0·0	$x^3-0{\cdot}5x = 0$	0·7	0·7	0·0	100	0
149	29·7	0·03	26·4	3·3	0·8	0·00	0·0	$x^3-0{\cdot}8x = 0$	0·9	0·9	0·0	100	0
150	26·8	0·03	23·7	3·1	0·7	0·00	0·0	$x^3-0{\cdot}7x = 0$	0·8	0·8	0·0	100	0
151	36·0	0·05	32·8	3·2	1·6	0·00	0·0	$x^3-0{\cdot}6x = 0$	1·3	1·3	0·0	100	0
152	19·4	0·02	17·3	2·1	0·3	0·00	0·0	$x^3-0{\cdot}3x = 0$	0·5	0·5	0·0	100	0
154	27·2	0·03	25·5	1·7	0·8	0·00	0·0	$x^3-0{\cdot}8x = 0$	0·9	0·9	0·0	100	0
156	34·2	0·04	30·0	4·2	1·2	0·00	0·0	$x^3-1{\cdot}2x = 0$	1·1	1·1	0·0	100	0
157	21·3	0·02	18·5	2·8	0·4	0·00	0·0	$x^3-0{\cdot}4x = 0$	0·6	0·6	0·0	100	0
159	35·2	0·05	29·7	5·5	1·5	0·00	0·0	$x^3-1{\cdot}5x = 0$	1·2	1·2	0·0	100	0
160	24·8	0·08	19·2	5·7	1·5	0·00	0·0	$x^3-1{\cdot}5x = 0$	1·2	1·2	0·0	100	0
161	37·0	0·07	29·0	8·1	2·0	0·00	0·0	$x^3-2{\cdot}0x = 0$	1·4	1·4	0·0	100	0
162	16·2	0·02	12·5	3·7	0·2	0·00	0·0	$x^3-0{\cdot}2x = 0$	0·4	0·4	0·0	100	0
163	18·2	0·01	14·5	3·7	0·1	0·00	0·0	$x^3-0{\cdot}1x = 0$	0·3	0·3	0·0	100	0
165	13·5	0·09	10·6	3·0	0·9	0·01	0·0	$x^3-0{\cdot}9x = 0$	0·9	0·9	0·0	100	0
166	15·2	0·08	11·8	3·5	0·9	0·01	0·0	$x^3-0{\cdot}9x = 0$	0·9	0·9	0·0	100	0
167	16·0	0·09	13·0	3·1	1·2	0·01	0·0	$x^3-1{\cdot}2x = 0$	1·1	1·1	0·0	100	0
168	19·1	0·07	15·6	3·6	1·1	0·00	0·0	$x^3-1{\cdot}1x = 0$	1·0	1·0	0·0	100	0
169	17·1	0·07	14·5	2·7	1·0	0·00	0·0	$x^3-1{\cdot}0x = 0$	1·0	1·0	0·0	100	0
									0·9	0·9	0·0	100	0

valent and bivalent ions expressed in m/l, have been calculated on the basis of the data of Tables 4 and 5. These mean values are shown in Table 6 together with the values for composition of the subsurface water samples. It must be noted that while in the case of well O the shale internal solution data are reliable, in well G the same data are less reliable owing to the difficulty of determining the chloride content at such low concentration values.

The comparison between the experimental and calculated data, in spite of the simplifications used in the calculation, is fairly good in the case of well O (experimental salinity about 4·3 – 4·9 m/l, calculated salinity 5·4 m/l) and less good in the case of well G (experimental salinity 0·5 m/l, calculated salinity 0·9 m/l). The main reason of these differences, in our opinion, is that the NaCl activity coefficients, external and inter-

TABLE 6. — COMPARISON OF THE EXPERIMENTAL AND CALCULATED VALUES OF THE WATER COMPOSITION IN WELL O AND WELL G

Sample	$Na^{+}+K^{+}$ eq/l.		$Ca^{++}+Mg^{++}$ eq/l.		Cl^{-} eq/l.		$\frac{M^{+}}{M^{+}+B^{++}}\times 100$		$\frac{B^{++}}{M^{+}+B^{++}}\times 100$	
	Det.	Cal.	Det.	Cal.	Det.	Cal.	Det.	Cal.	Det.	Cal.
1	0·5	0·9	0·0	0·0	0·5	0·9	100·0	100·0	0·0	0·0
2	3·0		1·6		4·3		63·5		36·5	
		4·2		1·2		5·4		77·8		22·2
3	3·1		1·7		4·9		63·9		36·1	

1. Well G—average of 11 analyses.
2. Well O—analysis carried out on a water sample taken during a drill-stem test (tested interval: 2480–2491 m, test time : 26 min).
3. Well O—analysis carried out on a water sample obtained by flushing with oil a sand core taken at reservoir conditions at 2450 m. Oil-base mud was used and the core was wax-coated soon after extraction.

nal to the shale, probably differ. Thus the calculation which assumed equal activity coefficients is less reliable, particularly at the low salinity values.

In conclusion the data seem to prove that the White (1965) classification could be extended to the shale pore solutions which are in thermodynamic equilibrium with those of the aquifer. The high chloride content, the abundance of bivalent cations, the complete absence in the shales of calcite, the good correlation between the base exchange capacity and the void volume, on the whole show that the interstitial waters in the shales —e.g. well O—are acid of a chloride-sodium-calcium type, i.e. of a concentrated type. On the contrary the monovalent cation dominance, the chloride low content, the presence of calcite, probably as cement, on the whole, show that the interstitial waters in well G shales are alkaline of a chlorine sodium type, i.e. are membrane filtered waters.

The flow of alkaline waters from the shales to the sands in well G is evidenced not only by the subsurface water analysis, but also by petrographic observations on the sandstones close to the shales. The presence of a calcitic cement among the quartz grains in the spots where a silica cement preexisted (see Fig. 7), confirms that the filtered waters have been subjected during geological time to a pH increase which has favoured quartz dissolution and calcite precipitation.

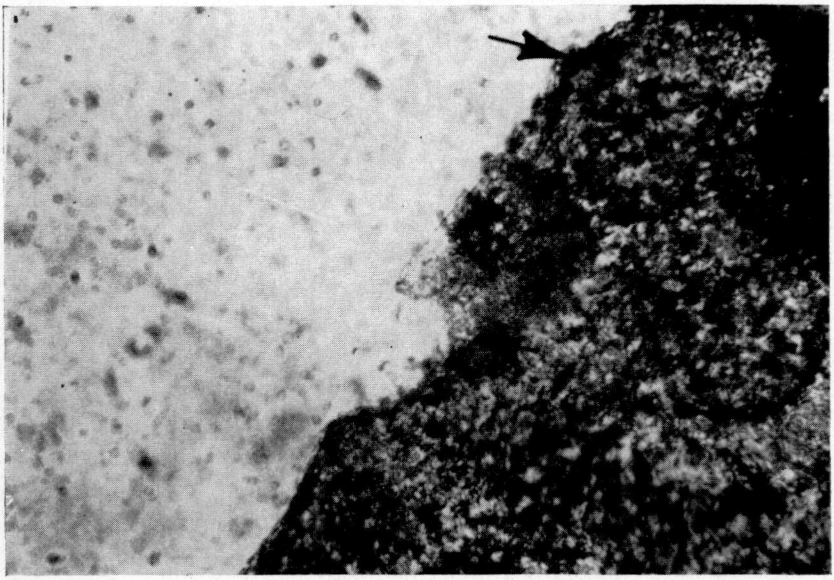

FIGURE 7. — Photo. 1. Well G sandstone core sample taken in the 2364–2372·50 m interval. Crossed nicols, ×200. Replacement, see arrows, of silica by calcite (dark) along the rim of a quartz crystal (bright).

Photo. 2. Well G sandstone core sample taken in the 2356–2364 m interval. Crossed nicols, ×80. Dissolution of the quartz crystals, see arrows, and precipitation of calcite.

Conclusions

In the present situation it is too early to draw definite conclusions from our work. We are convinced, however, that a thorough study of the pore solution of shales will contribute to the development of geochemistry in the various ways listed below.

1. The example of wells G and O and others now being studied, seems to indicate that it is possible to calculate the aquifer salinity starting from the composition of the shale pore solution. This would indicate, whenever cores are available in a basin, the distribution of the saline waters in various formations and indirectly give indications on the movement of the water in the subsurface. In particular, in a given aquifer, the migration of the water would occur from the deepest zones to structural highs, following the path of least resistance to reach the surface. The filtering properties of a shale would determine the variation of the composition of the water along an aquifer: the salinity should be greater in the high zone and gradually decrease laterally. Such a distribution has, indeed, been observed (but differently interpreted) by Jones (1960) in some Wyoming fields. Vertically, admitting that the water flow concerns a geological continuous series of permeable formations interbedded with shaly ones, the water salinity should, generally, be greater in the deepest aquifers and gradually decrease in the higher ones. As the movement of the hydrocarbons is believed to be associated with that of the water, the possibility can be seen that such geochemical studies are of practical interest for the determination of hydrocarbon migration trends.

In well G, for instance, the presence of alkaline water together with a sodium chloride content lower than that of the original marine water, can be explained by admitting that the water is filtered through thick shale beds. Since the Collesano formation in well G has a considerable thickness of perhaps thousands of metres, it is possible that the waters, that filter from great depths, dissolve the hydrocarbons (methane preferably) generated in the shales and that these, once arrived in less deep aquifers, come out of solution at the lower pressure. This process may explain the gas composition of G well: Methane = 90%, higher hydrocarbons = 8–9%.

On the other hand in the oil field, which produces oil from Triassic sands, transgressive on Palaeozoic sediments, we are dealing with a very concentrated water, whose salinity, in the range of 270 g/l, can be explained, since the Triassic shales above the oil have a small thickness, under 100 metres, by admitting that this water has formed in Palaeozoic sediments and has come late into the well O structure, together with the oil.

2. The degree of compaction reached by a shale formation depends upon the factors which have been able to modify the shale porosity during the geological time: clay mineral type, presence of a carbonate or silicate rigid matrix, temperature and pressure. Therefore, anomalies in the porosity values – such as the exceptionally low value for well G – can point to unusual pressures of tectonic origin.

3. The factors influencing the degree of compaction of the shales are the same that modify the molecular structure of the organic compounds owing to the removal of the lateral chains which hamper the flattening of the molecules (Erdman, 1965). As a consequence of the increase of temperature and pressure, hydrocarbon development and gradual transformation of the organic matter into graphite take place. The determination of the degree of compaction of the shale will thus contribute to determine the degree of evolution of the organic matter, which ends when the organic matter is completely transformed into graphite.

4. The composition of the shale pore solution in relation to the associated clay minerals and to their degree of compaction, appears to be the starting point for any further research on the diagenesis of clay minerals. If the principle of mobile equilibria is applied to the compacted shale system, we can easily understand how the expandable clay minerals tend to collapse, under the influence of the external pressure, if on their surface, cations capable of penetrating into the crystal lattice are present. This could explain the almost complete absence of pure montmorillonite in the compacted shales and instead the appearance of mixed-layers, an intermediate stage of the transformation of the montmorillonite into illite.

5. The knowledge of the pore solution composition of the shales appears very useful in interpreting the petrographical observations on samples of sandstone formations flushed by the water driven out of the shales. In well G, for instance, silification phenomena (due very likely to pressure solution), the dissolution of the silica and the precipitation of the carbonates can well be explained by the high pressures which have acted on the shales and by the alkaline water flow through shales and sandstones.

6. The complete or partial exclusion of anions from compacted shales is responsible for their behaviour as a more or less ideal membrane, which accounts for the origin of the static self-potential, SSP, which can be read an the electric logs. The presence of chloride ions, free to move in the shale pore solution, will lower the SSP value as the transference number

of the chloride ion in the shale will not be zero. Furthermore these chloride ions will lower the resistivity value of the shales, also measured on the electric logs.

The development of the studies which tend to correlate the electric characteristics of a shale formation with its pore solution composition seems to be very interesting since these studies would allow the extrapolation to the whole formation of results obtained from a few core samples.

Acknowledgements

The authors are grateful to AGIP for permission to publish this paper. Special thanks are due to T. Rocco, Exploration Manager for AGIP Direzione Mineraria, and to G. L. Morelli and G. C. Micheletti for X-ray and chemical analyses, respectively.

The authors are also indebted to J. G. Erdman, Phillips Petroleum Company, for the revision of the English translation.

References

Bates, T. F. and Comer, J. J. (1955) Electron microscoy of clay surfaces. In *Clays and Clay Minerals*, Natl. Acad. Sci. –Natl. Res. Council, pub. 395, pp. 1–9.

Bernstein, F. (1960) Distribution of water and electrolyte between homoionic clays and saturating NaCl solutions. In *Clays and Clay Minerals*, Natl. Conf. on Clay and Clay Minerals, 8th (Swineford, Ada, ed.), pp. 122–149.

Chebotarev, I. I. (1955) Metamorphism of natural waters in the crust of weathering. *Geochim. et Cosmochim. Acta*, Nos. 1–2, 22–48; No. 3, 137–170; No. 4, 198–212.

De Sitter, L. U. (1947) Diagenesis of oil-field brines. *Am. Assoc. Petroleum Geologists Bull.* 31, No. 11, 2030–2040.

Dickinson, G. (1951) Geological aspects of abnormal reservoir pressures in the Gulf Coast region of Louisiana, U.S.A. Third World Petroleum Congress, The Hague, Section 1.

Donnan, F. G. (1924) The theory of membrane equilibria. *Chem. Rev.* **1,** 73–90.

Dyal, R. S. and Hendricks, S. B. (1950) Total surface of clays in polar liquids as a characteristic index. *Soil Science*, **69,** 421–432.

Engelhardt, Wolf von, and Gaida, K. H. (1963) Concentration changes of pore solutions during the compaction of clay sediments. *Jour. Sed. Petrology*, **33,** 919–930.

Erdman, J. G. (1965) Petroleum-its origin in the earth. In *Fluids in Subsurface Environments* (A. Young and J. E. Galley ed.), published by A.A.P.G., Memoir 4, pp. 1–52.

Grahame, D. C. (1947) *Chem. Revs.* **41,** 441.

Grim, R. E. (1953) *Clay Mineralogy*, McGraw-Hill Book Company, Inc.

Jackson, M. L. (1958) *Soil Chemical Analysis.* Prentice-Hall Inc., Englewood Cliffs, N. J.

Jones, B. (1960) Electric log evaluation in exploring for Rocky Mountain petroleum traps. *Oil and Gas J.* **58,** No. 29, 142–145.

Harned, H. S., and Owen, B. B. (1958) *The Physical Chemistry of Electrolytic Solutions*. Reinhold Pub. Co., New York.

Kinter, E. B. and Diamond, S. (1958) Gravimetric determination of monolayer glycerol complexes of clay minerals. In *Clays and Clay Minerals*, Natl. Acad. Sci.-Nat. Res. Council, pub. 566, pp. 318–333.

Klotz, I. M. (1950) *Chemical Thermodynamics. Basic Theory and Methods*. New York, Prentice-Hall.

Kruyt, H. R. (1952) *Colloid Science*, vol. I. Elsevier Publishing Company.

Long, G., Neglia, S., and Favretto, L. (1964) Geochemical contribution to research for the reconstruction of the palaeogeography of a sedimentary basin. In *Advances in Organic Geochemistry*, pp. 239–259 (Colombo, U. and Hobson, G. D. ed.), Pergamon Press.

Mackenzie, R. C. (1951) A micromethod for determination of cation-exchange capacity of clay. *J. Coll. Sci.* **6**, 219–222.

McKelvey, J. G., and Milne, I. H. (1962) The flow of salt solution through compacted clay. In *Clays and Clay Minerals*, (Swineford, Ada, ed.), Natl. Conf. on Clays and Clay Minerals, 9th, pp. 248–259.

Moulenes, B. (1964) Origines des pressions anormales dans les gisements de petrole. Etude bibliographique. *Rev. Inst. Franc. du Petrole*, **XIX**, No. 2.

Norrish, K. (1954) The swelling of montmorillonite. *Discussion Faraday Soc.* **18**, 120–134.

Rowell, D. L. (1965) Influence of positive charge on the inter and intra-crystalline swelling of oriented aggregates of Na montmorillonite in NaCl solutions. *Soil Science*, **100**, No. 5, 340–347.

Van Olphen, H. (1954) Interlayer forces in bentonite. In *Clays and Clay Minerals*, Natl. Acad. Sci.-Natl. Res. Council, pub. 327, pp. 418–438.

Van Olphen, H. (1963) Compaction of clay sediments in the range of molecular particle distances. In *Clays and Clay Minerals*, Monograph No. 13, Pergamon Press.

Weaver, C. E. (1958) Geological interpretation of argillaceous sediments. Part. I Origin and significance of clay minerals in sedimentary rocks. *Bull. Amer. Ass. Petrol. Geol.* **42**, 254–271.

White, D. E. (1965) Saline waters of sedimentary rocks. In *Fluids in Subsurface Environments* (Young, A. and Galley, J. E., ed.), published by A. A. P. G., Memoir 4, Tulsa, Oklahoma.

DIFFUSION OF LIGHT PARAFFIN HYDROCARBONS IN WATER FROM 2° C TO 80° C

by

J. J. SAHORES

Société Nationale des Pétroles d'Aquitaine, France

and

P. A. WITHERSPOON

University of California, Berkeley

Introduction

The movement of hydrocarbons through water is a basic consideration in any attempt to understand the origin and migration of petroleum. Hydrocarbons are widely dispersed throughout recent and ancient sediments, and the pore spaces in these rocks are filled with waters of varying salinity. The migration of hydrocarbons through such sediments must, therefore, take place in an aqueous environment, but the exact process that leads to the accumulation of an oil or gas reservoir is not yet clear. In those instances where convection currents are negligible, the pertinent question thus arises, what part does diffusion play in this process.

Diffusion will not, of course, be the only factor that controls the movement of hydrocarbons through water-filled sediments. Adsorption phenomena will also be important, especially in view of the very large internal surface areas that are characteristic of porous media.

In as much as this hydrocarbon movement involves both diffusion and adsorption, it is necessary that the diffusion phenomena be studied independently. Unfortunately, only limited data on the diffusion of petroleum hydrocarbons in water are available. Kartsev *et al.*[4] have reported work credited to Antonov for the diffusion of methane, ethane, propane, and hexane in water. Witherspoon and Saraf[17] have reported diffusion coefficients for methane, ethane, propane and normal butane at temperatures ranging from 25° to 43° C. Gubbins *et al.*[3] have reported data for methane at 25°, 45°, and 65° C. They also have reported data on the diffusion of methane in aqueous electrolytes. With regard to other hydrocarbons, diffusion data have been reported only for acetylene in water at 18° C[10] and ethylene, propylene, and butene-1 in water between 6° and 65° C.[11]

Since petroleum is composed of several homologous groups of hydrocarbons, the diffusion results cited above constitute only a bare beginning. In approaching the problem of oil migration, one will have to know diffusion coefficients not only for the paraffins and their isomers, but the naphthenic and aromatic hydrocarbons as well. Furthermore, the effects of temperature, pressure and salinity on diffusion will also have to be studied in order to understand the role of diffusion in the migration of petroleum.

A project has therefore been undertaken at the University of California at Berkeley to study the general problem of the diffusion of hydrocarbons in water. The first results[17] reported included methane, ethane, propane and n-butane between 25° and 43° C. In the present work, we have extended the temperature range from 2° to 80° C for these same hydrocarbons and also obtained data on n-pentane and neopentane.

Experimental

The capillary – cell method of measuring diffusion coefficients perfected by Wang[12, 13] and coworkers[14, 15] and Saraf *et al.*[8] has been used for this work. Details of the method have been described elsewhere,[7, 8] and only a few new features will be mentioned here.

One of the problems has been that of preparing and maintaining standradized aqueous solutions of the individual hydrocarbons under study. We have found that for gaseous hydrocarbons, solutions were easily prepared using standard bubbling techniques; but in order to prevent loss of hydrocarbons over long periods of time, the solutions had to be kept in glass containers that were immersed in a temperature-controlled water bath. This prevented loss of hydrocarbons by diffusion through joints, plastic tubing, etc., that occurs rather easily if the container is kept in an air bath.

The preparation of solutions of hydrocarbons that are normally liquids presented another problem in that repetitive measurements of the initial concentration showed a significantly larger scatter than in the case of gaseous hydrocarbons. This is attributed to the tendency of liquid hydrocarbons (no matter how "pure") to form dispersions rather than true solutions in water. To avoid this difficulty, solutions were first prepared at temperatures above the normal boiling point where the usual bubbling technique could be employed. After saturation was achieved, the atmosphere over the solution was partially replaced by air and the container was then cooled to the temperature of interest and kept immersed in a water bath.

In general, all solutions were kept slightly undersaturated at any particular temperature to avoid oversaturation during transfer to other containers, diffusion cells, etc.

Discussion of results

The results of the diffusion measurements are summarized in Table 1. The data represent averages of a number of repetitive runs, with capillary lengths varying from 15 to 30 mm and diffusion times ranging from 20 to 170 hours. The precision shown is the standard deviation of the arithmetic mean.

TABLE 1. — EXPERIMENTAL RESULTS FOR DIFFUSION COEFFICIENTS OF HYDROCARBONS IN WATER

Hydrocarbon	$D \times 10^5$ cm²/sec at temperature of						
	2° C	4° C	6° C	20° C	40° C	60° C	80° C
Methane	0·82±0·01	0·85±0·02	0·91±0·01	1·49±0·04	2·43±0·05	3·43±0·17	4·5±0·50
Ethane	0·64±0·02	0·69±0·01		1·20±0·06	1·95±0·08	2·94±0·12	3·9±0·40
Propane	0·51±0·01	0·55±0·03	0·63±0·03	0·93±0·05	1·58±0·03	2·40±0·10	
n-Butane		0·50±0·02		0·93±0·02	1·56±0·03	2·26±0·16	
n-Pentane		0·30±0·03		0·72±0·07	1·17±0·13	1·57±0·06	
Neopentane		0·37±0·01		0·86±0·02	1·45±0·05	1·90±0·09	2·7±0·20

The data for methane, ethane, propane and n-butane are compared with the results of earlier work by Witherspoon and Saraf[17] as shown in Fig. 1. The data are in reasonably good agreement for methane, ethane, and propane, but the new results for n-butane are about 20% higher than those of the earlier work. It will also be noted that the temperature dependence of diffusion is significantly higher for our data than for the results obtained by Witherspoon and Saraf.[17] No explanation for this discrepancy has been found. Gubbins *et al.*[3] have recently published results for the diffusion of methane at 25°, 45° and 65° C. These data are included on Fig. 1 and are in good agreement with our results.

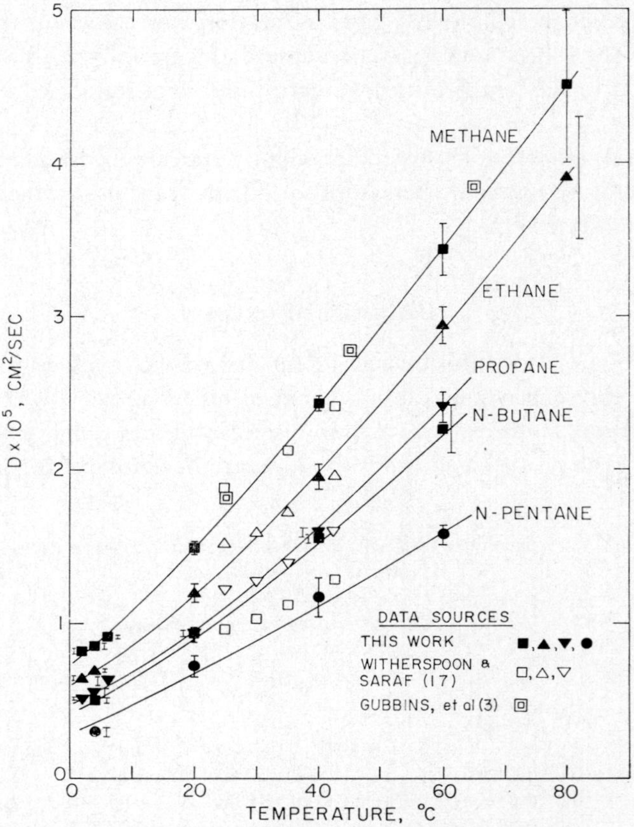

FIGURE 1. — Diffusion data versus temperature.

Witherspoon and Saraf[17] have proposed a linear relationship between log *D* and carbon number for the paraffin hydrocarbons that they investigated. The data of Table 1 are plotted in this manner as shown in Fig. 2, and the results confirm this semi-log relationship over a much wider temperature range than has previously been studied. One notes from the correlations shown on both Figs. 1 and 2 that the diffusion coefficients for n-butane are consistently too high by about 15%, which suggests experimental error.

On the basis of the correlations shown on Fig. 2, one can compute diffusion coefficients for the homologous series of methane through pentane at any temperature from 2° through 80° C from:

$$D_n = 1{\cdot}22\ D_{n+1}\dagger \tag{1}$$

As has been pointed out previously,[17] such an equation is not in agreement with correlations that have been proposed by Antonov.[1]

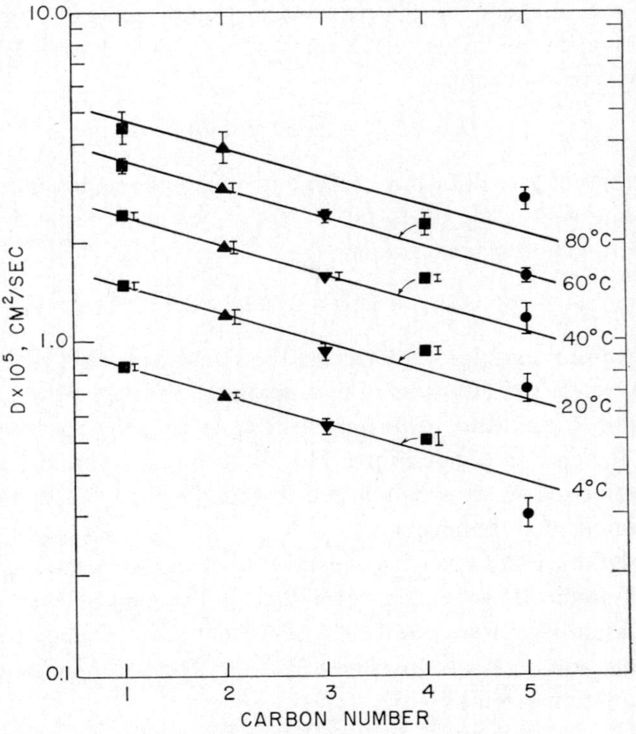

FIGURE 2. — Diffusion data versus carbon number.

One of the purposes of this study was to examine diffusion coefficients in the vicinity of the temperature where water has a maximum density. At 3·98° C, water has its maximum density of 0·999973 g/cc, and at this temperature the liquid structure would presumably be in its most compact state. One might therefore expect some effect if the structure of water is an important factor in controlling diffusion.

† All terms are defined in the nomenclature at the end of this paper.

For this purpose, experiments were made with methane, ethane and propane at 2°, 4° and 6° C. It can be seen in Fig. 1 that, within the limits of our precision, there does not appear to be any significant change in slope as one passes from the higher temperatures down to the minimum temperatures studied.

In fact, one notes from the data of Fig. 1 that diffusion coefficients increase with temperature in a predictable fashion. Although the temperature dependence is obviously non-linear, one can make a fairly good approximation of the diffusion coefficients for methane, ethane, propane, and n-butane by using the values given for 4° C on Table 1 in the following simple linear relationship:

$$D_t = D_4[1+0{\cdot}059\,(t-4)] \tag{2}$$

Equation (2) will give diffusion coefficients for methane, ethane, propane, and n-butane within 8% up to 60° C. For n-pentane a slightly different equation will give the same results:

$$D_t = D_4[1+0{\cdot}077(t-4)] \tag{3}$$

The above relationships demonstrate the significant effect that temperature can have on the diffusion of hydrocarbons through water. As equations (2) and (3) indicate, diffusion coefficients will be doubled for every 15 degree increase in temperature. This is an important effect that must not be overlooked by those interested in the movement of hydrocarbons through water-filled sediments.

A considerable effort has been expended by many workers in developing methods of predicting coefficients of diffusion. The methods that are based on theory usually employ idealized systems that are not likely to apply to the aqueous solutions of this investigation. For example, the Stokes-Einstein equation given by:

$$D = \frac{kT}{6\pi\eta_{\mathrm{B}}r_{\mathrm{A}}} \tag{4}$$

predicts the behavior of small spherical molecules of solute A in solvent B. The increasing chain length as one passes from methane to pentane obviously results in further and further departures from a sphere.

On should therefore not expect the Stokes-Einstein equation to apply to the diffusion of paraffin hydrocarbons in water. We examined the results of using such an equation very briefly. One of the problems was to determine the appropriate radius to employ in equation (4). As a first approach, we computed approximate radii from viscosity data using half the separation between two molecules at zero potential. Curves of diffusion coefficients computed from the Stokes-Einstein equation for such radii are compared to the experimental results as shown in Fig. 3.

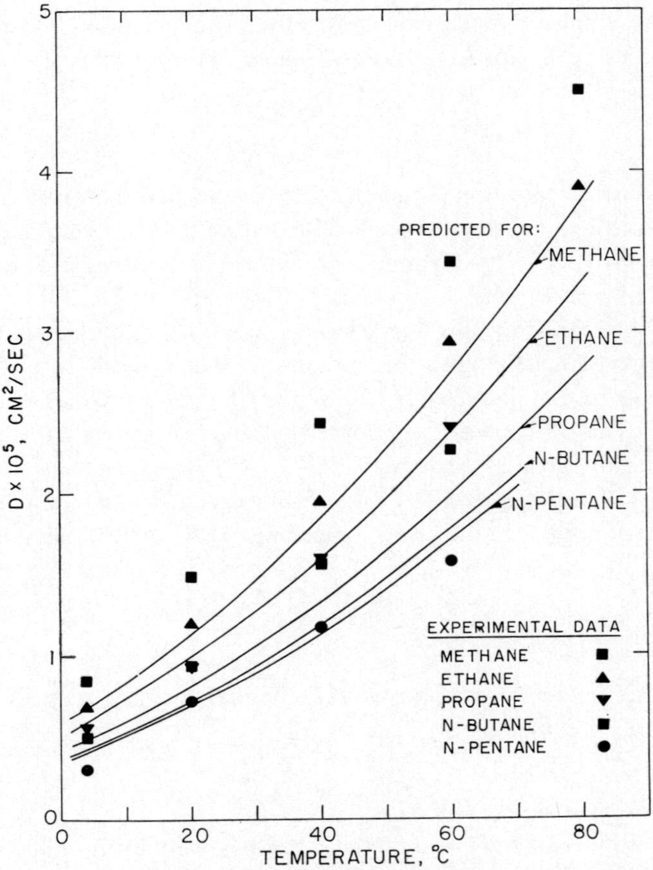

FIGURE 3. — Diffusion data versus predictions from Stokes-Einstein equation.

It will be noted that although the temperature effect is predicted rather well by the Stokes-Einstein equation, the computed coefficients are generally too low. One might anticipate that if the Stokes-Einstein equation were to apply at all, it should do so for methane, but Fig. 3 shows the greatest discrepancies for this particular hydrocarbon. If one chooses to determine the hypothetical radius from molecular volumes, the computed values are even lower than those obtained by using viscosity data.

We also investigated the theoretical equation developed by Eyring[2] and a modification of this equation that has been proposed by Olander.[5] Both approaches assume idealized spherical systems, and they too cannot be expected (nor were they found) to apply to the aqueous solutions of hydrocarbons that we investigated.

Various empirical equations have also been proposed for predicting coefficients of diffusion. The Wilke–Chang[16] formula:

$$D = 7{\cdot}4 \times 10^{-8} \frac{M_B \phi_B)^{1/2}\, T}{\eta_B\, V_A^{0{\cdot}6}} \tag{5}$$

where ϕ_B is an association parameter that is equal to 2·6 for water, has been reported to predict diffusion coefficients within ±10%. Results computed from the Wilke–Chang formula are compared with the experimental results shown in Fig. 4.

It may be seen from this figure that, in contrast to the results obtained by the Stokes-Einstein equation, which predicted coefficients that were too low, the results predicted by the Wilke-Chang formula are consistently too high. The differences vary somewhat but, in general, the predicted coefficients are 10 to 20% too high.

Scheibel[9] and Othmer and Thakar[6] have proposed two other empirical formulae for predicting diffusion coefficients. The Scheibel formula is given by:

$$D = \frac{8{\cdot}2 \times 10^{-8}[1+(3V_B/V_A)^{2/3}]\, T}{\eta_B\, V_A^{1/3}} \tag{6}$$

When water is the solvent, the Othmer-Thakar formula reduces to:

$$D = \frac{14 \times 10^{-5}}{\eta_B^{1{\cdot}1}\, V_A^{0{\cdot}6}} \tag{7}$$

Results computed from the Othmer-Thakar formula are also compared with the experimental data as shown in Fig. 4. Results computed using the Scheibel formula were found to be intermediate between those of the Wilke-Chang and Othmer-Thakar formulae and are therefore not included on this figure.

It is apparent from Fig. 4 that the results predicted by the Othmer-Thakar formula are considerably closer to our experimental results than the coefficients predicted by the Wilke-Chang formula. Above 20° C, the computed values are generally within 5% of the experimental results. Below 20° C, the results predicted by the Othmer-Thakar formula differ somewhat more from the experimental data and are generally 10 to 15% too high. It is apparent, however, that these three empirical methods of predicting diffusion coefficients[6, 9, 16] are not completely satisfactory.

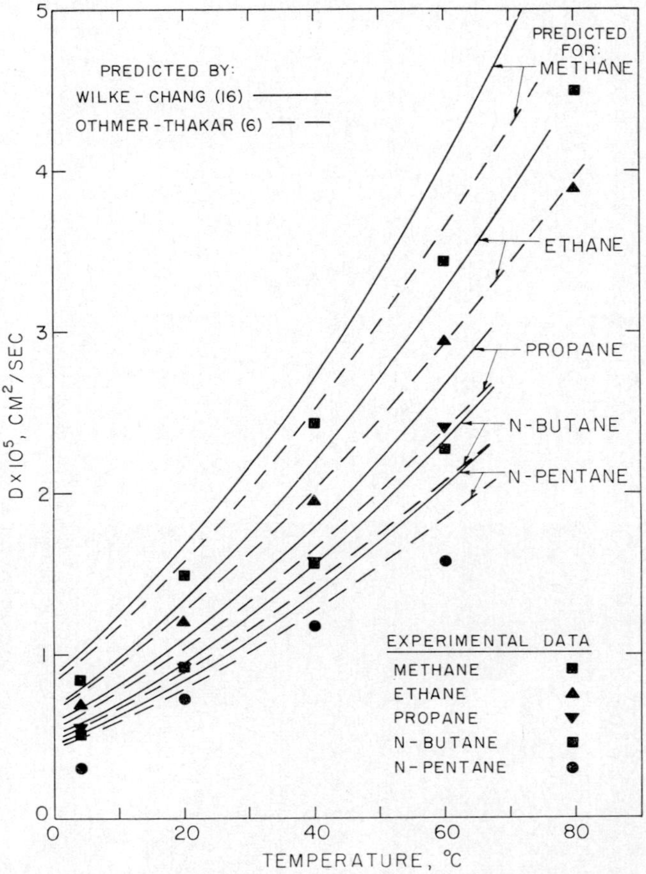

FIGURE 4. — Diffusion data versus predictions from empirical equations.

As an approach to the effect on diffusion of a different structured molecule, we chose neopentane, because it is a gas under normal conditions and has the most compact structure of all the pentanes. A comparison of the structures and boiling points of n-pentane, isopentane, and neopentane is given in Table 2.

Figure 5 shows a photograph of three-dimensional Stuart models of these three pentanes from which one can obtain a much better idea of the spatial relationships for the structures given in Table 2. It is apparent that

TABLE 2. — COMPARISON OF STRUCTURES AND NORMAL BOILING POINTS OF PENTANES

Name	Structure	Normal boiling point °C
n-pentane	$CH_3—CH_2—CH_2—$ $—CH_2—CH_3$	36·2
isopentane	$CH_3—CH_2—CH—CH_3$ CH_3	27·8
neopentane	CH_3 $CH_3—C—CH_3$ CH_3	9·5

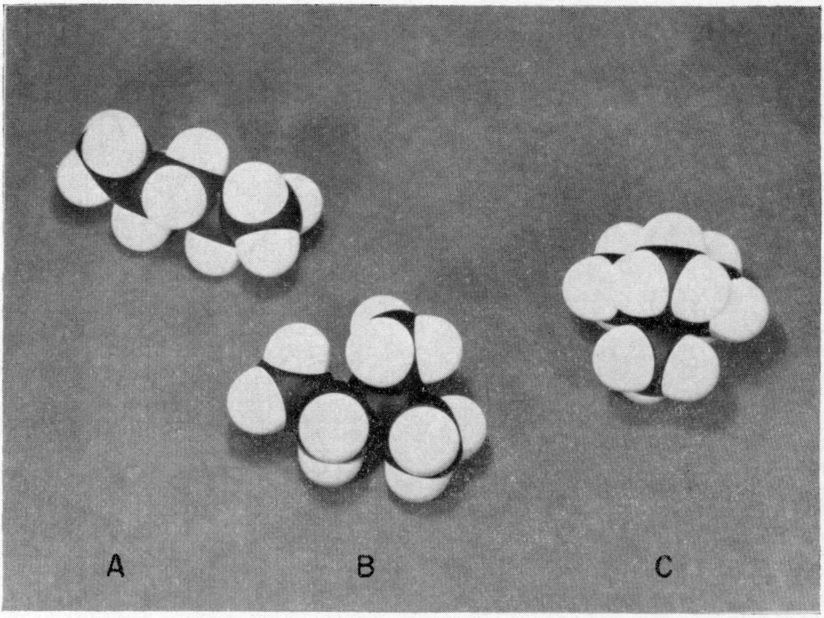

FIGURE 5. — Stewart models of pentane structures: A, n-pentane, B, isopentane, and C, neopentane.

as the degree of branching increases, the boiling point is lowered, and, of course, volatility is increased. Thus, on the basis of the results obtained for the normal paraffins, one would expect that the diffusion coefficient for neopentane should be higher than that of n-pentane. From the standpoint of the structural configurations shown in Fig. 5, one would also expect this result because neopentane is a more compact and more nearly a spherical molecule than n-pentane. It will be noted on Table 1 that diffusion coefficients for neopentane are consistently about 20% higher than the results obtained for n-pentane.

Summary

Diffusion coefficients have been measured for methane, ethane, propane, n-butane, n-pentane, and neopentane in pure water at temperatures ranging from 2° to 80° C. A previously reported linear relationship between log D and caıbon number for paraffin hydrocarbons has been found to hold over this range in temperature. Diffusion coefficients were found to be approximately doubled for every 15-degree increase in temperature. A number of theoretical and empirical methods of predicting diffusion coefficients were examined in the light of these results, but none was found to be entirely satisfactory.

Acknowledgments

This work was made possible through a grant from the National Science Foundation.

Nomenclature

- D diffusion coefficient, cm^2/sec;
- D_n diffusion coefficient of a given paraffin hydrocarbon at a given temperature, cm^2/sec;
- D_{n+1} diffusion coefficient of homologous paraffin hydrocarbon of next higher molecular weight, cm^2/sec;
- D_t diffusion coefficient of a given paraffin hydrocarbon at temperature of interest, cm^2/sec;
- D_4 diffusion coefficient of a given paraffin hydrocarbon at 4° C, cm^2/sec;
- k Boltzmann constant, erg/° C;
- M_B molecular weight of solvent;

r_A radius of dispersed sphere, cm;
T absolute temperature, °K;
t temperature, °C;
V_A liquid molal volume of solute at its normal boiling point, cm^3/g-mol;
V_B liquid molal volume of solvent at its normal boiling point, cm^3/g-mol;
η_B viscosity of solvent, cp;
ϕ_B association parameter of solvent.

References

1. P. L. Antonov, The diffusion permeability of some argillaceous rocks. *Geokhim. Metody Poiskov Neft. i Gaz. Mestorozhd.* An SSSR, Tr. Soveshch, Moscow, 1958.
2. S. Glasstone, K. T. Laidler, and H. Eyring, *The Theory of Rate Processes*, ch. 9, McGraw-Hill, New York, 1941.
3. K. E. Gubbins, K. K. Bhatia, and R. D. Walker, Diffusion of gases in electrolytic solutions, *A.I.Ch.E. J.* **12,** 548–552, 1966.
4. A. A. Kartsev, Z. A. Tabasaranskii, M. I. Subotta, and G. A. Mogilevskii, *Geochemical Methods of Prospecting and Exploration for Petroleum and Natural Gas*, p. 69. Univ. California Press, 1959.
5. D. R. Olander, Mutual diffusion in dilute binary systems, *A.I.Ch.E. J.* **9,** 207–210, 1963.
6. D. E. Othmer, and M. S. Thakar, Correlating diffusion coefficients in liquids, *Ind. Eng. Chem.* **45,** 589–593, 1953.
7. D. N. Saraf, Diffusion of ethane, propane and butane in water, *MS.* Thesis, Univ. California, Berkeley, 1963.
8. D. N. Saraf, P. A. Witherspoon, and L. H. Cohen, Diffusion coefficients of hydrocarbons in water, *Science*, **142,** 955–956, 1963.
9. E. G. Scheibel, Liquid diffusivities, *Ind. Eng. Chem.* **46,** 2007–2008, 1954.
10. G. Tamman, and V. Jessen, Über die Diffusionskoeffizienten von Gasen in Wasser und ihre Temperaturabhängigkeit, *Z. Anorg, Allgem. Chem.* **179,** 125–144, 1929.
11. J. E. Vivian, and C. J. King, Diffusivities of slightly soluble gases in water, *A.I.Ch.E. J.* **10,** 220–221, 1964.
12. J. H. Wang, Self-diffusion and structure of liquid water. I. Measurement of self-diffusion of liquid water with deuterium as tracer. *J. Amer. Chem. Soc.* **73,** 510–513, 1951.
13. J. H. Wang, Tracer-diffusion in liquids. I. Diffusion of tracer amount of sodium ion in aqueous potassium chloride solution, *J. Amer. Chem. Soc.* **74,** 1182–1186, 1952.
14. J. H. Wang, C. V. Robinson, and I. S. Edelman, Self-diffusion and structure of liquid water. III. Measurement of the self-diffusion of liquid water with H^2, H^3, and O^{18} as tracers, *J. Amer. Chem. Soc.* **75,** 466–470, 1953.
15. J. H. Wang, C. B. Anfinsen, and F. M. Polestra, The self-diffusion coefficients of water and ovalbumin in aqueous ovalbumin solutions at 10°, *J. Amer. Chem. Soc.* **76,** 4763–4765, 1954.
16. C. R. Wilke, and P. Chang, Correlation of diffusion coefficients in dilute solutions, *A.I.Ch.E. J.* **1,** 264–270, 1955.
17. P. A. Witherspoon, and D. N. Saraf, Diffusion of methane, ethane, propane and n-butane in water from 25° to 43°, *J. Phys. Chem.* **69,** 3752–3755, 1965.

ORIGIN OF TRACE METAL ENRICHMENT IN BITUMINOUS SHALES

by
MARGARETHA BRONGERSMA-SANDERS
Geologisch en Mineralogisch Instituut, Leiden, Holland

IN bituminous shales there is a marked enrichment of certain trace metals. The chief metals concerned are V, Co, Ni, Cu, Zn, Mo, Ag, Pb, U. The concentrations (absolute concentrations as well as ratios between different metals) are variable, but again and again this same assemblage is enriched.

Factors influencing the metal content of a sediment are:

(a) supply of metals to bottom layer of the sea,
(b) "precipitation" of metals in bottom layer (removal mechanisms),
(c) dispersal of metals by sediment, a high rate of sedimentation counteracting the enrichment of a metal.
(Freely adapted after Manheim, 1961.)

In the literature most attention has been paid to the removal mechanisms, the authors often losing sight of the fact that besides a removal mechanism, there must also be a more or less continuous supply.

Bituminous shales are deposited in places where the bottom water is devoid of oxygen, and contains H_2S, whether intermittently or permanently. Removal mechanisms in such waters are (Krauskopf, 1956):

(1) precipitation of metal sulphides,
(2) co-precipitation,
(3) precipitation of insoluble compounds, due to change in valence ($U^{6+} \rightarrow U^{4+}$),
(4) adsorption,
(5) biological processes.

Krauskopf showed that only part of the trace metals characteristic of bituminous shales are precipitated by one of the mechanisms (1)–(4);

V, Co, Ni, and Mo were stated not to precipitate. Afterwards Mo had to be struck from this list, since it appeared that the concentration of this metal in sea water is higher than had been accepted previously; moreover Mo co-precipitates with iron sulphide. Wedepohl (1964) calculated that the concentration of V, Ni, and Co in sea water should be at least a factor ten higher to be precipitated in sea water containing H_2S (concentration of H_2S 10^{-5} to 10^{-4} mol/l., i.e. the concentration in the bottom water of stagnating Norwegian fjords); the concentration for nickel in sea water is given as 0·5 mg/m^3, to be precipitated this concentration should be at least 3 mg/m^3.

For lack of an alternative possibility Krauskopf suggests that V, Ni, and Co are removed from the water by a biological process, whereas the other metals are removed by one of the mechanisms (1)–(4). This explanation is not satisfactory. We may consider that the biological process is indeed important, but not so much by furnishing a new removal mechanism as by influencing the metal supply: in some places of the sea the biological process leads to an increase in the concentration of certain trace metals in the lower water layers.

Processes affecting the concentrations of trace metals in the oceans are (Schutz and Turekian, 1965):

(a) supply from the hinterland or undergound,

(b) internal modification of concentration, i.e. shifting of metals from one part of the oceans to another,

(c) removal.

The supply from the hinterland or underground is not important for the case in question, since bituminous shales developed in different parts of the world with varying hinterland and underground. However, the internal modification of concentration may be important indeed. How does this modification come about?

A number of trace metals are concentrated by the living plankton in the surface water (the metal assemblage reminding one of that enriched in bituminous shales). After death the plankton bodies sink to deeper water layers, and this is followed by their decay, and the return to inorganic form and redissolution of the metals. The sinking of the dead bodies provides a quick transport from surface to bottom, and this may lead to an increase in concentration with depth. On the other hand the mixing of different water layers tends to smooth out the disparities in concentration (Schutz and Turekian, 1965). The change in concentration with depth is a function of:

(1) number and volume of concentrating organisms,

(2) concentration factor $\left(= \frac{\text{content of metal in organism}}{\text{content of metal in sea water}} \right)$,

(3) mixing rates in the sea.

If mixing times are short compared to the rate of removal from the surface layers and input into deeper water layers, the change along the vertical will be negligible. The rate of input or removal depends on the one hand on the concentration factor of a metal in the prevailing plankton, and on the other hand on the mass of the concentrating organisms. The concentration factors of the various metals differ widely, and vary from one organism to the other. Therefore, the nature of the prevailing plankton (or of prevailing micro-organisms) is important for the case in question, but space limitations prohibit consideration of this aspect of the problem.

The concentration factors of the metals are low as compared with that of phosphorus, which is a major organic constituent. A strong increase in concentration with depth is a general phenomenon in the case of phosphorus, but the trace metals generally do not show such an increase. A marked increase of the metals occurs only in places of highest organic productivity, where the number of concentrating organisms is high, e.g. in areas of upwelling water (Schutz and Turekian, 1965, p. 300). The change will be particularly great if high productivity is combined with a low mixing rate of the bottom water.

Besides contributing to an increase with depth the biological process may lead to an increase in the concentration of the whole water mass. Great changes may occur if the surface and subsurface layers flow in opposite directions. Redfield, Ketchum and Richards (1963) showed that such a countercurrent system may act as a trap wherein the trace elements, concentrated by the living plankton, accumulate. In this way there is a shifting of trace metals from one part of the oceans to the other, and geographical differences develop.

In the open ocean a countercurrent system is to be met with in regions of upwelling water; subsurface water from a depth of about 150–200 m flows in a nearly horizontal direction towards the coast, gradually ascending to the surface; near the coast the water overturns and flows along the surface in an offshore direction. Approaching the coast the subsurface water becomes more and more enriched; where this water reaches the well-lighted upper layers an enormous plankton production takes place.

Countercurrent systems are not confined to the open ocean; they also occur in semi-separated parts of the sea, notably in two-layered "estuaries" Estuaries are called positive when the evaporation is less than the fresh-water supply, and negative or inverse when the evaporation exceeds this supply. In positive estuaries of humid regions a wedge of sea water often underlies the outflowing river water. As the river water flows seaward it drags along some of the underlying sea water; to replenish the water that is dragged along a countercurrent of sea water sets in. This circulation leads to an accumulation of the trace elements, concentrated by plankton, in the inner part of the estuary.

In negative or inverse estuaries of dry regions an anti-estuarine circulation may develop: sea water flows in along the surface to replace water lost by evaporation, while water concentrated by the evaporation will sink and flow out along the bottom. This type is well exemplified in the Mediterranean Sea. The anti-estuarine circulation does not lead to an enrichment, but to an impoverishment of the basin, because the metals accumulate in the direction wherein the subsurface current is flowing. We may, however, stress the fact that in the dry sub-tropics there is still another possibility. If by winds of constant direction, such as the trade winds, the surface water is blown out of the basin, a circulation of an estuarine-like type may develop, again leading to an enrichment of trace metals in the inner part (Brongersma, 1966a, b). This wind-driven estuarine-like circulation is closely related to upwelling in the open ocean, the driving force in both cases being the trade winds.

Data on the geographical and depth distribution of Ag, Ni, and Co were given by Schutz and Turekian in 1965. These metals show relatively high concentrations in areas of high organic productivity (upwelling water) accompanied by an increase of concentration with depth. Of the five areas where high silver contents were found, only one was not associated with upwelling. The oceanic range for nickel found by these authors was 0·43 – 46·2 mg/m^3; the latter value occurred in the bottom water of the southeast Atlantic.

Data on the contents of Mo were given by Sugawara *et al.* (1960, 1962). Studying the vertical distribution of this metal in the nortwest Pacific these authors found a slight, but not definite tendency to increase with depth (1960, p. 96). Data on the contents of this metal and of vanadium in regions of upwelling are unknown to us. A large increase with depth and in the whole water mass possibly occurs only in waters where the concentration factor for these metals in the prevailing plankton (or microorganisms) is high; but this aspect of the problem will not be considered here.

The enrichment in the water does not necessarily imply an enrichment of the underlying sediment. For the latter there must be in addition a removal mechanism, the presence of a removal mechanism being a sine qua non. Now, the countercurrent systems lead, on the one hand, to trapping of trace metals and high productivity in surface waters, but on the other hand to a marked depletion of oxygen in the waters below, and occasionally even to complete de-oxygenation and development of H_2S in the bottom water. The depletion of oxygen is brought about by the decay of great masses of dead plankton sinking down from the fertile surface waters. Moreover, in the two-layered systems there is a sharp increase in density between surface and subsurface layers, limiting renewal of oxygen. In the open ocean complete de-oxygenation and the appearance of H_2S are rare and of short duration. In highly productive semi-separated parts of the sea these conditions may be of longer duration.

When H_2S appears in the free bottom water, a number of trace metals will be removed; even metals with a comparatively high solubility of the sulphide like Ni will be precipitated, since the concentration of this metal may be far above the required minimal value of 3 mg/m^3. In this way bituminous sediments with high metal contents may develop.

As high rate of sedimentation counteracts the enrichment of a metal in a sediment, in the deep sea environment of the open ocean (where anoxic conditions do not prevail, but where other removal mechanisms exist, e.g. the formation of manganese nodules) rather high metal contents may develop, even if the supply of metals per time unit is moderate. In the coastal area where the rate of sedimentation is much higher, the trace metal contents normally are low, but high contents may occur if the metal content of the overlying water is markedly enriched by organic interaction.

References

Brongersma-Sanders, Margaretha (1966a) The fertility of the sea and its bearing on the origin of oil. *Advancement of Science*, **23** (107), 41–46. (Paper presented on Sept. 3, 1965 at Cambridge meeting of Brit. Ass. Adv. Sci.)

Brongersma-Sanders, Margaretha (1966b) Metals of Kupferschiefer supplied by normal sea water. *Geol. Rundschau*, **55**, 365–375.

Krauskopf, K. B. (1956) Factors controlling the concentrations of thirteen rare metals in sea water. *Geochim. Cosmochim. Acta*, **9**, 1–32.

Manheim, F. T. (1961) A geochemical profile in the Baltic Sea. *Geochim. Cosmochim. Acta*, **25**, 52–70.

Redfield, A. C., Ketchum, B. H. and Richards, F. A. (1963) The influence of organisms on the composition of sea water. In *The Sea* (ed. M. N. Hill.), vol. II, pp. 26–77. New York/London, J. Wiley & Sons.

Schutz, D. F. and Turekian, K. K. (1965) The investigation of the geographical and vertical distribution of several trace elements in sea water using neutron activation analysis. *Geochim. Cosmochim. Acta*, **29**, 259–313.

Sugawara, K. and Okabe, S. (1960) Geochemistry of molybdenum in natural waters. *J. Earth Sci., Nagoya Univ.*, **8**, 93–107.

Sugawara, K., Terada, K., Kanamori, S., Kanamori, N. and Okabe, S. (1962) Different distribution of calcium, strontium, iodine, arsenic, and molybdenum in the northwestern Pacific, Indian and Antarctic Oceans. *J. Earth Sci., Nagoya Univ.*, **10**, 34–50.

Wedepohl, K. H. (1964) Untersuchungen am Kupferschiefer in Nordwestdeutschland. *Geochim. Cosmochim. Acta*, **28**, 305–364.

NITROGENOUS CONSTITUENTS OF DEEP-SEA SEDIMENTS

by

F. J. STEVENSON and S. N. TILO
Department of Agronomy, University of Illinois, Urbana, Illinois, U.S.A.

Introduction

Very little is known regarding the distribution of the forms of combined nitrogen in deep-sea sediments. Amino acids (Degens *et al.*, 1961, 1963, 1964; Emery *et al.*, 1964; Erdman *et al.*, 1956; Rittenberg *et al.*, 1963), amino sugars (Degens *et al.*, 1964; Rittenberg *et al.*, 1963), and purine and pyrimidine bases (Rosenberg, 1964) have been detected in hydrolysates of marine basin sediments using chromatographic techniques, but accurate quantitative data are lacking. In most studies, only a small fraction of the nitrogen has been accounted for in known compounds. Rittenberg *et al.* (1963), for example, recovered only 0·5 to 17% of the nitrogen in the experimental Mohole sediments as amino acids and concluded that—"the elucidation of the chemical nature of the non-amino acid organic nitrogen, which is the dominant one at all depths, is of great significance and constitutes a major geochemical problem." Emery *et al.* (1964) found that, in the San Diego Trough, the percentage of the nitrogen as amino acids decreased from 88% near the surface sediment to only 1% at 350–360 cm; for the Santa Barbara Basin the ratio was nearly constant at 12 to 16%. In an earlier study on amino acids in sediments of the Santa Barbara Basin, Degens *et al.* (1961) reported that not more than 5% of the total nitrogen occurred in these compounds.

Failure to account for a major fraction of the nitrogen in deep-sea sediments may be due to one, or both, of the following factors. First, yields of nitrogenous biochemicals may be low because of incomplete extraction. Sellers (1966) found that large quantities of amino compounds were released from some Miocene rocks by heating at elevated temperatures under aqueous conditions, and concluded that a hydrothermal treatment might reveal the presence of amino acids which are not released by acid hydrolysis. Second, some of the nitrogen which is regarded as "organic" may exist in an inorganic form, namely, as NH_4^+ held within the lattice structures of clay minerals. Terrestrial soils (Bremner, 1959, 1965a; Dhariwal and Stevenson, 1958; Hinman, 1964; Rodrigues, 1954; Schachtschabel, 1960; Stevenson, 1959; Stevenson and Dhariwal, 1959; Walsh and Murdock, 1960; Young, 1962) and sedimentary rocks (Stevenson, 1962) contain significant amounts of lattice-bound NH_4^+, and marine sediments should also contain nitrogen in this form, particularly those that are rich in clay minerals (montmorillonite, illite) capable of retaining NH_4^+ within their structures. In soil science terminology, the NH_4^+ irreversibly bound to clay minerals is referred to as "fixed NH_4^+," and this convention will be employed herein.

A variety of methods have been developed for determining the various forms of nitrogen in terrestrial soil, including the estimation of fixed NH_4^+ (Bremner, 1959, 1965a; Dhariwal and Stevenson, 1958; Rodrigues, 1954; Schachtschabel, 1960), amino acids (Bremner, 1950, 1965b; Ferguson and Sowden, 1966; Lynch *et al.*, 1959; Stevenson, 1954, 1956a, 1956b, 1965a; Young and Mortensen, 1958), and amino sugars (Bremner, 1958, 1965b; Sowden, 1959; Stevenson, 1957, 1965b). Unfortunately, most of these methods cannot be applied directly to deep-sea sediments. The concentration of nitrogen compounds in deep oceanic basin sediments is several orders of magnitude lower than that found in terrestrial soil, and, in addition, the amount of material available for experimental work is limited. Samples from oceanic drillings are often available in gram quantities only.

The present paper deals mainly with the distribution of the forms of nitrogen in the experimental Mohole sediments. In addition, microanalytical techniques are described for the determination of fixed NH_4^+-N, amino acid-N, and amino sugar-N.

Descriptions of samples

The sediment samples were from the experimental Mohole drillings of 1961 off Guadalupe Island, Mexico (Riedel *et al.*, 1961; Rittenberg *et al.*, 1963). Eight samples from hole EM 8 were provided for the study (by W. R. Riedel acting for the Mohde organization). They were stored at about 2 to 4° C until used for analysis.

According to Riedel *et al.* (1961) the entire sedimentary section of the Guadalupe site is late Tertiary in age, with the topmost sample possibly Pliocene and those directly above the basalt layer late (or middle) Miocene. The sedimentation rate was calculated by Dymond (1966) to be 1·1 cm per 1000 years. As shown in Fig. 1, the samples examined in this study ranged in ages from about 3·2 to 12·3 million years.

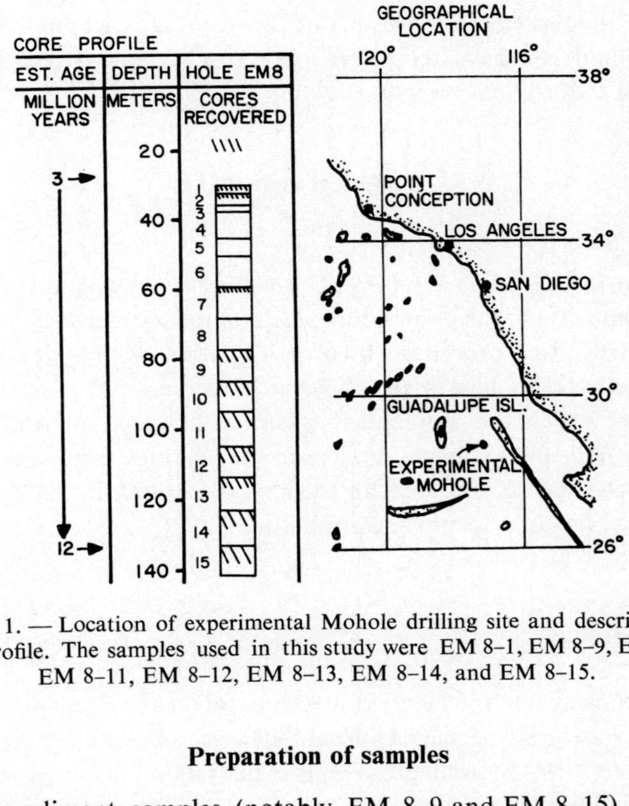

FIGURE 1. — Location of experimental Mohole drilling site and description of core profile. The samples used in this study were EM 8–1, EM 8–9, EM 8–10, EM 8–11, EM 8–12, EM 8–13, EM 8–14, and EM 8–15.

Preparation of samples

Several sediment samples (notably EM 8–9 and EM 8–15) contained high amounts of carbonates. The removal of these constituents would greatly simplify the extraction and subsequent characterization of nitrogen compounds. Preliminary studies on the amounts of nitrogen extracted from the samples with acidified 1 N KCl (pH 1·0) confirmed the finding of Rittenberg et al. (1963) that the sediments contained negligible quantities of soluble nitrogen coumpounds; therefore, the samples were treated with acidified KCl prior to further analysis. For this purpose, the moist sample was dispersed in a 1 N KCl solution (10 ml/g moist sediment) following which the pH of the mixture was adjusted to 1·0 by addition of 1 N or 6 N HCl, depending upon the carbonate content. The sample was then filtered on a Büchner funnel and washed first with copious amounts of acidified 1 N KCl (pH 1·0) and then with a small quantity of 0·01 N HCl (to remove exchangeable K). Finally, the residue was washed with reagent grade methanol and dried in a vacuum desiccator. Since only negligible quantities of nitrogen could be detected in acidified KCl extracts of the sediments, the results obtained on the carbonate-free preparations could be translated directly in terms of the composition of the nitrogen in the original samples.

Analytical methods

Total Nitrogen

Nitrogen in the original sediments and in the carbonate-free samples was determined by the sealed-tube technique described by Stevenson (1960). Briefly, this procedure involves digestion of the specimen with concentrated H_2SO_4 in a sealed tube at a temperature of 420° C for 2 hours, after which the ammonia is distilled in a microKjeldahl steam distillation unit and estimated titrimetrically, or colorimetrically, depending upon the nitrogen content of the sample. A 200- to 300-mg sample of sediment was used for this determination.

Fixed NH_4^+*-N*

(a) PRINCIPLES. Of the several methods that have been proposed for determining naturally occurring fixed NH_4^+ in soil, only the procedure of Bremner (1959) showed promise for analysis of deep-sea sediments. Other procedures were elminated from consideration for one or more of the following reasons: (1) the sample was subjected to drastic chemical treatment prior to extraction of fixed NH_4^+, (2) high concentrations of HF were used for extraction, and (3) distillation of NH_4^+ was performed under strongly alkaline conditions. All of these procedures can bring about the deamination of nitrogenous organic compounds, thereby giving high values for fixed NH_4^+. For example, in the method of Rodrigues

(1954), the soil is treated with a 4 : 1 (v/v) mixture of 40% HF and 50% H_2SO_4, following which the NH_4^+ is recovered by distillation with alkali after removal of HF by heating with H_2SO_4. Bremner (1959) found that this procedure led to extensive decomposition of organic nitrogen compounds to NH_4^+. Several attempts (Bremner, 1965a; Dhariwal and Stevenson, 1958) have been made to remove labile compounds from soil prior to HF extraction by treating the sample with hot KOH solution. However, the ability of this hot KOH treatment to remove organic nitrogen compounds which are intimately bound to clay minerals (perhaps within their lattice structures) is suspect. The desirability of using mild extraction and distillation procedures for determining fixed NH_4^+ in marine sediments was emphasized by preliminary trials which indicated that a significant fraction of the organic nitrogen is so intimately associated with mineral matter that it cannot be removed by hydrolysis with 6 N HCl; on the other hand, an appreciable amount of this organic nitrogen was released when the clay was destroyed by HF.

In Bremner's (1959) method of determining fixed NH_4^+, a 5 g sample of finely ground soil (60 mesh) is shaken with 100 ml of a N-HF : N-HCl solution for 24 hours, after which the NH_4^+ is recovered by distillation with MgO at 25° C in a microdiffusion unit, or by steam distillation with pH 8·8 borate buffer in a microKjeldahl distillation assembly. Amino acids, amino sugars, and some proteins were found to be stable in the HF mixture, as well as under the condition employed for distillation of NH_4^+. Stevenson *et al.* (1967) found that this extraction procedure gave low recoveries of fixed NH_4^+ from clayey subsoils; this problem was solved by increasing the concentration of HF in the extractant at the expense of HCl. The acid mixture they adopted for recovering fixed NH_4^+ was a 2·5 N-HF : 0·1N-HCl solution, using a solution/soil ratio of 20. The total concentration of acid in this mixture is only slightly greater than that employed in Bremner's method, and, despite the higher concentration of HF, the risk of degrading organic forms of nitrogen to NH_4^+ was no greater. The following method, adapted from Stevenson *et al.* (1967), was found to give satisfactory values for fixed NH_4^+ in marine sediments:

(b) PROCEDURE. A 125-mg sample of sediment material previously ground in a Spex mixer mill† so that 80% passed through a 300 mesh sieve is added to a 50-ml polyethylene centrifuge tube, 5 ml of 2·5N-HF : 0·1N-

† Spex Industries, Inc., Scotch Plains, New Jersey.

HCl solution is added, and the sample is shaken for a 16-hour period. Several drops of methyl red indicator solution are then added, and the sample is titrated to a red color (pH of about 6·0) by cautious addition of 5 N NaOH. This neutralization procedure is necessary because HF is a volatile reagent and it interferes with the determination of NH_4^+ by distillation procedures. The sample is stirred vigorously after each addition of base in order to minimize decomposition of organic nitrogen compounds in localized areas of high NaOH concentration. Finally, the mixture is swirled gently for several minutes to complete the adsorption of any volatile HF in the tube, after which the sample is transferred quantitatively to a 100 ml microKjeldahl distillation flask. The transfer can best be accomplished using a long-stem medicine dropper. Several drops of phenolphthalein indicator solution are then added, and the sample is titrated to a pink color (pH of about 9) by step-wise addition of 1 N NaOH. The flask is swirled gently during this operation, and, when the addition of base is completed, 2 g of pH 8·8 buffer, prepared by grinding 8 g boric acid with about 50 g sodium borate, is added. The flask is attached immediately to a microKjeldahl distillation assembly of the type described by Bremner (1965b), and the NH_4^+ is recovered and analyzed colorimetrically with Nessler's reagent.

For the direct analysis of calcareous sediments, carbonates should be destroyed prior to extraction with HF. In this case, the sample is suspended in 5 ml of 1 N KCl, following which 1 N HCl is added a drop at a time until effervescence ceases. The residue from the acid treatment is recovered by centrifugation and extracted with HF as described above.

The sediments examined in this study had been treated with acidified KCl prior to analysis; therefore, any exchangeable NH_4^+ would have been removed. The most convenient method for estimating exchangeable NH_4^+ in deep-sea sediments is by analysis of KCl extracts by the steam distillation technique described above.

Amino Acids

(a) PRINCIPLES. Conventional methods for the determination of amino acids are based on the classical ninhydrin reaction. The ninhydrin-CO_2 method has been used extensively for determining amino acids in soil hydrolysates (Bremner, 1949; Stevenson, 1954, 1956b; Young and Mortensen, 1958), and, more recently, the ninhydrin-NH_3 method has been applied (Bremner, 1965b; Ferguson and Sowden, 1966; Keeney and Bremner, 1964). These methods, unless suitably modified, would appear to be unsuitable for analysis of deep-sea sediments, because of the low quantities of amino acids normally present.

Several investigators (Degens *et al.*, 1961, 1963, 1964; Emery *et al.*, 1964; Erdman *et al.*, 1956; Rittenberg *et al.*, 1963), have estimated amino acids in marine sediments by paper partition chromatography, using ninhydrin as the spray reagent. However, this is a tedious procedure which yields results of questionable accuracy.

Two methods, both based upon the ninhydrin color test, were used to determine amino acid-N in hydrolysates of the experimental Mohole sediments. In the first method, the hydrolysate was desalted prior to the colorimetric analysis by displacement from an ion exchange resin; in the second method, the hydrolysate was analyzed directly after removal of ammonia and amino sugars.

(b) EXTRACTION. Amino acids in sediments occur largely in combined forms. They exist in stable linkages with humic substances, such as humic acids, and they are bound to clay minerals. There is no evidence for the occurrence of intact proteins in sediments.

The conventional method of recovering amino acids from sediments is by heating with 6 N HCl (reflux conditions). Experience obtained in this laboratory has shown that this procedure does not give quantitative release of organic nitrogen compounds. In brief, it was found that considerably less nitrogen was solubilized by hydrolysis with 6 N HCl than by extraction with a 2·5N-HF : 0·1N-HCl solution at room temperature. Figure 2 shows that only from 32 to 50% of the nitrogen in three experimental Mohole samples was dissolved by hydrolysis with 6 N HCl for 16 hours, whereas from 65 to 90% was extracted with 2·5N-HF : 0·1N-HCl at room temperature. Slightly higher recoveries of nitrogen were obtained when the HF-extracted residues were hydrolyzed with 6 N HCl. Further studies showed that an HF pretreatment enhanced the release of both amino acids and amino sugars by acid hydrolysis. For example, the amounts of amino acid-N recovered by direct hydrolysis of samples EM 8–9, EM 8–13, EM 8–14, and EM 8–15 were 67·0, 17·7, 10·0, and 55·9%, respectively, of the values obtained by hydrolysis of the HF-treated samples.

In view of the results obtained above, an HF pretreatment has been incorporated in the hydrolysis procedure. Amino acids, and to a lesser extent amino sugars, are stable towards heating with dilute solutions of HF; nevertheless, HF is removed by freeze-drying prior to hydrolysis with 6 N HCl. This freeze-drying technique served as a convenient method for removing HF from the experimental sample.

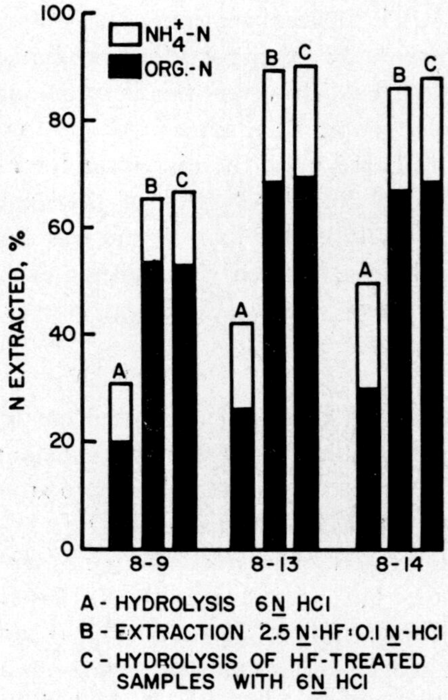

FIGURE 2. — Organic nitrogen and NH_4^+-N recovered from some experimental Mohole sediments by three extraction methods. A, Hydrolysis with 6 N HCl B, Extraction with 2·5N-HF: 0·1N-HCl. C, Hydrolysis of HF-treated samples with 6 N HCl.

The reason for the low recoveries of organic nitrogen compounds by direct acid hydrolysis of sediments is unknown. It is suspected that the treatment with HF resulted in the liberation of amino acids and other nitrogen compounds which occurred at inaccessible sites of clay minerals. Research on the sorption of organic compounds by minerals indicates that sediments may contain low-molecular-weight substances (such as amino acids) that are held within the expanded layer lattices of certain clays. An alternate explanation is that the HF enhanced the release of nitrogen constituents from humic acids and other heteropolycondensates.

(c) COLORIMETRIC ESTIMATION OF AMINO ACIDS IN DESALTED HYDROLYSATES. A 1-g sample of sediment is placed into a 50-ml polyethylene centrifuge tube, 10 ml of 5N-HF : 0·1N-HCl is added, and the sample is shaken on a reciprocal shaker for 16 hours. The HF and HCl are then removed by freeze-drying, following which the residue is heated under reflux with 10 ml of 6 N HCl for 16 hours. Any insoluble material is removed by filtration, and the resulting solution is evaporated to near-dryness *in vacuo* at 40° C. This residue is dissolved in a minimum amount of water and added to a 1×5 cm column of Dowex-50 (200 to 400 mesh, H-form). The column is leached first with deionized water until the effluent is free of chlorides and then with 0·1 M piperidine solution (Buchanan, 1957). The first 50 ml of effluent is discarded, and the next 200 ml, containing the amino acids, is reduced to a volume of about 5 ml by evaporation at 40° C *in vacuo*. If a precipitate forms (from peptized Al or Si) it is removed by centrifugation, dissolved in dilute acid, and transferred to a second column of Dowex 50. Any amino acids coprecipitated with mineral matter are subsequently recovered by elution with 0·1 M piperidine in the manner described above. This effluent is also reduced to a volume of about 5 ml and combined with that from the initial resin treatment.

Before the preparations obtained above can be analyzed for amino acids any amino sugars have to be destroyed. For this purpose, 0·5 ml of 1 N NaOH is added, and the sample is heated in a boiling water bath for 20 to 30 minutes with a stream of air blowing over the surface of the liquid. The sample is then acidified with 1 N HCl, following which the volume of the solution is adjusted to a volume of 10 ml. Aliquots of this solution are then analyzed for amino acids by the colorimetric method of Troll and Cannan (1953).

(d) DIRECT COLORIMETRIC ANALYSIS OF ACID HYDROLYSATES. This procedure was devised from the observation (Cheng and Stevenson, 1967) that interference due to metal cations in hydrolysates of clayey marine sediments can be eliminated by performing the ninhydrin reaction in the presence of sodium citrate, a strong chelating agent. A complete description of the method is being prepared for publication elsewhere; consequently, only the essential details are given below.

A 125-mg sample of HF-treated sediment is hydrolyzed with 6 N HCl as described above. After removal of HCl, the solution is made alkaline and ammonia and amino sugars are removed. Sufficient 6 N HCl is then added to dissolve metallic hydroxides (mixture assumes a pale yellow color), following which aliquots are analyzed for amino acids by the method of Yemm and Cocking (1955), except that excess sodium citrate is applied to chelate metals. The standard curve used for the calculations was obtained by adding known amounts of amino acid-N to 125 mg montmorillonite clay and following through with the entire procedure.

Results obtained by the two colorimetric methods were in excellent agreement. The data reported for amino acid-N represent average values secured on three separate hydrolysates, two of which were analyzed without prior desalting.

Amino Sugars

Amino sugars in acid hydrolysates of the sediments were determined by the procedure described by Stevenson (1957). This method involves removal of metallic ions and dark-colored humin-like substances by treatment of the hydrolysate on an anionic resin (carbonate form), followed by colorimetric analysis of the desalted preparation. Because the supply of sediment material was limited, only a 1-g sample was used. The sample was treated with HF prior to hydrolysis with 6 N HCl (see section on *Amino Acids)*, and the acidified effluent from the anionic resin was reduced to a volume of 10 ml by evaporation of 40° C *in vacuo* prior to analysis of 1-ml aliquots for amino sugars.

Two methods were used for the colorimetric estimation of amino sugars, namely, the classical Elson and Morgan (1933) method and the newer indole-HCl procedure of Dische and Borenfreund (1950). Identical results were obtained by both procedures.

Nonhydrolyzable-N

Not all of the organic nitrogen in the experimental Mohole sediments was solubilized by the extraction procedure used for recovering amino acids and amino sugars (hydrolysis of HF-treated samples with 6 N HCl). This nitrogen, hereafter designated as "nonhydrolyzable-N," was determined by two methods, namely, from an estimation of total nitrogen in hydrolysates from one of the amino acid experiments (in which case nonhydrolyzable-N was taken as the difference between nitrogen in the sediment and that recovered in the hydrolysate) and by direct analysis of the hydrolyzed residues (two sets available). Table 1 shows that, except for samples EM 8–1 and EM 8–10, there was excellent agreement in the

TABLE 1. — NONHYDROLYZABLE-N IN EXPERIMENTAL MOHOLE SEDIMENTS AS ESTIMATED BY TWO METHODS
(Results given in $\mu g/g$, carbonate-free samples)

Sample	By difference	Analyses of hydrolyzed residues		
		I	II	Average
EM 8–1	36	123	121	122
EM 8–9	214	195	200	198
EM 8–10	143	73	78	76
EM 8–11	124	95	91	93
EM 8–12	108	117	117	117
EM 8–13	53	59	60	60
EM 8–14	31	43	35	39
EM 8–15	175	210	219	215

values obtained by the two methods. The low value obtained for EM 8–1 by analysis of the hydrolysate was probably due to NH_3 contamination, because the preparation used for this determination was found to contain considerable amounts of NH_3 which could not be accounted for as being derived from fixed NH_4^+. The direct method was used to calculate nonhydrolyzable-N (last column of Table 1).

Unknown Acid-soluble Organic Nitrogen Compounds (Undetermined-N)

The organic nitrogen which occurred in unknown acid-soluble compounds (undetermined-N) can be defined as that nitrogen which was not recovered in the other forms, specifically, as fixed NH_4^+, amino acids, amino sugars, or nonhydrolyzable-N. Thus, this estimation is subject to errors inherent in all of the above-mentioned determinations. Nevertheless, the values reported are believed to be reasonably accurate.

Results and discussion

Organic Nitrogen and Fixed NH_4^+ Contents

Total-, organic-, and fixed NH_4^+-N in the experimental Mohole sediments, together with other pertinent data, are presented in Table 2. A result of particular interest is that a significant fraction of the total nitrogen, ranging from 10·0 to 21·5 percent, occurred as fixed NH_4^+. The variation in the absolute amount of fixed NH_4^+-N (from 40 to 114 $\mu g/g$) can be

TABLE 2. — TOTAL-, ORGANIC-, AND FIXED NH_4^+-N IN EXPERIMENTAL MOHOLE SEDIMENTS

Sample	Depth	Water content	Ca^{++}- and Mg^{++}-carbonates*	Total N*	Organic N*	Fixed NH_4^+-N*	Fixed NH_4^+-N / Total N	Total N in carbonate-free sample†
	m	%	%	μg/g	μg/g	μg/g	%	μg/g
EM 8–1	28·3	97·0	—	598	493	105	17·6	795
EM 8–9	76·5	98·9	48·4	418	366	52	12·4	637
EM 8–10	85·6	118·8	—	511	406	105	20·6	496
EM 8–11	95·3	121·3	8·1	442	369	73	16·5	456
EM 8–12	104·5	82·3	27·5	336	269	67	19·9	612
EM 8–13	113·4	110·5	—	531	417	114	21·5	545
EM 8–14	122·4	140·5	1·6	333	268	65	19·5	279
EM 8–15	132·0	105·9	53·0	400	360	40	10·0	627

* Moisture-free basis. Organic nitrogen was taken as the difference between total N and fixed NH_4^+-N. Trace quantities of soluble nitrogen were ignored in making the calculations. The values for Ca^{++} -and Mg^{++}-carbonates were estimated from data reported by Rittenberg *et al.* (1963) on a separate section of each core. Samples EM 8-1,8-10, and 8-13 appeared to have low carbonate contents.

† Air-dried basis.

attributed, at least in part, to differences in carbonate levels, indicating that the clay-sized fraction of each sediment was somewhat similar in NH_4^+-fixing capacity. Along this line, it is appropriate to mention that kaolinitic clays and colloidal amorphorus (allophanic) materials do not retain NH_4^+ in non-exchangeable forms whereas vermiculitic and illitic clays have high fixing capacities; montmorillonitic clays fix very little NH_4^+. The fixed NH_4^+ content of terrestrial soils has been found to be related to the kinds and amounts of clay minerals they contain (Stevenson and Dhariwal, 1959).

The data for fixed NH_4^+ emphasize that the Kjeldahl method does not provide a true measure of organic nitrogen in deep-sea sediments, and that the C/N ratio should no longer be considered a characteristic of the organic matter. With regard to the latter point, the reviews of Waksman (1933) and Arrhenius (1950) show that the C/N ratio of marine sediments generally ranges from about 5 to 13, although values below 5 are not uncommon. Mohamed (1949), for example, found that the C/N ratios of some sediments from the northern Red Sea ranged from 2·0 to 8·2, with a mean of 5·2. The frequent observation that sediments low in organic matter generally have very low C/N ratios can be explained on the basis that the percentage of the nitrogen as fixed NH_4^+ will be highest where organic matter has been diluted with mineral matter (clay).

Emery *et al.* (1964) observed a drastic increase in the percentage of the Kjeldahl nitrogen occurring in "resistant decomposition products" with increasing depth in sediment cores from three California offshore basins. A plausible explanation for this result is that a greater fraction of the nitrogen in the deeper layers occurred as fixed NH_4^+.

Inspection of Table 2 shows that the organic nitrogen contents of the sediments varied considerably, with no consistent change being evident with increasing depth. The variability cannot be accounted for entirely by differences in Ca^{++}- and Mg^{++}- carbonate contents, which ranged from a negligible amount to about 50%. The organic nitrogen content of sample EM 8–14, for example, was considerably lower than samples EM 8–10 and EM–13, even though all three had low carbonate contents. It should be pointed out, however, that the acid-leached preparations from the sediments rich in carbonates generally had higher nitrogen contents than those from the sediments low in carbonates. The presence of excess Ca^{++} and Mg^{++} may have led to better preservation of nitrogenous constituents through the formation of insoluble salts and chelate complexes.

At least three reasons can be given for the unusually low organic nitrogen content of sample EM 8–14: (1) the detrital material in the runoff waters was unusually low in organic matter because of unsatisfactory conditions for biosynthesis reactions on the continent, (2) conditions in the surface sediment had been particularly favorable for mineralization of organic matter by microorganisms, and (3) the detrital material from the continent had been diluted to a considerable extent with volcanic ash.

The nitrogen contents of some of the carbonate-free preparations are not in complete agreement with values expected on the basis of their Ca^{++}- and Mg^{++}- carbonate levels. However, it should be mentioned that the data for carbonates were obtained on a separate section of each core; thus, the values given are only approximations. The weight of material recovered by acid treatment of sediments EM 8–9 and EM 8–15 exceeded that which was anticipated from published carbonate values (see footnote to Table 2).

Forms of Organic Nitrogen

(a) AMINO ACIDS AND AMINO SUGARS. Figures 3 and 4 give the distribution of amino acid-N and amino sugar-N in the experimental Mohole sediments, respectively. It is apparent that the percentage of the organic nitrogen which occurred in these compounds decreased progressively with increasing depth (age). Thus, approximately 28% of the organic

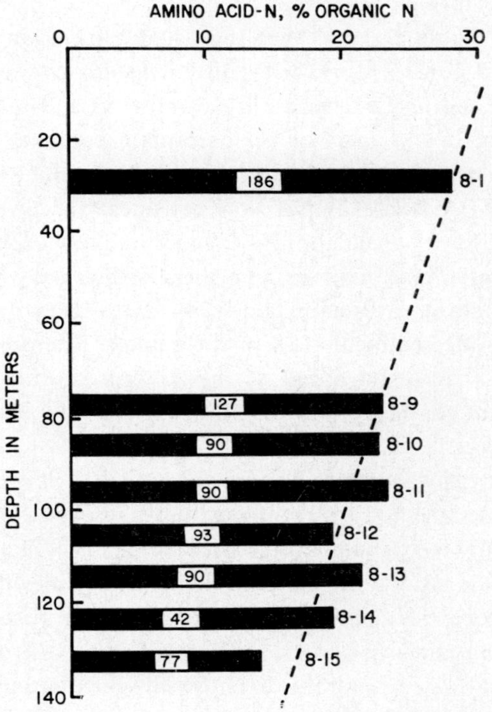

FIGURE 3. — Distribution of amino acid-N in experimental Mohole sediments. The values in the solid portion of the bars represent the amount of amino acid-N in the carbonate-free preparations (μg/g). The amounts of amino acid-N in the original sediments (dry weight basis) were: EM 8–1, 140 μg/g; EM 8–9, 83 μg/g; EM 8–10, 93 μg/g; EM 8–11, 87 μg/g; EM 8–12, 51 μg/g; EM 8–13, 88 μg/g; EM 8–14, 50 μg/g; EM 8–15, 49 μg/g.

nitrogen in sample EM 8–1 (about 3·2 million years old) existed in amino acids, while only 13·6% of that in sample EM 8–15 (about 12·3 million years old) existed in this form. The comparable figures for amino sugars are 6·9 and 1·2%, respectively.

The decrease with depth in the proportion of the organic nitrogen which occurred as amino acids was more gradual than was the case with amino sugars. Consequently, the ratio of amino sugar-N to amino acid-N de-

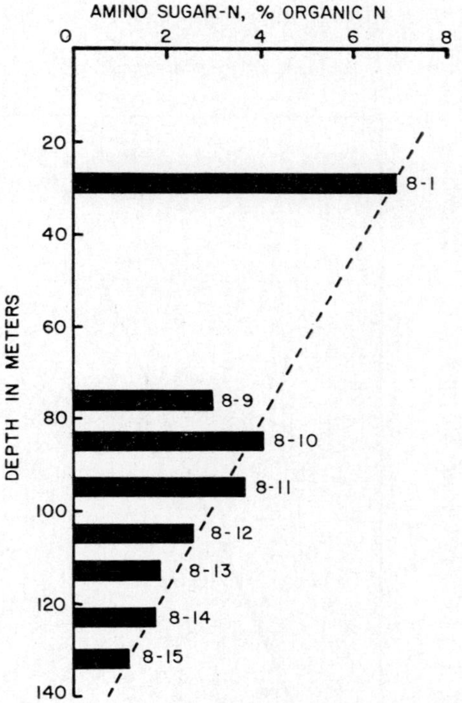

FIGURE 4. Distribution of amino sugar-N in experimental Mohole sediments. The amounts of amino sugar-N in the original samples (dry weight basis) were: EM 8–1, 35 μg/g; EM 8–9, 11 μg/g; EM 8–10, 16 μg/g; EM 8–11, 14 μg/g; EM 8–12, 7 μg/g; EM 8–13, 6 μg/g; EM 8–14, 5 μg/g; EM 8–15, 4 μg/g.

creased in a general way with increasing depth (Fig. 5). Although this result may reflect variations in production rates of amino acids and amino sugars during the time the sediments were deposited, it is more likely that the amino sugars had undergone greater chemical destruction since deposition. Amino sugars are less stable than amino acids; therefore, they would not be expected to persist for equivalent geological periods of time.

The absolute amounts of amino acid-N and amino sugar-N in the sediments did not follow any depth relationships, a result which was expected in view of the wide variation in organic nitrogen levels (see previous section). In the case of amino acid-N, the amounts ranged from 140 μg/g for sample EM 8–1 to about 50 μg/g for samples EM 8–12, EM 8–14, and EM 8–15 (see Fig. 3).

The proportion of the nitrogen recovered as amino acids was somewhat higher than reported elsewhere for the experimental Mohole sediments (Degens *et al.*, 1964; Rittenberg *et al.*, 1963), and for some California

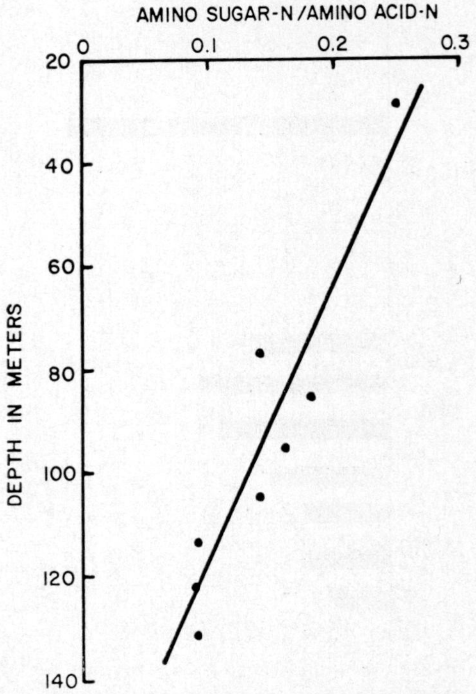

FIGURE 5. — Change in ratio of amino sugar-N to amino acid-N in experimental Mohole sediments.

offshore basin sediments (Emery *et al.*, 1964; Degens *et al.*, 1961, 1963, 1964). The higher recoveries obtained in this study can be attributed, at least in part, to improved methods of isolation.

(b) NONHYDROLIZABLE-N AND UNKNOWN ACID-SOLUBLE NITROGEN COMPOUNDS. The major fraction of the organic nitrogen in the experimental Mohole sediments occurred in the two vague groups designated as non-hydrolyzable-N and undetermined-N, the latter being defined as the acid-soluble nitrogen in the hydrolysates which could not be accounted for as fixed NH_4^+, amino acids, or amino sugars.

Figure 6 shows that there was considerable variation in the proportion of the organic nitrogen which occurred as nonhydrolyzable-N. The relative level of Ca^{++}- and Mg^{++}- carbonates in the sediments is indicated on the diagram because there appears to be a relationship between carbon-

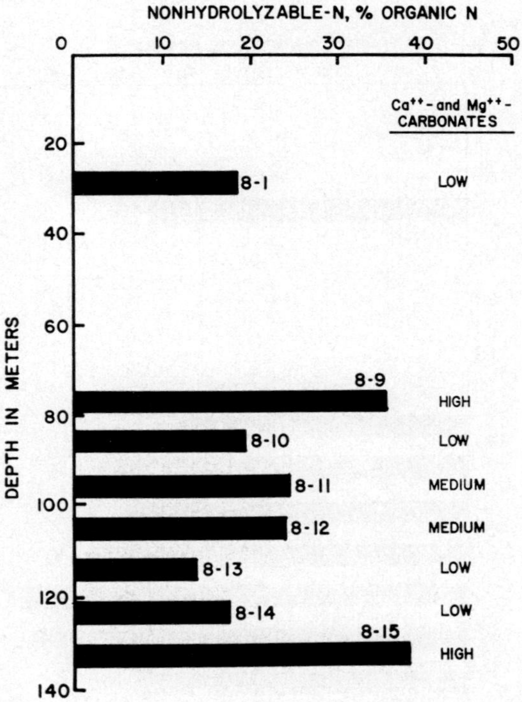

FIGURE 6. — Distribution of nonhydrolyzable-N in experimental Mohole sediments. The approximate level of Ca^{++}- and Mg^{++}- carbonates is shown in the column on the right: high, 45 to 55%; medium, 8 to 28%; low, less than about 5%. Additional data on carbonates are given in Table 1.

ate content and the percent of the organic nitrogen which occurred in this form. For example, over 35% of the organic nitrogen in samples EM 8–9 and EM 8–15, which have Ca^{++}- and Mg^{++}- carbonate contents of about 50%, occurred as nonhydrolyzable-N, whereas less than 20% of the organic nitrogen in samples EM 8–1, EM 8–10, EM 8–13, and EM 8–14, with negligible amounts of carbonates, was recovered in this form. Two explanation for this result are: (1) the presence of excess Ca^{++} and Mg^{++} led to the selective preservation of humic acids (as insoluble salts), and (2) greater polymerization of nitrogenous constituents had taken place in the calcareous sediments due to a more favorable pH. Results discussed below indicate that the latter was probably the case.

Figure 7 shows that the percentage of the organic nitrogen which occurred as unknown acid-soluble compounds (undetermined-N) tended to

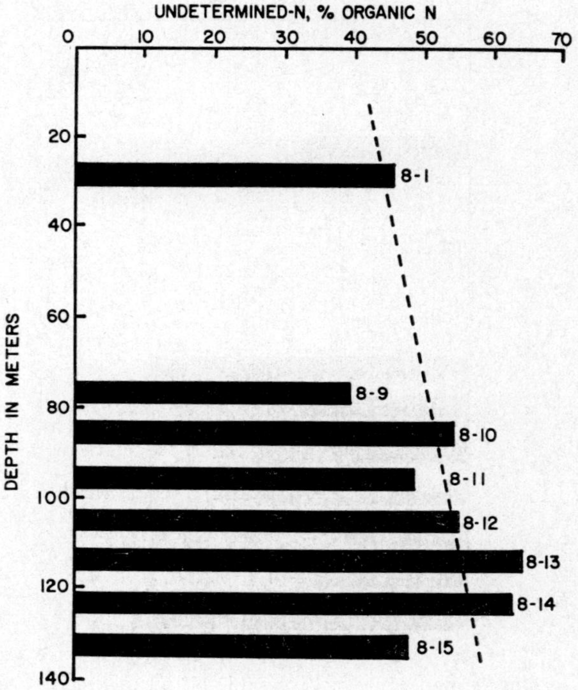

FIGURE 7. — Nitrogen recovered in unknown acid-soluble compounds (undetermined-N).

increase with increasing depth. However, the relationship was not distinct because of the variation introduced by differences in carbonate content. In contrast to the results shown for nonhydrolyzable-N (see Fig. 6), a larger fraction of the organic nitrogen in the noncalcareous sediments was recovered in unknown acid-soluble compounds. This finding suggested that a plot of the sum of the nitrogen occurring in the two fractions against depth would reflect the overall changes that had occurred in organic nitrogen during the long history of the experimental Mohole sediments. Figure 8 definitely shows that there was an inverse relationship between the amounts of nitrogen recovered in the two fractions. Although definite proof is lacking, the evidence indicates that polymerization of nitrogenous compounds had taken place since deposition, the reaction being favored by conditions existing in the calcareous sediments. From 61·8 to 82·5% of the organic nitrogen in the sediments was recovered in the two fractions, with the

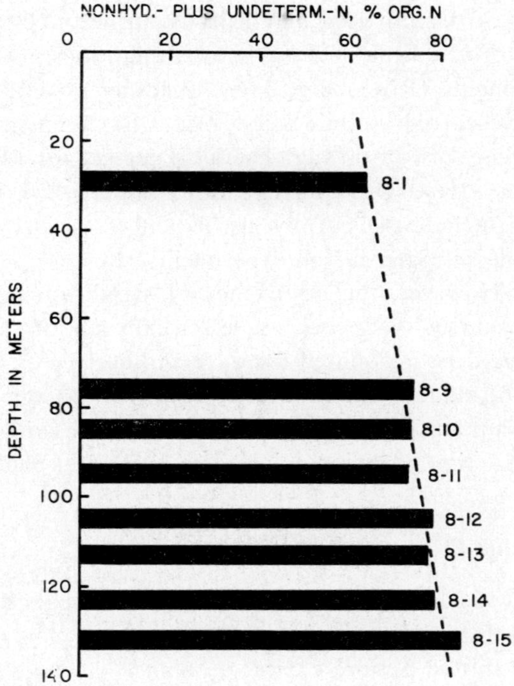

FIGURE 8. — Nitrogen recovered in unknown acid-soluble compounds plus nonhydrolyzable-N.

percentage increasing systematically with depth. The increase in the amount recovered with depth can be accounted for by losses of amino acids and amino sugars (compare Fig. 8 with Fig. 3 and 4).

Extensive research will be required to characterize the organic nitrogen which occurred in nonhydrolyzable forms and as unknown acid-soluble compounds. It is highly probable that the nonhydrolyzable-N occurred as a structural component of humic acids. Information on this point was obtained by extracting several hydrolyzed residues with 0·5 N NaOH and precipitating out the humic acids by addition of mineral acid. Accurate quantitative data were not possible because of an inadequate supply of sediments; however, it was evident that samples EM 8–9 and EM 8–15 contained higher amounts of humic acids per unit of organic nitrogen than the other samples. Bremner (1955) found that from 20 to 60% of the nitrogen in a variety of humic acids could not be solubilized by acid hydrolysis.

With regard to the unknown acid-soluble compounds (undetermined-N), a small fraction of this nitrogen may exist as purine and pyrimidine bases; these compounds have been detected in trace quantities in the experimental Mohole sediments (Rosenberg, 1964). Another possibility is that part of the nitrogen occurred as amines, because it has been shown that these compounds can be formed through chemical destruction of certain amino acids (Vallentyne, 1964). Other nitrogenous compounds of known geological significance include the porphyrins. Rittenberg *et al.* (1963) suggested that some of the nitrogen in the experimental Mohole sediments might occur as urea. However, the occurrence of significant amounts of this compound appears unlikely, because essentially all of the NH_3 in the hydrolysates could be accounted for as being derived from fixed NH_4^+. The point will be made in the next section that the sediments may contain acid-soluble compounds in the form of intermediate condensation products from the reaction of amino acids with sugars and phenols.

Distribution of the Forms of Nitrogen

Results of the fixed NH_4^+, amino acid, and amino sugar determinations, together with information obtained on other types of organic nitrogen compounds, permitted a more complete balance sheet to be made of the forms of nitrogen in the experimental Mohole sediments. The general findings, summarized in Table 3, show that considerable progress has been made. However, only from about one-fourth to one-half of the nitrogen has been accounted for in known compounds (as fixed NH_4^+, amino acids, and amino sugars). Characterization of the remaining nitrogen constitutes a major problem in organic geochemistry.

Diagenetic Changes in Organic Nitrogen Compounds

According to Rittenberg *et al.* (1963), most of the detritus in the experimental Mohole sediments was derived from the nearby continent. Some nitrogenous compounds may have been preserved intact through fixation to clay minerals and high-molecular-weight humic substances, but it is

TABLE 3. — SUMMARY OF DATA RELATIVE TO KINDS AND AMOUNTS OF NITROGEN COMPOUNDS IN EXPERIMENTAL MOHOLE SEDIMENTS

Form of N	% of total N	% of organic N	Comments
Fixed NH_4^+-N	10·0 to 21·5	—	This nitrogen is considered to occupy interstices formed by hexagonal oxygen rings within the lattice structures of clay minerals.
Amino acid-N	12·3 to 23·4	13·6 to 28·0	The fraction of the organic nitrogen in the form of amino acids decreased progressively with increasing depth (age).
Amino sugar-N	1·1 to 5·8	1·2 to 6·9	The proportion of the organic nitrogen in this form decreased sharply with increasing depth (age).
Nonhydrolyzable-N	10·9 to 34·2	14·0 to 37·8	This nitrogen was not solubilized by acid hydrolysis and may be associated with humic acid. A direct relationship existed between the percent of the organic nitrogen which occurred in this form and the carbonate content of the sediments.
Unknown acid-soluble compounds	34·0 to 50·0	38·3 to 64·7	This fraction is the acid-soluble nitrogen not accounted for as fixed NH_4^+, amino acids, or amino sugars. The hydrolysates contained very little ammonia which could be attributed to deamination of organic compounds.

more reasonable to assume that most of the organic nitrogen had been recycled numerous times through living tissues as a consequence of intense microbial activity in the top sediment. The result of this turnover would be a decline in the organic nitrogen content of the detritus. Greatest loss of organic nitrogen would occur in the first few centimeters of the sediment. After burial to the extent of only a few decimeters, the loss would be greatly reduced, and at greater depths, where low Eh values are attained, the loss would be negligible. Following deep burial, further changes occur through strictly chemical reactions.

A major difference between the distribution of the forms of organic nitrogen in the experimental Mohole sediments and terrestrial soil is that the latter contains large quantities of labile compounds which yield ammonia by hydrolysis (see Bremner, 1949). Thus, if it is assumed that the organic nitrogen associated with the original detrital material was similar in composition to that of present-day soils, the conclusion would have to be reached that the labile compounds (as yet unidentified) were the first to have been lost during diagenesis (within three million years). This was followed by the gradual disappearance of amino sugars, and, finally, the amino acids. Information on the distribution of nitrogenous constituents may provide useful information concerning the depositional environment of sediments.

According to Rittenberg *et al.* (1963), each successive centimeter of sediments in the drilling site of the experimental Mohole is about 1000 years older than the previous centimeter. This is borne out by data published recently by Dymond (1966). The results of the amino sugar study indicates that, under conditions existing in the experimental Mohole, a period of about $1{\cdot}5\times10^{17}$ years would be required for their complete chemical destruction.

As far as the chemical destruction of amino acids is concerned, Vallentyne's (1964) thermal stability study suggests that serine and threonine would be extensively altered during the long history of the experimental Mohole sediments. However, the loss of amino acids during the nine million years between the deposition of samples EM 8–1 and EM 1–15 (see Figure 1) cannot be accounted for entirely by simple chemical oxidation. Consequently, other mechanisms must have been responsible for their disappearance. One possibility is that the amino acids had reacted with reducing sugars and phenols. The latter compounds, because of their ubiquity and ability to combine with nitrogen-containing biochemicals, should receive considerable attention in ascertaining the fate of amino acids and related substances in geological environments. Both sugars and phenols were detected in the experimental Mohole sediments by Degens *et al.* (1964) and Rittenberg *et al.* (1963).

With regard to the reaction between amino compounds and reducing sugars (the Maillard or browning reaction), the first step involves addition of the amine to the carbonyl group of the sugar, with the formation of the N-substituted glycosylamine. This compound undergoes the Amadori rearrangement to form the N-substituted 1-amino-1-deoxy-2-ketose, which is highly reactive and undergoes dehydration and fragmentation to yield unsaturated intermediates (Hodge, 1953). In the final stages of browning, the intermediates polymerize into brown nitrogenous polymers and copolymers. The overall reactions are illustrated in Fig. 9. Browning occurs slowly at normal temperatures in an aqueous media. However, this factor

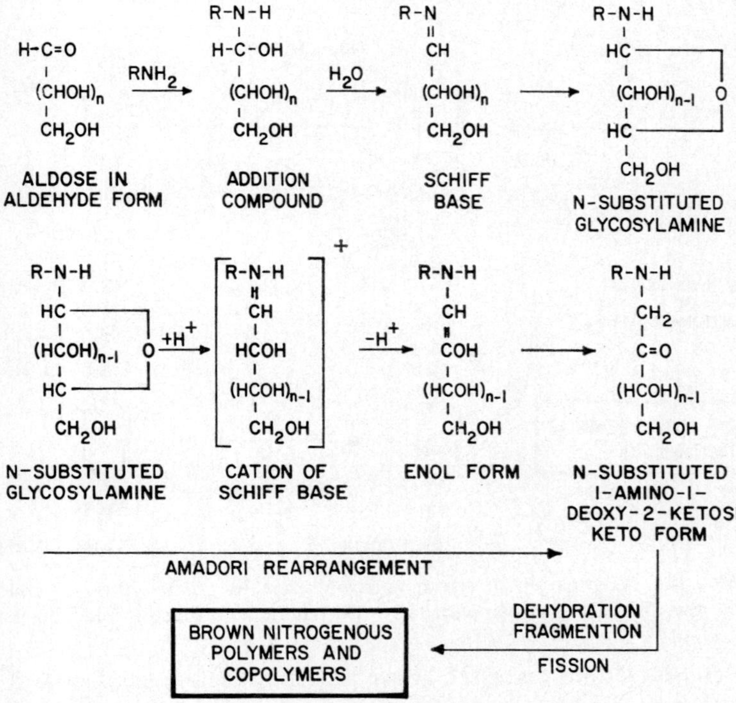

FIGURE 9. — Formation of brown nitrogenous polymers by sugar-amine condensation reactions.

assumes a minor role when one considers the long period of time available for the reaction to take place. The intermixing of reactants with mineral matter having catalytic capabilities may facilitate browning.

Attention has been given recently to the importance of the reaction between amino acids and polyphenols in the formation of humic substances (Swaby and Ladd, 1962). Simultaneous oxidation of the polyphenol to the corresponding quinone, such as by the polyphenol oxidases, is involved. The addition product readily polymerizes into brown nitrogenous polymers, according to the general sequence illustrated in Fig. 10.

Some of the nitrogen which was recovered in unknown acid-soluble compounds may have occurred as intermediate condensation products from the reaction of amino compounds with sugars and phenols. The inverse relationship between the amount of nitrogen occurring in this fraction and nonhydrolyzable-N can be explained by enhanced polymerization of intermediates in the calcareous sediments due to a more favorable pH.

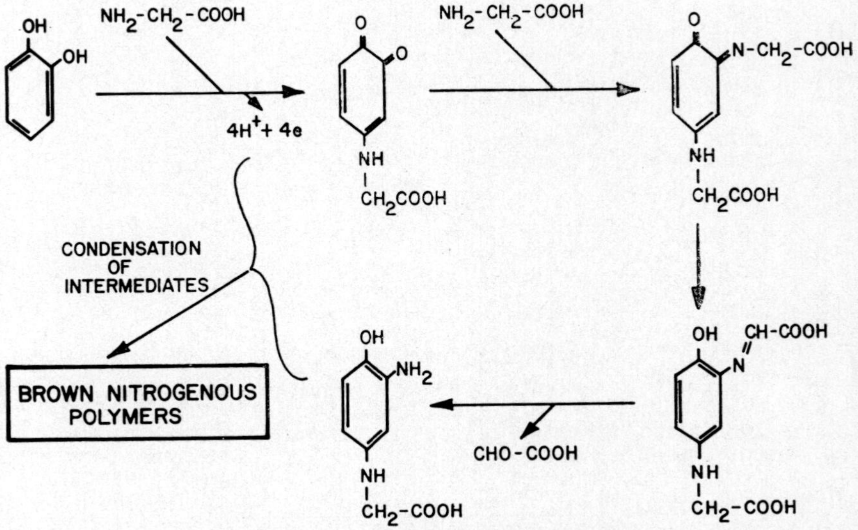

FIGURE 10. — Formation of nitrogenous polymers by condensation of amino acids and phenols, as exemplified by the reaction of glycine and catechol.

The considerations discussed above have far reaching implications. For example, the concept that humic acids in marine sediments are derived exclusively from the lignin of land plants may require modification. Kerogen, the insoluble residue of ancient sediments, may also originate from the condensation of simple organic molecules.

Acknowledgements

This investigation was supported by the National Science Foundation (Grants GP 1938 and GA 348).

References

Arrhenius, G. (1950) Carbon and nitrogen in subaquatic sediments. *Geochim. et Cosmochim. Acta*, **1**, 15–21.

Bremner, J. M. (1949) Studies on soil organic matter. I. The chemical nature of soil organic nitrogen. *J. Agr. Sci.* **39**, 183–193.

Bremner, J. M. (1950) The amino-acid composition of the protein material in soil. *Biochem. J.* **47,** 538–542.

Bremner, J. M. (1955) Studies on soil humic acids: I. The chemical nature of humic nitrogen. *J. Agr. Sci.* **46,** 247–256.

Bremner, J. M. (1958) Amino sugars in soil. *J. Sci. Food Agr.* **9,** 528–532.

Bremner, J. M. (1959) Determination of fixed ammonium in soil. *J. Agr. Sci.* **52,** 147–160.

Bremner, J. M. (1965a) Inorganic forms of nitrogen: *Methods of Soil Analysis*: Volume II. Edited by C. A. Black *et al.*: American Society of Agronomy, Madison, Wisconsin, pp. 1179–1237.

Bremner, J. M. (1965b) Organic forms of nitrogen: *Methods of Soil Analysis*: Volume II. Edited by C. A. Black *et al.*: American Society of Agronomy, Madison, Wisconsin, pp. 1238–1255.

Buchanan, D. L. (1957) Desalting amino acid solutions by displacement with piperidine. *Anal. Chem.* **29,** 1877–1878.

Cheng, C., and Stevenson, F. J. (1967) Colorimetric determination of amino acids in marine sediments. Unpublished data.

Degens, E. T., Prashnowsky, A., Emery, K. O., and Pimenta, J. (1961) Organic mate rials in recent and ancient sediments. Part II. Amino acids in marine sediments of Santa Barbara Basin, California. *N. Jb. Geol. Pal., Mhft.*, 1961 (8), pp. 413–426.

Degens, E. T., Emery, K. O., and Reuter, J. H. (1963) Organic materials in recent and ancient sediments. Part III. Biochemical compounds in San Diego Trough, California. *N. Jb. Geol. Pal., Mhft.* 1963 (5), pp. 231–248.

Degens, E. T., Reuter, J. H., and Shaw, K. N. F. (1964) Biochemical compounds in offshore California sediments and sea waters. *Geochim. et Cosmochim. Acta*, **28,** 45–66.

Dhariwal, A. P. S., and Stevenson, F. J. (1958) Determination of fixed ammonium in soils. *Soil Sci.* **86,** 343–349.

Dische, Z., and Borenfreund, E. (1950) A spectrophotometric method for the microdetermination of hexosamines. *J. Biol. Chem.* **184,** 517–522.

Dymond, J. R. (1966) Potassium-argon geochronology of deep-sea sediments. *Science,* **152,** 1239–1240.

Elson, L. A., and Morgan, W. T. J. (1933) A colorimetric method for the determination of glucosamine and chondrosamine. *Biochem. J.* **27,** 1824–1828.

Emery, K. O., Stitt, C., and Saltman, P. (1964) Amino acids in basin sediments. *J. Sed. Petr.* **34,** 433–437.

Erdman, J. G., Marlett, E. M., and Hanson, W. E. (1956) Survival of amino acids in marine sediments. *Science*, **124,** 1026.

Ferguson, W. S., and Sowden, F. J. (1966) A comparison of methods of determining nitrogen fractions in soils. *Can J. Soil Sci.* **46,** 1–6.

Hinman, W. C. (1964) Fixed ammonium in some Saskatchewan soils. *Can. J. Soil Sci.* **44,** 151–157.

Hodge, J. E. (1953) Chemistry of browning reactions in model systems. *Agr. Food Chem.* **1,** 928–943.

Keeney, D. R., and Bremner, J. M. (1964) Effect of cultivation on the nitrogen distribution in soils. *Soil Sci. Soc. Amer. Proc.* **28,** 653–656.

Lynch, D. L., Hughes, D. H., and Rhodes, Y. E. (1959) Pressure and gradient elution in ion exchange chromatography of the amino acids in soils. *Soil Sci.* **87,** 339–344.

Mohamed, A. F. (1949) The distribution of organic matter in sediments from the northern Red Sea. *Amer. J. Sci.* **247,** 116–127.

Riedel, W. R., Ladd, H. S., Tracey, J. I., and Bramlette, M. N. (1961) Preliminary drilling phase of Mohole project. II. Summary of coring operations. *Bul. Amer. Assoc. Petrol Geologists*, **45,** 1793–1798.

Rittenberg, S. C., Emery, K. O., Hülsemann, J., Degens, E. T., Fay, R. C., Reuter, J. H., Grady, J. R., Richardson, S. H., and Bray, E. E. (1963) Biogeochemistry of sediments in experimental Mohole. *J. Sed. Petr.* **33,** 140–172.

Rodrigues, G. (1954) Fixed ammonia in tropical soils. *J. Soil Sci.* **5,** 264–274.

Rosenberg, E. (1964) Purine and pyrimidines in sediments from the experimental Mohole. *Science*, **146,** 1680–1681.

Schachtschabel, P. (1960) Fixieter Ammoniumstickstoff im Löss- und Marschböden. *Trans. 7th Intern. Congr. Soil Sci.* **2,** 22–27.

Sellers, G. A. (1966) Hydrothermal experiments on the thermal stability of amino substances in sediments. Ph.D. thesis, California Institute of Technology. (University Microfilms, Inc., No. 66–2197, Ann Arbor, Michigan).

Sowden, F. J. (1959) Investigations on the amounts of hexosamines found in various soils and methods for their determination. *Soil Sci.* **88,** 138–143.

Stevenson, F. J. (1954) Ion exchange chromatography of the amino acids in soil hydrolysates. *Soil Sci. Soc. Amer. Proc.* **18,** 373–377.

Stevenson, F. J. (1956a) Isolation and identification of some amino compounds in soils. *Soil Sci. Soc. Amer. Proc.* **20,** 201–204.

Stevenson, F. J. (1956b) Effect of some long-time rotations on the amino acid composition of the soil. *Soil Sci. Soc. Amer. Proc.* **20,** 204–208.

Stevenson, F. J. (1957) Investigations of aminopolysaccharides in soils. I. Colorimetric determination of hexosamines in soil hydrolysates. *Soil Sci.* **83,** 113–122.

Stevenson, F. J. (1959) Carbon-nitrogen relationships in soil. *Soil Sci.* **88,** 201–208.

Stevenson, F. J. (1960) Microdetermination of nitrogen in rocks and silicate minerals by sealed tube digestion. *Anal. Chem.* **32,** 1704–1706.

Stevenson, F. J. (1962) Chemical state of nitrogen in rocks. *Geochim. et Cosmochim. Acta*, **26,** 797–809.

Stevenson, F. J. (1965a) Amino acids: *Methods of Soil Analysis*: Volume II. Edited by C. A. Black *et al.:* American Society of Agronomy, Madison, Wisconsin, pp. 1437–1451.

Stevenson, F. J. (1965b) Amino sugars: *Methods of Soil Analysis*: Volume II. Edited by C. A. Black *et al.*: American Society of Agronomy, Madison, Wisconsin, pp. 1429–1436.

Stevenson, F. J., and Dhariwal, A. P. S. (1959) Distribution of fixed ammonium in soils. *Soil Sci. Soc. Amer, Proc.* **23,** 121–215.

Stevenson, F. J., Kidder, G., and Tilo, S. N. (1967) Extraction of organic nitrogen and ammonium from soil with hydrofluoric acid. *Soil Sci. Soc. Amer. Proc.* **31,** 71–76.

Swaby, R. J., and Ladd, J. N. (1962) Chemical nature, microbial resistance, and origin of soil humus. *Trans. Intern. Soc. Soil Sci.* Comm. IV, pp. 3–8.

Troll, W., and Cannan, R. K. (1953) A modified photometric ninhydrin method for the analysis of amino and imino acids. *J. Biol. Chem.* **200,** 803–811.

Vallentyne, J. R. (1964) Biogeochemistry of organic matter-II. Thermal reaction kinetics and transformation products of amino compounds. *Geochim. et Cosmochim. Acta*, **28,** 157–188.

Walsh, L. M., and Murdock, J. T. (1960) Native fixed ammonium and fixation of applied ammonium in several Wisconsin soils. *Soil Sci.* **89,** 183–193.

Yemm, E. W., and Cocking, E. C. (1955) The determination of amino-acids with ninhydrin. *Analyst*, **80,** 209–213.

Young, J. L. (1962) Inorganic soil nitrogen and carbon: nitrogen ratios of some Pacific Northwest soils. *Soil Sci.* **93,** 397–404.

Young, J. L., and Mortensen, J. L. (1958) Soil nitrogen complexes. I. Chromatography of amino compounds in soil hydrolysates. *Ohio Agr. Exp. Sta. Res. Circ.* No. 61, pp. 1–18.

Waksman, S. A. (1933) On the distribution of organic matter in the sea bottom and the chemical nature and origin of marine humus. *Soil Sci.* **36**, 125–147.

ÉTUDE DES PREMIERS STADES D'ÉVOLUTION DE LA MATIÈRE ORGANIQUE DANS DES MARES EN MILIEU RÉCIFAL (Polynésie française)

par

JEAN TRICHET

Laboratoire de Géologie de l'École Normale Supérieure
24, rue Lhomond,
Paris, — France

Les formations coralliennes en général et les atolls en particulier sont des hauts lieux de l'activité organique. Construits par des êtres vivants coloniaux, ils servent de support au développement d'une faune et d'une flore extraordinaires en quantité et en qualité : du côté de la mer poussent des algues à squelette calcaire, croissent des éponges, des échinodermes. Quant aux lagons, leurs eaux constituent une véritable arche de Noé de la vie marine : poissons, chéloniens, échinodermes, vers, coelentérés, éponges, foraminifères, algues, etc. C'est dans le lagon, piège à sédiment naturel que s'accumulent les débris organiques de ces êtres vivants, Il n'apparaît pas étonnant à la suite de cette énumération sommaire que, si les conditions de conservation et d'accumulation de la matière organique s'y prêtent, l'on trouve liées à certaines formations récifales anciennes des accumulations considérables de produits d'origine organique (produits bitumineux en particulier). Dès 1846, Silliman notait que des lipides n'étaient pas seulement présents dans la mince pellicule de coraux vivants mais bien aussi dans tout le récif fossile et en 1940, Bergmann et Lester précisaient les possibilités pour ces matières de constituer des sources de pétrole. Sickenberger enfin, observait en 1892 l'activité organique considérable qui caractérise certaines mares ou certains bassins : la consommation (par les invertébrés) de la matière disponible y est beaucoup plus faible que la synthèse de la matière elle-même et partant, le sédiment du fond de ces mares contient de notables quantités de matière organique.

C'est cet aspect de la sédimentation dans les mares que nous avons étudié dans plusieurs atolls de l'Archipel des Tuamotous (Polynésie Française).

Les figures 1 et 2 représentent deux coupes d'atoll où l'on peut placer deux types de mares bien différentes : dans la première (figure 1), la mare se trouve située immédiatement derrière les cordons littoraux sableux qui entourent en général les lagons. De faible profondeur (quelques dizaines

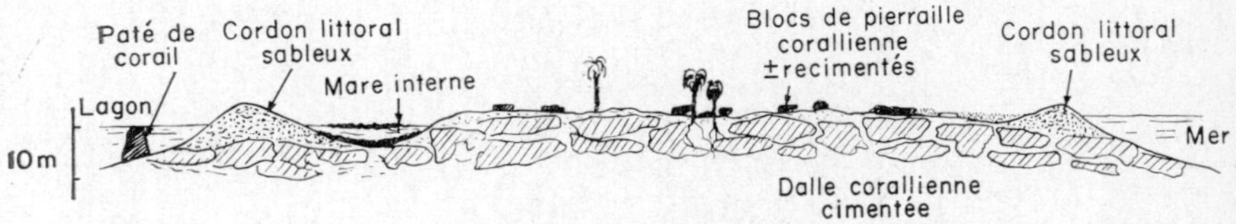

FIGURE 1. — Coupe transversale schématique d'un ilôt a l'endroit où l'atoll domine le niveau de la mer et porte une certaine végétation.

de centimètres), elles renferment une eau salée ou saumâtre; leur niveau suit fidèlement celui de l'eau du lagon à travers le cordon littoral. Ces mares sont très variables et lorsqu'elles persistent un temps suffisant pour le permettre, elles sont le siège d'un développement abondant d'algues bleues et vertes qui flottent à leur surface.

La figure 2, au contraire, montre une mare interne située au cœur de la couronne de l'atoll ; en général de beaucoup plus grande taille que les précédentes, les mares de ce type sont aussi beaucoup plus profondes

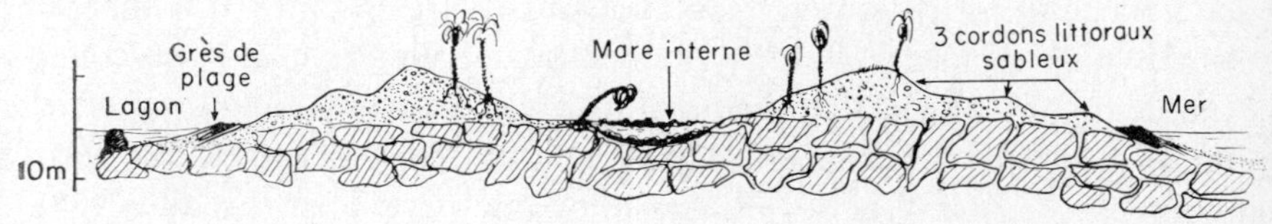

FIGURE 2. — Coupe transversale schématique passant par une mare d'eau douce.

(plus de 2 mètres dans certains cas). Elles reposent ainsi directement sur la dalle corallienne. Leur origine ne sera pas discutée ici du point de vue géomorphologique (Trichet, à paraître, 1966).

Elles contiennent de l'eau douce et leur niveau est en équilibre avec celui de la nappe d'eau douce de l'atoll (lentille de Gyben-Herzberg). À la surface de l'eau se développent ici aussi des algues bleues et vertes mais la source du matériel organique qu'elles reçoivent est essentiellement la végétation spontanée (ou plantée) de l'atoll. Les associations végétales qui bordent ce type de mare ne seront pas étudiées ici (Trichet, à paraître, 1966).

A. Les matières organiques accumulées dans les mares d'eau douce

Nous commencerons par l'étude de ce type d'accumulation en eau douce. Afin de pouvoir suivre l'évolution de la matière organique de façon simple, nous avons choisi une mare entourée d'une végétation monospécifique, constituée ici d'un arbre fréquent dans les atolls «Hibiscus tiliaceus» (le Purau en langue polynésienne). Nous avons recueilli le bois et les feuilles de cet arbre, les algues flottant à la surface de l'eau et avons pratiqué des sondages superficiels dans le sédiment déposé dans la mare. Ce sédiment est brun et constitué essentiellement de matière végétale plus ou moins décomposée mélangée à du sable aragonitique dû à l'érosion de la dalle corallienne.

I. *Traits généraux de l'évolution*

Nous suivrons l'évolution des parties hydrolysables (sucres, acides aminés), des lipides et des pigments et l'apparition des matières humiques.

La figure 3 montre la constitution du matériel originel : dans les trois cas, les parties hydrolysables sont abondantes, les lipides et pigments présents de façon importante dans les feuilles et les algues ; les matières humiques sont absentes, la lignine manque dans les algues ; quant à la cellulose, elle forme une part importante du bois et des feuilles ; dans le cas des algues, sa nature n'est pas nettement définie, il semble bien plutôt que cette partie soit constituée de mucopolysaccharides et de muco-polypeptides qui forment les membranes des algues bleues (Höcht *et al.*, 1965).

La figure 4 est à comparer à la précédente : elle montre l'évolution des matières organiques accumulées en fonction de la profondeur dans deux types de stations différentes. Le premier correspond à une «tourbe» de feuilles (étant donnée la quantité relativement faible de lignine dans les horizons de surface), le second à une «tourbe» de bois (riche en lignine au sommet du dépôt). Dans les deux cas, la proportion des parties solubles ou hydrolysables est élevée dans les horizons de surface et diminue en profondeur. Cette proportion élevée dans les horizons de surface est à relier à la disparition presque totale de la cellulose dans les mêmes horizons. La chute très brusque de la quantité de cellulose (elle a lieu dans les

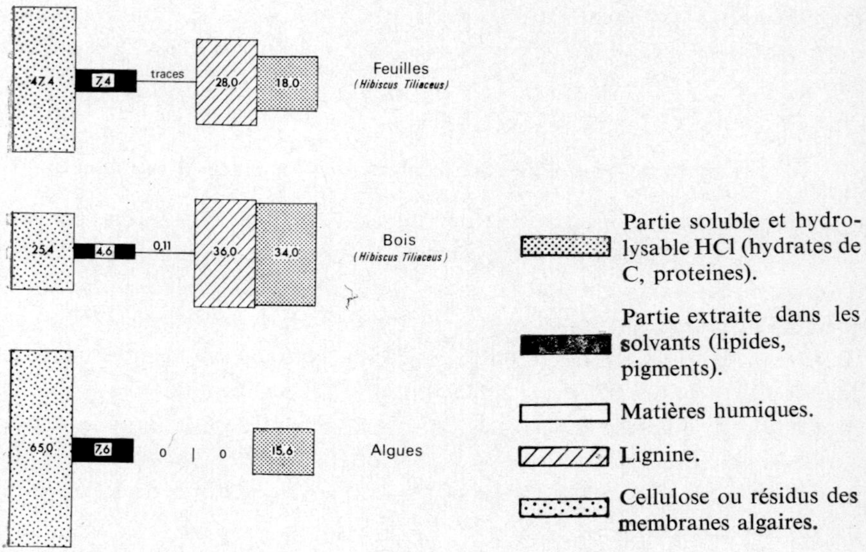

FIGURE 3. — Spectre de la composition des éléments végétaux s'accumulant dans les mares d'eau douce.

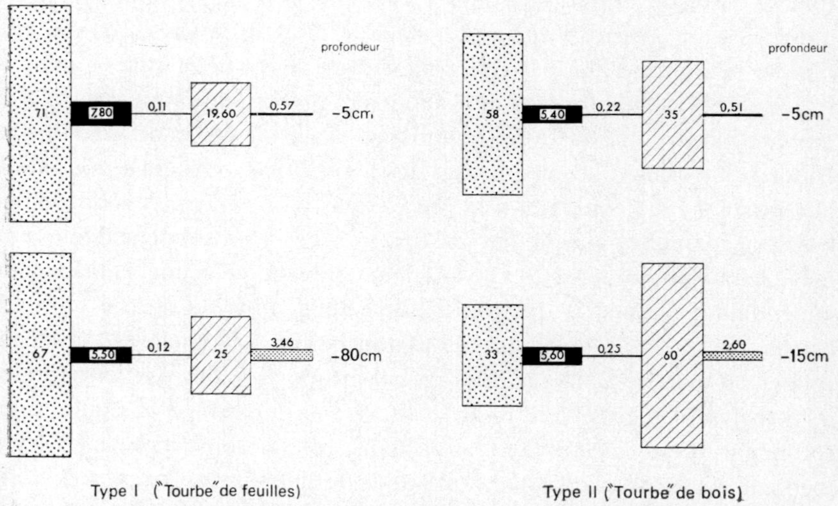

FIGURE 4. — Influence de la profondeur sur la composition des matières enfouies.

premiers centimètres du dépôt) est évidemment le fait majeur de l'évolution du dépôt organique. Un phénomène parallèle a été observé dans d'autres types de tourbes et Bailey (1953) l'attribue à l'activité bactérienne cellulolytique. Breger (1949) le décrit également comme le résultat d'une activité bactérienne aérobie. L'abondance des sucres réducteurs dans les extraits confirme l'attaque de la cellulose (quantités toujours supérieures à 10 % du poids sec de matière organique et pouvant aller jusqu'à 50 % dans certaines tourbes de bois). Quant à l'augmentation de la teneur en cellulose dans les horizons inférieurs, elle est relative aux variations des autres constituants dont la proportion diminue. (Si l'on se reporte à la teneur en lignine en la considérant comme constante, la variation de la teneur en cellulose par rapport à la lignine peut être aussi due à une teneur différente en cellulose d'un dépôt différent.)

À côté de ce trait majeur, nous noterons la permanence des fractions extraites par les solvants (lipides, pigments). La lignine enfin se montre très résistante dans ces premiers stades d'évolution et la quantité de matières humiques reste constante et d'ailleurs difficilement appréciable.

Ce type de composition et d'évolution est parallèle à ceux décrits par d'autres auteurs (Zhuraoleva, 1954).

II. *Étude de la fraction soluble dans l'eau et hydrolysable par les acides*

Les techniques chromatographiques classiques nous ont permis de séparer et d'étudier les acides aminés et les sucres obtenus dans la fraction soluble dans l'eau et hydrolysable dans les acides.

1) *Les acides aminés* trouvés sont réportés dans le tableau de la figure 5. Beaucoup d'entre eux ont une origine commune (les feuilles, le bois ou les algues). Il s'agit de l'alanine, de l'acide aspartique, de la glycine, de la leucine, de la sérine et de la valine. Les feuilles constituent la source essentielle de l'hydroxyproline, de la phénylalanine et de la thréonine. Le bois d'Hibiscus apporte la cystéine et les algues bleues la grande partie de l'arginine.

Deux problèmes se posent en ce qui concerne ces amino-acides : celui de leur présence et celui de leur stabilité. La présence des amino-acides dans les sédiments récents a été beaucoup étudiée et des listes tout-à-fait comparables à celle obtenue ici sont reportées dans les articles rédigés par Swain et Abelson dans l'ouvrage de Breger (1963).

Quant au problème de la stabilité, on peut déduire de la lecture du tableau précédent les indications suivantes : huit acides aminés semblent très stables (alanine, acide aspartique, glycine, leucine, phénylalanine, sérine, thréonine, valine) ; au contraire l'arginine, la cystéine, l'hydroxyproline et la lysine montrent une assez grande instabilité. Cette comparaison indique que les acides aminés acides et neutres sont d'une manière générale beaucoup plus stables que les acides aminés basiques (arginine, lysine). (Dans cet ordre d'idées, l'existence d'histidine dans les horizons de profondeur et dans eux seuls, demeure inexplicable.)

	Ala	Arg	Asp	Cys	glu	Gly	His	Hpro	Leu	Lys	Phe	Ser	Thr	Val
Feuilles	+		+		+	+		+	+		+	+	+	+
Bois	+	(+)	+	+	+	+			+	(+)		+		+
Algues	+	+	+			+			+			+		+
I - 5 cm	+		+		+	+		+	+		+	+	+	+
I - 80 cm	+		+			+	+		+		+	+	+	+

FIGURE 5. — Les acides aminés dans le matériel sédimenté dans les mares d'eau douce et leur évolution avec la profondeur (cas de la «tourbe» du type I).

Si l'on compare la stabilité observée ici avec celle que rapportent d'autres auteurs pour des sédiments récents (Erdman, Marlet et Hanson, 1956), on retrouve six des huits acides aminés reconnus comme stables dans les neufs indiqués par ces auteurs (valine, leucine, alanine, acide aspartique, glycine, phénylalanine). Il s'agit d'acides aminés acides ou neutres. Notons également que l'acide glutamique se comporte apparemment comme un acide peu stable.

2) *L'étude des sucres* a été réalisée de la même façon. Le tableau de la figure 6 indique leur présence et suggère leur stabilité. Les feuilles, le bois et les algues sont des sources communes pour le glucose, l'arabinose et le xylose, qui sont tous stables dans les conditions de dépôt. Le fructose semble provenir en grande partie des algues, mais n'est pas conservé dans ce type de mares. Le galactose et le ribose sont présents seulement dans les lits superficiels. Un point intéressant (non indiqué dans le tableau) est l'existence de mannitol dans le bois et les algues. Or les sucres obtenus par Höcht *et al.* (1965) lors de l'hydrolyse des membranes d'une autre algue bleue (Tolyphotrix) sont essentiellement le mannose, le glucose et le xylose. Il semble bien que le glucose, le xylose et le mannitol aient pour une part, une telle origine dans nos sucres hydrolysés. La raison pour

laquelle le mannose se serait transformé en mannitol reste posée. Il faut noter que le fructose lui aussi disparaît dans le sédiment et que sa réduction a pu être à l'origine d'une certaine quantité de mannitol. Ce mannitol se retrouve dans les horizons supérieurs du sédiment. En ce qui concerne la présence de mannitol dans le bois, rappelons que l'association de glucanes-mannanes-xylanes-galactanes et arabanes caractérise les bois tendres,

		Glucose	Galactose	Fructose	Arabinose	Xylose	Ribose	Rhamnose
Feuilles		+	+		+	+	+	+
Bois		+	+	+	+	+		
Algues		+		+	+	+		
I	- 5cm	+	+		+	+	+	
	- 80cm	+			+	+	(+)	

FIGURE 6. — Les sucres dans les matériaux sédimentés dans les mares d'eau douce et leur évolution en fonction de la profondeur («tourbe» du type I).

les bois durs étant plus spécialement munis en glucanes et en xylanes (Gustafson *et al.* in Polglase, 1955). Le bois d'Hibiscus tiliaceus se présente comme un bois tendre.

Du point de vue quantitatif, notons la grande abondance de sucres réducteurs et le fait que leur quantité diminue avec la profondeur (50 % en surface et 7 % à 1 mètre de profondeur).

III. *Étude de la fraction extraite par les solvants organiques*

1) Nous avons recherché la profondeur à laquelle la chlorophylle pouvait encore être reconnue. L'absorption photométrique de l'extrait alcoolique de quantités équivalentes de sédiment organique a été enregistrée : la diminution d'intensité du pic d'absorption de la chlorophylle est telle que au-delà de 1 mètre de profondeur la présence de ce pigment n'est plus appréciable (figure 7). Sur la même figure sont reportés les spectres d'absorption des extraits alcoolique de feuilles et de bois d'Hibiscus.

2) L'étude de la fraction lipidique extraite par l'éther a été réalisée à l'aide des techniques chromatographiques.

La figure 8 donne l'image du chromatogramme sur plaque de silice du mélange des corps gras extraits par l'éther à partir de quantités équivalentes de feuilles, de bois et d'algues d'une part et de sédiments organiques prelevés à deux profondeurs différentes d'autre part. Tous les constituants se retrouvent (paraffines, lipides neutres, phospholipides). Rap-

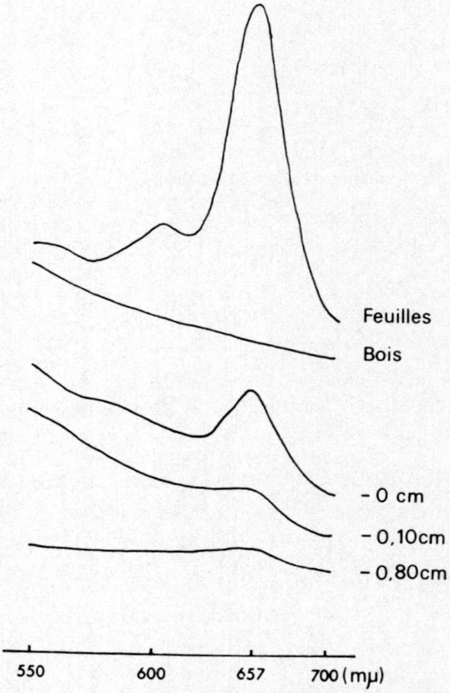

FIGURE 7. — Diminution de l'intensité du pic d'absorption de la chlorophylle avec la profondeur (mares d'eau douce) (extrait alcoolique).

pelons que la quantité totale de corps gras reste toujours élevée (⩾ 5 %) aux profondeurs testées. L'étude des acides gras par chromatographie de leurs esters méthyliques en phase gazeuse nous a conduit à y déceler un type d'évolution possible. Le tableau supérieur de la figure 9 montre la variété de ces acides dans les feuilles, le bois décomposé et les algues d'une part et dans le sédiment prélevé à deux profondeurs différentes d'autre part. La lecture du tableau conduit à l'observation suivante: les quantités de tous les acides en C18 diminuent avec la profondeur alors que les quantités des acides en C16 augmentent. Or l'on sait qu'un type de dégradation biologique important pour les acides gras est la

β-oxydation, processus de diminution de la longueur des chaînes d'acides par section d'un groupement acétique contenant le groupement carboxylique de l'acide initial.

$$CH_3—CH_2—\ldots \underset{\beta}{CH_2} \updownarrow \underset{\alpha}{CH_2}—COOH$$

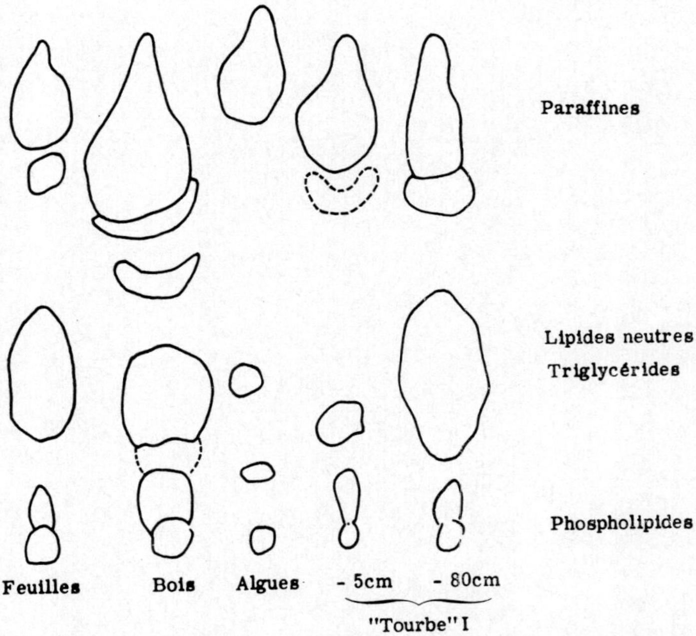

FIGURE 8. — Chromatographie sur plaque de silice des extraits solubles dans l'éther, issus des matériaux accumulés dans les mares d'eau douce.

Si l'on reporte face à face (figure 9) la diminution de la quantité d'un acide en C18 et l'augmentation de celle de l'acide en C16 ayant le même nombre de doubles liaisons, on peut observer qu'elles se correspondent approximativement, compte-tenu de la précision des mesures ($\Delta C = \pm 5\,\%$).

De toutes façons, plus la chaîne d'acide est longue et plus elle possède de doubles liaisons, plus l'acide est instable et sensible à la dégradation. C'est bien ce qui est observé pour la série des acides en C18. Les acides en C16 ne semblent pas encore affectés de la même manière ; leur quantité est d'ailleurs constamment augmentée par la dégradation des acides en C18. Il est certain que cette relation entre les acides possédant le même nombre de doubles liaisons serait à confirmer par la méthode d'oxydation permettant de localiser ces doubles liaisons.

	C 14	C 16	C 16 : 1	C 16 : 2	C 16 : 3	C 18	C 18 : 1	C 18 : 2	C 18 : 3
Feuilles	9,5	32,6	3,1	1,0	5,1	6,6	7,0	12,2	27,7
Bois en décomposition	12,7	17,6			17,6	6,3	10,1	35,2	
Algues	6,8	25,9	10,4			10,6	28,4	17,4	
- 80cm	12,3	22,1	5,8	3,0	19,6	10,6	11,5	13,8	

Proportions relatives des différents acides gras dans les parties végétales accumulées dans les mares d'eau douce et dans un échantillon de "tourbe" situé à - 0,80m.

	C 14	C 16	C 16 : 1	C 16 : 2	C 16 : 3	C 18	C 18 : 1	C 18 : 2	C 18 : 3
Matériel de surface	12,1	31,8	3,9	0,2	10,1	7,1	11,6	20,7	11,3
- 80cm	12,3	22,0	5,8	3,0	19,6	10,6	11,5	13,8	0

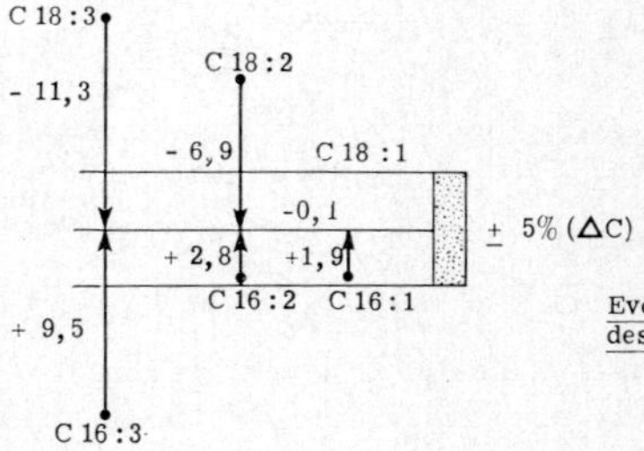

Evolution des proportions relatives des acides gras avec la profondeur.

FIGURE 9.

IV. *Études des matières humiques*

L'extraction par le pyrophosphate de Na à pH 9 des sédiments bruns accumulés dans la mare à permis d'y déceler la présence de faibles quantités de matières humiques (0,1 à 0,2 %). Leur dosage est incertain et la séparation en acides fulviques et en acides humiques par précipitation en milieu acide montre que les acides fulviques, légers, se trouvent de préférence dans les horizons superficiels.

En conclusion, ce premier type d'évolution de la matière organique en eau douce est caractérisé par : une rapide disparition de la cellulose et des hydrates de carbone — une libération abondante de produits légers (acides aminés, sucres) — une disparition progressive des pigments chlorophylliens — la présence simultanée d'acides gras dus à la dégradation par β-oxydation des chaînes les plus longues, d'autant plus que celles-ci sont plus insaturées.

Les conditions physico-chimiques (pH voisin de la neutralité, eau riche en bicarbonate de calcium au contact d'aragonite) permettent de rapprocher ce sédiment de celui des tourbières alcalines. L'absence de bryophytes remplacée ici par la présence d'algues est une différence importante entre les deux formations. La nature de la végétation arborée qui ceinture chaque mare est aussi cause de variété. Du point de vue botanique, la fréquence d'espèces à feuilles luisantes et couvertes de cires favorise l'accumulation de corps gras.

B. Les matières organiques accumulées dans les mares d'eau saumâtre

I. *Traits généraux de l'évolution*

Situées derrière le, cu les cordons littoraux sableux qui ceinturent le lagon, les mares d'eau saumâtre sont remarquables par leur variabilité. Le niveau de l'eau y varie en fonction des mouvements du lagon (dus aux petites marées, aux tempêtes). Dès que la persistance de l'eau est suffisante des algues bleues s'y développent. Un voile mucilagineux couvre la surface de l'eau, fruit d'une croissance active des algues, jusqu'au moment où, le niveau de l'eau venant à baisser, le voile se dépose et vient ajouter un lit organique supplémentaire au sédiment du fond de la mare. Ce sédiment est ainsi constitué de tels lits organiques aux couleurs encore vives mélangés à des sables aragonitiques et calcitiques. Une coupe tranche de tels lits sur des épaisseurs de plusieurs dizaines de centimètres. C'est là un premier type de dépôt possible pour les algues. Mais il en est un second que l'on peut observer aussi fréquemment : les mares sont alors envahies par des gastéropodes (*Cerithium patiens* Bayle) (figure 10) suspendus au voile d'algues qui leur sert de nourriture. Le dépôt au fond de l'eau est alors constitué essentiellement des coprolites de ces gastéropodes (figure 10). À peine peut-on distinguer quelques restes algaires déposés directement. Des couches de coprolites de plusieurs centimètres peuvent être reconnues et témoignent d'une permanence remarquable de cette consommation.

Il était alors intéressant de comparer les deux voies d'évolution possibles de ce matériel algaire, par dépôt direct ou à travers le tube digestif d'un gastéropode. Les algues considérées sont essentiellement des algues bleues-vertes du genre Phormidium mélangées à quelques algues vertes unicellulaires.

Du point de vue sédimentaire, les dépôts obtenus dans les deux cas sont comparables par leur richesse en matière organique (40 à 50 % du sédiment global).

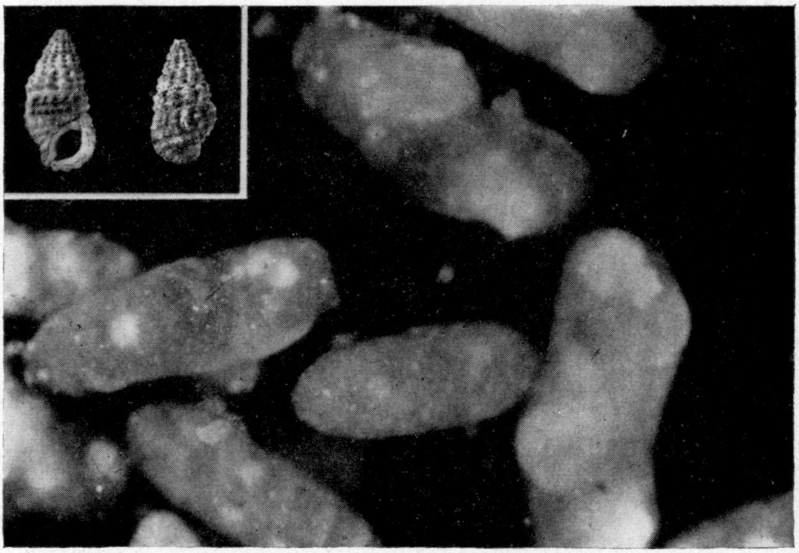

FIGURE 10. — Dépôt de coprolites dans les mares d'eau saumatre (×30). En cartouche, *Cerithium patiens* Bayle (×1).

La figure 11 montre le devenir des différentes fractions végétales dans les deux cas envisagés. Dans l'un et l'autre cas, les parties solubles et hydrolysables par les acides restent importantes; elles augmentent même de façon significative dans le cas d'algues consommées. Les quantités de lipides et de pigments augmentent eux aussi, relativement à la dernière fraction constituée par les résidus de membranes des algues. La diminution de la quantité de cette fraction, sensible dans le type de dépôt direct, devient le trait dominant de l'évolution dû au passage à travers le tube digestif des gastéropodes. La comparaison de ces spectres de composition montre que l'activité digestive dû gastéropode consiste essentiellement à attaquer la fraction algaire constituée de muco-polysaccharides et de muco-polypeptides. L'attaque enzymatique de cette fraction libère des acides aminés, des sucres etc. qu'on retrouve en partie dans la fraction soluble et hydrolysable par les acides. Il semble bien, en

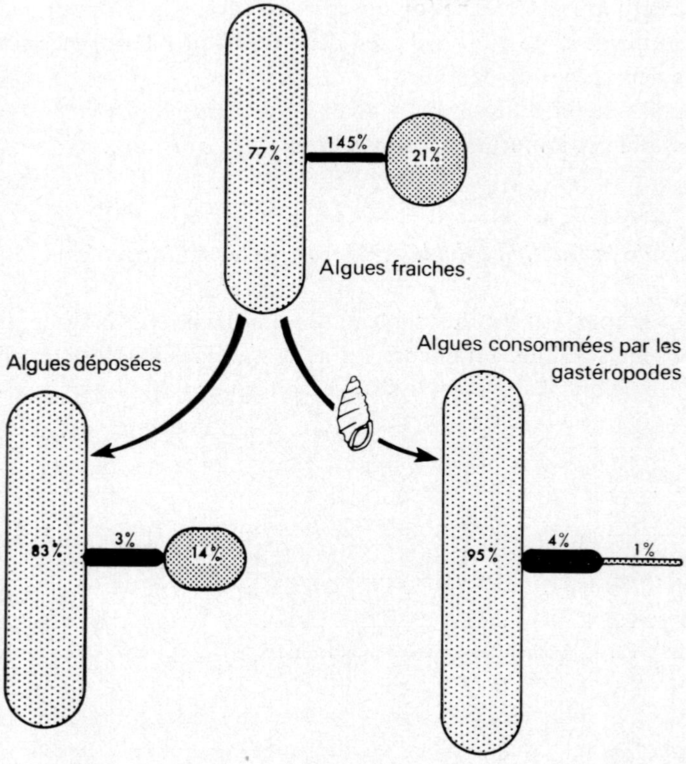

Partie soluble et hydrolysable HCl (hydrates de C, proteines, etc.).
Partie extraite par les solvants (lipides, pigments, etc.).
Residus des membranes algaires.

FIGURE 11. — Deux types d'évolution de la matière organique des algues des mares d'eau saumâtre.

effet, que les gastéropodes vivent là dans des conditions de croissance luxueuses. Notons par ailleurs que le résultat de cette hydrolyse des membranes algaires est celui recherché en utilisant la technique classique d'action du lyzozyme, enzyme des sucs digestifs de l'escargot, pour accéder au contenu cellulaire des algues bleues par attaque des membranes mucilagineuses.

La diminution de cette fraction muco-polysaccharidique et muco-polypeptidique dans le cas de dépôt direct (7 % environ) est due à une possible activité bactérienne (on observe, en effet, à la surface du sédiment, le développement de films colorés dus à une prolifération bactérienne, dans les deux types de dépôt).

Quant à l'augmentation de la quantité des corps gras et des pigments, elle est rapide et importante. Nous l'étudierons plus loin.

II. *Étude de la fraction soluble et hydrolysable par les acides*

1) LES ACIDES AMINÉS. Le tableau de la figure 12 indique leur présence et leur stabilité. Ici encore les acides aminés stables sont l'alanine, l'acide aspartique, la glycine, la sérine, soit quatre acides aminés «acides»

Acide aminé	Ala	Arg	Asp	Cys̍	Cys	Hypro	glu	gly	Ile	Leu	Phe	Ser	Val
Algues fraîches	+	+	+	+	+	+		+	+	+		+	+
Algues déposées	+	+	+	+	+			+				+	
Algues consommées par les gastéropodes	+		+				+	+		+	+	+	

FIGURE 12. — Les acides aminés dans les dépôts des mares d'eau saumâtre.

qu'Erdman *et al.* avaient reconnus comme stables. La cystéine, la cystine et l'arginine sont consommées ou détruites lors du passage à travers le tube digestif des gastéropodes. L'isoleucine, la valine et l'hydroxyproline se montrent instables. La présence d'acide glutamique et de phénylalanine dans les seuls coprolites demeure inexpliquée. (La libération de l'acide glutamique à partir de membranes d'algues bleues a été obtenue par Höcht (1965) précisément sous l'action du lyzozyme. Il aurait été ici aussi plus complétement libéré que dans le cas du dépôt direct, l'hydrolyse acide n'ayant pas atteint cette fraction des membranes ?)

2) LES SUCRES. En se reportant au tableau de la figure 13 on peut avoir une idée de la nature des sucres présents dans les dépôts algaires, de leur quantité approximative (indiquée par une croix simple ou redoublée), enfin, de la teneur en sucres réducteurs.

Le glucose et le xylose dominent nettement en quantité. Il sont stables dans le dépôt simple et sont encore présents de façon importante dans les coprolites des gastéropodes. Le fructose semble instable au cours du passage à travers le tube digestif, à l'inverse de l'arabinose.

Le trait remarquable de cette évolution est l'augmentation considérable de la quantité des sucres réducteurs due au passage à travers le tube digestif des mollusques. Comme nous l'avons dit plus haut, ce fait est

	Glucose	Fructose	Xylose	Arabinose	sucres réducteurs %
Algues fraîches	++	+	++	+	2,0
Algues déposées	+	+	+	+	3,2
Algues consommées par les gastéropodes	++		++	+	14,5

FIGURE 13. — Les sucres dans les dépôts des mares d'eau saumâtre.

le résultat de l'attaque enzymatique des réserves glucidiques des algues (amidon des cyanophytes, Hough *et al.*, 1962 ; Frederick, 1957). Cet amidon a une structure de glucane branchu, c'est-à-dire d'une amylopectine. Sa digestion libère de grandes quantités de glucose. En ce qui concerne les membranes et leurs enveloppes mucilagineuses, rappelons que Höcht *et al.* (1965) ont trouvé le glucose, le xylose et le mannose dans les hydrolysats de membranes de l'algue bleue Tolyphotrix. Or, comme dans le cas des dépôts en eau douce, nous retrouvons le mannitol dans les dépôts des deux types en eau saumâtre. Ce mannitol peut provenir de la réduction du mannose (que nous n'avons pas trouvé dans les algues fraîches) mais bien aussi du fructose qui disparaît dans le tube digestif et dont la quantité est encore notable dans le dépôt direct. Les mares d'eau saumâtre constituent d'ailleurs un milieu très réducteur, riche en acide sulfhydrique.

III. *Étude de la fraction extraite par les solvants organiques*

Cette fraction qui contient essentiellement les corps gras et les pigments représente une partie importante de la fraction organique du sédiment (3 %). Le passage à travers le tube digestif des gastéropodes augmente encore cette proportion.

1) La permanence *des pigments* dans les divers lits organiques frappe l'observateur. Après extraction par l'acétone on a pu obtenir le spectre d'absorption du mélange des pigments chlorophylliens et des caroténoïdes. La saponification des chlorophylles a permis de séparer les caroténoïdes et d'enregistrer leur spectre d'absorption.

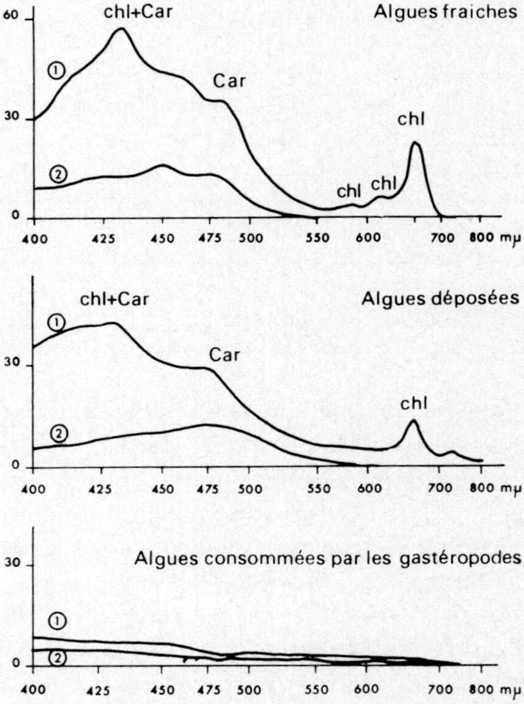

FIGURE 14. — Spectres d'absorption des pigments dans les dépôts des mares d'eau saumâtre (chl, chlorophylle; car, carotenoïdes). ① pigments totaux; ② caroténoïdes.

La figure 14 reproduit ces deux spectres pour les trois matériaux étudiés. La comparaison des deux diagrammes supérieurs montre la permanence des pigments chlorophylliens et des caroténoïdes dans le dépôt simple; les spectres d'absorption dans le second cas illustrent la disparition des pigments au cours du passage à travers le tube digestif des mollusques. (Le sédiment coprolithique est effectivement gris-blanchâtre.) On sait que les invertébrés peuvent synthétiser ou transformer les caroténoïdes et que les mollusques possèdent des pigments, souvent dans leur manteau, toujours dans leurs ocelles.

La stabilité des pigments est remarquable dans un tel milieu de sédimentation. Les caroténoïdes sont, en effet, très sensibles à la chaleur et à la lumière. Or ces deux facteurs peuvent agir intensément à la surface des mares au moment où elles se dessèchent. Il semble bien que le mélange de sable aragonitique blanc à cette matière pigmentée assure à cette dernière une protection efficace. La surface du dépôt n'est, en effet, jamais colorée;

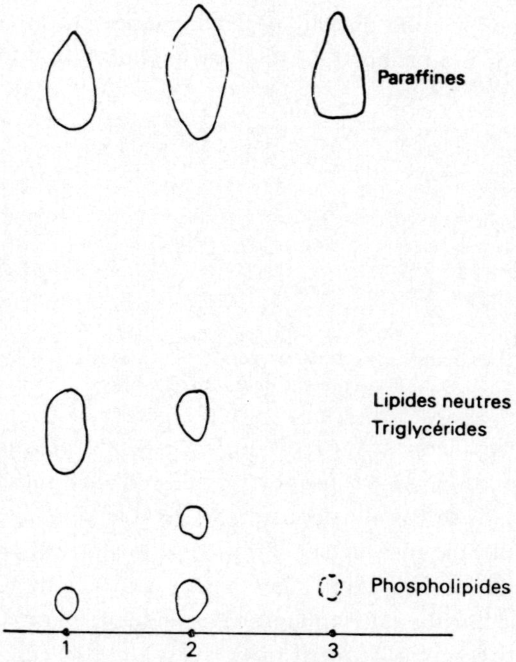

FIGURE 15. — Chromatographie sur plaque de silice des corps gras extraits par l'éther dans les algues fraiches (1); les algues déposées (2); les algues consommées par les gastéropodes (3).

elle est constituée de sables fins répandus à la surface du sédiment par les mouvements de l'eau.

2) *La fraction lipidique* est abondante. La figure 15 montre un chromatogramme sur plaque de silice du mélange des corps gras extraits par l'éther. Entre les algues fraîches et les algues déposées, peu de différence, les corps gras semblent stables. Le point intéressant est la disparition de de la partie neutre des lipides (triglycérides et lipides neutres) et sans doute d'une partie des phospholipides au cours du passage à travers le tube digestif des gastéropodes. Ce fait, intéressant pour la connaissance

des besoins alimentaires de ces animaux, l'est aussi du point de vue géochimique. La fraction des corps gras (4 % environ) stable dans ce type de dépôt est constituée essentiellement de paraffines (leur étude est en cours).

La chromatographie en phase gazeuse des esters méthyliques des acides gras a permis d'établir la liste consignée dans le tableau de la figure 16. Les acides trouvés sont rencontrés classiquement dans les algues, en particulier les acides insaturés.

Le passage par le tube digestif des mollusques entraîne la disparition de l'acide C18 : 2. Les proportions relatives des autres fractions ne semblent

	C 12	C 14	C 16	C 16 : 1	C 18	C 18 : 1	C 18 : 2
Algues fraîches	5,0	7,0	27,0	11,0	11,0	20,0	18,0
Algues déposées	20,0	22,5	18,7		8,2	12,0	18,5
Algues consommées par les gastéropodes	11,5	8,2	32,8	13,6	9,5	24	

FIGURE 16. – Les acides gras dans les dépôts des mares d'eau saumâtre (en % de la quantité totale d'acides gras).

pas modifiées de façon à suggérer un mécanisme possible d'évolution Seule l'augmentation de la proportion des acides saturés d'ordre bas (C12 et C14) dans le cas du dépôt direct semble significative. Du point de vue global, il faut insister sur le fait que, si l'on rapporte à la même quantité de carbone la quantité des acides gras, la différence entre les deux types de sédiments est frappante : la somme de l'aire des pics obtenus sur le chromatogramme des acides gras est de 200 cm^2 pour le dépôt direct et de 20 cm^2 pour le dépôt digéré par les gastéropodes. Le chromatogramme sur plaque de silice avait d'ailleurs montré la disparition de la fraction des lipides neutres.

CONCLUSION. Si l'on compare l'importance des mares d'eau douce (décrites plus haut) à celle des mares d'eau saumâtre, l'avantage est très nettement aux dernières. Les premières sont figées du point de vue géomorphologique et ne sont pas soumises aux cycles fréquents de croissance et de sédimentation de la matière organique. Il n'est pas rare de rencontrer le souvenir de mares d'eau saumâtre sous forme de quelques centimètres ou de quelques dizaines de centimètres de lits organiques plus ou moins décolorés et décomposés. Ce type de sédiments, riche en corps gras, amino-acides, gaz ... et tout-à-fait comparable au sapropèle décrit par Kuznetsov (1951). (Ici l'origine algaire explique l'absence de matières humiques.)

Conclusions

La production de matériel organique dans les formations coralliennes est le fait de plusieurs milieux, parfois même de micromilieux dont les deux types de mares que nous venons de décrire constituent deux exemples. Ils sont de faible étendue par rapport aux autres sources de matériel organique (le récif corallien lui-même et les eaux du lagon et du bord récifal). Mais un de leur caractère essentiel, dû aux conditions climatiques, est une activité constante dans la production organique ; c'est la poursuite incessante de cette synthèse qui leur confère sans doute une grande importance, l'avantage revenant d'ailleurs aux mares d'eau saumâtre. (La part de la végétation ligneuse serait indiquée par l'abondance de matières humiques dans le sédiment.) Les récifs coralliens se présentent ainsi comme des réservoirs pour les produits de décomposition des êtres qui les on construits (Bergmann et Lester, 1940) et aussi pour les produits exogènes dus aux activités biologiques auxquelles ils servent de support (Bergmann, 1963).

Remerciements

Je tiens à remercier M. Louis et Mme Califet, de l'Institut Français du Pétrole, qui ont bien voulu m'accueillir dans leur laboratoire et m'ont ainsi permis d'entreprendre cette étude. Je remercie également M. Mazliak et M. Bomsel, du Laboratoire de Physiologie végétale appliquée de la Sorbonne qui m'ont prodigué conseils et aide pour les études chromatographiques ainsi que M. Salvat du Laboratoire de malacologie du Museum National d'Histoire Naturelle qui m'a très aimablement fourni la détermination du *Cerithium patiens.*

Bibliographie

Bailey, R. W. (1953) Composition of the organic fraction of the Waikato and Hauraka peats. *N.Z. J. Sci. Tech.* **A 34** , 421–429.

Bergmann, W. et Lester, D. (1940) Coral reefs and petroleum formation. *Science*, **92**, 452–453.

Bergmann, W. (1963) Geochemistry of lipids. In Breger, I. A. (1963)

Breger, I. A. (1949) Depositional environment of source materials forming organic sediments. *Am. J. Biol.* **36,** 382.

Breger, I. A. (1963) *Organic Geochemistry*. Pergamon Press.

Frederick, J. F. (1956) Physico-chemical studies of the phosphorylating enzymes of oscillatoria princeps. *Physiol. Plantarum*, **9,** 446–451.

Höcht, H., Martin, H. H. et Kandler, O. (1965) Zur Kenntnis der chemischen Zusammensetzung der Zellwand der Blaualgen. *Z. Pflanzenphysiologie*, **53,** 39–57.

Hough, L., Jones, J. K. et Wadman, W. H. (1952) An investigation of the polysaccharide components of certain fresh-water algae. *J. Chem. Soc.* 3393–3399.

Kuznetsov, S. I. (1951) Influence of micro-organisms in forming sapropel deposits. *Mikrobiologiya*, **20,** 245–255.

Polglase, W. J. (1955) Polysaccharides associated with wood cellulose. *Adv. Carbohydr. Chem.*, **10,** 283–333.

Sickenberger, (1892) Zur Entstehung des Erdöls. *Chem. Z.*, **15,** 1582–1583.

Silliman, B. Jr. (1846) On the chemical composition of calcareous corals. *Am. J. Sci.* **I,** (2), 189–199.

Trichet, J. (1966) À paraître in *C.R. Soc. Biogéogr.*

Zhuraoleva, M. M. (1954) Chemical composition of peat. *Torfyanaza Prom.*, **31,** n°. 5, 22–24.

STRUCTURAL CHARACTERISTICS OF HUMIC SUBSTANCES IN RECENT LAKE SEDIMENTS

by

RYOSHI ISHIWATARI

Department of Chemistry
Faculty of Science
Tokyo Metropolitan University
Tokyo, Japan

Introduction

In soils, sediments and sedimentary rocks, there exist the yellow to dark-brown coloured organic substances termed "humic substances" or "humus", which, because of the peculiarity of their nature, cannot be related to any of the existing groups of organic chemistry. These substances often predominate in the organic substances present in natural formations. However, the origin and nature of these substances are not yet well investigated. In particular, those in sediments and sedimentary rocks have been less investigated, as compared with those in soils.

According to Kononova, soil humic acid, which is the main component of humic substances, is a high-molecular-weight compound, which is formed as a result of the synthesis of simpler, chemically individual substances formed by the decomposition of tissue components of dead organisms. In other words, soil humic acid is considered to be a heteropolycondensate of aromatic compounds with products of protein decomposition, with the possible participation of substances of carbohydrate nature.(1)

Consequently, it is very interesting for geochemists, ecologists, palaeobiologists and investigators on petroleum and coal genesis, to study the nature, origin and transformation of "humic substances" in bottom sediments of water bodies in relation to the individual organic substances, such as chlorophyll derivatives, amino-compounds, carbohydrates, organic acids and hydrocarbons. The latter have been comparatively well studied by various authors.

The starting point of the present study is based on the assumption that the chemical nature of "humic substances" must depend on the nature of the source materials and biological, chemical and physical factors acting in their transformation, especially in its early stage, and therefore, humic substances formed in the bottom sediments of bodies of water may differ from those formed in soils. This thought is not new, but was already pointed out by some authors. For example, Karavaev and Budyak[(2)] reported the aliphatic character of humic acids from fresh-water lake sediments, on the basis of the presence of aliphatic organic acids in the oxidation products of humic acids.

The present study was undertaken to obtain structural information on the humic substances in lake sediments as the first step in a study of the geochemical transformation of humic substances in the bottom sediments of water bodies. For this purpose, humic substances were extracted from the bottom sediment samples of seven fresh-water lakes, and their various physico-chemical characters were examined. In particular, the study was focused on the chemical structure of humic acid. Furthermore, in order to clarify the characteristics of humic substances from lake sediments, a comparative study was undertaken of humic substances from lake sediments, soils, bog peat and coals.

Experimental

Sample Collection and Preparation of Humic Substances

SAMPLE COLLECTION. The location and description of the lakes from which samples were collected are given in Fig. 1 and Table 1. In these lakes, the source material of the organic substances in bottom sediments is thought to be mainly phytoplankton. The samples of the lake sediments studied were taken with an Ekman-Berge bottom sampler. The samples of bog peat and bay sediment were collected at Ishikari-gun, Hokkaido, and in Minamata Bay, Kumamoto-ken, Kyushu. The core sample of sedimentary rock (2877 m depth at Kasukabe, Saitama-ken) was kindly given by the Geological Survey of Japan.

EXTRACTION AND PREPARATION OF HUMIC SUBSTANCES. The extraction method for the humic substances is based on that presented by Tyurin.[1]

The air-dried mud sample was ground in a porcelain dish and passed through a 16–100 mesh sieve. The organic solvent extraction was carried out from 10–100 g sample in a Soxhlet apparatus, using 1:1 ethanol–benzene mixture or 1:1:2 methanol–acetone–benzene mixture. The extraction was continued for about 40 hours until the solvent no longer showed

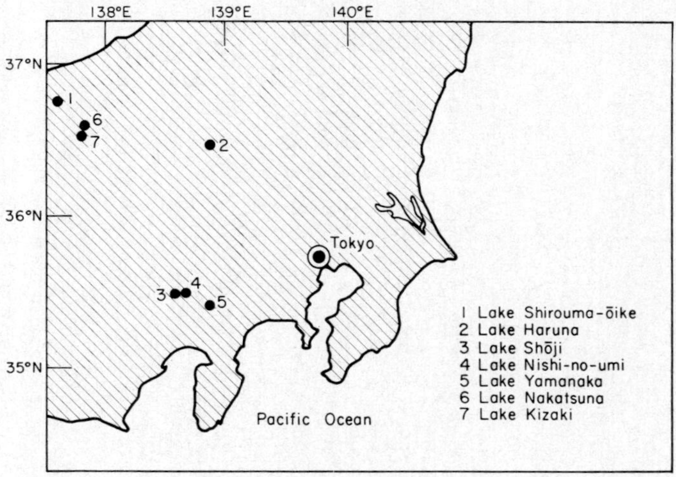

FIGURE 1. — Location of lakes studied.

any colour. In this extraction, chlorophyll derivatives in the sample were completely removed.

When the extraction was completed, the thimble containing the sample was removed from the apparatus, the sample transferred to a porcelain dish and the solvent completely evaporated.

Known volumes of the organic solvent extract were taken for the determination of carbon and chlorophyll derivatives. The other part of the extract was evaporated to dryness on a steam bath. The dried extract was subjected to the measurements of its physical properties.

After extraction with organic solvent, the sample was transferred to a beaker, covered with 200 ml 0·1 N sodium hydroxide solution, and allowed to stand with occasional mixing for a whole day. The supernatant liquid was separated from the sample by centrifugation, and the sample was again covered with alkali solution and allowed to stand for a whole day.

TABLE 1. — DESCRIPTION OF LAKES[3]

Name of Lake	Location (Prefecture)	Altitude (m)	Area (km^2)	Max. depth (m)	Volume (km^3)	Type	Origin*
1. Shirouma-ôike	Nagano	2379	0·06	13·5	0·0004	Oligotrophic	a
2. Haruna	Gumma	1084	1·23	13·0	0·01	Mesotrophic	b
3. Shôji	Yamanashi	896	0·65	11·2	0·0024	Mesotrophic	c
4. Nishi-no-umi	Yamanashi	898	2·1	66·5	0·073	Oligotrophic	c
5. Yamanaka	Yamanashi	982	6·5	13·2	0·06	Mesotrophic	c
6. Nakatsuna	Nagano	815	0·14	12·0	0·0008	Mesotrophic	d
7. Kizaki	Nagano	764	1·4	29·5	0·025	Mesotrophic	e

*(a) Lake, which has connection with recent vulcanism, although the exact origin of lake is still unknown.
(b) Lake filled up by the eruption of central cones in the caldera.
(c) Lakes formed by a barrier constituted by a volcano or group of volcanoes across a valley
(d) Lake held by rockslide dams.
(e) Fluviatile dam holding lake, due to deposition by a lateral tributary, either temporarily, of more sediment than the main stream can remove.

These treatments were repeated for about 1 month until the alkali solution had a slightly yellow colour.

After each extraction, the solution obtained after centrifugation was immediately transferred to a separate beaker, concentrated sulphuric acid was added to make the solution below pH 2, and it was placed on a steam bath for 0·5–1 hour. The separation of the precipitate of humic acids from the clear solution of fulvic acids was carried out by centrifugation. The precipitate of humic acids was dissolved in a 0·1 N sodium hydroxide solution, and known volumes of this solution were taken for the carbon determination. The carbon content of the fulvic acids was determined for the supernatant liquid. The carbon determination was made by the iodic acid decomposition method.[4]

In order to obtain the humic substances readily soluble in 0·1 N sodium hydroxide solution, short-time extraction of humic substances was carried out. The method consists of mixing the mud sample with 0·1 N sodium hydroxide solution in a separating funnel for 3 hours. This procedure was repeated twice.

For purification of humic acid, the crude preparation was dissolved in 0·1 N sodium hydroxide solution, and gel-filtered through Sephadex G-25 for several samples, and precipitated with dilute sulphuric acid, washed with distilled water, and dried in air. The ash content of the final preparation was below 3%. The purified preparation of humic acid was subjected to measurement of infrared spectra, X-ray diffraction, density and elemental composition.

Chromatographic separation of the organic solvent extract into n-heptane, benzene and methanol fractions was made by the method of Smith.[5]

Determination of Properties

ULTRAVIOLET AND VISIBLE ABSORPTION. The absorbances of organic solvent extract, humic acid and fulvic acid extracts were measured with a Hitachi EPS recording spectrophotometer. The organic solvent extract was subjected to the measurement in a methanol solution, humic acid in a 0·1 N sodium hydroxide solution, and fulvic acid in a 0·05 N sulphuric acid solution.

INFRARED ABSORPTION. Infrared spectra were recorded, using a Shimazu IR 27A type double-beam spectrophotometer. Pellet mixtures of sample concentration of 1% by weight in potassium bromide were prepared

Films were used for some organic solvent extracts and were prepared by spreading the sample on an absorption cell (KR-6). Carbon tetrachloride and carbon disulphide were used for the fractionated samples of the organic solvent extracts. A 0·5 mm cell was used for solution measurements.

DENSITY. Density measurements were made by a pycnometric method for solid material in water at 27–30° C. In order to obtain the true density of solid materials having an internal surface, the samples were ground and suspended in water under water-pump vacuum pressure in a pycnometer. The submerged particles were kept under vacuum in water until no more air bubbled. Duplicate determinations on a 400–1000 mg sample agree within one percent.

ELEMENTAL ANALYSIS. Microanalyses for carbon, hydrogen, nitrogen and ash were performed. The oxygen was estimated by difference.

THIN-LAYER CHROMATOGRAPHY. The molecular weights of humic acids were estimated by the Sephadex gel-filtration method, using the method employed by Andrews on proteins.[6]

Sephadex G-100 gel particles (40–120 μ) were swollen in distilled water for 3 days before use. Thin gel layers were prepared on glass plates (20 cm × 20 cm) with an apparatus manufactured by Yazawa Chem. Co. Ltd. The thickness of the gel layer was kept at 0·5 mm, and the angle of tilt of the glass plates was 10° from horizontal. The flow of buffer solution through the layer was maintained for one day before the application of samples. The flow rate was about 3 cm/hr with respect to cytochrome c, which was used as the reference substance. Samples (20–50 μg in 2–4 μl) were applied to the plates in a 0·05 M phosphate buffer solution (pH 7) containing 0·1 M potassium chloride. After interruption of the run, the plates were dried at 60° C in an oven. The location of spots of proteins used as the references was determined with a ninhydrin spray. As reference substances for the estimation of molecular weight of humic acids, cytochrome c (M.W. 12,400), chymotrypsinogen (M.W. 25,000), ovalbumin (M.W. 45,000) and serum albumin (bovine, M.W. 67,000) were used. These samples were kindly given by Dr. S. Sasakawa and Dr. T. Masuda of Tokyo Metropolitan University.

The standard curve of migration distance against the molecular weight of protein, shown in Fig. 2, was used for the estimation of the molecular weight of the humic acid.

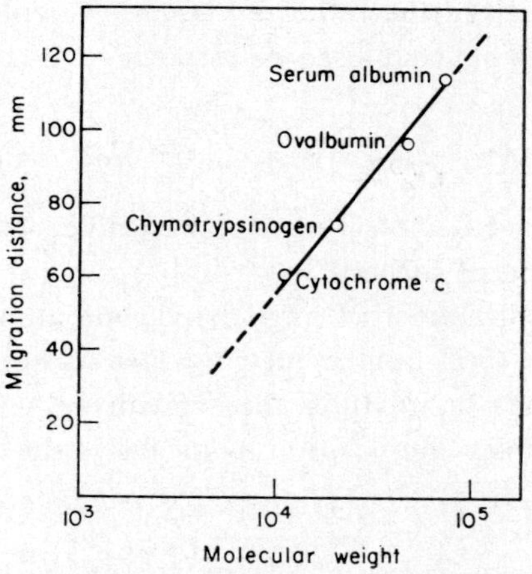

FIGURE 2. — Plot of migration distance against molecular weight of protein.

X-RAY DIFFRACTION. X-ray diffraction patterns were recorded by a Toshiba self-recording X-ray diffractometer (Model ADX-103). The powdered samples were packed into an acryl resin holder. A nickel filter was used to achieve monochromatic CuKα radiation.

COMPOSITIONAL ANALYSIS. Analyses of nitrogen compounds and carbohydrates were performed on the hydrolysates of humic acid. For hydrolysis, humic acid was refluxed with inorganic reagents, such as sulphuric and hydrochloric acids and sodium hydroxide. Total nitrogen in the samples was determined by a micro-Kjeldahl method, ammonium nitrogen by a micro-distillation of ammonia, and total amino nitrogen photometrically with ninhydrin. Carbohydrates were determined by a phenol-sulphuric acid method.[7]

ESTIMATION OF RING STRUCTURE. In order to determine the structural character of humic acid, the densimetric method for ring analysis developed by van Krevelen was applied.[8] The method is based on the Traube's result on the molar volume of a liquid, which is given by:

$$V_M = \sum_i n_i \cdot V_i - K_M + \Phi_M$$

where n_i = number of atoms of the ith kind per molecule;
V_i = volume of these atoms per gram atom;
K_M = molar volume contraction caused by structural factors;
Φ_M = molar free space due to end groups.

Van Krevelen derived the following equation applicable to high polymeric resins, by using cellulose, polystyrene and graphite as standard substances:

$$M_c/d = 9{\cdot}9+3{\cdot}1\ H/C+3{\cdot}75\ O/C+1{\cdot}5\ N/C-(9{\cdot}1-3{\cdot}65\ H/C)\ R/C$$

where M_c = reduced molecular weight (molecular weight per gram atom of carbon);

R/C = total number of rings per carbon atom.

Accordingly, the total number of rings per carbon atom in aliphatic-aromatic compounds or mixtures thereof can be derived from data on density and elementary composition by means of the next equation:

$$R/C = \frac{0{\cdot}9+3{\cdot}1\ H/C+3{\cdot}75\ O/C+1{\cdot}5\ N/C-(12{\cdot}011+1{\cdot}008\ H/C+}{9{\cdot}1-3{\cdot}65\ H/C}$$

$$\frac{+16000\ O/C+14{\cdot}007\ N/C)\ 1/d}{9{\cdot}1-3{\cdot}65\ H/C}$$

where, H/C, O/C and N/C = atomic ratios.

Results and discussion

The Composition of Humic Substances in Lake Sediments

The results on the humus composition are given in Table 2.

The organic substances directly soluble in organic solvent amount to 4–7% of the total organic substances in lake sediments. This value is practically the same as those in soils and sediments presented by various authors. Of the organic solvent soluble substances, the chlorophyll derivatives (mainly chlorophyll a and pheophytin a) amount to 2·2–8·0%. Since the chlorophyll derivatives are the main components of pigments, about 90% of the organic solvent soluble substances may be accounted for by the non-pigments. By chromatographic analysis, it was revealed that the polar compounds (O–N–S compounds) amount to about 90% of the organic solvent soluble substances, as shown in Table 3.

Humic substances directly extractable with a dilute alkali solution amount to 41–53% of the total organic substances. And 43–76% of the extractable humic substances was passed into solution by short-time treatment with alkali solution. "Humin", or the organic substances remained after the alkali treatment, amounts to 43–51% of the total organic sub-

TABLE 2. — COMPOSITION OF HUMIC SUBSTANCES IN LAKE SEDIMENTS

Name of Lake	C (mg/g)	N (mg/g)	Pheophytin a	Composition (% of the total carbon in sediment)												
					6 hours extraction						A month extraction					
				(1)	Humic acid (A)	Fulvic acid (B)	(A)/(B)	PQ	(A) + (B)	Humin	Humic acid (A)	Fulvic acid (B)	(A)/(B)	PQ	(A) + (B)	Humin
1. Shirouma-ôike	38·2	3·37	0·10	4·6	14	22	0·6	39	36	58	19	34	0·6	36	53	43
2. Haruna	52·6	4·76	0·33	7·2	8	23	0·3	26	31	59	17	24	0·7	41	41	51
3. Shôji	41·9	4·44	0·52	6·5	18	12	1·5	60	30	63	23	28	0·8	45	51	43
4. Nishi-no-umi	30·5	3·21	0·22	5·5	18	8	2·1	68	25	70	17	29	0·6	37	46	48
5. Yamanaka	45·1	4·36	0·17	4·3	11	9	1·2	55	20	76	16	31	0·5	34	47	48
6. Nakatsuna	51·8	4·06	0·31	5·4	24	6	4·0	80	30	65	31	18	1·7	63	49	46
7. Kizaki	43·0	3·19	0·31	5·9	20	8	2·5	71	28	67	27	19	1·4	59	46	48

1) The organic substances soluble in organic solvent (The amount of pheophytin a is not subtracted).

TABLE 3. — CHROMATOGRAPHIC SEPARATION OF THE ORGANIC SUBSTANCES SOLUBLE IN ORGANIC SOLVENTS

Sample	Substances soluble in organic solvent		Group composition (% of substances soluble in organic solvent)			
	% of sediment	% of total organic content*	n-Heptane fraction	Benzene fraction	Methanol fraction	Remaining on silicagel
Lake sediment (L. Haruna)	0·85	8·2	5·9	2·8	57·5	33·9
Bay sediment	1·18	17	17·4	16·5	36·6	29·7
Sedimentary rock	0·057	3·5	8·1	15·1	58·4	18·4
Bog peat	4·86	14	0·7	3·8	45·2	50·5

* Total organic content was calculated, assuming that a half of organic substances in the sediment consist of elemental carbon.

stances. Accordingly, it can be seen that the ratio of organic solvent soluble substances : humic substances soluble in alkali : humin has a nearly constant value of $1:9\pm3:9\pm2$ in the sediment samples.

The composition of humic substances can be also characterized by the *PQ* value (the percentage of humic acid to the extractable humic substances). In the case of soils, this value increases with increasing degree of humification.[9] Evidently the *PQ* values on a month's extraction are lower than those on short-time extraction, as given in Table 2. The decrease of the *PQ* values on a month extraction may be attributed to the decomposition of humic acid to fulvic acid during alkali treatment. The only exception is the case of Lake Haruna.

In addition, it is interesting to point out that the *PQ* values of the sample from Lake Nakatsuna and Lake Kizaki are considerably higher than the others. This indicates that, according to Kobo's idea on soil humic substances,[9] the degree of humification in the sediment of the former lakes is higher than that of the latter. The *PQ* values of the samples of the latter lakes correspond to the lower values among those of soil humic substances in Japan (*PQ* : 34–88).[10]

Although the number of the sediment samples studied is limited, the following facts can be pointed out.

No essential difference in the humus composition can be observed between lake sediments and soils. The humus content of sediment is relatively high, and humic acids are predominantly in a form readily soluble in alkali solution. No relationship, however, can be seen between humic acid content and the total organic content, in contrast with the case of soils.[1] An inverse relationship seems to exist between the fulvic acid and humic acid contents. Judging from the *PQ* values, the humic substances of lake sediments correspond to the group of soil humus with a comparatively low degree of humification.

According to Tyurin's idea, derived from the examination on the cause of regular changes in the content and nature of the humic acids in the composition of the humic substances in soils, (1) the humus accumulation in soil is associated with the decomposition of a large amount of dead organisms under humid conditions, (2) the formation of humic acid is accelerated by the lower humidity, and (3) the decomposition of humic acid, with which the formation of fulvic acid is closely connected, increases under conditions of adequate moisture and higher temperature.[1]

By analogy with soil humic substances, the following interpretation may be given to the results on the composition of humus in lake sediments.

In lake sediments, the amount of organic substances accumulating in the sediment is small compared with the case of soils, because a fairly large part of the dead plankton appears to decompose in the course of deposition through water. However, the deposition of the less decomposable, therefore, well preservable parts of dead plankton in bottom sediment, and the surrounding physico-chemical conditions, such as low temperature and nearly reducting conditions through out a year, are favourable for the accumulation of the humic substances, in spite of the possible slow humification processes, but not for decomposition processes. On the other hand, under such conditions, the humic acid cannot develop to a high degree nor form a stable linkage with matrix materials, since the humic acid in sediment is always exposed to water. Therefore, the lake humic substances can be dissolved directly in dilute alkali solution without preliminary removal of exchangeable calcium and other ions.

The Optical Properties of Humic Substances

ULTRAVIOLET AND VISIBLE REGION. The data on light absorption of humic acid and fulvic acid are shown in Fig. 3 (1) – (2) and Table 4. Optical properties are expressed in *RF* value (absorption coefficient of extract at wave length 600 mμ and carbon concentration of 10 mg/ml)

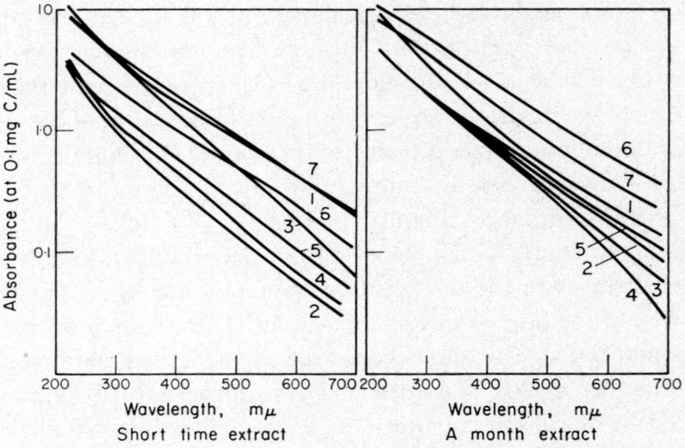

FIGURE 3 (1). — Ultraviolet and visible absorption spectra of humic acids from lake sediments.

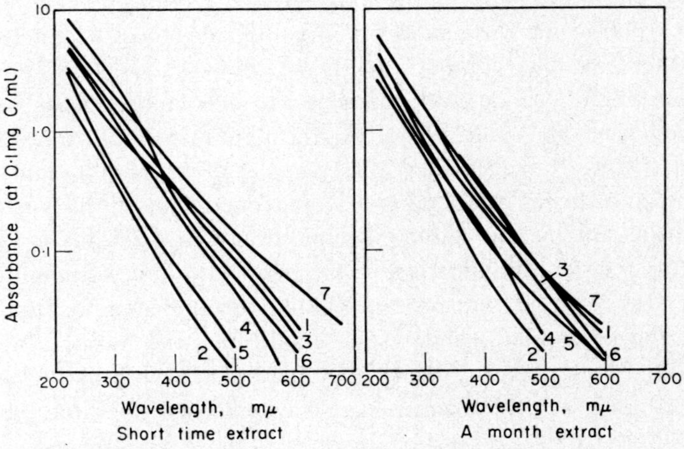

FIGURE 3 (2). — Ultraviolet and visible absorption spectra of fulvic acids from lake sediments.

and $\Delta \log k$ value (logarithm of the ratio of absorption coefficient at 400 mμ to that at 600 mμ). In soil humic substances, the *RF* value becomes higher and $\Delta \log k$ value becomes lower with increasing degree of humification.[9]

Absorption spectra of humic substances in the ultraviolet and visible regions, show no maximum, and differ little from those of soil humic substances. *RF* values and $\Delta \log k$ values are 11–62 and 0·61–0·84 for

TABLE 4. – OPTICAL PROPERTIES OF HUMIC SUBSTANCES FROM LAKE SEDIMENTS

Name of Lake	6 hours extraction				A month extraction			
	Humic acid		Fulvic acid		Humic acid		Fulvic acid	
	*RF**	*Δ* log *k***	*RF*	*Δ* log *k*	*RF*	*Δ* log *k*	*RF*	*Δ* log *k*
1. Shirouma-ôike	43	0·64	2·2	1·18	26	0·64	2·0	1·15
2. Haruna	6	0·79	~0·2	1·36	18	0·71	~0·2	1·53
3. Shôji	20	0·86	1·9	1·21	14	0·78	1·3	1·20
4. Nishi-no-umi	7	0·89	~0·2	2·04	11	0·84	~0·1	2·18
5. Yamanaka	12	0·78	~0·3	1·46	20	0·70	1·3	0·96
6. Nakatsuna	23	0·68	1·6	1·27	62	0·61	1·2	1·41
7. Kizaki	45	0·68	5·2	1·03	40	0·63	2·3	1·00

* RF, absorption coefficient of extract at wave length 600 mμ and carbon concentration of 10 mg/ml
** Δ log k, logarithm of the ratio of absorption coefficient at 400 mμ to that at 600 mμ

humic acids, 0·1–2·3 and 0·96–2·84 for fulvic acids, respectively. An inverse relationship between *RF* value and Δ log *k* value was observed, as in the case of soils. Clearly, the values for *RF* and Δ log *k* of the humic substances from Lake Nakatsuna and Lake Kizaki, are different from those from the other lakes. This agrees fairly well with the relationship with respect to the *PQ* value. If it is assumed, modifying Kononova's statement,[1] that the extinction coefficient of humic acids depends on the degree of condensation of the molecule, it could be concluded that the degree of condensation of the humic acid molecule from Lake Nakatsuna and Lake Kizaki is higher than that of the other lakes. Accordingly, the former is higher than the latter, with respect to the degree of humification.

INFRARED REGION. The absorption spectra of the humic substances from lake sediments are given in Fig. 4. The spectra of a typical green algae (scenedesmus), humic acid from bay sediment, bog peat and sedimentary rock are also given, for comparison, in Figs. 4 (1) – (3). The final interpretation of the spectra is listed in Table 5.

Practically no differences can be observed in the absorption spectra of the organic solvent soluble substances, humic acid and fulvic acid among the samples from the seven lake sediments.

The marked characteristics of the organic solvent soluble substances of lake sediments are the clear presence of a long chain of the type —$(CH_2)_n$—(n = 4 or more) and absence of the bands indicating the aromatic nature. These characteristics become more clear when the spectra

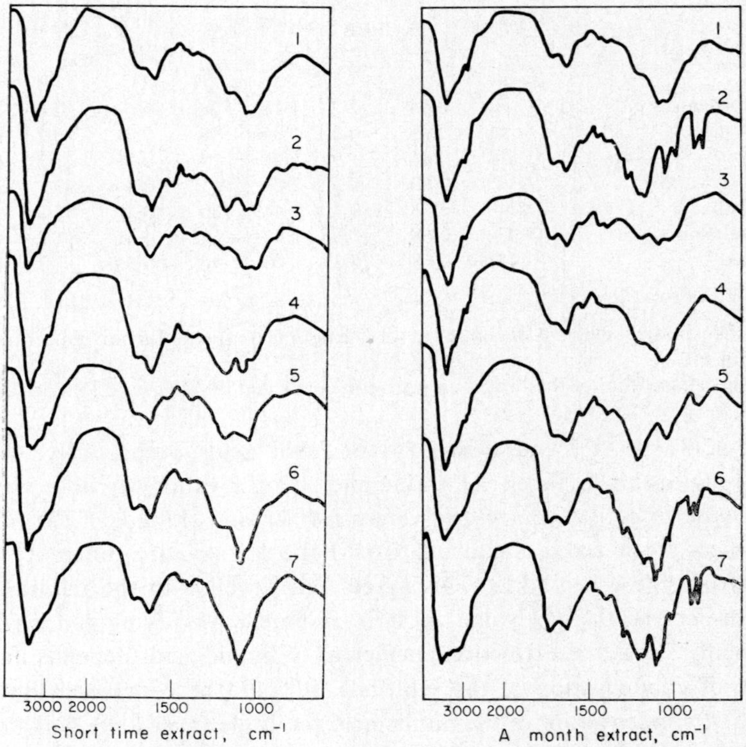

FIGURE 4 (1). — Infrared absorption spectra of humic acids from lake sediments.

of the fractionated samples of the organic solvent soluble substances are examined, as shown in Fig. 5. The absorption bands at 1615, 1600, 740 and 700 cm^{-1}, suggesting the presence of an aromatic structure, are absent in the sample from lake sediment, while the bands appear in the samples from bay sediment, bog peat and sedimentary rock.

The absorption spectra of humic acids from lake sediments show the bands at 3450, 1720, 1640, 1540, 1220 and 1100–1020 cm^{-1}. The peak at 1540 cm^{-1} and a part of the peak-height of 1640 cm^{-1} are thought to be due to the polypeptide-like bonds, and the peak at 1100–1020 cm^{-1} to polysaccharide-like substances. The main bases for interpretation of the 1640 and 1540 cm^{-1} bands are that: (1) relative intensities of 1640, 1540 cm^{-1} and the other bands of humic acid are similar to those of proteins and scenedesmus (abounding in proteins), and (2) there is a nearly straight-line relationship between absorbance of the 1540 cm^{-1} band and nitrogen

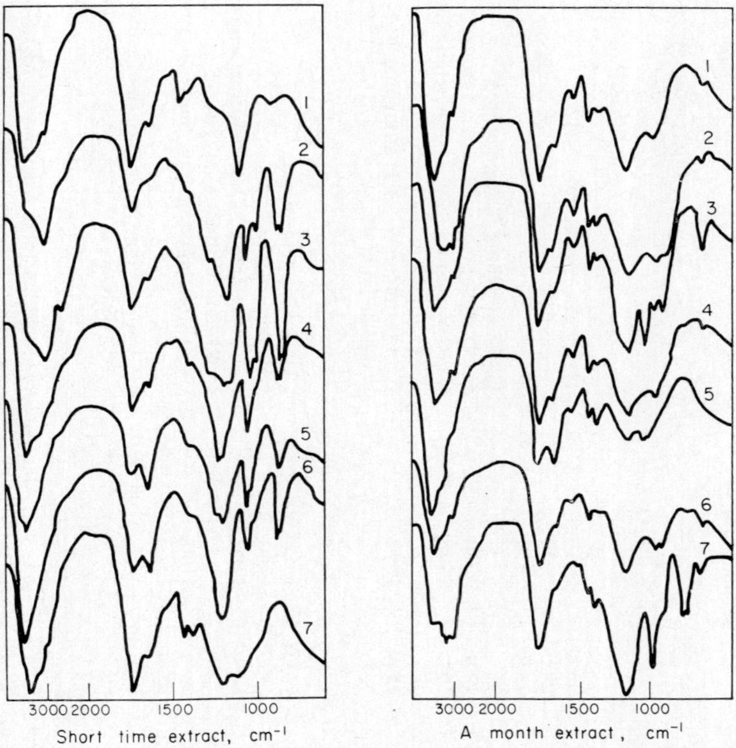

FIGURE 4 (2). — Infrared absorption spectra of fulvic acids from lake sediments.

content of the humic acid remaining after acid or alkali hydrolysis (see Fig. 6). The bases are described in more detail in previous publications.[11, 12] The presence of protein-like substances and carbohydrates will be mentioned later in this paper. It is also established that the humic acid from lake sediment contains no, or very few, aromatic bonds. Aliphatic CH, CH_2 and CH_3 groups, on the other hand, do occur. C=O groups, CN and NH groups due to polypeptide-like bonds, and C—O groups due to polysaccharides, are probably also present. On the contrary, humic acids from bay sediment, bog peat and sedimentary rock, have the 1620, 1505 and

TABLE 5. — INTERPRETATION OF IR-SPECTRA OF HUMIC SUBSTANCES AND SCENEDESMUS

Humic substances				Scenedesmus			Interpretation
(A)	Humic acid	Fulvic acid (1)	Fulvic acid (2)	(3)	(4)	(5)	
3400	3450	3450	3450	3300	3350	3450	Hydrogen bonded OH/NH
2930	2940	2960	2960	2925	2930	2925	Aliphatic CH_2 and CH_3 stretching
2850	2850	2850	2850	2850	2850	2850	
1740	—	1740	1740	—	—	—	C=O stretching (ester, ketone)*
1720	1720	—	—	1720	1720	—	C=O stretching (acid, ester, ketone)**
—	1640	1640	1640	1645	1645	1645	C=O stretching + C—N stretching (amide I)
							C=O stretching (acid, quinone?)
—	1540	—	1540	1530	1530	—	C=O stretching + CH stretching + NH deformation (amide II)
1470, 1380	—	—	—	—	—	—	CH_2 & CH_3 deformation
—	1450	1440	1440	1455	1450	1450	CH_2 & CH_3 deformation
—	1400	1400	1400	1400	1380	—	CH_3 deformation
—	1220	1220	1220	1240	1220	—	C—N stretching + NH deformation (amide III)
							C—O deformation (acid, ester)
1170	—	—	—	—	1160	—	C—O stretching (alcohol, linear & cyclic ester)
—	1100	—	1070	1090	—	1070	
	~1040		~1000	~1030		~1010	C—O stretching (polysaccharide, cellulose)
725	—	—	—	—	—	—	C—C skeletal (—$(CH_2)_n$—, n = 4 or more)

(A) Organic substances soluble in organic solvent

(1) Short time extract; (2) A month extract; (3) Whole body; (4) Alkali soluble, acid insoluble fraction; (5) Alkali and acid insoluble fraction

* For example, 5-membered-ring ketone, α-, β-, γ-keto ester, δ-lactone, Rs—COO—R and Ar—COO—Ar etc.

** For example, 6-membered-ring ketone, α-diketone, Rs—CO—Rs, Rs—COOH, R—COSH, Ar—COO—R and C=C—COOR. etc.

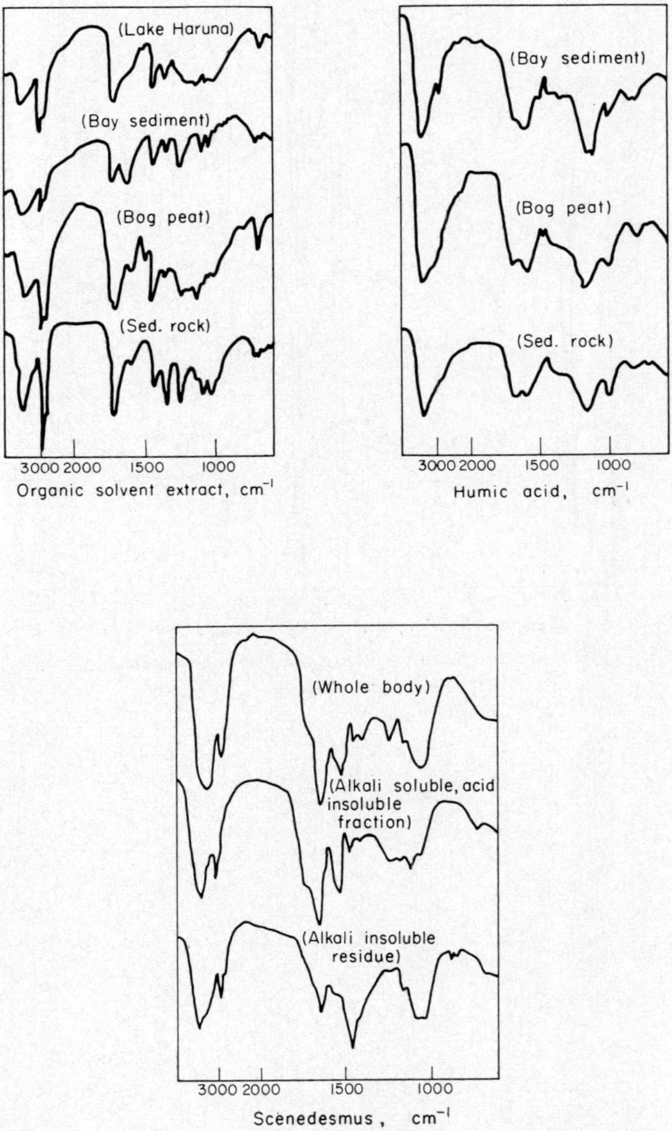

FIGURE 4(3). — Infrared absorption spectra of Scenedesmus and humic acids from various sediments.

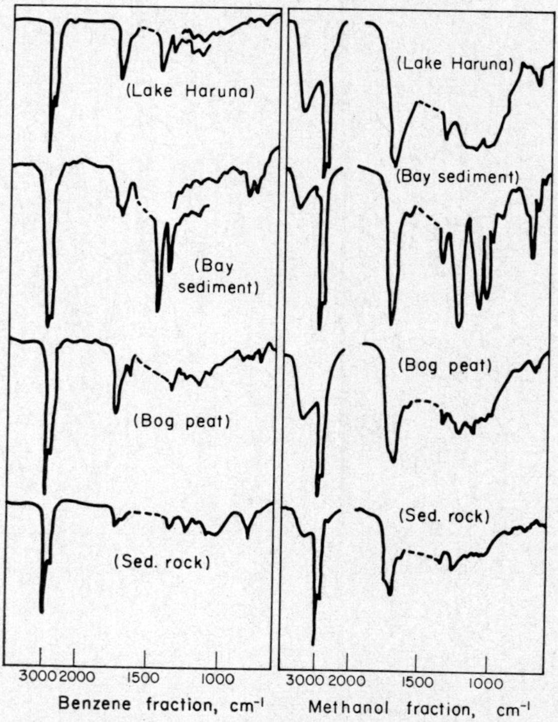

FIGURE 5. — Infrared adsorption spectra of the fractionated samples of the organic solvent extracts.

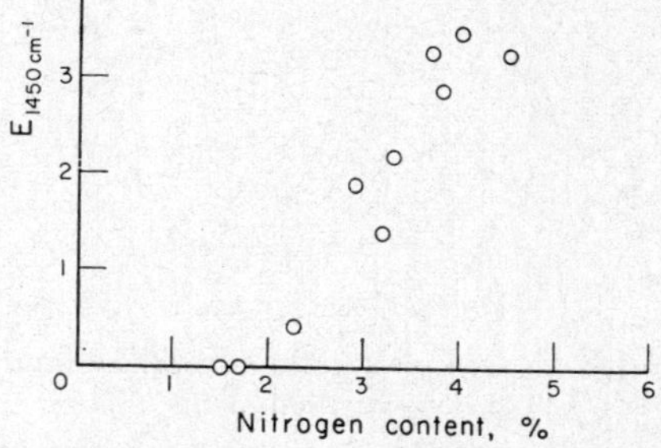

FIGURE 6. — Absorbance of 1540 cm^{-1} band against nitrogen content of the humic acid recovered after hydrolysis.

880–850 cm^{-1} bands, which are considered to be evidence of their aromatic nature.

The fulvic acid is distinguishable from the humic acid by the following points. (1) The band due to the aliphatic CH_2 and CH_3 groups of the fulvic acid, appears at 2960 cm^{-1} instead of 2940 cm^{-1}. This indicates that the CH_3 content of the fulvic acid is higher than that of the humic acid.[13] (3) The absorption spectra of the fulvic acid shows no sign of the presence of polypeptide-like bond and polysaccharides, with the exception of the fulvic acids obtained after one month's treatment with alkali. In the latter case, these substances passed into solution, probably as a result of the partial decomposition of humic acids or of proteins and polysaccharides in the sediment sample during long-time contact with alkali solution.

Structural Characteristics of Humic Acid

In studying the structural characteristics of humic acid, the author used the purified preparation extracted from Lake Haruna, and scenedesmus and humic acid from bog peat for comparison.

MOLECULAR WEIGHT. The result obtained by thin-layer chromatography is shown in Fig. 7. The width of the band shows graphically the relative amount of sample. Evidently, the molecular weight of the humic acid from lake sediment is distributed in the wide range of 4000–30,000. A very slight part of the humic acid reaches 100,000. The main part of the humic acid has a molecular weight of approximately 6400. Almost similar molecular weights were observed on the sample from scenedesmus (alkali soluble fraction coloured light-yellow, containing proteins). The

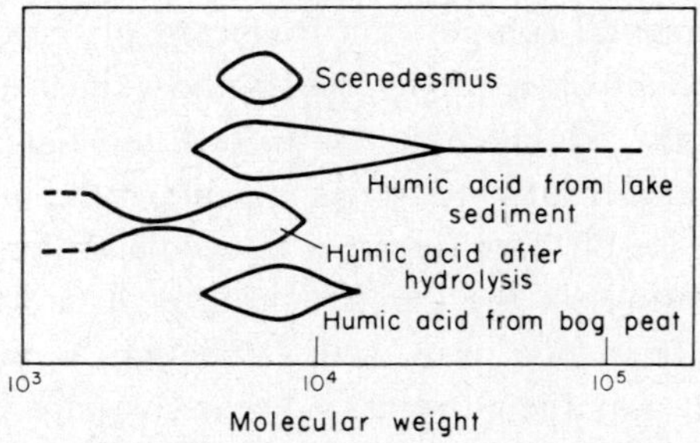

FIGURE 7. — Molecular weight distribution of Scenedesmus, humic acid from lake sediment and from bog peat.

molecular weight of the humic acid from bog peat (about 7600) is slightly higher than that from lake sediment. In the humic acid recovered after hydrolysis with 3 N sulphuric acid solution, the molecules over 9000 vanished, and on the contrary, those below 2000 appeared. But the molecules with about 6400 molecular weight did not disappear. This suggests that the unit of the humic acid molecules, if one exists, might have a molecular weight of about 6400.

As compared with the values of humic acids from various soils—1000–1200, 1350, 30,000–50,000,[1] 5000–100,000 (av. 25,000)[14]—the values obtained by the present author are not extraordinary. The present method using thin-layer chromatography, is simple and good for a comparative study of humic acid molecules.

DENSITY, ELEMENTAL COMPOSITION AND THE RING STRUCTURE. The results on the density, elemental composition and the ring structure are given in Table 6.

We can find no data on density of soil humic acids in the literature. The results show that humic acid from lake sediment has a density smaller than that of the humic acid from bog peat. However, the density of the humic acid from lake sediment is not specific, as compared with coals with wide variety of values (0·85–1·8).[15]

Data on the elemental composition of humic acid from lake sediment, have been also presented by Karavaev and Budyak.[2] The values have a strong resemblance to that obtained by the present author. A comparison of the data on elemental composition, shows a somewhat lower percentage of carbon and higher percentage of hydrogen and nitrogen in humic acid from lake sediment than in soil humic acids.[1]

The total number of rings per carbon atom was derived from the data on density and elemental composition by means of the equation given in the section *Determination of properties*. If the values obtained by thin-layer chromatography are taken as the average molecular weight of each humic acid, the number of rings per molecule could be calculated. On this basis the presence of three rings per molecule in the humic acid from lake sediments, 12 rings in the humic acid after hydrolysis and 61 rings in the humic acid from bog peat, was estimated. A differentiation between aromatic and naphthenic rings, however, cannot be made. Clearly the humic acid from lake sediment has a rather strongly non-cyclic structure. This is markedly distinguished from coals and probably humic acid

TABLE 6. — DATA ON THE DENSITY, ELEMENTAL COMPOSITION AND RING STRUCTURE

Sample	Elemental composition (%)				Atomic ratio			Empirical formula	Density g/cm^3	Number of rings per carbon atom (R/C)
	C	H	O	N	H/C	O/C	N/C			
Scenedesmus	51·88	7·47	31·43	9·22	1·716	0·454	0·152	$C_7H_{11}O_3N$	—	—
Humic acid from lake sediment	53·64	6·14	35·14	5·08	1·364	0·492	0·081	$C_{12}H_{17}O_6N$	1·396	0·01
Humic acid from lake sediment (after hydrolysis)	61·46	5·90	29·83	2·81	1·144	0·364	0·039	$C_{26}H_{29}O_9N$	1·332	0·04
Humic acid from lake sediment (after hydrolysis)	57·44	5·79	34·39	2·38	1·201	0·449	0·036	$C_{28}H_{34}O_{13}N$	—	—
Humic acid from lake sediment (after hydrolysis)	59·08	5·40	33·48	2·04	1·089	0·425	0·030	$C_{34}H_{37}O_{14}N$	—	—
Humic acid from lake sediment (Hydrolysable part)	46·3	6·81	36·7	10·2	1·753	0·595	0·189	$C_5H_9O_3N$	—	—
Humic acid from lake sediment (Hydrolysable part)	44.3	7·41	38·1	10·2	1·993	0·646	0·189	$C_5H_{10}O_3N$	—	—
Humic acid from bog peat	53·40	5·59	38·18	2·83	1·247	0·537	0·045	$C_{22}H_{27}O_{12}N$	1·498	0·18
Humic acid from lake sediment[2]										
(from Lake Galich)	56·21	5·84	31·40	6·55	1·238	0·419	0·100	$C_{10}H_{12}O_4N$	—	—
(from Lake Tolpolovskoje)	55·70	6·38	32·49	5·43	1·365	0·438	0·084	$C_{12}H_{16}O_5N$	—	—
(from Lake Disna)	52·0	6·0	—	—	1·375	—	—	—	—	—

from soils. Since the number of rings in the humic acid recovered after hydrolysis is slightly more than in the raw humic acid, the non-hydrolysable part—skeletal part(?)—of the humic acid may have a slightly cyclic structure. This result showing the non-cyclic nature of the humic acid from lake sediment, agrees with the result of the infrared analysis, and Karavaevs' conclusions.

The hydrolysable part of the humic acid also has clearly an aliphatic character, as indicated by elemental analysis (see Table 6).

X-RAY DIFFRACTION PATTERN. The X-ray diffraction patterns of scenedesmus, humic acids from lake sediment and from bog peat are shown in Fig. 8. These patterns with 1–3 diffuse peaks are typical of the amorphous state. Scenedesmus and the raw humic acid from lake sediment show similar lines at spacing near 9·8 Å and near 4·6 Å. On the basis that these two peaks are not present in the humic acid recovered after hydrolysis and the humic acid from bog peat, and by analogy with the X-ray pattern

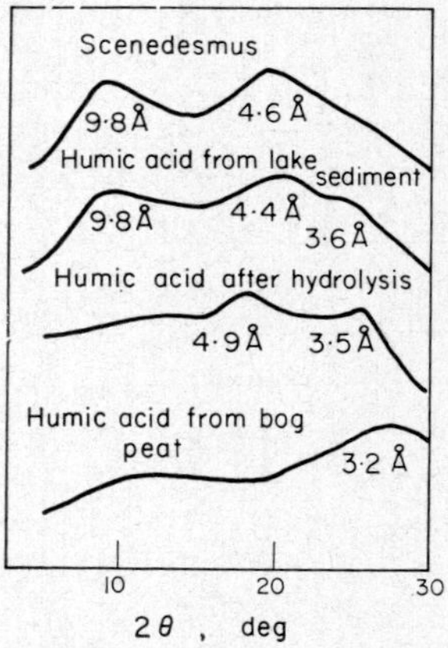

FIGURE 8. — X-ray diffraction pattern of Scenedesmus, humic acid from lake sediment and from bog peat.

given by the denaturation of β-proteins,[16, 17] these two peaks are considered to show the presence of the denatured β-configuration of proteins. The regular crosslinks (hydrogen bonds) between polypeptide chains are much damaged, but the 4·6 Å and 9·8 Å distances characteristic of the β-configuration are retained, which shows that short-range order is still present. The illustration is given in Fig. 9. The above estimation is not so strange, because the presence of protein-like substances in the humic acid has been inferred from the infrared and compositional analysis. However, the possibility that the 4·4–4·6 Å peak corresponds with the spacing of the amorphous polyethylene-type bond cannot be denied.[18]

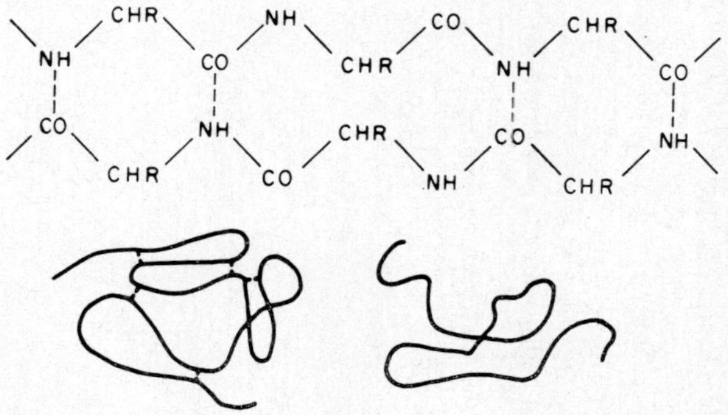

FIGURE 9. — β-configuration of polypeptide chain and its denaturation.

If a molecule has a graphite-type structure, the (002)-spacing at 3·5 Å, which is accepted as representing the spacing between the layers of a condensed aromatic system, will appear as in graphite, carbon blacks, coals petroleum asphaltenes and some soil humic acids. The peak near 3·5 Å, observed on the humic acid from lake sediment, corresponds with the above mentioned peak. However, it is hard to regard the peak as the (002)-spacing, because the non-cyclic structure and the non-aromatic character of the humic acid have been indicated by the densimetric analysis and the infrared analysis. This peak is probably due to a certain intramolecular diffraction, other than graphite-type spacing. Further study should be made on this peak.

COMPOSITION OF THE HYDROLYSABLE PART. The analytical results on the hydrolysate of humic acid from lake sediment are shown in Table 7.

The hydrolysable part of the humic acid reaches about 50% of its total weight. Total carbohydrate content of humic acid was estimated to be 2–4% (measured as "glucose-equivalent"). These values correspond to 4–10% of the hydrolysable part. The carbohydrate content of the humic acid from lake sediment does not differ so much from that of soil humic acid.[1]

About 70–90% of the nitrogen in the humic acid passed into solution on acid hydrolysis. At the same time, the infrared absorption band of humic acid at 1540 cm^{-1} vanished (see Fig. 6). About 40–50% and 20%

TABLE 7. — COMPOSITION OF THE HYDROLYSABLE PART OF HUMIC ACID

Condition of hydrolysis		Hydrolysed organic matter (% of humic acid)	Composition of hydrolysate (% of humic acid)					
				Nitrogen compound				
Medium	Duration (hr)		Carbohydrates	Total N		Amino N	Ammonium N	Unaccounted N
					(1)			
0·1 N H_2SO_4	17	34·1	3·61	3·48	21·8	1·30	0·79	1·39
0·1 N H_2SO_4	38	36·7	3·17	3·76	23·5	1·91	0·82	1·03
3 N H_2SO_4	25	49·7	2·04	4·50	28·2	—	0·80	—

(1) Values calculated as protein.

of the nitrogen in the hydrolysate is present in the α-amino and ammonium forms respectively. The form of the remaining nitrogen is unknown. These results suggest that the nitrogen in the hydrolysate is derived from substances closely similar to proteins. The hydrolysable nitrogen percentage of the total nitrogen seems to be a little higher in the humic acid from sediment than in soil humic acids. This is in accordance with the conclusion derived from the elemental analysis of the hydrolysate of humic acid, the X-ray diffraction and infrared absorption analysis of humic acid, as already mentioned. If it is assumed that the hydrolysable part of the total nitrogen consists of protein-like substances, it could be concluded that these substances together with carbohydrates occupy approximately 60–75% of the hydrolysate of humic acid, and correspond to 25–30% of the humic acid. The remaining organic substances probably consist of organic acids and others, broken up into small fractions by hydrolysis of the end parts of humic acid molecules.

To make clear whether or not the nitrogen in the humic acid molecule is an accidental admixture, the chromatographic fractionation of the humic acid was carried out through Sephadex G-25. As shown in Fig. 10, practically no difference could be observed among the various fractions of humic acid.

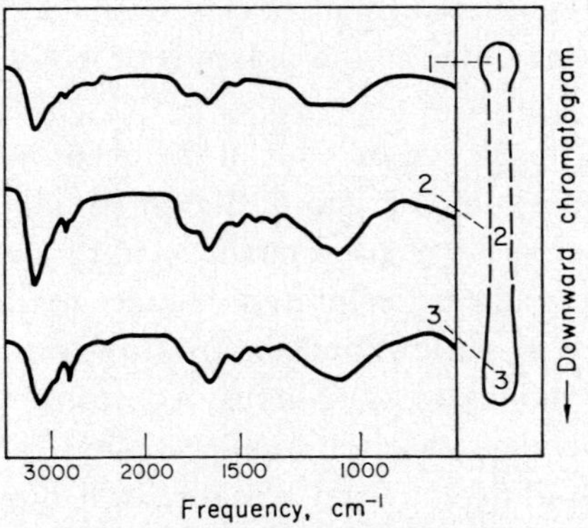

FIGURE 10. — Infrared absorption spectra of the humic acid fractionated through Sephadex G-25.

Conclusions

On the basis of the present study, the following characterization of the humic substances in lake sediments can be made.

(1) There are no essential differences between lake-sediment humus

and soil humus, with respect to humus composition and the ultraviolet and visible light absorption properties.

(2) However, the non-aromatic nature of the humic substances as a whole is most characteristic compared with soil humic substances.

(3) Humic substances in lake sediments belong to the group with a relatively low degree of humification.

As to the structure of the humic acid from lake sediment, it was established that;

(1) *General structure.* Density 1·396; empirical formula $C_{12}H_{17}O_6N$; molecular weight —6400 (4000—100,000); number of rings per carbon atom 0·01; mostly amorphous; hydrolysable part 30–50%; non-hydrolysable part 50–70%; functional groups OH/NH, C=O (ketone, carboxyl), C=O (quinone?), CONH (amide), C—O (alcohol, carboxyl).

(2) *Non-hydrolysable part.* Density 1·332; empirical formula $C_{26\sim34}H_{29\sim37}O_{9\sim14}N$; molecular weight 2000–6400 (<9000); number of rings per carbon atom 0·04; mostly amorphous; functional groups OH/NH, C=O (ketone, carboxyl), C=O (quinone?).

(3) *Hydrolysable part.* Empirical formula $C_5H_{9\sim10}O_3N$; carbohydrates 4–11%; Protein-like substances 56~64%; unknown compounds 25–40%.

(4) *Form of nitrogen.* 70–90% of the total nitrogen is hydrolysable. This nitrogen is present in substances closely similar to denatured β-proteins. The short-range order of the polypeptide bond is considered to be present.

Finally, a rough image can be built of the chemical structure of humic acid from lake sediment as a group of high-molecular-weight compounds of specific character, having an aliphatic nature and partially proteinaceous components and carbohydrates. Humic substances on the earth have been discussed by various authors from the point of view regarding them as substances similar to soil humus. According to the present study, however, it became clear that the humic substances in sediments under water are markedly different from so-called soil humus.

The author agrees, in general, with Kumada's statements that "The so-called browned substances (including humic substances) can be regarded as a kind of stable form (or metastable form) of organic substances, which preserve themselves against oxidation in the oxygen-rich condition on the earth" and "Because of their general distribution on the earth, the humic substances should be given an important position in the geochemical cycle of carbon compounds."[19]

Acknowledgments

I am very grateful to Prof. T. Hanya for his guidance during the course of the work, and for a critical reading of the manuscript. I also wish to thank Mr. M. Kosaka for his help in carrying out the work.

References

1. M. M. Kononova. *Soil Organic Matter, Its Nature, Its Role in Soil Formation and in Soil Fertility*. Translated from the Russian by T. Z. Nowakowski and G. A. Greenwood. Pergamon Press, 1961.
2. N. M. Karavaev and N. F. Budyak. Investigation on the so-called humic acids from fresh-water sapropels. *Doklady Akademii Nauk SSSR* **132,** 192–194, 1960.
3. S. Horie. Morphometric features and the classification of all the lakes in Japan. *Mem. Coll. Sci., Univ. Kyoto*, Ser. B, **29,** 191–262, 1962.
4. S. Ohashi. Determination of carbon in organic compounds by the iodic acid decomposition method. *Bull. Chem. Soc. Japan* **28,** 585–588, 1955.
5. P. V. Smith, Jr. The occurrence of hydrocarbons in recent sediments from the Gulf of Mexico. *Science*, **116,** 437–439, 1952.
6. P. Andrews. Estimation of the molecular weights of proteins by Sephadex gel-filtration. *Biochem. J.* **91,** 222–233, 1964.
7. M. Dubois *et al.* Colorimetric method for determination of sugars and related substances. *Anal. Chem.* **28,** 350, 1956.
8. D. W. van Krevelen and H. A. G. Chermin. Chemical structure and properties of coal. I. Elementary composition and density. *Fuel,* **33,** 79–87, 1954.
9. K. Kobo and R. Tatsukawa. Composition and forms of soil humus. *Kagaku*, **31,** 318–323, 1961.
10. K. Kobo, R. Tatsukawa and Y. Oba. The nitrogen distribution of soil organic matter and its relation to the type of humus. Extrait des rapports présentés au VIe Congrès International de la Science du Sol (Paris, 1956,) pp. 485–491.
11. R. Ishiwatari, M. Kosaka and T. Hanya. Compositional and optical characteristics of humic substances extracted from recent lake sediments. *J. Chem. Soc. Japan* (*Pure chemistry section*) **87,** 557–566, 1966.
12. R. Ishiwatari. Infrared absorption band at 1540 cm^{-1} of humic acid from a recent lake sediment. *Geochem. J.* **1,** 61–70, 1967.
13. L. J. Bellamy. *The Infra-red Spectra of Complex Molecules*. John Wiley & Sons Inc., 1958.
14. N. C. Mehta, P. Dubach und H. Deuel. Untersuchungen über die Molekulargewichtsverteilung von Huminstoffen durch Gelfiltration an Sephadex. *Z. Pflanzenernahr. Dung., Bodenkunde* **102,** 128–137, 1963.
15. J. P. Forsman. Geochemistry of kerogen. Organic Geochemistry Monograph No. 16, Earth science series, pp. 148–182 edited by I. A. Breger, Pergamon Press, 1963.
16. B. K. Vainshtein. *Diffraction of X-rays by Chain Molecules*, Elsevier Pub. Co., 1966.
17. *Chemistry of Proteins*, Vol. 2 edited by S. Mizushima and S. Akabori, Kyoritsu Shuppan Co., 1954.
18. *X-ray Crystallography*, Vol. 1 edited by I. Nitta, Maruzen Co., 1959.
19. K. Kumada. A consideration on humification. *J. Sci. Soil. Man. Japan*, **32,** 466–474, 1961.

BOTANICAL, CHEMICAL AND MICRO-BIOLOGICAL STUDIES OF PEAT ACCUMULATION PROCESSES IN THE EVERGLADES OF FLORIDA

by

P. H. GIVEN, W. SPACKMAN, A. COHEN, J. IMBALZANO, L. E. CASIDA and T. HISCOTT

Pennsylvania State University, University Park, Pa., U.S.A.

Abstract

The southern section of the Florida peninsula contains large areas of swamp or slough in which peat is accumulating. A number of distinctive environments are recognizable, in each of which the plant populations are relatively simple. As a relatively stable sub-tropical area of accumulation of decayed plant tissue, it is believed that southern Florida represents a suitable model system for studying some of the problems of coal formation.

A technique has been devised for sampling sections of organic sediment without disturbance and mounting them for microscopic study. Sections of anatomical components of living plants have been differentially stained and examined microscopically. Direct comparison of living with decayed tissue from the sediment shows that certain types of cell wall and cell contents (e.g. palisade cells in red mangrove leaves) rapidly disappear, while others (e.g. certain xylem cells, dense cell fillings that are believed to represent flavonoid pigments) are fairly well preserved. Surprisingly, lignified cell walls in the rootlets of red mangrove appear to be less well preserved than unlignified.

In a preliminary survey of the chemistry and microbiology of the peat accumulations, it was found that pH is uniformly near neutrality, and Eh values in most cores indicated positive values even at depths of 2–3 m.; it is not clear that the latter values should be interpreted as proving oxidizing conditions, since the thermodynamic reversibility of reactions in these organic media is questionable. Measurement of dehydrogenase activity showed a high level of microbiological activity in the top 20 cm of three different cores, which at greater depths decayed to a fairly constant rather low activity. Plating techniques indicated the probable presence of some unidentified fungal hyphae and protozoa.

Simple proximate analyses showed low humic acid contents, except in peat derived from sawgrass (*Mariscus*). A systematic trend towards higher C^{13} contents with increasing depth was found in the humic acids from a sawgrass peat core; this may be associated with the fact that the source of the peat changed from sawgrass to red mangrove in about the middle of the core, indicating that the surface environment had changed during the accumulation of the peat.

It appears that peat microorganisms, unlike the well-known microflora in soil, do not plate satisfactorily on to glass sheets, so that isolation of organisms for study is not easy. Moreover, attempts to culture the organisms in samples removed for the sediment indicate that it is the adventitious inclusions that multiply most freely, not the true, active, peat microflora. However, a large indigenous microflora has been found in centrifuged slurries of peat by staining and ultra-violet fluorescence microscopy; this resembles but is not identical with the recently discovered "Q-form" microflora in soils, and nothing is known of its physiology.

A specially designed probe was used for measuring the pH, ER and conductivity values *in situ* in the peat.

FATTY ACIDS IN SEDIMENTS INCLUDING THE GREEN RIVER SHALE (EOCENE) AND SCOTTISH TORBANITE (CARBONIFEROUS)

by

A. G. DOUGLAS,[†] K. DOURAGHI-ZADEH, G. EGLINTON, J. R. MAXWELL and J. N. RAMSAY

Department of Chemistry, The University of Glasgow, Glasgow W.2, Scotland

Introduction

Fatty acids occur very widely distributed in nature, in simple lipids usually as esters of glycerol, cholesterol or long chain alcohols, and also in more complex lipids and lipoproteins. This very ubiquity, and abundance, is one of the reasons why the distribution of fatty acids in the plant and animal kingdom is difficult to use taxonomically. Despite this, however, one may make some rather general correlations as shown in Table 1.

Considering both the relative stability and widespread occurrence of fatty acids, it is perhaps not surprising that they are found in organic detrital remains. Early in this century several workers had established the presence of fatty acids in various fossil environments, and their isolation and identification has been recorded in soils (Schreiner and Shorey, 1908, 1910), petroleums (Tanaka and Kuwata, 1928), boghead coal (Stadnikov and Kaschtanov, 1929), recent marine sediments (Trask and Wu, 1930), peats and lignites (Cawley and King, 1945), mud (Shabarova, 1954) and freshwater lakes (Goryunova, 1952). Since 1960 their occurrence has been established in montan wax (Hewett *et al.*, 1961; Wollrab *et al.*, 1962), ocean waters (Williams, 1961; Slowey *et al.*, 1962), fossil brines (Cooper, 1962), meteorites (Nagy and Bitz, 1963), and recent and ancient sediments.

[†] Present address: Organic Geochemistry Unit, The University, St. Thomas Street, Newcastle upon Tyne 1, England.

TABLE 1. — DISTRIBUTION OF FATTY ACIDS IN NATURE*

Acid type	Carbon number	Range	Typical sources	Distribution	Concentration
n-saturated	even	C_2–C_{36}	natural fats and waxes	wide	high
	odd	C_3–C_{25}	ruminant fats	wide	low
		C_7–C_{17}	human hair fat	narrow	low
		C_{17}–C_{33}	sugar cane wax	narrow	low
	even	C_{10}–C_{24}	milk fats	wide	high
		C_{10}–C_{34}	seed oils and waxes	wide	high
n-mono-		C_{16}–C_{24}	aquatic organisms	wide	high
unsaturated	odd	C_{11}–C_{17}	human hair fat	narrow	low
		C_{15}–C_{23}	milk fat	wide	low
		C_{10}	flowers	narrow	low
n-poly-	even	C_{10}–C_{24}	aquatic organisms	wide	high
unsaturated	odd	C_{17}	bacteria	narrow	low
		C_{10}–C_{28}	woolwax	narrow	high
	even	C_{14}–C_{18}	bacilli	narrow	low
		C_{14}–C_{18}	ruminant fat	narrow	low
iso-					
		C_{13}–C_{17}	ruminant fat	narrow	low
	odd	C_{13}–C_{19}	fish oil	narrow	low
		C_{15}–C_{19}	bacilli	narrow	low
		C_9–C_{31}	woolwax	narrow	high
anteiso-	odd	C_{13}–C_{17}	ruminant fat	narrow	low
		C_{15}, C_{17}	fish oil	narrow	low
	even	C_{26}–C_{36}	bacilli	narrow	low
		C_{12}, C_{14}	preen gland	narrow	low
multi-					
branched	odd	C_{27}–C_{25}	bacilli	narrow	low
		C_{11}, C_{15}	preen gland	narrow	low
	even	C_{20}	butterfat	narrow	low
		C_{20}	animal sera	narrow	low
isoprenoid		C_{20}	in Refsum's syndrome	narrow	high
	odd	C_{19}	butterfat	narrow	low
	even	C_6–C_{14}	seed fats	narrow	low
		C_{16}, C_{18}	seed fats	narrow	high
cyclic					
	odd	C_{17}, C_{19}	bacilli	narrow	high
		C_{19}	seed fat	narrow	high

In recent sediments reports by Cooper (1962), Cooper and Bray (1963), Abelson and Parker (1962), Parker and Leo (1965), Leo and Parker (1966) and Kvenvolden (1966), are, with one exception, on the occurrence of unbranched fatty acids ranging from about C_{10} to C_{34}. With respect to this work on sediments it is noteworthy that the relative abundance of fatty acids containing odd numbers of carbon atoms is greater in sediments than in biological systems, and that as the age of the sediment increases the even/odd carbon number ratio (carbon preference index: CPI) usually falls to a value near 1·0. Thus Kvenvolden (1966), has shown that the CPI value for fatty acids of a modern sediment from the San Nicolas Basin, California is 7·0, whereas that of a Lower Cretaceous shale is 1·05. Similarly, Cooper and Bray (1963), have examined recent and ancient sediments and oilfield waters, and have shown that for these the CPI values are of the order 15, 1·6 and 1·1, respectively.

Results, taken from the literature, for some ancient sediments are shown in Table 2. An interesting result is Kvenvolden's finding that the distribution of fatty acids can differ in similar lithologies of the same geological unit. He suggests that this does not necessarily imply differences of exposure in terms of time, temperature or pressure, but may reflect a different chemical environment or different chemical composition of the original deposit. The predominant n-acids found by the authors listed are all, with one exception, in the $C_{16}-C_{22}$ range, and this is noteworthy in view of our results for the total fatty acids of Green River Shale, in which the predominant acids appear in the range $C_{22}-C_{32}$.

It is evident, that in the analysis of a shale for fatty acids the results will in some measure reflect the particular extraction procedures used, and the extent to which the rock sample has been ground. Thus Hoering (1965), obtained the results shown in Table 3 when he investigated the extractable organic matter in the Precambrian Fig Tree shale of Swaziland from South Africa. Although Hoering concluded that much of the organic matter was loosely bound at grain surfaces, it is apparent that considerably more was recoverable on fine grinding and demineralizing.

In shales, fatty acids may be present as free acids, simple lipids, esters

---

* This table is not complete, but is meant rather as a guide for the geochemist.

† For simplicity high may mean that only some of the acids in the range indicated occur in large proportion, e.g. in woolwax, of the *anteiso-* acids in the range C_9 to C_{31}, only the C_9, C_{11}, C_{29} and C_{31} acids occur in large amount.

Information compiled largely from Hilditch, T. P. and Williams, P. N. (1964), and Shorland, F. B. (1962).

TABLE 2. – DISTRIBUTION OF FATTY ACIDS IN ANCIENT SEDIMENTS

Epoch	Shale	Range of n-acids	Predominant acid	Even/Odd ratio	Authors
Eocene	Green river	C_{12}–C_{20}	C_{18}	high	Abelson and Parker (1962)
Eocene	Green river	C_{10}–C_{35}	C_{18}	high	Lawlor and Robinson (1965)
Eocene	Green river	C_{12}–C_{18}	C_{16}	high	Leo and Parker (1966)
Upper cretaceous	Eagle ford	C_{11}–C_{34}	C_{16}	1·74	Cooper and Bray (1963)
Upper cretaceous	Navesink	C_{14}–C_{30}	C_{16}, C_{18}	high	Nagy and Bitz (1963)
Lower cretaceous	Skull creek	C_{8}–C_{35}	C_{18}	1·45	Cooper and Bray (1963)
Lower cretaceous	Mowry	C_{8}–C_{34}	C_{20}	1·42	Cooper and Bray (1963)
Lower cretaceous	Mowry	C_{13}–C_{34}	C_{28}	2·49	Kvenvolden (1966)
Lower cretaceous	Mowry	C_{11}–C_{33}	C_{16}, C_{18}, C_{20}	1·05	Kvenvolden (1966)
Lower cretaceous	Thermopolis	C_{13}–C_{33}	C_{20}, C_{22}	1·05	Kvenvolden (1966)
Lower cretaceous	Thermopolis	C_{12}–C_{35}	C_{20}, C_{22}	1·32	Kvenvolden (1966)
Mississippian	Chattanooga	C_{8}–C_{28}	C_{16}	1·57	Cooper and Bray (1963)
Cambrian	Alun	C_{12}–C_{18}	C_{18}	high	Abelson and Parker (1962)

TABLE 3. — PARTICLE SIZE VS. EXTRACTABLE ORGANIC MATTER (HOERING, 1965)

Particle size of shale	Extract* (p.p.m.)
$\frac{1}{2}$ inch	13
$-60+30$mesh	+30
3–10 μ	+15
Demineralized (HF/HCl)	+ 6
Total	64

* I.e. extract at each stage, the same material being used for the three stages subsequent to the first.

bound to the kerogen matrix or as salts bound to the inorganic matrix; the acids and other lipid materials may also be present interstitially or be held loosely bound on the grain surfaces. The authors in Table 1 used a variety of procedures for extracting the fatty acids including,

1. Soxhlet extraction with or without pretreatment of the shale with mineral acid.
2. Alkaline digestion of the powdered rock before, and after, extraction with organic solvents.
3. Chemical 'refining' of the extract to reduce the amount of tars present.

The subsequent separation of the normal from the branched acids has involved the use of urea adduction followed by gas chromatography of the methyl esters. Identification of the methyl esters by gas chromatographic retention times, infrared spectroscopy and mass spectrometry has been reported; quantitative data have been obtained from peak heights or areas of the gas chromatograms, and from mass spectroscopic parent ion abundances. This latter method should be used with caution in view of the non-linearity of the plot of parent peak intensities against carbon number of methyl esters as reported by Ryhage and Stenhagen (1959).

Experimental

In the work to be described, the crushing and extraction procedure adopted (with the exceptions noted) is shown in Fig. 1. In all procedures, precautions were taken to minimize contamination (Burlingame *et al.*, 1965). Typically, the rock was broken into about 1 inch pieces, and those pieces bearing freshly exposed surfaces were then cleaned by ultrasonic washing in benzene/methanol (3 : 1) for about 10 min. The drained and dried pieces were pulverized in a star beater mill (*ca.* 15 min) and the gritty powder was finely powdered in a disc mill (*ca.* 15 min) so that all passed through a 200-mesh sieve. The finely powdered rock (50 g) was then demineralized by allowing a suspension in HF/HCl (300 cc; 3 : 1) to stand for *ca.* 7–10 days; the residue was filtered off (water pump) and washed with distilled water until acid free. A glass centrifuge bottle containing the residue (1 part) and benzene/methanol (1 : 1; 3 parts) was placed in an ultrasonic tank; three successive extractions of 30 min duration recovered most of the soluble organic matter. The suspension resulting from the ultrasonic extraction was centrifuged and the clear supernatant solution was removed and evaporated to give what we term the "crude extract";

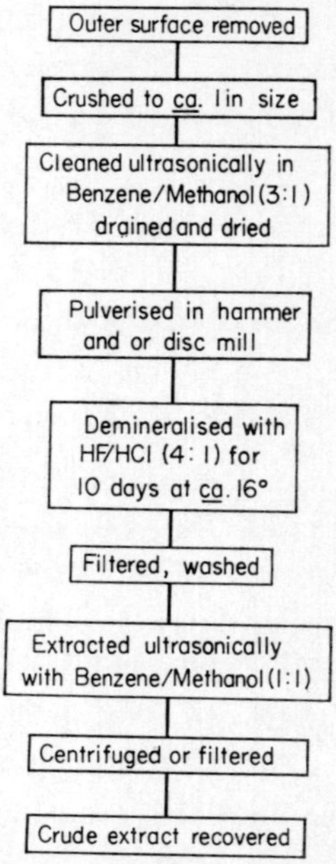

FIGURE 1. — Flow chart for the isolation of crude lipid extracts from demineralized and powdered shale.

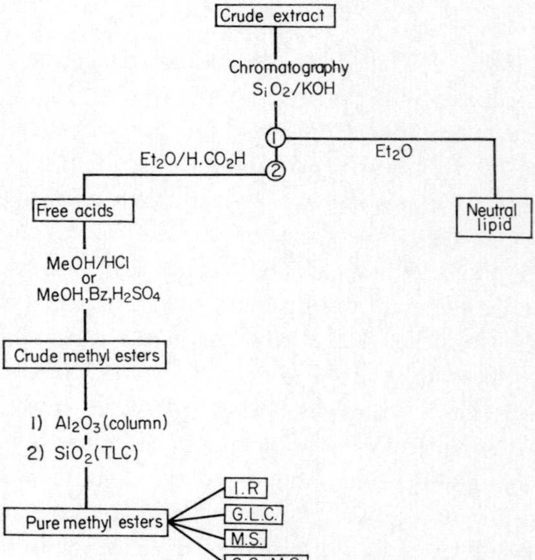

FIGURE 2. — Flow chart for the separation and identification of the "free" fatty acids from crude lipid extract.

in some experiments the suspension was filtered rather than centrifuged. The crude extract, obtained as a dark gum, would now contain the fatty acids which had been free in the rock, or bound as salts to the inorganic matrix. Recovery of the acids from this crude extract was accomplished by chromatographing the mixture on a silica column containing potassium hydroxide (McCarthy and Duthie, 1962) according to the scheme shown in Fig. 2. The acids were methylated with anhydrous methanolic/hydrochloric acid or a methanol/benzene/sulphuric acid mixture, and the crude methyl esters were then purified by column chromatography on alumina, followed by preparative T.L.C. on silica gel. The fatty acid methyl esters were then examined further by gas liquid chromatography (g.l.c.), infrared spectroscopy (i.r.), mass spectrometry (m.s.) and combined gas chromatography–mass spectrometry (gc–ms).

Results

Green River Shale (Eocene)

This shale is the largest single known hydrocarbon deposit in the world, and has been studied both geologically (Bradley, 1964) and chemically (Robinson *et al.*, 1963, 1965; Cummins *et al.*, 1964; Eglinton *et al.*, 1964). It is believed to be the consolidated organic rich sediment of large inland lakes of Eocene age, and contains remains of algae and other aquatic species. Recent studies of Green River Shale from the Mahogany Zone have shown the presence of the normal C_{10} to C_{34} alkanes, with a marked dominance of the odd carbon numbers, especially C_{27},C_{29} and C_{31}, a characteristic of most plant waxes (Eglinton and Hamilton, 1963; Douglas and Eglinton, 1966), also the presence of pristane, phytane, other isoprenoid hydrocarbons (Cummins and Robinson, 1964), steranes and triterpanes (Burlingame *et al.*, 1965).

Normal fatty acids have been reported in this shale (Table 1) and more recently Leo and Parker (1966) have recorded the presence of branched-chain fatty acids of the *iso*- and *anteiso*-series. Since branched acids are not plentiful in nature their presence in sediments raises the question of their origin. Parker and Leo (1966) state that in view of the similar biological and chemical stabilities of branched and normal fatty acids the sedimentary branched acids will reflect the original lipid of the organisms which contributed to the organic detritus. Since common marine organisms contain a ratio of branched/normal acids of about 1 : 100 to 1 : 500 (Ackman and Sipos, 1965), whereas in the sediment they found a ratio of 1 : 1 to 1 : 20, Parker and Leo concluded that such organisms are not major sources of the fatty acids in Green River Shale. Similar conclusions have been reached with respect to the fatty acid content of a number of species of blue green algae and phytoplankton in which little or no branched acids occur. Both bacteria and protozoa live in the upper sediment, deriving food from organic detritus, and there is evidence for large proportions of *iso*- and *anteiso*- acids in some bacteria (Kaneda, 1963) and some ciliated protozoa (Erwin and Bloch, 1963); Leo and Parker therefore suggest bacterial lipid as a likely source.

We have recently reported the occurrence of the isoprenoid phytanic and norphytanic acids in Green River Shale (Eglinton *et al.*, 1966) and wish now to report the extension of this series from C_{14} to C_{21} (with the exception of a C_{18} acid).

A sample of Green River Shale (200 g) taken from the 1100-ft level at Sulphur Creek was powdered, demineralized, extracted and chromatographed to give a crude extract from which the free acids were obtained and converted to their methyl esters (12 mg). The infrared spectrum (in CCl_4)

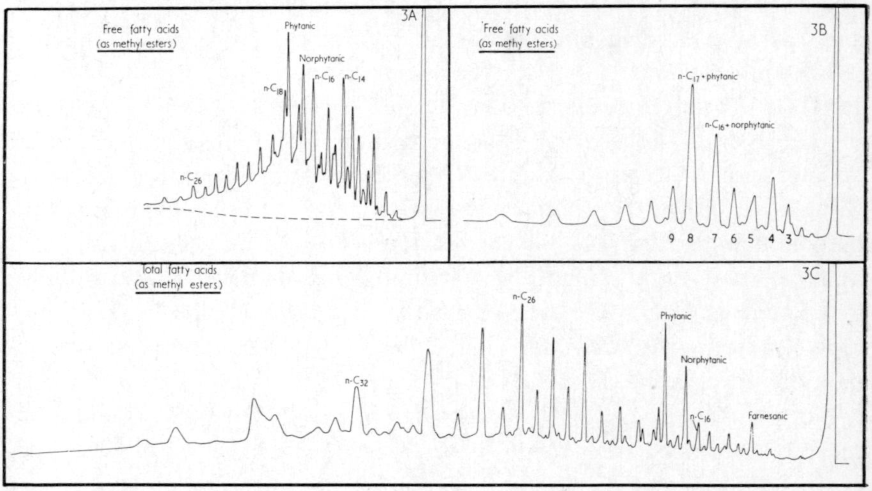

FIGURE 3. — Green River Shale (Eocene) 1100 ft. Gas–liquid chromatograms of the methyl esters: (A) "Free" fatty acids. Column conditions: 10 ft × $\frac{1}{8}$ in., 1% SE–30 on 100–120 mesh gas chrom P (DMCS); 20 ml/min nitrogen; temp. programmed at 8°/min from 125°–300°; injector temperature 280°. (B) "Free" fatty acids. Column conditions: 6 ft × $\frac{1}{4}$ in., 10% PEGA on 100–120 mesh chromosorb W; 60 ml/min helium; temp. programmed at 4°/min from 150°–210°; injector temp. 265°. (C) Total fatty acids. Column conditions: 10 ft × 1/16 in., 3% SE–30 on 100–120 mesh gas chrom Z; nitrogen, 30 p.s.i.; temp. programmed at 5°/min from 100°–280°; injector temp. 280°.

was consistent with an alkanoic acid ester fraction [ν(C=O), 1740 cm^{-1}, ν(C—O), 1175 cm^{-1}]. No attempt was made to separate the normal from the branched/cyclic esters, but the whole was gas chromatographed on silicone gum rubber (SE–30) to give the tracing shown in Fig. 3A. There is here an apparent even/odd predominance which was established on characterization of the peaks. Co-injection of authentic phytanic, norphytanic, n-C_{14}, n-C_{16} and n-C_{18} methyl esters gave peak enhancement of the peaks so labelled. Fractions 3 to 9 trapped from a preparative, programmed, gas chromatograph (Fig. 3B) were analysed by combined gas chromatography–mass spectrometry, using a 1% silicone/gum (SE–30) column in a commercial combined instrument (LKB 9000); this phase clearly resolved the individually collected peaks, and showed that the major components in cuts 3, 4, 5, 6, and 9 were the n-C_{12}, n-C_{13}, n-C_{14}, n-C_{15} and n-C_{18} normal saturated fatty acid esters, and in cuts 7 and 8, although the corresponding n-C_{16} and n-C_{17} esters were present, the major components were the methyl esters of norphytanic (C_{19}) and phytanic (C_{20}) acids, respectively. These proofs of structure were made by comparison of their mass spectra with those of authentic esters. For the other isoprenoid acid esters, where no authentic esters were available for comparison, the structures for the compounds isolated were assigned by examination of the mass spectra, and comparing them with literature values where they were available. Thus the C_{14} and C_{15} isoprenoid acid esters compared well with those reported by Cason and Graham (1965); structural assignments for the C_{16}, C_{17} and C_{21} esters have been made by adopting the method of Ryhage and Stenhagen (1963) to rationalize the structures of branched acids.

Similar findings to the above have been obtained with a sample of Green River Shale from the 1900-ft level, with some difference in the amount of fatty acids present. Thus a 200 g sample of shale gave only 3·5 mg of pure esters, and the infrared spectrum (in CCl_4) showed less methyl branching than did that of the ester fraction from the 1100-ft level, the ratios of optical densities for $\frac{\nu(C{=}O)\ 1740\ cm^{-1}}{\nu(CH)\ 2850\ cm^{-1}}$ being 1·09 and 1·49, respectively. Gas chromatography of the esters on SE–30 and polyethylene glycol adipate (PEGA) gave traces similar to those obtained for the 1100-ft sample, with the exceptions that a very large peak equivalent to an n-C_{12} acid ester had appeared, and the amount of branched acids was much reduced. Again, the even/odd predominance was evident. Samples collected from a programmed preparative gas chromatograph were re-injected into the gas chromatograph–mass spectrometer as before, and the presence of n-alkanoic acids ranging from C_{12}–C_{18}, and of the isoprenoid acids was established. This evidence, combined with coinjection measurements for standard n-C_{14}, n-C_{16} and n-C_{18} acid methyl esters, established the identity of the acids mentioned; gas chromatography indicated the presence of normal acids ranging from C_{11} to C_{30}.

In experiments designed to establish the total fatty acids present in Green River Shale the powdered and demineralized shale was digested with methanolic potassium hydroxide for 3 days, and the fatty acid esters were recovered in essentially the same manner as previously described.

Preliminary findings are that the amount of the total fatty acids is more than ten times that for the free fatty acids at the 1100-ft level. A gas chromatogram of the methyl esters of the total fatty acids is shown in Fig. 3c; the methyl esters of the normal acids ranging from C_{13} to C_{32} have been identified by examination of their mass spectra. The even-carbon-numbered long chain normal acids extending from about C_{24} to C_{32} are evident in large amount. Presumably they appear in the total fatty acid fraction as a result of hydrolysis of lipid esters or kerogen-bound esters. The isoprenoid acids ranging from C_{14} to C_{21} are present in the total fatty acid fraction, as revealed by the mass spectra of their methyl esters. A number of branched or cyclic acids, and higher molecular weight acids are evidently present, as can be seen from the gas chromatogram (Fig. 3c). The results obtained for the 'free' and total fatty acids of the Green River Shale are listed in Table 4.

TABLE 4. — FATTY ACIDS FROM THE GREEN RIVER SHALE (EOCENE)

	"Free" fatty acids		Total fatty acids 1100 ft
	1100 ft	1900 ft	
mg/100 g shale (as methyl esters)	4·8	1·5	65·0
*Range of n-acids	(a) C_{14}–C_{26} (b) C_{10}–C_{29}	C_{12}–C_{18} C_{11}–C_{30}	C_{13}–C_{32} C_{12}–C_{34}
Isoprenoid acids	C_{14}–C_{21}	C_{14}–C_{21}	C_{14}–C_{21}

* (a) Determined by mass spectrometry.
(b) Assessed from gas chromatograms.

Torbanite (*Carboniferous*)

This material was described as a boghead or cannel coal by Macgregor (1938), but more recently Dulhunty (1943) has stated that general opinion favours the view that Torbanite and cannel coal form distinct classes, because their respective peat stages resulted from the accumulation of unlike vegetable debris. Torbanite is from the Carboniferous Limestone Series of the Scottish Lothians and has an interesting morphology, for in addition to the presence of scales, teeth and spines of freshwater fish there is present a large amount of alveolar "yellow bodies", believed by Temperley (1936) to be the remains of an alga which does not differ in any material respect from the now living alga *Botryococcus braunii*. The high organic content of Torbanite is reflected in the high yield of oil obtained on pyrolysis (90–130 gal/ton), and indeed, this material saw the beginnings about 100 years ago of the now defunct Scottish Oil Shale industry.

Two different extraction procedures have given different results which are described below. In one procedure the finely powdered shale was extracted ultrasonically with benzene/methanol, and in another procedure the powdered shale was demineralized for 10 days with a hydrofluoric/hydrochloric acid mixture (3 : 1), the residue from which was extracted ultrasonically as for the previous method. Thus for the demineralizing treatment 250 g of powdered Torbanite gave 160 g of residue which was thoroughly extracted three times ultrasonically in benzene/methanol (1 : 1). Acids recovered from the gum (3·1 g) were methylated and chromatographed on alumina to give a crude ester fraction (13·5 mg) which was purified by thin-layer chromatography. The infrared spectrum showed a strong ester band at 1740 cm^{-1}, and the optical density ratio of $\frac{\nu(C{=}O)\ 1740\ cm^{-1}}{\nu(CH)\ 2850\ cm^{-1}} = 1{\cdot}24$ (stearic acid, as standard = 1·00) indicated a considerable amount of branching. For the sample of Torbanite which had not been demineralized, but which had otherwise been treated as above, the amount of gum and ester was somewhat less as shown in Table 5, and the optical density ratio of $\frac{\nu(C{=}O)\ 1740\ cm^{-1}}{\nu(CH)\ 2850\ cm^{-1}} = 1{\cdot}17$ indicated less branching than for the above.

These findings were borne out by inspection of the gas chromatograms of the methyl esters which are shown in Figs. 4A and 4B. Although it was deemed unlikely that the treatment with hydrofluoric acid would lead to

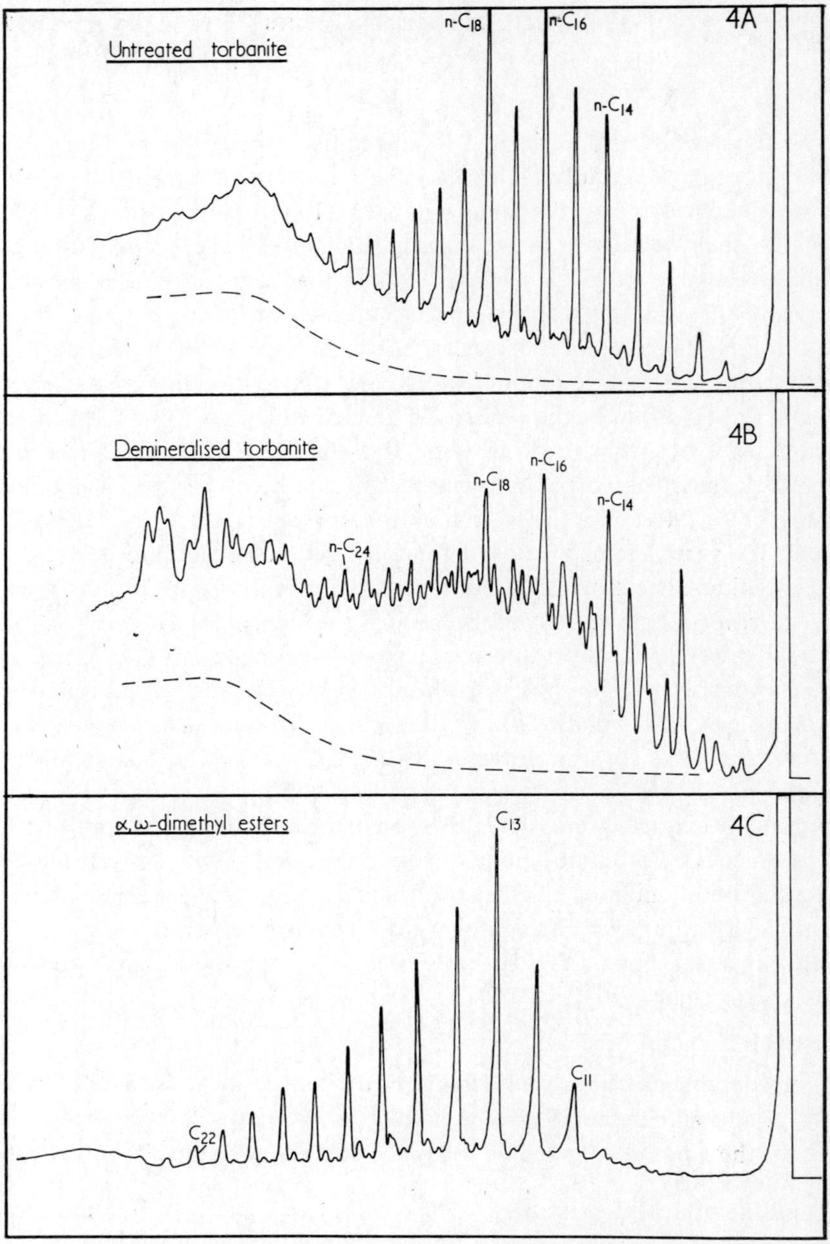

FIGURE 4. — Torbanite — Torbane Hill (Carboniferous). Gas-liquid chromatograms of the methyl esters: (A) "Free" fatty acids from the untreated material. Column conditions: 6 ft × $\frac{1}{8}$ in., 3% SE–30 on 100–120 mesh gas chrom P(DMCS); 30 ml/min nitrogen; temp. programmed at 4°/min from 125°–300°; injector temperature 280°. (B) "Free" fatty acids from demineralized material. Column conditions: 6 ft × $\frac{1}{8}$ in., 3% SE–30 on 100–120 mesh gas chrom P(DMCS); 30 ml/min nitrogen; temp. programmed at 4°/min from 150°–300°; injector temperature 280°. (C) Total α, ω-dibasic acids from demineralized material. Column conditions: 10 ft × 1/16 in., 3% SE–30 on 100–120 mesh gas chrom Z; nitrogen, 30 p.s.i., temp. programmed at 5°/min from 130°–300°; injector temp. 280°.

TABLE 5. — ISOLATION OF "FREE" FATTY ACIDS FROM TORBANITE (CARBONIFEROUS)

Torbanite	Wt. (g)	Gum (g)	Methyl esters (mg)	O.D. $\frac{\nu(C{=}O)\ 1740\ cm^{-1}}{\nu(CH)\ 2850\ cm^{-1}}$
Untreated	250	1·5	8·0	1·17
Demineralized HF/HCl	250	3·1	12·0	1·24
				Stearic acids = 1·00

any alteration of the structure of the fatty acids to produce branched or cyclic from normal acids, an experiment designed to establish this was carried out. Thus, pure stearic acid when left in contact with hydrofluoric acid for 10 days (and sonicated from time to time) was recovered unchanged as was shown by comparing the gas chromatographic patterns of the methyl esters of the starting acid and the acid which had been treated with hydrofluoric acid.

Preliminary conclusions are, therefore, that a range of normal acids ranging from C_{10} to C_{28} are present in Torbanite, and that there is a marked dominance of the C_{16} and C_{18} acids. The identity of the normal C_{14}, C_{16} and C_{18} acids was established by coinjection of standard esters, and by comparing their mass spectra with those of standard esters and with spectra appearing in the literature (Ryhage and Stenhagen, 1959). Resolution on SE–30 was clearly much better than on PEGA, and the minor components appearing between the large peaks of the normal acids (from C_{11} to C_{25}) must be branched or cyclic acids. Mass spectra of some of these minor components have been measured, and there appears to be a series of acids with a methyl substituent in the 2-position and another series of acids of unknown structure.

The total fatty acid fraction of Torbanite was obtained in a manner similar to that already described, the shale being hydrolysed by heating with alkali: i.e. the powdered and demineralized shale was hydrolysed by heating with methanolic potassium hydroxide, and the fatty acids were recovered and methylated as already described. Results obtained so far have shown that there is a range of normal acids extending from C_9 to C_{21}, the structures of which have been established by mass spectrometry and by coinjection of authentic esters, and normal acids of higher carbon number are apparent from the gas chromatogram. By careful thin-layer chromatography a second series of acids were isolated whose mass spectra are consistent with their structure being α, ω-dicarboxylic acids. These spectra are in agreement with the available published spectra (Ryhage and Stenhagen, 1959b), and with those of two authentic esters. A gas chromatogram of these α, ω-dicarboxylic acid esters, showing a smooth envelope of peaks extending from C_8 to C_{22} is shown in Fig. 4c. The resulte obtained for the 'free' and total fatty acids of Torbanite are illustrated in Table 6.

TABLE 6. — FREE AND TOTAL FATTY ACIDS FROM TORBANITE (CARBONIFEROUS)

	"Free" fatty acids		Total fatty acids
	Untreated	Demineralized	Demineralized
mg/100 g shale (as methyl esters)	3·2	4·8	11·0
* Range of n-acids	(a) C_{16}, C_{18} (b) C_{10}–C_{28}	C_{14}, C_{16}, C_{18}, C_{24} C_{10}–C_{28}	(a) C_9–C_{21} (b) C_9–C_{27}
Range of di-acids			(a) C_8–C_{21} (b) C_8–C_{25}

* (a) Determined by mass spectrometry.
(b) Assessed from gas chromatograms.

Discussion

The results described pose a number of questions. Previous workers have provided chemical evidence of the biological history of the Green River Shale in showing the predominance of odd-numbered hydrocarbons and even-numbered fatty acids, as is typical of plants and animals. Further support for this biological history is revealed by the presence of *iso-* and *anteiso-* fatty acids (Leo and Parker, 1966), and by the present findings of the architecturally specific polyisoprenoid acids. Comparative abundances of n-paraffins and n-fatty acids having one more carbon atom have been established for Green River and other shales (e.g. Kvenvolden, 1966; Lawlor and Robinson, 1965; Cummins and Robinson, 1964), and it has been proposed (Cooper and Bray, 1963) that decarboxylation of an acid, during diagenesis, provides the corresponding alkane. Thus Lawlor and Robinson (1965) give very convincing graphical evidence for such an acid/hydrocarbon relationship (in the C_{24}–C_{34} range) for Green River Shale, as does Kvenvolden (1966) for a sample of Mowry shale. In the present work the abundances of the isoprenoid hydrocarbons pristane (C_{19}) and phytane (C_{20}) are parallelled by the abundances of the isoprenoid norphytanic (C_{19}) and phytanic (C_{20}) acids. Thus it is clear that any genetic relationship between these compounds cannot arise solely through a mechanism involving the loss of CO_2. Also, the presence of substantial amounts of the C_{15} isoprenoid acid would suggest the presence of the corresponding C_{14} hydrocarbon, but this substance was not found among the isoprenoid hydrocarbons isolated by Cummins and Robinson (1965). There exists the possibility that these isoprenoid acids are indigenous to the original organic deposit, but they are rare in nature today. Phytanic acid has been reported as being a constituent of ruminant fats and sera (e.g. Hansen *et al.*, 1966), rumen bacteria (Hansen, 1966) and in persons afflicted with Refsum's syndrome (Steinberg *et al.*, 1965), while norphytanic acid has been reported in butterfat (Hansen, 1964).

The appearance of α, ω-dicarboxylic acid in the total fatty acid fraction of Torbanite requires some comment, although it is not possible to be specific as to their origin. Some such acids occur free in nature, in the range from the common oxalic acid (C_2) to about C_{23}, but the higher molecular weight acids are not abundant. Glycerides of dibasic acids have been reported, as in Japan Wax, in which two molecules of glycerol are linked by an α,ω-dicarboxylic acid containing 22 or 23 carbon atoms, and hydrolysis of cork (suberin) is reported to give a C_{22} α,ω-dicarboxylic acid (phellogenic acid). Bacterial oxidation of hydrocarbons or mono-carboxylic acids to give α,ω-diacids is known (Kester and Foster, 1963), and it may be that the appearance of these acids in Torbanite does not reflect the organic matter indigeneous to the original deposit, but rather microbial oxidation of the appropriate substrate.

Bendoraitis *et al.* (1962) suggested the phytyl side chain of chlorophyll as the source of pristane and phytane found in petroleum, and it would be convenient to think that this was also the source of the isoprenoid acids. In view of our finding that a C_{21} isoprenoid acid occurs in the Green River Shale and as the same acid occurs in Serpiano Oil Shale (Blumer, 1966) one has to seek other natural products which contain the appropriate carbon skeleton, assuming of course that the acids themselves are not indigenous to the organic matter deposited. Molecules containing polyisoprenoid moeties include the menaquinones (Vit. K_2 series) in which C_{20} to C_{45} unsaturated chains are present, e.g.

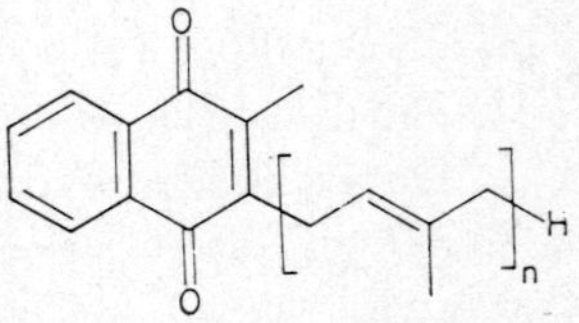

FIGURE 5.

where $n = 4, 6, 7, 8, 9$, and it is interesting to note that these menaquinones are present in the bacteria living on putrid fish meal (Morton, 1965). Similarly, the ubiquinones in which the $(C_5H_8)_n$ chain occurs (n = 6–10) are present in yeasts, bacteria and animal tissue, and plastoquinone with a $C_{45}H_{73}$ polyisoprenoid chain occurs in chloroplasts. It is noteworthy that this compound is sometimes present in amounts equal to about 10% of the chlorophyll content of the chloroplast (Henninger *et al.*, 1963). The isoprenoid alcohol solanesol $C_{45}H_{73}OH$ occurs in tobacco and other plants with possibly its C_{50} and C_{55} isoprenologues (Morton, 1962), and prenols have been reported having as many as 22 isoprenoid units. There is also the C_{30} triterpenoid squalene and C_{40} tetraterpenoids (e.g. lycopene) which have, respectively, C_{15} and C_{20} isoprenoid units linked tail to tail. Finally, in certain halophilic bacteria the major phosphatide is present as a C_{20} isoprenoid di-ether (Kates *et al.*, 1965), and in certain bacterio/chlorophylls the esterifying alcohol is not phytol but farnesol (Rapoport and Hamlow, 1961).

The prominence of the normal C_{12} fatty acid in the free fatty acid fraction from the 1900-ft level of Green River Shale and in the total fatty acids of Torbanite is not easily rationalized. The possibility of post-geological contamination, e.g. with a soap containing large amounts of lauric acid must be considered, but the appearance of this acid variously in the present work, in deep ocean waters (Slowey *et al.*, 1962) and in another Carboniferous Oil Shale at present being examined by us suggests that this is not the answer. Although lauric acid does not appear ubiquitously in large amounts, it does occur in the endospermic seed fats of some Palmae (45–50% of fatty acid content) and Lauraceae (80–90% of fatty acid content) species, and it may be that some of the plant types of the Eocene and Carboniferous periods contained lipids which were rich in lauric acid. It may be, however, that this acid was produced by degradation of a suitable precursor, i.e. a molecule having a point of attack at the appropriate carbon atom; thus certain Umbelliferae seeds contain Δ^6-octadecenoic (petroselenic) acid ($CH_3(CH_2)_{10}CH=CH(CH_2)_4CO_2H$) and the corresponding acetylenic acid (tariric) is known, both of which could conceivably lead to the C_{12} acid by an oxidative process.

The isoprenoid acids have not been found in the Torbanite free fatty acids, although they may be present in small amounts; this seems rather surprising, since Torbanite is held to have originated from *Botryococcus braunii*, which is an alga of the order Chlorophyceae (Belcher and Fogg, 1955).

A number of methods for examining the lipid fraction of geological materials exist, but we still need better methods for separation and identification. Progress will depend (too) on some standardization of procedure whereby results from different laboratories can be compared. On the first point, the use of the new technique of combined gas chromatography–mass spectrometry is of obvious advantage, for it allows separation and spectra to be obtained on submicrogram amounts. The method however, in our opinion, is only as good as the gas chromatographic technique available, and here a search for phases of greater stability and higher resolving power would be worthwhile. Further, before subjecting material to gas chromatography or any of the spectroscopic techniques it is clearly useful to have done as much preliminary separatory work as possible and the use of molecular sieves, clathrating and molecular complexing agents should be pursued. The classical techniques of organic chemistry whereby the conversion of one compound to another for which physical separatory methods are available, should be fruitful when they are possible on a micro-scale; thus the conversion of fatty acids to hydrocarbons followed by separation of the normal from the branched and cyclic products by 5 Å molecular sieve.

Acknowledgements

We thank Dr. W. E. Robinson and Dr. W. D. I. Rolfe for samples of Green River Shale and Torbanite, respectively. We are also indebted to the Science Research Council (J.R.M.), Natural Environment Research Council (A.G.D.), National Aeronautics and Space Administration (N.s.G. 101-61, K.D-Z.) and Shell International for personal and general financial support.

References

Abelson, P. H. and Parker, P. L. (1962) Fatty acids in sedimentary rocks (Annual Report, Director Geophys. Lab.). *Carnegie Inst. Wash. Yr. Book*, **61**, 181–184.

Ackman, R. G. and Sipos, J. C. (1965) Isolation of the saturated fatty acids of some marine lipids with particular reference to normal odd-numbered fatty acids and branched-chain fatty acids. *Comp. Biochem. Physiol.* **15**, 445–456.

Belcher, J. H. and Fogg, G. E. (1955) Biochemical evidence of the affinities of *Botryococcus*. *New Phytologist*, **54**, 81–83.

Bendoraitis, J. G., Brown, B. L. and Hepner, L. S. (1962) Isoprenoid hydrocarbons in petroleum. *Anal. Chem.* **34**, 49–53.

Blumer, M. (1966) Personal communication.

Bradley, W. H. (1964) Aquatic fungi from Green River formation of Wyoming. *Amer. J. Science*, **262**, 413–416.

Burlingame, A. L., Haug, P., Belsky, T. and Calvin, M. (1965) Occurrence of biogenic steranes and pentacyclic triterpanes in an Eocene Shale (52 million years) and in an early Precambrian Shale (2·7 billion years): A preliminary report. *Proc. Nat. Acad. Sci., U.S.* **5**, (5), 1406–1412.

Cason, J. and Graham, D. W. (1965) Isolation of isoprenoid acids from a California petroleum. *Tetrahedron*, **21**, 471–483.

Cawley, C. M. and King, J. G. (1945) Ester waxes from British lignites and peat. *Chem. and Ind.* 237–242.

Cooper, J. E. (1962) Fatty acids in recent and ancient sediments and reservoir waters. *Nature*, **193**, 744–746.

Cooper, J. E. and Bray, E. E. (1963) A postulated role of fatty acids in petroleum formation. *Geochim. et Cosmochim. Acta*, **27**, 1113–1127.

Cummins, J. J. and Robinson, W. E. (1964) Normal and isoprenoid hydrocarbons isolated from oil-shale bitumen. *J. Chem. Eng. Data*, **9**, 304–307.

Douglas, A. G. and Eglinton, G. (1966) The distribution of alkanes. In *Comparative Phytochemistry*, T. Swain (Editor), 57–71. Academic Press, London.

Dulhunty, J. A. (1943) Classification and origin of the New South Wales Torbanites. D. Sc. Thesis, University of Sydney, Australia.

Edwards, V. A., Kipping, P. J. and Jeffrey, P. G. (1963) Composition of Montan Wax. *Nature*, **199**, 171–172.

Eglinton, G., Scott, P. M., Belsky, T., Burlinghame, A. L., Richter, W. and Calvin, M. (1964) Isoprenoid alkanes in a Precambrian sediment. In *Advances in Organic Geochemistry 1964*, G. D. Hobson and M. C. Louis (Editors), 41–74. Pergamon Press, London.

Eglinton, G., Douglas, A. G., Maxwell, J. R., Ramsay, J. N. and Ställberg-Stenhagen, S. (1966) Occurrence of isoprenoid fatty acids in the Green River Shale. *Science*, **153**, 1133–1135.

Eglinton, G. and Hamilton, R. J. (1963) The Distribution of Alkanes. In *Chemical Plant Taxonomy*, T. Swain, (Editor), 187–217. Academic Press, London.

Erwin, J. and Bloch, K. (1963) Lipid metabolism of ciliated protozoa. *J. Biol. Chem.* **238**, 1618–1624.

Goryunova, S. V. (1952) Characteristics of dissolved organic substances in the water of Glubokoe Lake. Cited from *Chem. Abstracts*, **47**, 8293h.

Hansen, R. P. (1966) Occurrence of phytanic acid in rumen bacteria. *Nature*, **210**, 841.

Hansen, R. P., Shorland, F. B. and Prior, I. A. M. (1966) The fate of phytanic acid when administered to rats. *Biochim. Biophys. Acta*, **116**, 178–180.

Henninger, M. D., Dilley, R. A. and Crane, F. L. (1963) Restoration of ferricyanide reduction in acetone-extracted chloroplasts by β and γ tocopherol quinones. *Biochem. Biophys. Res. Commun.* **10,** 237–242.

Hewett, D. R., Kipping, P. J. and Jeffrey, P. G. (1961) Separation, identification and determination of the fatty acids of Montan Wax. *Nature*, **192,** 65.

Hilditch, T. P. and Williams, P. N. (1964) *The Chemical Constitution of Natural Fats.* Chapman and Hall, London.

Hoering, T. C. (1965) The extractable organic matter in Precambrian rocks and the problem of contamination. *Carnegie Inst. Wash. Yr. Book*, **64,** 215–218.

Kaneda, T. (1963) Biosynthesis of branched fatty acids. I. Isolation of fatty acids from *B. subtilis. J. Biol. Chem.* **238,** 1222—1228. II. Microbial synthesis of branched long-chain fatty acids from certain substrates. *Ibid.* 1229–1235.

Kates, M., Yengoyan, L. S. and Sastry, P. S. (1965) A diether analogue of phosphatidyl glycerophoshpate. In *Halobacterium cutirubrum. Biochem. Biophys. Acta*, **98,** 252–268.

Kester, A. S. and Foster, J. W. (1963) Diterminal oxidation of long-chain alkanes by bacteria. *J. Bacteriol.* **85,** 859–869.

Kvenvolden, K. A. (1966) Molecular distributions of normal fatty acids and norma hydrocarbons in some Lower Cretaceous sediments. *Nature*, **209,** 573–577.

Lawlor, D. L. and Robinson, W. E. (1965) Fatty acids in Green River Formation Oil-Shale. *Div. Pet. Chem. Amer. Chem. Soc.* (Detroit Meeting), 5–9.

Leo, R. F. and Parker, P. L. (1966) Branched-chain fatty acids in sediments. *Science*, **152,** 649–650.

McCarthy, R. D. and Duthie, A. H. (1962) A rapid quantitative separation of free fatty acids from other lipids. *J. Lipid Res.* **3,** 117–119.

Macgregor, M. (1938) Oil-Shales and cannel coals of Scotland. In *Oil Shale and Canne Coal*, 6–17. Institute of Petroleum, London.

Morton, R. A. (1962) Lipid co-factors in biochemistry. Preprint from *The Advancemen of Science*, **19,** 1–12.

Morton, R. A. (1965) Quinones as biological catalysts. *Endeavour*, **24,** 81–86.

Nagy, B. and Bitz, M. C. (1963) Long-chain fatty acids from the Orgueil meteorite *Arch. Biochim. Biophys.* **101,** 240–248.

Parker, P. L. and Leo, R. F. (1965) Fatty acids in blue-green algal mat communities. *Science*, **148,** 373–374.

Rapoport, H. and Hamlow, H. P. (1961) *Chlorobium* – Chlorophyll – 660; esterifying alcohol. *Biochem. Biophys. Res. Comm.* **6,** 134–137.

Robinson, W. E., Lawlor, D. L., Cummins, J. J. and Fester, J. I. (1963) Oxidation of Colorado Oil Shale. *U.S. Dept. of Interior, Bureau of Mines*, No. 6166.

Robinson, W. E., Cummins, J. J. and Dineen, G. U. (1965) Changes in Green River Shale paraffins with depth. *Geochim. et Cosmochim. Acta*, **29,** 249–258.

Ryhage, R. and Stenhagen, E. (1959a) Mass spectrometric studies. I. Methyl esters of saturated normal chain carboxylic acids. *Arkiv. Kemi.* **13,** 523–542.

Ryhage, R. and Stenhagen, E. (1959b) Mass spectrometric studies. III. Esters of saturated dibasic acids. *Arkiv. Kemi.* **14,** 497–504.

Ryhage, R. and Stenhagen, E. (1960) Mass spectrometric studies. IV. Esters of monomethyl-substituted long-chain carboxylic acids. *Arkiv. Kemi.* **15,** 291–315.

Ryhage, R. and Stenhagen, E. (1963) Mass spectrometry of long-chain esters. In *Mass Spectrometry of Organic Ions.* F. W. McLafferty (Editor), 399–452. Academic Press, London.

Schreiner, O. and Shorey, E. C. (1908) The isolation of dihydroxystearic acid from soils. *J. Amer. Chem. Soc.* **30,** 1599–1607.

Schreiner, O. and Shorey, E. C. (1910) Some acid constituents of soil humus. *J. Amer. Chem. Soc*. **32,** 1674–1680.

Shabarova, N. T. (1954) The organic matter of marine sediments. *Usp. Sovr. Biol.* **37,** 203–208.

Shorland, F. B. (1962) Comparative aspects of fatty acid occurrence and distribution. In *Comparative Biochemistry*, M. Florkin and H. S. Mason (Editors), Vol. III, Part A, 1–102. Academic Press, London.

Slowey, J. F., Jeffrey, L. M. and Hood, D. W. (1962) Fatty acid content of ocean water. *Geochim. et Cosmochim. Acta*, **26,** 607–616.

Stadnikov, G. and Kaschtanov, L. (1929) Die Chemische Struktur der Bogheadkohle. *Brennstoff—Chem.* **10,** 417–419.

Steinberg, D., Avigan, J., Mize, C., Eldjarn, L., Try, K. and Refsum, S. (1965) Conversion of U—C_{14} phytol to phytanic acid and its oxidation in Heredopathia Atactica Polyneuritiformis. *Biochem. Biophys. Res. Comm.* **19,** 783–789.

Tanaka, Y. and Kuwata, T. (1928) Higher fatty acids in petroleum. *J. Faculty Eng. Tokyo Imp. Univ.* **17,** 293–303.

Temperley, B.N. (1936). *Botryococcus* and the algal coals. Part II–The boghead controversy and the morphology of the boghead algae. *Trans. Roy. Soc. Edin.* **58,** 855–868.

Trask, P.D. and Wu, C.C. (1930), Does petroleum form in sediments at time of deposition? *Bull. Am. Ass. Petrol. Geol.* **14,** 1451–1463.

Williams, J. R. (1961) Organic acids in Pacific Ocean waters. *Nature*, **189,** 219–220.

Wollrab, V., Streibl, M. and Sorm, F. (1962) Gas-chromatographic analysis of the wax components of Montan Wax. *Chem. Ind.* 1762.

EVIDENCE FOR TRANSFORMATIONS OF NORMAL FATTY ACIDS IN SEDIMENTS

by

KEITH A. KVENVOLDEN*

Mobil Oil Corporation
Field Research Laboratory
Dallas, Texas

GEOCHEMICAL evidence for fatty acid transformations in sediments has been obtained from a study of six modern unconsolidated sediments and eleven ancient rocks. The six modern sediments came from four different environments–basin, delta, bay, and lake. The eleven ancient rocks included shales and limestones ranging in age from Devonian to Pliocene. Speculations regarding fatty acid transformations in sediments are bounded by distributions of normal fatty acids and possibly related normal paraffins found in modern sediments on one hand, and distributions of these same compounds in ancient sediments on the other. A principal assumption is that originally deposited fatty acids and paraffins in ancient sediments were distributed similarly to fatty acids and paraffins found in modern sediments. Differences in distributions of normal fatty acids and paraffins in modern and ancient sediments suggest that during geological time chemical transformations have taken place.

Previous Work

The presence of normal fatty acids and paraffins in sediments has been noted by several previous investigators. Normal fatty acids in modern and ancient sediments have been reported by Cooper (1962), Abelson and Parker (1962) and Cooper and Bray (1963). Normal paraffins in sediments have been described by Bray and Evans (1961). Lawlor and Robinson

* Present address: National Aeronautics and Space Administration, Ames Research Center, Moffett Field, California 94035.

(1965) described both normal fatty acids and paraffins in a sample of oil shale from the Green River Formation. Later, Kvenvolden (1966) contrasted distributions of normal fatty acids and paraffins in some Lower Cretaceous sediments with distributions of these same compounds in a modern sediment. These investigations have shown that in sediments normal fatty acids and paraffins may have a variety of distributions. Even-carbon-numbered normal fatty acids usually dominate odd-carbon-numbered fatty acids, but in some ancient sediments normal fatty acids have almost equal abundances of even- and odd-carbon-numbered molecules. Normal paraffins, on the other hand, usually show a preference for odd-carbon-numbered molecules or nearly equal distributions of odd- and even-carbon-numbered molecules.

Analytical Procedure

An analytical procedure was designed so that from a single sample both normal fatty acids and normal paraffins from about C_{15} to about C_{34} could be recovered. This procedure has been modified at various times during the course of the examination of samples reported here. Modifications have not affected the internal consistency of the results concerning the relative molecular weight distributions of fatty acids or paraffins in a sample; they have, however, improved the total recovery of fatty acids and paraffins, and have reduced the content of unwanted materials.

Normal fatty acids and paraffins were obtained from dried, crushed sediments by extraction with a solution of 10% potassium hydroxide in methanol. The extraction solution was recovered from the sediment; the concentrations of fatty acids and paraffins reported were adjusted to compensate for incomplete solution recovery. Fatty acids and paraffins were extracted from the concentrated saponification solution after the addition of water and hydrochloric acid. Normal fatty acids were converted to methyl esters which were separated along with normal paraffins from other lipid material by urea adduction. Methyl esters of normal fatty acids were separated from normal paraffins by silica-gel chromatography. The fraction containing methyl esters of normal fatty acids was further purified by alumina chromatography. Identification of normal fatty acids (their methyl esters) and normal paraffins was based on the isolation procedure, infrared spectrometry, and gas chromatography. In some earlier work identifications were confirmed by mass spectrometry (Kvenvolden, 1966). Relative concentrations of fatty acids and paraffins were calculated frompeak heights on gas chromatograms. Typical gas chromatograms are shown in Figs. 1 and 2. Calibrations were obtained from the gas chromatographic peak heights of standard solutions containing methyl esters of C_{18} and C_{26} normal fatty acids and normal paraffins C_{20} and C_{28}. Internal standards containing the same compounds were added to half of the original saponification solution for compound

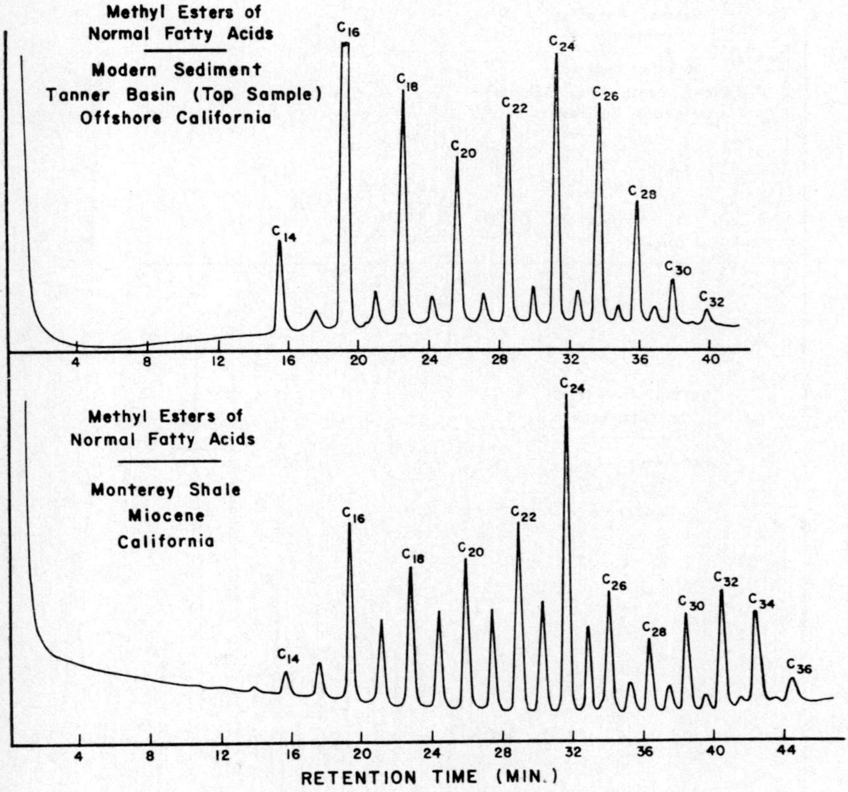

FIGURE 1. — Typical gas chromatograms of methyl esters of normal fatty acids in a modern and ancient sediment. F. & M. model 500 gas chromatograph; 48 in. by $\frac{1}{4}$ in. column packed with 4% SE 30 on 60–80 mesh chromasorb W DMCS; temperature programmed from 100° C to 350° C at 5·6° C/min.

identification purposes, and to determine losses occurring during the isolation procedures. Infrared spectrometry established the presence of methyl esters of normal fatty acids and normal paraffins. Figure 3 compares typical infrared spectra of methyl esters of normal fatty acids from several samples with a standard solution containing nine methyl esters of normal fatty acids. Figure 4 compares infrared spectra of normal paraffins. In this figure two spectra, Mud Lake and Green River oil shale, show some extraneous absorptions, especially near 10·4μ. The cause of these absorptions was not determined; unsaturation in some carbon chains is a possibility.

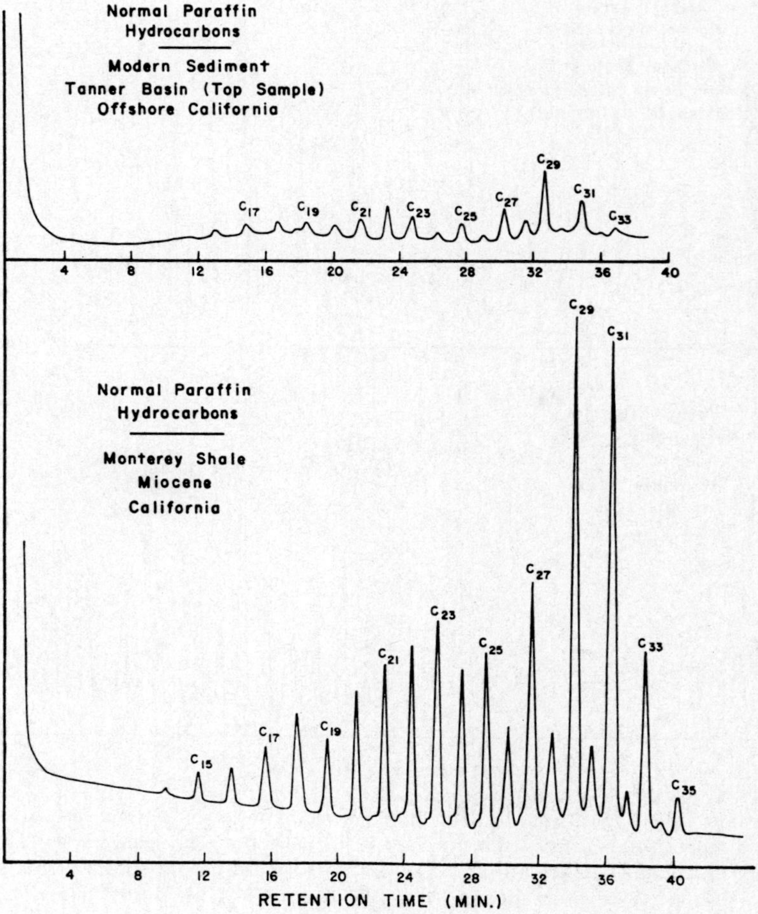

FIGURE 2. — Typical gas chromatograms of normal paraffin hydrocarbons in a modern and ancient sediment. Chromatographic conditions as Fig. 1.

This analytical procedure is better suited for recovering fatty acids than for recovering paraffins from sediments. The methanolic-KOH solution is a poor solvent for paraffins; 3 ml of methanolic-KOH were used for each gram of sediment in order to promote solution of paraffins by providing a large volume of solvent. The concentrations of paraffins recovered by this procedure were usually less than the concentrations of paraffins recovered by a benzene-methanol extraction as described by Meinschein and Kenny (1957). For example, from a sample of New

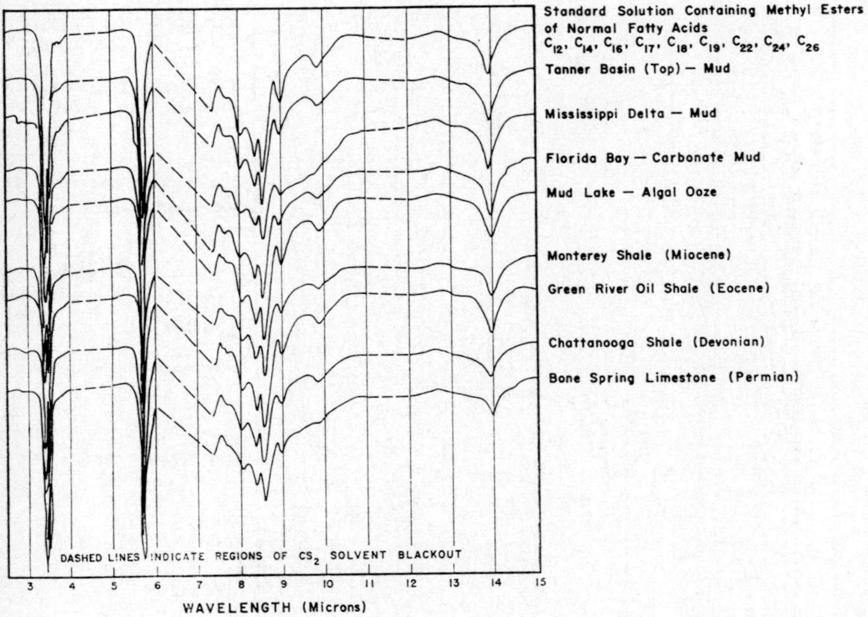

FIGURE 3. — Infrared spectra of chromatographic fractions containing methyl esters of normal fatty acids from modern and ancient sediments.

Albany shale the benzene-methanol method recovered about 2·5 times more paraffins than did the methanolic-KOH method. However, relative molecular weight distributions of normal paraffins obtained by the two procedures on a given sample were about the same. Therefore, the molecular weight distributions of normal paraffins reported here closely reflect the distributions present in the sediments; the total concentrations of normal paraffins reported are usually smaller than would be obtained by a benzene-methanol extraction method, except in cases where the amount of paraffins in the sediment is small.

In the analytical procedure silica gel chromatography was used to separate normal paraffins from methyl esters of normal fatty acids. Chromatographic fractions containing esters usually were yellow-brown or brownish-black in color. Color was removed from the methyl ester fraction by alumina chromatography. In this step the methyl ester fraction was eluted over a column of Woelm alumina (activity 2) with a mixed solvent composed of heptane and benzene (8 : 2). Methyl esters of normal

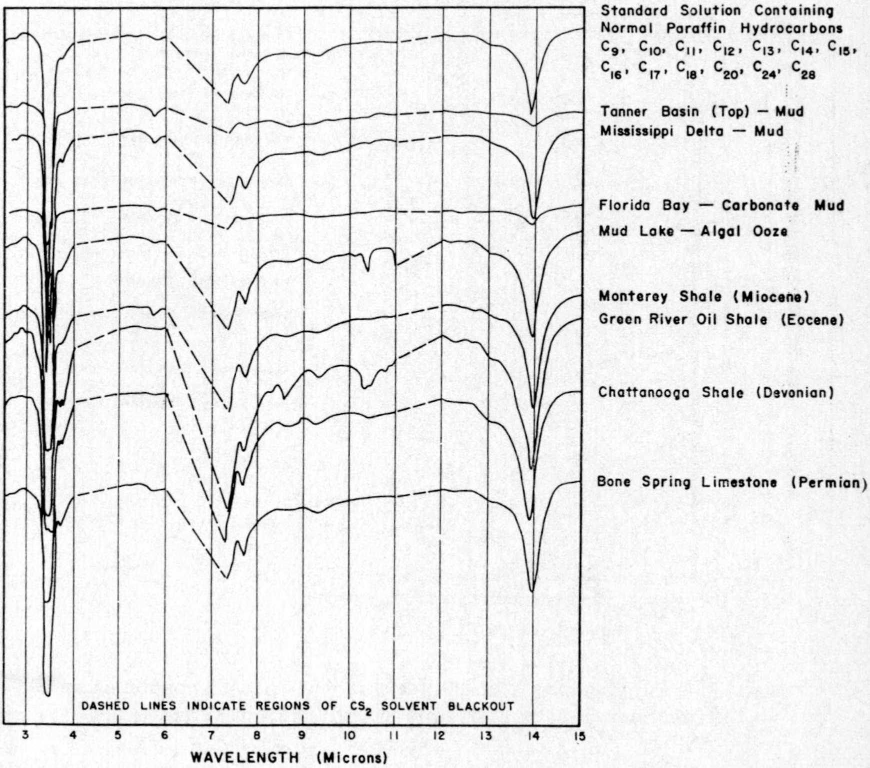

FIGURE 4. — Infrared spectra of chromatographic fractions containing normal paraffin hydrocarbons from modern and ancient sediments.

fatty acids came out free of color in a predetermined chromatographic cut. The presence of unsaturation in methyl ester fractions from some modern sediments was noted by the strong absorption at 10·4 microns on infrared spectra and by gas chromatographic peaks between peaks assigned to methyl esters of normal fatty acids. Fatty acid esters with unsaturation were removed from the fractions by bromination and reduction.

Display of Data

Distributions of normal fatty acids and normal paraffins for each sample considered here will be shown by a pair of graphs with fatty acids plotted above and paraffins below. The percentage of each fatty acid based on total fatty acids (corrected for losses during processing) and the percentage of each paraffin based on total paraffins (corrected for losses during processing) are plotted on the respective graphs at appropriate carbon numbers. For fatty acids the carbon numbers range from 14 to 34; for paraffins from 13 to 33. Graphs permit rapid comparison of various distributions. To characterize various distributions of fatty acids and paraffins, carbon preference indices will be used (Cooper and Bray, 1963; Kvenvolden 1966). CPI_A will represent the carbon preference index for normal fatty acids and CPI_p will represent the carbon preference index for normal paraffin hydrocarbons. The formulae for calculating these indices are:

$$CPI_A = \frac{1}{2}\left(\frac{\Sigma \text{ concentrations of even-carbon fatty acids } C_{16} \text{ to } C_{30}}{\Sigma \text{ concentrations of odd-carbon fatty acids } C_{15} \text{ to } C_{29}} + \frac{\Sigma \text{ concentrations of even carbon fatty acids } C_{16} \text{ to } C_{30}}{\Sigma \text{ concentrations of odd-carbon fatty acids } C_{17} \text{ to } C_{31}}\right),$$

$$CPI_p = \frac{1}{2}\left(\frac{\Sigma \text{ concentrations of odd-carbon paraffins } C_{17} \text{ to } C_{31}}{\Sigma \text{ concentrations of even-carbon paraffins } C_{16} \text{ to } C_{30}} + \frac{\Sigma \text{ concentrations of odd-carbon paraffins } C_{17} \text{ to } C_{31}}{\Sigma \text{ concentrations of even-carbon paraffins } C_{18} \text{ to } C_{32}}\right).$$

The limits used in determining these indices are dictated by the analytical method and the concentrations of molecular species. In this study fractionation of both fatty acids and paraffins containing less than about 16 carbon atoms is severe due to evaporation and inefficiency of adduction in the analytical scheme. Concentrations of molecular species with more than about 33 carbon atoms are low and approach the limits of detection of the analytical method. Use of carbon preference indices to describe molecular distributions of fatty acids and paraffins is less satisfactory but more convenient than the use of graphs. In calculating carbon preference indices limits will be noted when they are different from those specified in the formulae. The limits in the formulae were chosen because in most samples concentrations of fatty acids and paraffins could be determined within these limits.

Data related to six modern, unconsolidated sediments and eleven ancient sedimentary rocks are shown in Tables 1 and 2. In Table 1 the type

TABLE 1. — NORMAL FATTY ACIDS AND PARAFFINS IN MODERN SEDIMENTS

Sample	Age Years	Location	Normal Fatty Acids				Normal Paraffins			
			μg/g recovered	μg/g corrected	CPI_A	Range found	μg/g recovered	μg/g corrected	CPI_p	Range found
Mud 18–60 in.	<22,000	San Nicolas Basin, California 33°3′N, 119°W	41	64	8·39	C_{12}–C_{34}	2·2	3·9	4·78	C_{14}–C_{35}
Mud 10–80 cm.	<10,000	Tanner Basin, California 32°50′N, 119°33′W	7·0	28	9·00	C_{14}–C_{32}	0·19	0·61	2·57	C_{16}–C_{33}
Mud 500–555 cm.	<30,000	Tanner Basin, California 32°50′N, 119°33′W	17	26	8·40	C_{14}–C_{32}	1·5	2·3	2·13	C_{15}–C_{35}
Mud 0–32 in.	<5000	Mississippi Delta 28°33′N, 89°50′W	20	27	4·46	C_{12}–C_{36}	1·7	2·2	3·98	C_{14}–C_{35}
Carbonate Mud 0–68 in.	<2000	Florida Bay, Florida 25°05′N, 80°38′W	4·1	77	5·58	C_{12}–C_{38}	0·09	0·96	1·80 (C_{18}–C_{30})	C_{17}–C_{31}
Algal Ooze	<1500	Mud Lake, Florida 19 mile NE Ocala	570	—	6·57	C_{12}–C_{34}	35	—	5·19	C_{20}–C_{33}

TABLE 2. — NORMAL FATTY ACIDS AND PARAFFINS IN ANCIENT SEDIMENTS

Sample	Age	Location	Normal Fatty Acids				Normal Paraffins			
			μg/g recovered	μg/g corrected	CPI_A	Range found	μg/g recovered	μg/g corrected	CPI_P	Range found
Repetto mudstone	Pliocene	Los Angeles Co., California	0·83	1·9	3·18	C_{14}–C_{34}	0·68	1·5	1·16 (C_{20}–C_{32})	C_{19}–C_{34}
Monterey shale	Miocene	Santa Barbara Co., California	13	19	2·22	C_{13}–C_{36}	2·6	3·6	2·18	C_{14}–C_{35}
Green River oil shale	Eocene	Garfield Co., Colorado	13	23	1·85	C_{10}–C_{33}	29	34	2·38	C_{12}–C_{33}
Skull Creek shale	Cretaceous	Weston Co., Wyoming	13	23	2·05	C_{12}–C_{32}	6·7	8·9	2·21	C_{13}–C_{35}
Mowry shale	Cretaceous	Converse Co., Wyoming	28	32	1·06	C_{9}–C_{33}	32	43	1·10	C_{12}–C_{35}
Heath shale	Mississippian	Fergus Co., Montana	27	54	1·10	C_{10}–C_{32}	19	32	1·13	C_{13}–C_{32}
Chattanooga shale	Devonian	Cherokee Co., Oklahoma	5·6	21	1·23	C_{10}–C_{32}	14	52	1·05 (C_{16}–C_{30})	C_{12}–C_{31}
New Albany shale	Devonian	Rowan Co., Kentucky	3·8	5·5	1·26	C_{9}–C_{31}	23 60*	29 71*	1·13 (C_{16}–C_{30})	C_{11}–C_{31}
Heath limestone	Mississippian	Fergus Co., Montana	0·59	1·8	4·34	C_{14}–C_{33}	1·2	1·2	1·01	C_{13}–C_{33}
Bone Spring limestone	Permian	Culberson Co., Texas	0·45	0·73	3·45 (C_{15}–C_{29})	C_{12}–C_{30}	1·7	2·9	0·90	C_{14}–C_{36}
Lamar limestone	Permian	Culberson Co., Texas	9·1	10	1·11	C_{10}–C_{34}	8·8	12·0	0·92	C_{14}–C_{37}

* Determined by benzene-methanol extraction

of sedimentary material is followed by a measure of distance of the sample from the sediment-water interface. The ages of these samples were determined by radiocarbon dating. In Table 2 the geological name and a lithological term identify the samples; geological ages of the samples are listed next, followed by location. The concentrations of normal fatty acids and normal paraffin hydrocarbons in micrograms per gram of dried sediment are reported in two ways. First are listed the concentrations of total fatty acids or total paraffins recovered from the sediments by the analytical procedure. In the adjacent column are the concentrations of total fatty acids or total paraffins corrected for losses, as determined from internal standards, which occurred during the analytical procedure. Carbon preference indices for both fatty acids and paraffins are shown. Finally the largest and smallest normal fatty acid and paraffin detected in each sample are listed.

Normal Fatty Acids and Paraffins in Modern Sediments

Six modern, unconsolidated sediments were examined for fatty acids and normal paraffins. Five of these samples are from marine environments. Three samples came from deep basins off the coast of southern California. One sample was collected in the Gulf of Mexico from sediments of the Mississippi delta. Another sample is from shallow, carbonate bank deposits of Florida Bay. Algal ooze from Mud Lake, Florida, represents the one sample from a non-marine environment.

San Nicolas Basin (*Fig. 5*)

A sample was obtained from San Nicolas Basin, one of several basins lying off the coast of southern California, at a water depth of almost 6000 ft. The sample was composited from the bottom 18 to 60 in. of several cores. Distributions of fatty acids and paraffins are shown; even-carbon-numbered fatty acids are dominant as are odd-carbon-numbered paraffins. The most pronounced normal fatty acids are C_{16} and C_{26}, while the most abundant normal paraffin is C_{29}. The ratio of total fatty acids to paraffins, as obtained by the analytical procedure, is about 16 to 1.

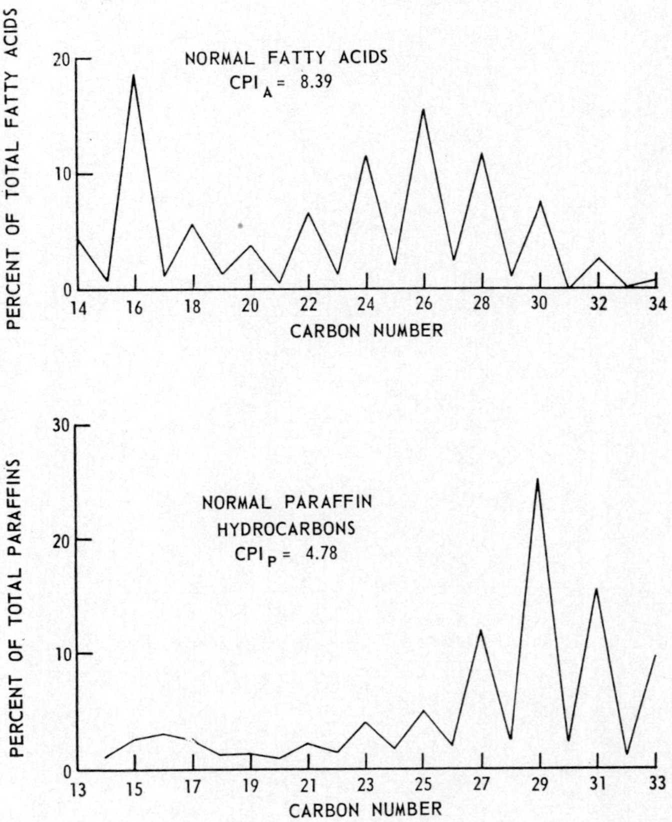

FIGURE 5. — Normal fatty acids and paraffins in San Nicolas Basin sediment.

Tanner Basin (Figs. 6 and 7)

Fatty acids and paraffins are reported for two samples from the top and bottom of a long core of sediment taken at a depth of 4500 feet from Tanner Basin, another of the basins off the coast of southern California. The core was provided by Dr. D. S. Gorsline, University of Southern California. The distributions of fatty acids and paraffins are generally similar in the two samples with a few exceptions. In both cases C_{16} is the dominant normal fatty acid. The second most prominent fatty acid is C_{24} at the top of the core and C_{26} at the bottom. The distribution of fatty acids in the bottom sample is similar to that of the sample from San Nicolas Basin. As in the San Nicolas Basin sample, the most prominent paraffin in the Tanner Basin samples is C_{29}. Tanner Basin

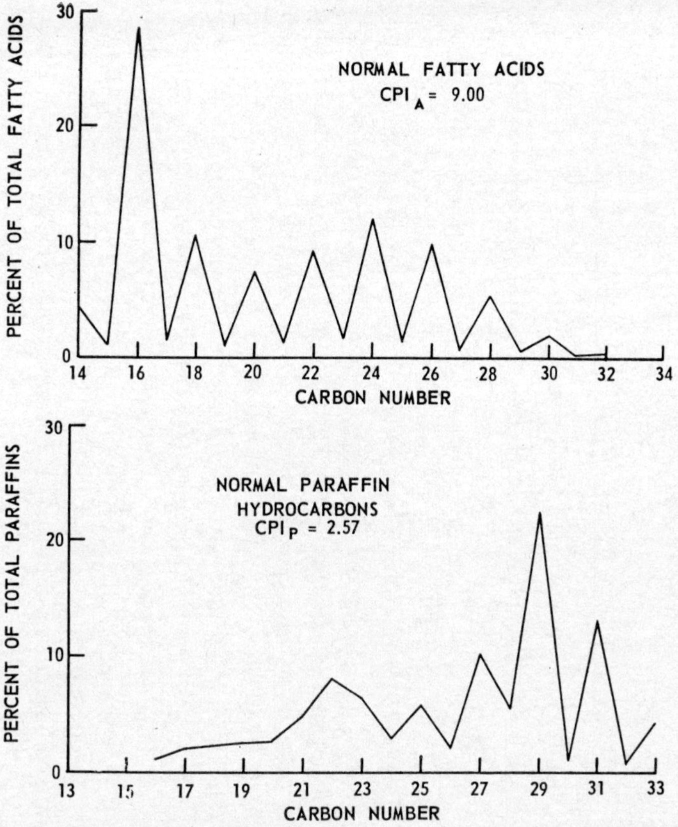

FIGURE 6. — Normal fatty acids and paraffins in Tanner Basin (top) sediments.

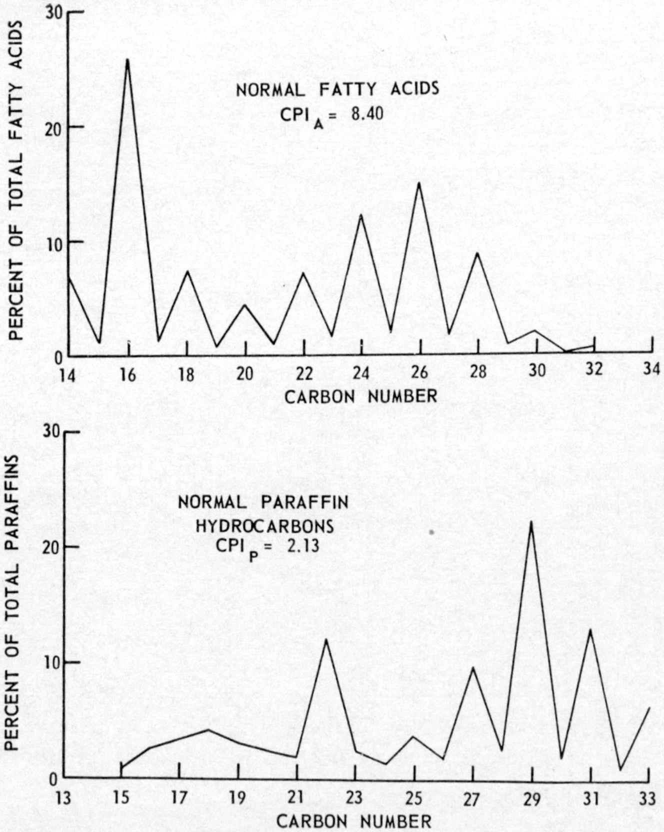

FIGURE 7. — Normal fatty acids and paraffins in Tanner (bottom) sediment.

samples differ from the San Nicolas sample in that the C_{22} normal paraffin is pronounced in the Tanner Basin sediments and becomes enhanced with depth. A pronounced even-carbon-numbered normal paraffin in a modern sediment is rare. The concentrations of normal fatty acids and paraffin in Tanner Basin sediments are about half as large as in the sediments of San Nicolas Basin.

Mississippi Delta (Fig. 8)

Normal fatty acids in a mud from the Mississippi delta resemble those in the California basin sediments. C_{16} normal fatty acid is dominant. In the high molecular weight range C_{28} is most abundant. The distribution of normal paraffins in the delta sample resembles most closely the

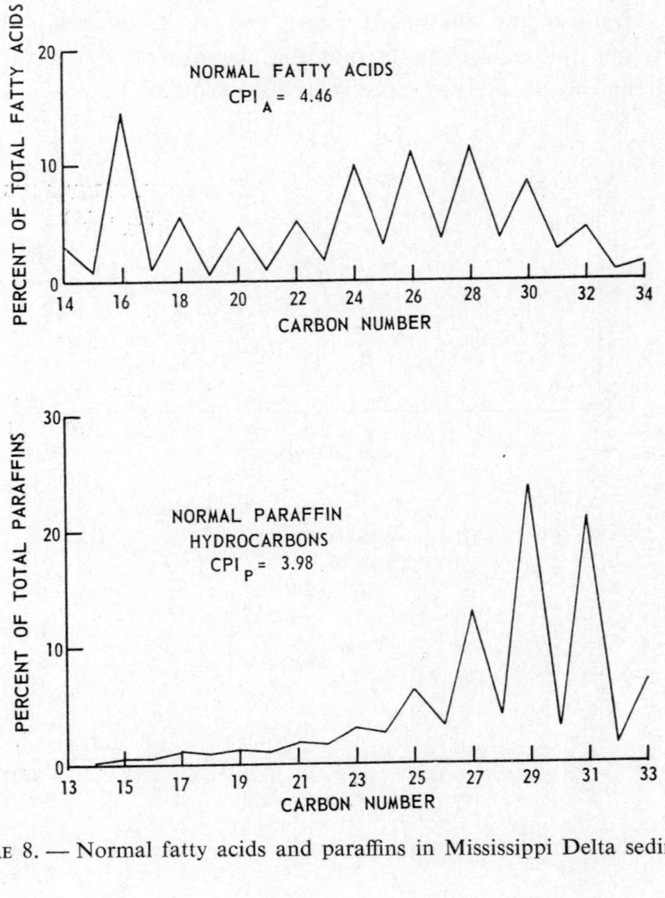

FIGURE 8. — Normal fatty acids and paraffins in Mississippi Delta sediments.

distribution of paraffins in San Nicolas Basin. The Delta sample was composed of the first 32 in. of a core collected south of Grand Isle, Mississippi at a depth of 1540 ft. Concentrations of normal paraffins and acids are similar to concentrations in Tanner Basin samples.

Florida Bay (Fig. 9)

A 68 in. piston core from a carbonate mud bank in Florida Bay was collected in 0·5 ft of water. The two prominent normal fatty acids present are C_{16} and C_{24}. In contrast to the previous samples C_{25} normal paraffin is largest instead of C_{29}. An unusually large correction factor was applied when total concentrations of both fatty acids and paraffins were adjusted for losses sustained by the internal standards (Table 1). Experiments were repeated on the sample and in all cases the internal standards fractionated badly. However, when limestones, which are composed of lithified carbonate material, were analyzed, this problem of such great loss of internal standards did not arise.

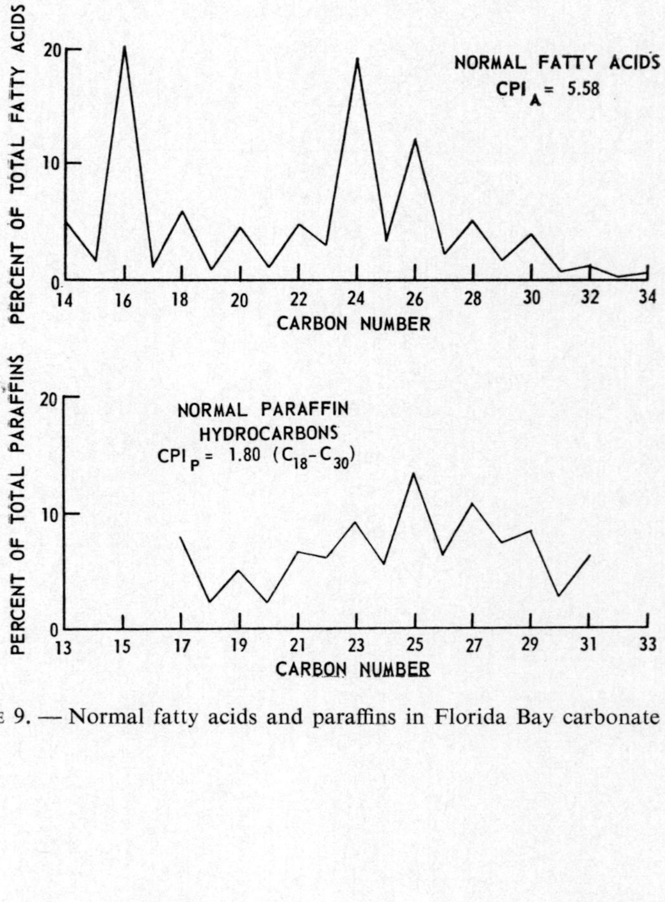

FIGURE 9. — Normal fatty acids and paraffins in Florida Bay carbonate mud.

Algal Ooze (Fig. 10)

This sample was obtained by W. H. Bradley, U.S. Geological Survey, from the middle of Mud Lake at the sediment-water interface. The lake is about 19 miles northeast of Ocala, Florida. The algal ooze of this lake is believed to be like the precursor of much of the organic material of Green River oil shale (Bradley, 1965). The concentration of normal fatty acids recovered is 570 μg per gram of which 40% is C_{16}. Higher molecular

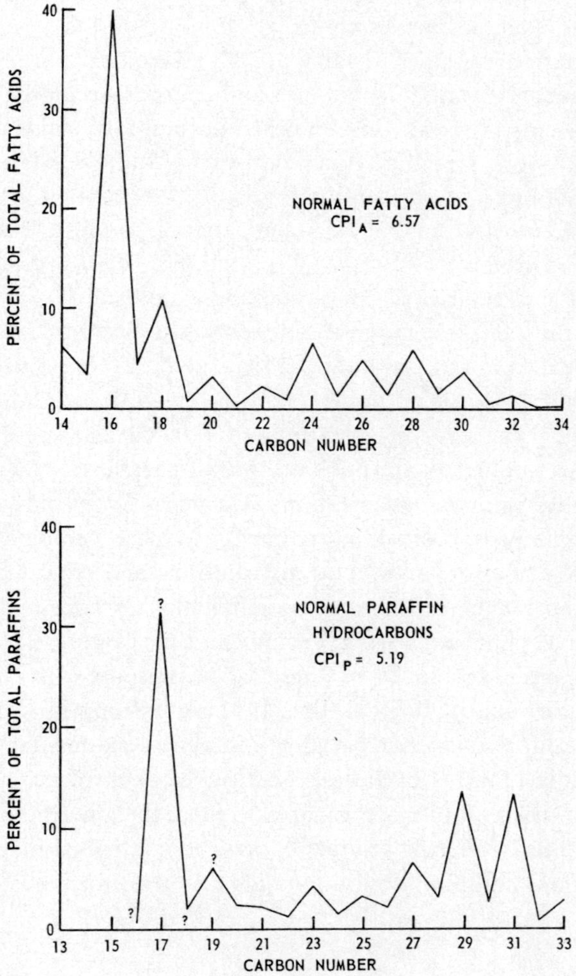

FIGURE 10. — Normal fatty acids and paraffins in Mud Lake algal ooze.

weight fatty acids are present with C_{24} and C_{28} most abundant. Trask and Wu (1930) examined a sample from Mud Lake and reported only C_{26} normal fatty acid. Perhaps they were observing a mixture composed dominantly of fatty acids C_{24}, C_{26}, and C_{28}. The paraffins in this sample have an odd-carbon preference, and for molecules larger than C_{20}, the most abundant normal paraffin is C_{29}. Assignment of gas chromatographic peaks in the region C_{16} to C_{20} is not certain because of the large number of compounds with retention times in this range. If peak assignment has been correct, C_{17} normal paraffin is most abundant. Insufficient sample prevented further study.

The distributions of normal fatty acids and normal paraffins in these modern sediments from different environments are remarkably similar. In these sediments even-carbon-numbered normal fatty acids are as much as nine times more abundant than odd-carbon-numbered normal fatty acids. Distributions of fatty acids are bimodal with the primary mode at C_{16}, always the most abundant fatty acid, and a secondary, less prominent mode, in the region of $C_{24}-C_{26}-C_{28}$. Paraffin distributions are also similar, but the similarity is not as striking as in the case of fatty acids. Normal paraffins in these modern sediments are dominantly of odd-carbon number with the ratio of odds to evens being as large as 5·19. Normal paraffins of odd-carbon numbers in the range C_{27} to C_{31} are usually most abundant. The total concentrations of fatty acids are 16 to 80 times larger than the total concentrations of normal paraffins.

The distributions and concentrations of normal fatty acids and normal paraffins originally deposited in sediments in past geological time are assumed to be somewhat similar to distributions and concentrations that have been reported here for modern sediments. Of course the kinds of organisms have changed during geological time, but the distributions of fatty acid building blocks of these organisms are assumed to have been similar through geological time. If these assumptions are correct, a study of the fatty acids and paraffins in ancient sedimentary rocks can show the results of transformations that have taken place in geological time. If these assumptions are incorrect, then the distribution of fatty acids or paraffins in ancient sediments may be controlled more by initial fatty acid-normal paraffin distributions than by transformation processes.

Normal Fatty Acids and Paraffins in Ancient Sediments

In ancient sediments normal fatty acids and paraffins have a variety of molecular distributions. In some rocks these distributions are like those in modern sediments in that even-carbon-numbered normal fatty acids and odd-carbon-numbered normal paraffins predominate, but the predominance of these molecules is seldom as great as in modern sediments. In other rocks the distributions of normal fatty acids and paraffins differ greatly from the distributions of these molecular species in modern sediments. The predominances of even-carbon-numbered normal fatty acids become smaller; that is, the ratios of even-to odd-carbon-numbered normal fatty acids approach 1·0. At the same time the normal paraffin distributions show less preference for odd-carbon-numbered molecules. In fact, the ratios of odd- to even-carbon-numbered normal paraffins approach 1·0, and may become less than 1·0, indicating that even-carbon-numbered paraffins predominate in the rock.

In an earlier paper (Kvenvolden, 1966) parallelism in the distributions of normal fatty acids and normal paraffins with one less carbon atom than the fatty acids was used as evidence that fatty acids served as precursors for many paraffins in ancient sediments. A lack of parallelism of these distributions in a modern sediment was used as evidence that fatty acids and paraffins in modern sediments do not have the same precursor-product relationship as in ancient sediments. Among the samples of ancient sediments described in this paper some show the parallelism of fatty acid-normal paraffin distributions observed earlier, but others do not.

Normal fatty acids and paraffins have been obtained from eleven ancient sedimentary rocks ranging from Devonian to Pliocene. The rocks were collected in Kentucky, Oklahoma, Texas, Colorado, Wyoming, Montana and California. Ten of the sediments were deposited in a marine and one in a non-marine environment. The marine sedimentary rocks include seven shales or mudstones and three limestones.

Repetto mudstone-Pliocene (Fig. 11)

A sample of Repetto mudstone was collected from a fresh exposure in the Repetto Hills of the Los Angeles Basin. The distribution of normal fatty acids is similar to the distributions of fatty acids obtained from modern sediments in basins off southern California or the Mississippi delta. (Figs. 5–8). Some transformation of fatty acids has taken place because

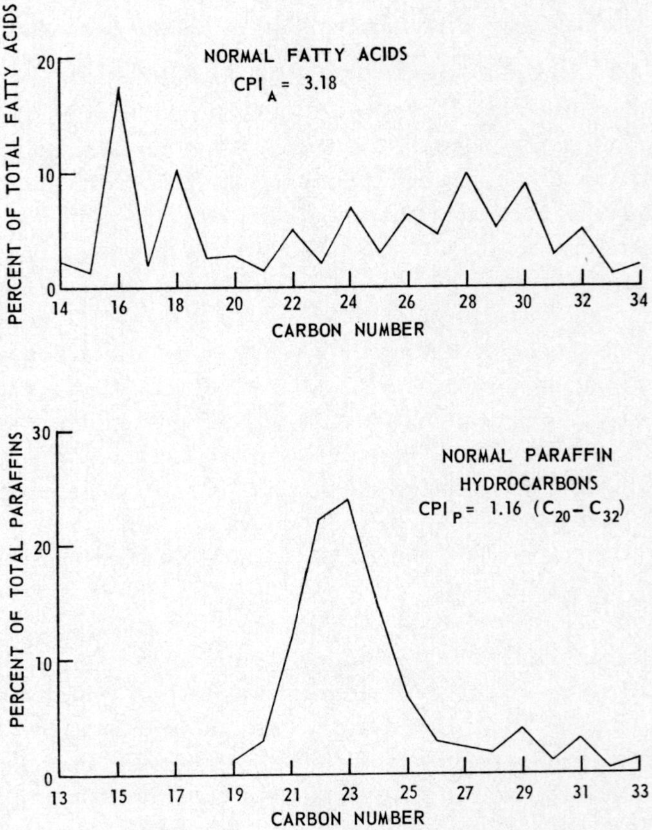

FIGURE 11. — Normal fatty acids and paraffins in Repetto mudstone (Pliocene).

the ratios of even- to odd-carbon-numbered fatty acids are smaller in the Repetto sample than in the modern sediments. Normal paraffins in the Repetto mudstone are distributed differently from those in the modern sediment. Although odd-carbon numbered paraffins are more abundant than even-carbon-numbered paraffins at C_{29}, C_{31}, C_{33}, below C_{28} the distribution of paraffins becomes smooth with C_{23} being present in the largest concentration. In this sample few if any of the normal paraffins could have been derived from normal fatty acids. If normal paraffins had been derived from fatty acids, the distribution of these molecular species should have been parallel as explained earlier. Most of the paraffins in the range C_{19} to C_{28} probably come from sources other than transformed fatty acids and other than originally deposited normal paraffins.

Monterey shale-Miocene (Fig. 12)

Normal fatty acids in this sample have a predominance of even-carbon-numbered molecules, but the distribution is different from any in the modern sediments. Normal fatty acid C_{24} is most abundant, with C_{16} second.

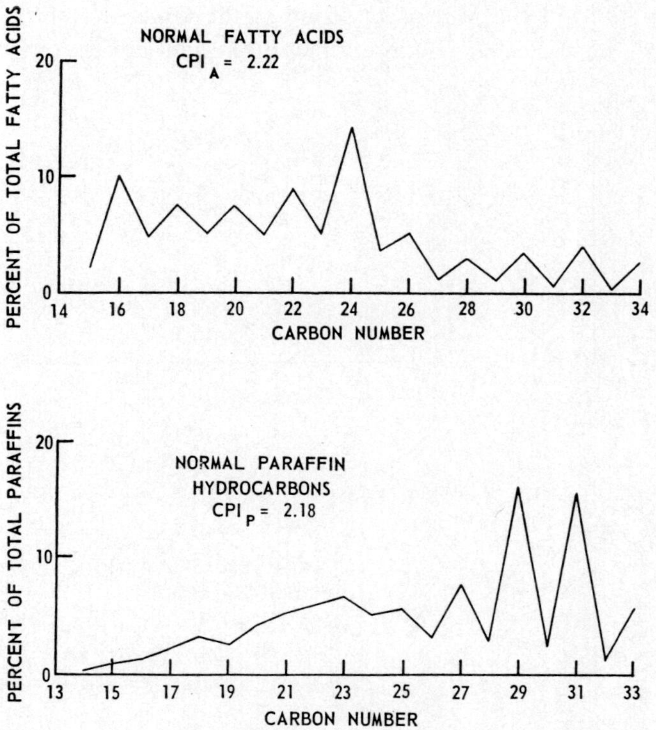

FIGURE 12. — Normal fatty acids and paraffins in Monterey shale (Miocene).

On the other hand, the normal paraffin distribution is somewhat like that in modern sediments, with prominent molecules at C_{27}, C_{29}, C_{31} and C_{33}. If transformation of fatty acids to paraffins has taken place in this sediment, its effect is small. Perhaps some of the paraffins in the range from C_{15} to C_{25} came from transformations involving some kind of decarboxylation of fatty acids C_{16} to C_{26}. This siliceous shale sample was collected on the sea cliff at Gaviota Beach State Park near Santa Barbara, California.

Green River oil shale-Eocene (Fig. 13)

Green River oil shale has been examined previously for normal fatty acids (Lawlor and Robinson, 1965), and for normal paraffins (Cummins and Robinson, 1964). The analyses reported now show distributions of normal fatty acids and paraffins similar to those obtained by others. The sample came from the Mahagony Zone of the Green River Formation, which was deposited in Eocene time in a non-marine environment.

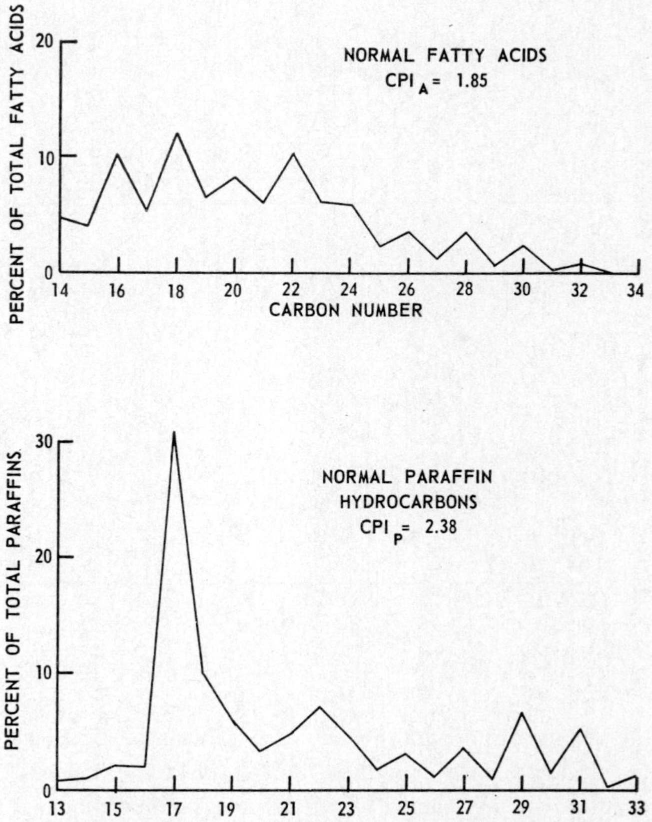

FIGURE 13. — Normal fatty acids and paraffins in Green River oil shale (Eocene).

Although even-carbon-numbered fatty acids are dominant, their distribution is unlike any found in modern sediments. A comparison of normal fatty acids of Green River oil shale with normal fatty acids in the Mud Lake algal ooze, which is thought to be like the precursor material for the oil shale, shows no direct relationship between the two fatty acid populations. The distribution of fatty acids in the Green River oil shale suggests that some transformation of fatty acids probably has taken place. Normal paraffin distributions in the Green River oil shale and Mud Lake algal ooze are similar, especially in that both distributions show an unusually high abundance of C_{17}. The assignment of normal paraffin C_{17} in the Mud Lake sample is not certain, however. Lawlor and Robinson (1965) compared the distribution of normal fatty acids, C_{24}–C_{34}, in Green River oil shale with the distribution of normal paraffins C_{23}–C_{33} and observed a parallelism. They wrote that the parallelism of these distributions resulted from early transformations of organic material.

Skull Creek shale-Cretaceous (Fig. 14)

A sample of Skull Creek shale was obtained from an outcrop near Upton, Wyoming. Normal fatty acids in this rock still retain a strong preference for even-carbon-numbered molecules, but the preference is less

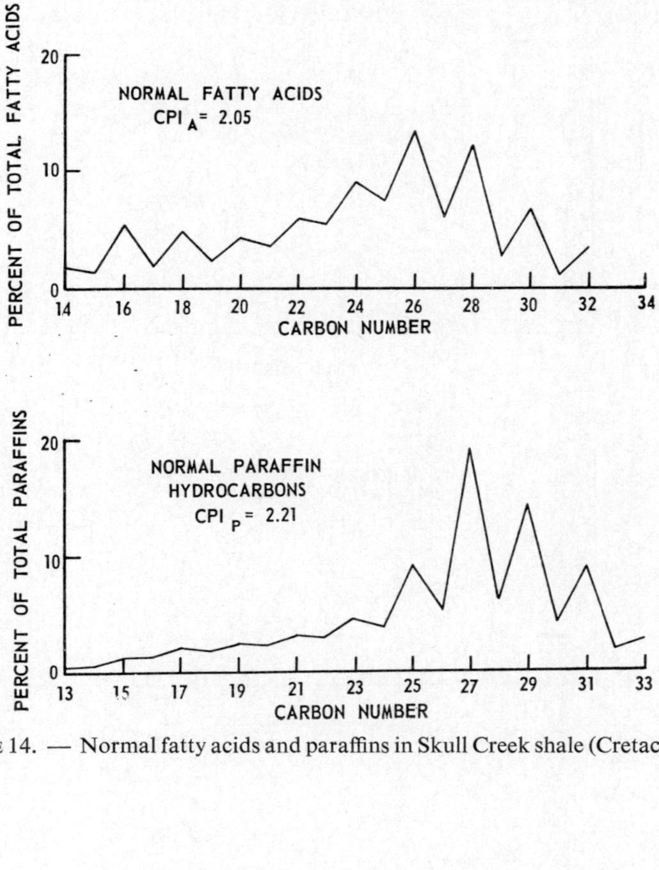

FIGURE 14. — Normal fatty acids and paraffins in Skull Creek shale (Cretaceous).

than that observed in modern sediments. Even-carbon-numbered fatty acids at C_{24}, C_{26}, C_{28}, and C_{30} are most abundant, corresponding to the secondary mode in the distributions of fatty acids in modern sediments. The fact that C_{16} is not more abundant is puzzling. If C_{16} normal fatty acid were transformed by decarboxylation to C_{15} normal paraffin, this molecule should be prominent in the paraffin distribution; it is not. The distribution of normal paraffins in this rock is like that observed in modern sediments from San Nicolas Basin or the Mississippi Delta (Figs. 5 and 8), except that C_{27} instead of C_{29} is dominant, and CPI_p is less for the Skull Creek sample. Some paraffins apparently have been added to this sediment to reduce the CPI_p. Some of these paraffins may have come from the transformation of fatty acids. However, fatty acid transformation has not been great, because the CPI_A is still relatively large.

Mowry shale-Cretaceous (Fig. 15)

This sample was collected from an outcrop south of Douglas, Wyoming. Distributions reported previously (Kvenvolden, 1966) for this sample were not corrected for losses sustained during processing as determined

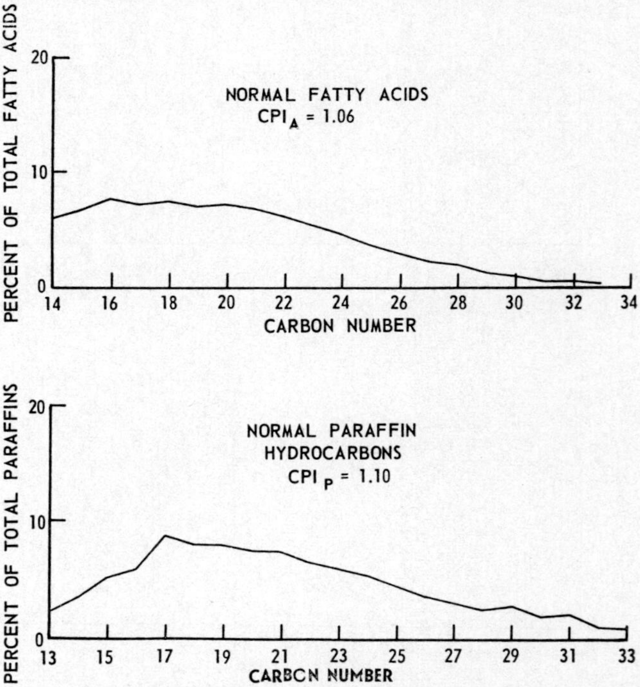

FIGURE 15. — Normal fatty acids and paraffins in Mowry shale (Cretaceous).

from internal standards. The distributions reported here differ only slightly from those reported earlier. Both the distributions of normal fatty acids and of normal paraffins are quite different from those observed in modern sediments. In this sample of Mowry shale even- and odd-carbon-numbered normal fatty acids are almost equally abundant; odd- and even-carbon-numbered normal paraffins are also almost equally abundant. Transformations of fatty acids must have taken place. The parallelism of distributions of normal fatty acids and paraffins suggests that a large amount of the paraffins present must have come from fatty acid transformations. One way to account for these smooth distributions, which probably came from original distributions having dominantly even-carbon-numbered fatty acids and odd-carbon-numbered paraffins, is to call upon ideas expressed earlier by Cooper (1962). He postulated that original fatty acids degraded by means of a free radical decarboxylation scheme in which at each step a lower-carbon-numbered fatty acid and paraffin are formed. Each newly formed acid participates in the same reaction. Paraffins generated by the scheme collect with paraffins originally present. The absolute concentrations of fatty acids diminish. "Under conditions favoring acid formation over paraffin formation a secular equilibrium would be attained eventually in which no preference between even- or odd-numbered paraffins or acids exists" (Cooper and Bray, 1963). The distribution of fatty acids and paraffins in this sample of Mowry shale can be explained by this scheme.

Mowry shale and Skull Creek shale are both of Cretaceous age, but Mowry shale is slightly younger geologically than Skull Creek shale. Yet fatty acids in Mowry shale are apparently further transformed than are fatty acids in Skull Creek shale. Time undoubtedly is important in these transformations, but the chemical and physical environments for transformations must be more important in establishing differences in the distributions of normal fatty acids and paraffins in sediments.

Heath shale-Mississippian (Fig. 16)

The distributions of normal fatty acids and paraffins in this sample are similar to distributions observed in the Mowry sample. A reason for showing these distributions is to point out that rocks of very different geological ages may have similar distributions of fatty acids and paraffins. Mowry shale is Mesozoic, and the sample came from an outcrop in the Powder River Basin of Wyoming. Heath shale is Palaeozoic, and its sample came from a core hole on the Big Snowy Uplift in Montana. In terms of total concentrations of fatty acids and paraffins is these sediments (Table 2) they are very much alike. Lithologically they differ in that Heath shale is

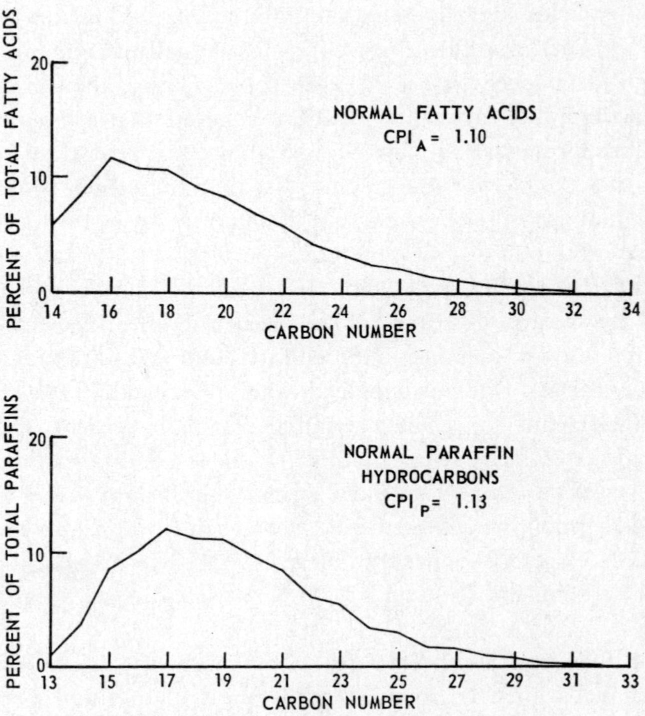

FIGURE 16. — Normal fatty acids and paraffins in Heath shale (Mississippian).

non-siliceous while Mowry shale is usually siliceous. On the other hand, Mowry shale and Monterey shale, mentioned earlier, are lithologically very similar, but the distributions of fatty acids and paraffins in these shale samples are quite different.

Chattanooga shale-Devonian (Fig. 17)

The oldest samples of this collection are Chattanooga shale and equivalent New Albany shale of Devonian age. The sample of Chattanooga shale was collected in Cherokee County, Oklahoma. The distributions of normal fatty acids and paraffins are similar to those obtained for Mowry and Heath shales. All of these samples have normal fatty acids which seem to be the residual acids of some transformation processes. Normal paraffins seem to be the products of the same transformation processes.

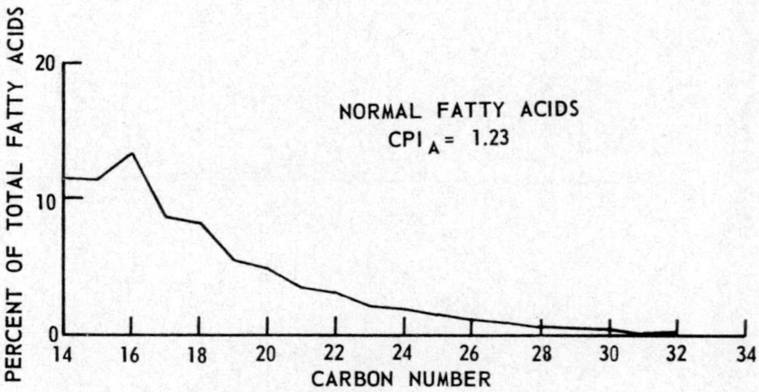

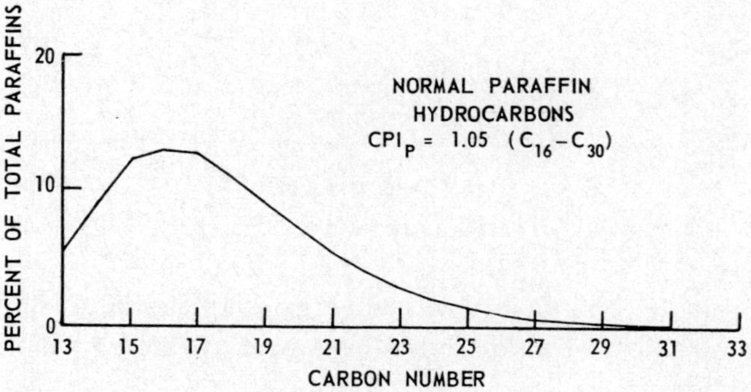

FIGURE 17. — Normal fatty acids and paraffins in Chattanooga shale (Devonian).

New Albany shale-Devonian (Fig. 18)

A new Albany shale sample came from the east side of the Cincinnati Arch in Kentucky. The distributions of normal fatty acids and paraffins in this sample are much like those that have been described earlier (Figs. 15, 16, and 17). The distributions of normal fatty acids in the New Albany and Chattanooga shales are similar; the distributions of paraffins differ slightly. A remarkable similarity is observed when the distribution of normal fatty acids in the New Albany shale is compared with the distribution of normal paraffins (with one less carbon atom than the acids) in

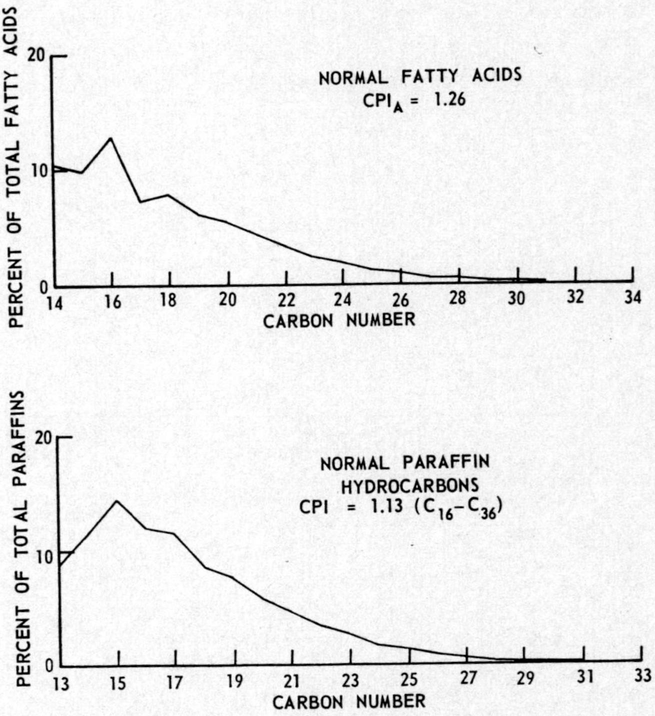

FIGURE 18. — Normal fatty acids and paraffins in New Albany shale (Devonian).

the same sample. This favorable comparison is believed to indicate that in this sample most of the paraffins have been generated from fatty acid transformations.

Heath limestone-Mississippian (Fig. 19)

The last three samples that will be discussed are limestones. The distributions of fatty acids and paraffins in these samples differ from distributions found in the other samples.

Heath limestone comes from the Heath Formation of the Big Snowy Uplift of Montana, as did the Heath shale sample. The distribution of normal fatty acids shows a dominance of C_{16} and C_{18}. Higher carbon numbered fatty acids are present, but their concentration diminishes rapidly with increasing molecular sizes. Even-carbon-numbered fatty acids dominate. This distribution differs from those found in modern sediments in that the secondary mode of the acid distribution at C_{24}–C_{26}–

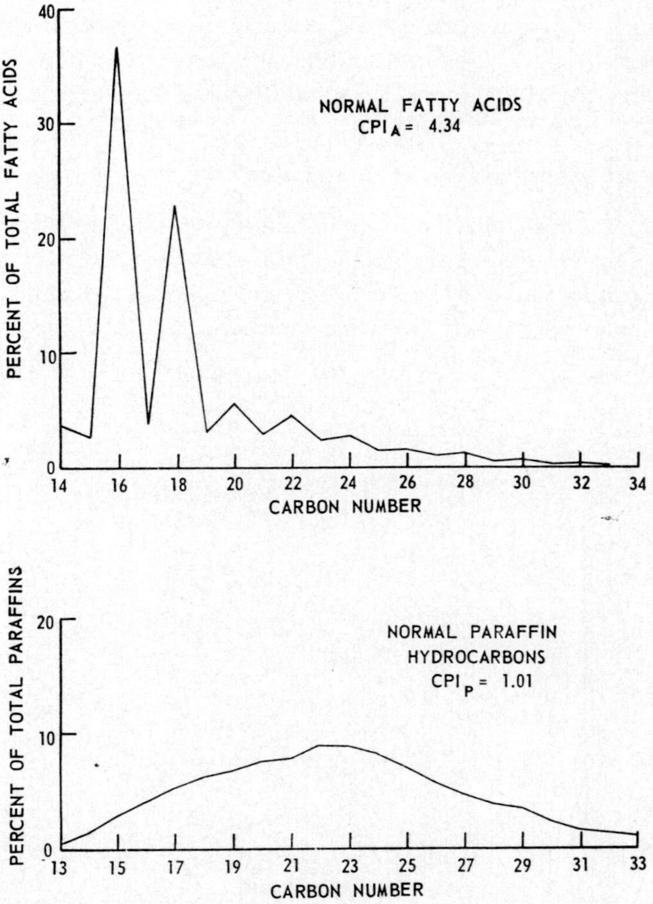

FIGURE 19. — Normal fatty acids and paraffins in Heath limestone (Mississippian)

C_{28} is absent. If this mode is missing because the fatty acids in this molecular weight range have been transformed, then normal paraffins resulting from this transformation should be present in approximately the same range. The paraffin distribution (Fig. 19) is smooth, with most of the paraffins in the range that would be expected if these paraffins came from fatty acids in the range of the secondary mode shown by the distributions of fatty acids in modern sediments. The reason that C_{16} and C_{18} normal fatty acids have survived the postulated transformations is not known at this time. At first contamination was suspected as the cause of the high concentration of C_{16} and C_{18}, but when other samples, including ones from the subsurface of different age and locality, were found to have similar distributions of fatty acids and paraffins, the idea of contamination causing the large relative abundances of C_{16} and C_{18} was discounted.

Bone Spring limestone-Permian, Leonardian (Fig. 20)

This sample was collected from an outcrop on the western flank of the Delaware Basin, Texas. The distribution of normal fatty acids is about the same as that found in the Heath limestone. Normal paraffins are distributed over approximately the same molecular weight range as they are

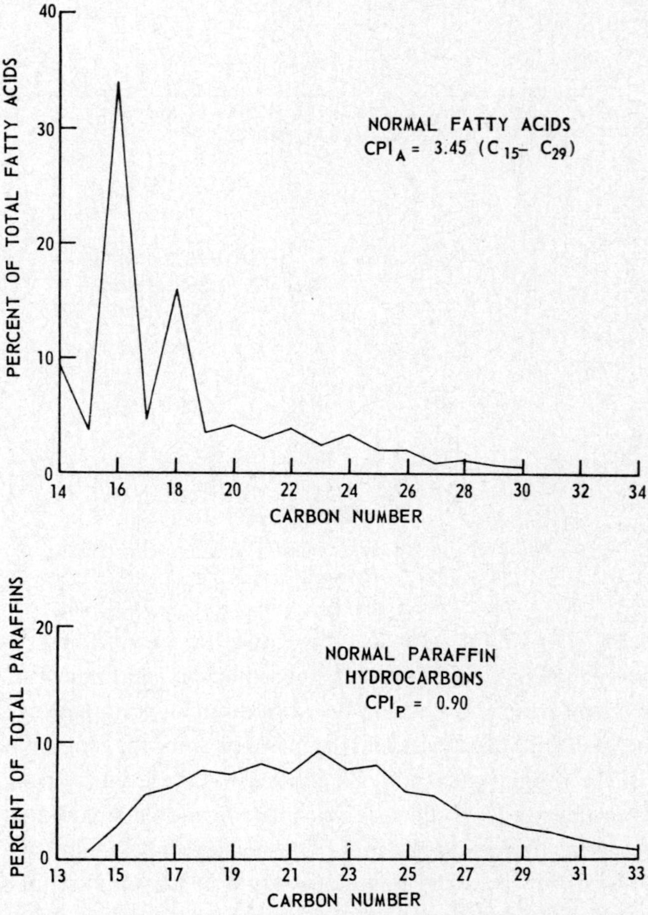

FIGURE 20. — Normal fatty acids and paraffins in Bone Spring limestone (Permian)

in the Heath limestone. An important difference in the paraffin distribution is that paraffins in the Bone Spring limestone have a definite preference for even-carbon-numbered molecules. Such distributions of paraffins are unusual. Therefore, in this sample both normal fatty acids and normal paraffins have an even-carbon preference. Distributions of normal paraffins with even carbon preferences cannot result from fatty acid transformations following the scheme of Cooper (1962). If, however, Cooper's step-by-step decarboxylation scheme is modified so that reduction (hydrogenation) of a small number of carboxyl groups to methyl groups (without loss of carbon atoms) occurs concurrently, mixtures of normal paraffins with even-carbon-numbered molecules slightly predominant can be accounted for. Geochemical evidence for hydrogenation occurring in sediments was obtained by Blumer (1965) in his studies of fossil porphyrins. The geochemical history of fatty acids and paraffins in the Bone Spring limestone must have been similar to that in the Heath limestone. However, in the Bone Spring limestone normal paraffins may have resulted from some kind of transformation most likely involving decarboxylation and hydrogenation.

Lamar limestone-Permian, Guadalupian (Fig. 21)

Lamar limestone contains normal fatty acids and paraffins distributed differently from those found in the two previously discussed limestone samples. The Lamar sample came from the west flank of the Delaware Basin in West Texas. Normal fatty acids larger than C_{22} are distributed so that neither even- nor odd-carbon-numbered molecules are more abundant. This kind of distribution of fatty acids has been observed previously in Mowry, Heath, Chattanooga and New Albany shales. Normal fatty acids smaller than C_{20} show a preference for even-carbon-numbered acids. This preference is not as dramatic as in the Heath and Bone Spring limestones. The distribution of normal paraffins in Lamar limestone is like that in Bone Spring limestone in that both distributions have more even- than odd-carbon-numbered molecules. The distributions differ in that the most abundant normal paraffin in the Bone Spring limestone is C_{22}, while the most abundant in the Lamar limestone is C_{26}, with C_{28} next most abundant. The geochemical history of this limestone may be like that experienced by the other limestones, except that the C_{16} and C_{18} acids apparently have disappeared and did not generate normal paraffins during their disappearance.

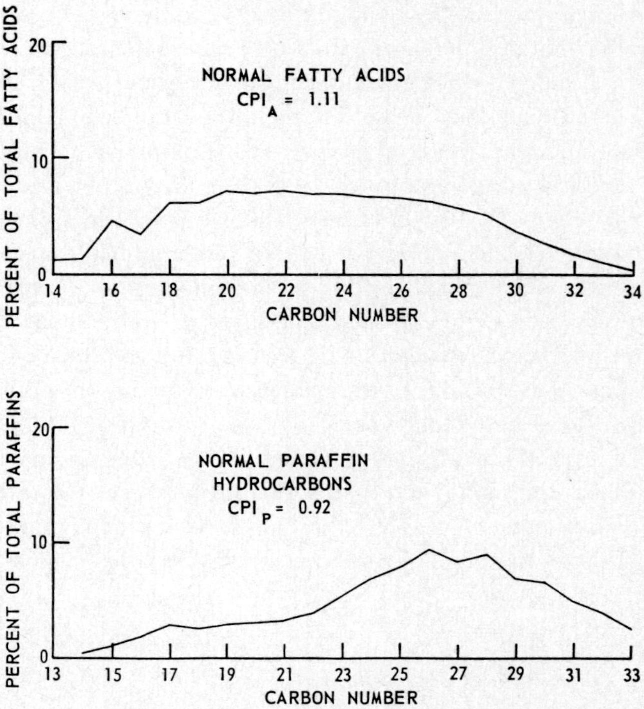

FIGURE 21. — Normal fatty acids and paraffins in lamar limestone (Permian).

Summary

Normal fatty acids in sediments show a variety of distributions, as do normal paraffin hydrocarbons. In the modern sediments studied the normal fatty acids are remarkably similar. This similarity shows that chemical and geochemical degradations which take place during sedimentation produce similar mixtures of normal fatty acids, at least in the modern sediments studied here. Normal fatty acid C_{16} is always most abundant; next most abundant are even-carbon-numbered fatty acids in the range from C_{24} to C_{28}. Normal paraffin distributions in modern sediments are also similar, but not as much so as are distributions of fatty acids. Odd-carbon-numbered normal paraffins are most abundant, especially in the range C_{27} to C_{31}. It is assumed that initially deposited sediments in ancient geological time had distributions of fatty acids and paraffins like those found today in modern sediments.

In ancient sediments normal fatty acids must have undergone some kind of transformation, because distributions of fatty acids range from those similar to what is found in modern sediments to distributions in which the even- and odd-carbon-numbered normal fatty acids are equally abundant. Samples examined here suggest that the older the rock, the more likely is the distribution of fatty acids to be smooth (show equal abundances of even- and odd-carbon-numbered molecules). But time is not as important as favorability of environment for chemical transformation in determining the distribution of fatty acids in a rock. This conclusion comes from the fact that of two Cretaceous samples the one that is geologically younger has smooth distributions of both fatty acids and paraffins. The older sample still shows a predominance of even-carbon-numbered fatty acids and odd-carbon-number paraffins. Both of these samples, which came from outcrops in the Powder River Basin, had experienced about the same geological history of temperature and pressure, so far as could be ascertained from field evidence. The dominant fatty acids in Precambrian Nonesuch Shale reported by Hoering and Abelson (1964) are C_{14}, C_{16}, C_{18}, C_{20} and C_{22}. Even in a rock of this antiquity fatty acids have not altered to give mixtures of equal abundances of even- and odd-carbon-numbered fatty acids. Apparently the chemical and physical environments in this rock were not favorable for these changes.

The transformations which take place in sediments tend to increase the relative abundances of odd- compared to even-carbon-numbered normal fatty acids. At the same time the relative abundances of even- compared to odd-carbon-numbered normal paraffins tend to increase. How this transformation takes place is not yet known. Ideas expressed by Cooper (1962) and cited earlier are the best explanations offered so far, but they do not account for all the observed variations in the distributions of fatty acids and paraffins. Undoubtedly many reactions take place, including simple decarboxylations, step-by-step decarboxylation (Cooper, 1962), hydrogenation and cleavage. The work of Jurg and Eisma (1964) suggests that some sort of carbon addition may also occur. The evidence is clear that a complicated sequence of chemical events must take place in sediments. This sequence may be different in shales than in limestones. The next step is to find out what has actually happened in sediments to account for the variety of distributions of normal fatty acids and paraffins.

Acknowledgement

The laboratory work involved in this paper was done by Mr. Jack T. Wall, to whom I am very grateful.

References

Abelson, P. H., and Parker, P. L. (1962) Fatty acids in sedimentary rocks. *Carnegie Institution of Washington Yearbook* **61,** 181–184.

Blumer, M. (1965) Organic pigments: their long-term fate. *Science*, **149,** 722–726.

Bradley, W. H. (1965) Tropical lakes, Copropel, and oil shale. Presidential Address, Annual Meeting, Geological Soc. Am., Kansas City, Mo., Nov. 4–6.

Bray, E. E., and Evans E. D. (1961) Distribution of *n*-paraffins as a clue to recognition of source beds. *Geochim. Cosmochim. Acta*, **22,** 2–15.

Cooper, J. E. (1962) Fatty acids in recent and ancient sediments and petroleum reservoir waters. *Nature*, **193,** 744–746.

Cooper, J. E., and Bray, E. E. (1963) A postulated role of fatty acids in petroleum formation. *Geochim. Cosmochim. Acta*, **27,** 1113–1127.

Cummins, J. J., and Robinson, W. E. (1964), Normal and isoprenoid hydrocarbons isolated for oil-shale bitumen. *J. Chem. Engr. Data*, **9,** 304–307.

Hoering, T. C., and Abelson, P. H. (1964) Chemicals from the Nonesuch shale of Michigan. *Carnegie Institution of Washington Yearbook* **63,** 262–264.

Jurg, J. W., and Eisma, E. (1964) Petroleum hydrocarbons: generation from fatty acids. *Science*, **144,** 1451–1452.

Kvenvolden, K. A. (1966) Molecular distributions of normal fatty acids and paraffins in some Lower Cretaceous sediments. *Nature*, **209,** 573–577.

Lawlor, D. L., and Robinson, W. E. (1965) Fatty acids in Green River formation oil shale. Preprint, Detroit Meeting, Am. Chem. Soc., Div. Petroleum Chem., p. 5–9.

Meinschein, W. G. and Kenny, G. S. (1957) Analyses of a chromatographic fraction of organic extracts of soils. *Anal. Chem.* **29,** 1153–1161.

Trask, P. D., and Wu, C. C. (1930) Does petroleum form in sediments at time of deposition? *Am. Assoc. Petroleum Geologists Bull.* **14,** 1451–1463.

THE MECHANISM OF THE GENERATION OF PETROLEUM HYDROCARBONS FROM A FATTY ACID

by

J. W. JURG and E. EISMA

Koninklijke/Shell Exploratie en Produktie Laboratorium, Rijswijk, The Netherlands

Abstract

During the International Congress of Organic Geochemistry in Paris (1964), we presented a paper concerning the generation of petroleum hydrocarbons from a fatty acid. Since that time we have done some complementary experiments. The results of these and the earlier experiments lead us to the suggestion of a reaction mechanism.

We have already reported on the generation from a fatty acid of n-alkanes with carbon chains both shorter and longer than that of the original acid. We have now analysed the fatty acids formed during the heating of behenic acid in the presence of clay and water.

Gas-chromatographic analyses indicate that the heating of behenic acid in the presence of clay and water yields fatty acids with carbon numbers ranging from C_{15} to C_{24}.

The generation of series of both fatty acids and n-alkanes with carbon chains shorter and longer than that of the original acid suggests that the formation of these compounds is brought about by related processes.

We suggest that the scheme of reactions involved is as follows:

(A)		RCOOH (1)	→ .R	(2) $+CO_2+$.H
(B)	.R	(2) + RCOOH	(1) → RH	(3) + .(R)COOH (4)
(C)		.(R)COOH (4)	↗ $R_{\bar{1}}^{=}$	(5) + .R_2COOH (6)
			(α–β) olefin	
			↘ .R_3	(7) + $R_{\bar{4}}^{=}$COOH (8)
				(ω–ϕ) unsat. acid
(D)	.R_3	(7) + RCOOH	(1) → R_3H	(9) + .(R)COOH (4)
(E)	.R	(2) + $R_{\bar{1}}^{=}$	(5) → .(R + R_1)	(10)
	.(R + R_1)	(10) + RCOOH	(1) → (R + R_1)H	(11) + .(R)COOH (4)

(F)	$R_1^{=}$	(5)$+.R_3$	(7) → $.(R_1+R_3)$	(12)
	$.(R_1+R_3)$	(12)+RCOOH	(1) → $(R_1+R_3)H$	(13) +.(R)COOH (4)
(G)	.R	(2)$+.R_3$	(7) → $(R+R_3)$	(14)
(H)	$.R_2COOH$	(6)+RCOOH	(1) → R_2HCOOH	(15)+.(R)COOH (4)
(I)	.R	(2)+$R_4^{=}COOH$	(8) → $.(R+R_4)COOH$	(16)
	$.(R+R_4)COOH$	(16)+RCOOH	→ $(R+R_4)HCOOH$(17)+.(R)COOH	(4)
(J)	.R	(2)$+.R_2COOH$ (6)	→ $(R+R_2)COOH$	(18)

The initiation reaction is the decarboxylation of the fatty acid (1) resulting in the alkyl radical (2). This intermediate will react with the original fatty acid, which is present in the relative highest concentration, to give the n-alkane (3) and the secondary radical of the fatty acid (4). This secondary radical can split up by β-scission into four products.

(a) An (α–β) olefin (5) and a primary radical of a fatty acid (6)
(b) A primary alkyl radical (7) and a (ω–ϕ) unsaturated fatty acid (8).

By combination of these intermediates the following products can be formed:

(A) n-Alkanes with carbon chains shorter and longer than the original acid.
(B) Fatty acids with carbon chains shorter and longer than the original acid.

The presumed reaction scheme is supported by the following facts:

1. The C_{21} n-alkane (corresponding to product (3) from reaction B) is always the most predominant n-alkane formed during the reaction of behenic acid. This indicates that decarboxylation is an important step in this reaction. That decarboxylation is an essential step for the generation of the n-alkanes with a carbon chain longer than the original acid follows from an experiment in which we used n-hexadecane instead of behenic acid. After heating n-hexadecane in the presence of clay we were unable to find hydrocarbons with a carbon chain longer than the n-C_{16}.
2. In general the absolute amount of unsaturated low-molecular-weight hydrocarbons decreases with increasing heating time.
3. In an experiment in which we heated behenic acid in the presence of 2-2′-azopropane, an initiator for radical reactions, for 65 hours at 200° C we also found that n-alkanes with longer and shorter carbon chains than that of the original acid were formed.

The results of these experiments indicate that from one fatty acid a homologous series of fatty acids and n-alkanes can be derived. Generation of such a mixture in a sediment may be a clue to the explanation of smooth distribution of n-alkanes and fatty acids in petroleum and in petroleum reservoir waters respectively.

THERMAL ALTERATION OF THE ORGANIC MATTER IN SEDIMENTS

by

A. G. DOUGLAS*, G. EGLINTON and W. HENDERSON

The Chemistry Department, The University, Glasgow, Scotland

Introduction

It is now widely accepted that sediments of all ages, including those from the Precambrian era as old as 3000 million years, contain organic matter. This organic matter is mostly insoluble in common solvents and reagents, and is termed kerogen, but a variable fraction is soluble and can be shown to contain identifiable compounds such as hydrocarbons (Douglas and Eglinton, 1966), fatty acids (Douglas *et al.*, 1966) and porphyrins (Thomas and Blumer, 1964; Barghoorn *et al.*, 1965; Eisma and Gransch, 1966). Current studies aim to isolate and identify these constituents, and among other things to use their distributions to rationalise the possible methods of petroleum origin and migration, and also to correlate their distributional patterns with those of identical or related compounds in present-day living things. Such correlations presuppose *inter alia* that (1) biochemical patterns existing in past times were similar to those now prevalent, (2) the chemical changes undergone by these constituents are explicable in terms of straightforward chemical reactions not dependent on geological time for their operation, and (3) the distribution patterns of the compounds cannot be explained entirely by the operation of abiogenic processes, such as the Fischer–Tropsch conversion of CO and H_2 into alkanes, alkenes, etc. (Lovelock, 1965; Meinschein, 1963).

* Present address: Organic Geochemistry Unit, The University, St. Thomas St., Newcastle upon Tyne 1, England.

The present paper represents an interim report on our attempts to investigate some aspects of point (2); that is, the diagenesis of organic matter in sediments under the influence of elevated temperatures. (In Fig. 1 we have used the term "geogenesis"; this is acceptable to the organic chemist for it implies the formation of compounds under earth conditions, thereby paralleling "biogenesis" in living things). Our aim is to establish the pattern of change undergone by individual organic compounds, classes of organic compounds and types of organic debris, when present in a sediment exposed to raised temperatures. This has relevance to the geo-

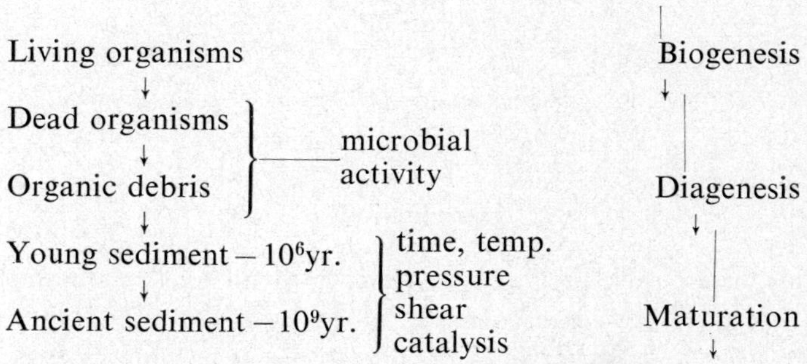

FIGURE 1. — Organic geochemistry—geogenesis of organic compounds.

logical experience of the higher temperatures occasioned by great depth of burial, earth movements and igneous intrusions. Definitive results should in turn lead to the use of this approach as a guide to the past biological, chemical and physical history of a sediment. Commercially, shales are heated to distil out 'shale oil' (Stewart and Forbes, 1938; de Nevers, 1966). In the realm of the space sciences, certain lunar and planetary probes are planned to provide automated controlled pyrolysis of surface samples, followed by instrumental analysis and telemetry (Rea, 1966).

We have chosen two materials for our initial experiments – the Eocene non-marine Green River Shale from Colorado and the Carboniferous Torbanite from the Lothians of Scotland.

The Green River Formation is held on geological and palaeontological grounds to be the accumulated sediments from very extensive shallow inland lakes. It is believed that at no time has this formation been deeply buried or subjected to high temperatures (Milton and Eugster, 1959); Jones and Vallentyne (1960) suggest, from its alanine content, that it has been subjected to a maximal continuous temperature of 74°. It therefore provides a useful model for thermal alteration experiments, since the compounds identifiable in the unaltered shale can be compared with those present in the same shale after laboratory thermal treatment. Torbanite, from the Carboniferous Limestone series of the Scottish Lothians, was described as a boghead or cannel coal by Macgregor (1938), but more recently Dulhunty (1943) has stated that Torbanite and cannel coal form distinct classes, because their respective peat stages resulted from the accumulation of unlike vegetable debris. Its microscopic structure reveals a large amount of alveolar "yellow bodies" which resemble the now living alga *Botryococcus braunii* (Blackburn and Temperley, 1936), and this in turn is regarded as the precursor of the recent rubbery deposit Coorongite and related materials (Blackburn and Temperley, 1936; Belcher, 1958). Pyrolysis of Torbanite gives a very high yield of oil (90–130 gal/ton), which reflects its high organic content.

Chemical studies on the Green River Shale have resulted in the identification of biologically significant alkanes such as a range of isoprenoids, including farnesane (C_{15}), pristane (C_{19}) and phytane (C_{20}) (Eglinton *et al.*, 1966) and a number of steranes and triterpanes (Burlingame *et al.*, 1965), one of which has been conclusively identified as gammacerane (Hills and Whitehead, 1966). The n-alkanes show the appropriate marked dominance of odd-numbered molecules, particularly n-C_{31}, C_{29}, C_{27}, and C_{17}, while the n-carboxylic acids are dominantly even-numbered (Douglas *et al.*, 1966; Ramsay, 1966). In another paper at this conference we have reported the identification of the C_{14}, C_{15}, C_{16}, C_{17}, C_{19}, C_{20} and C_{21} isoprenoid acids, while Leo and Parker (1966) have reported the presence of *iso*-acids in this shale. In summary, the alkanes and fatty acids already reported reflect the biological history of the shale to an impressive extent.

Similar chemical studies with Torbanite have resulted in the identification of phytane and pristane; there were, however, no steranes or triterpanes apparent. The n-alkanes showed no marked predominance of odd-numbered molecules; in fact, the carbon preference index (Cooper and Bray, 1963) was almost unity (Maxwell, 1967). The fatty acids ranged from n-C_{10} to n-C_{28}, with a marked dominance of the n-C_{16} and n-C_{18} acids (Ramsay, 1966; Douglas *et al.*, 1966). The remaining acids showed no even/odd predominance.

The composition of a room-temperature solvent extract of shale will serve as a basis for comparison with those of the pyrolysates obtained by heating it to elevated temperatures. We can conduct comparable experiments by using a single batch of pulverized shale. Several types of heating equipment are being used, for example, heating in a closed vessel in an atmosphere of nitrogen as described by Welte (1965), and heating in a sublimation apparatus under vacuum; in the latter experiment compounds which distil or sublime out are not subjected to further heat treatment. The results from the pyrolysis experiments, i.e. heating in vacuum at 500° C, can be considered under the following headings:

(i) An indication of what might be expected from igneous contact with sedimentary formations.

(ii) Similarities in distribution patterns and in the compounds identified, with those for the 200° C sealed-vessel experiment would encourage the belief that lower temperatures might result in similar but much less extensive changes. Coalification is accepted as being a temperature-dependent process (Teichmüller 1962), and although high temperatures are not believed to be involved in the genesis of petroleum (Hanson, 1959), it is likely that mild thermal processes obtain in its formation (Hoering and Abelson, 1963). Thus it might be possible to establish a relationship between our experiments and the maturation process.

(iii) As a comparison with the experiments suggested for the automated planetary probes.

Experimental

All solvents were re-distilled immediately before use, and in the experiments with *Botryoccocus braunii*, the bentonite clay was extracted in a soxhlet apparatus with a mixture of benzene/methanol/chloroform (1 : 1 : 1). Green River Shale and Torbanite were prepared for examination by first washing the exterior of the pieces of rock with a mixture of benzene/methanol (1 : 1), and then pulverizing in a disc mill for 15 min. to give a powder which passed a 100-mesh sieve. Portions of the pulverized shale were then used in a pyrolysis experiment, in which the shale was heated at 100°, 200°, 300°, 400° and 500° C in vacuum. Another portion was repeatedly extracted ultrasonically over a period of 5 hours, using a mixture of benzene/methanol (1 : 1), and acted as the control experiment. Finally, a third portion was re-compressed (at *ca.* 30 tons/in.^{2}) into pellets (2 g) using a hydraulic press; thus, the pellet was homogenous (at least down to the mesh size used). The pellets were then heated in a nitrogen atmosphere in a Parr bomb (stainless steel; teflon seal) at 200° C for 10 days. The brownish suspension of *Botryococcus braunii* was mixed with bentonite clay (2 : 1, by weight) and the mixture was freeze-dried. It was similarly compressed into pellets (2 g) which were heated (nitrogen atmosphere) in a Parr bomb for 10 days at 150° and at 200° C.

The pellets used in the prolonged thermal treatment of the shales and of the *Botryococcus braunii*-bentonite mixtures were afterwards extracted ultrasonically with a mixture of benzene/methanol (3 : 1). The suspension was centrifuged and the supernatant solution was evaporated to give a crude extract. The pyrolysates, which collected in a cold trap from the pyrolysis of the shales, were recovered using a solvent mixture of hexane, benzene and ethyl acetate, and then evaporated to furnish the crude extract.

The analytical techniques used in the separation and work-up of the different classes of compounds included column chromatography with neutral alumina, thin-layer chromatography (T.L.C.) using silica gel impregnated with silver nitrate (Morris, 1964), both on analytical and preparative scales, and finally gas–liquid chromatography (G.L.C.). The spectroscopic techniques used included ultraviolet (U.V.) and infra-red (I.R.) spectroscopy and mass spectrometry (M.S.). The final identification step usually involved the technique of combined gas chromatography–mass spectrometry (G.C.–M.S.). A diagram of one commercial instrument (L.K.B. 9000) is shown in Fig. 2.

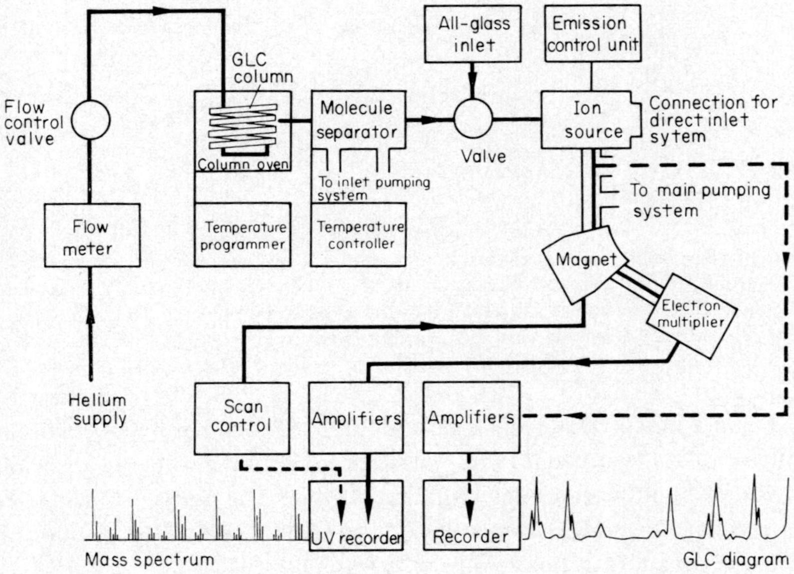

FIGURE 2. — Diagrammatic representation of a commercial combined gas chromatograph–mass spectrometer (G.C.–M.S.) (L.K.B. Model 9000).

The separation and analytical procedure for the hydrocarbon fractions from the shale extracts and pyrolysates is shown in Fig. 3. Column chromatography was carried out using neutral alumina activated at 120° C for 30 min; the eluant was normally benzene or xylene. Evaporation of the eluate furnished the total hydrocarbon extract, which was then separated into alkanes, alkenes and aromatic hydrocarbons by preparative

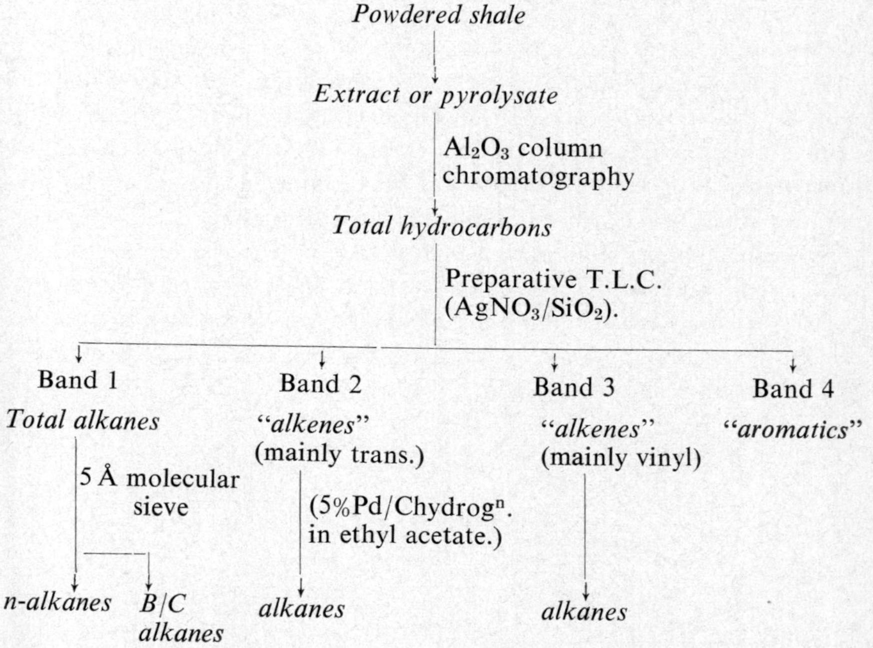

FIGURE 3. — Flow diagram showing the general separatory and analytical techniques used for the isolation and analysis of hydrocarbons. The alkane fractions derived by hydrogenation of the alkene fractions can be further separated into normal and branched-cyclic alkanes by the action of 5 Å sieve. All the alkane fractions can then be examined by G.L.C. and G.C.–M.S. for identification of individual hydrocarbons.

T.L.C. on 1-mm layers of silica gel containing 10% silver nitrate. Development with hexane/benzene (9 : 1) afforded four bands which were 'cut-out' and eluted from the silica gel with diethyl ether. The purity of each extract was verified by analytical thin-layer chromatography (Fig. 4), infrared spectroscopy and gas–liquid chromatography.

The alkenes were hydrogenated using 5% palladium charcoal as catalyst and ethyl acetate as solvent. Hydrogenations were carried out in a micro-

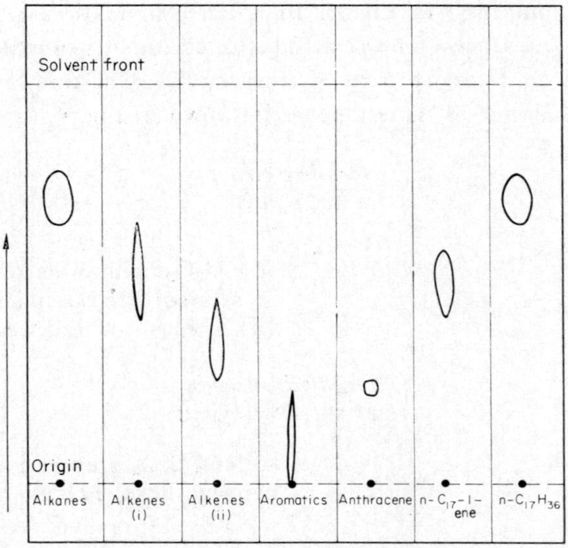

FIGURE 4. — Drawing of an analytical $AgNO_3$—Silica Gel T.L.C. plate showing the alkane, alkene and aromatic fractions isolated from Green River Shale by solvent extraction at 20° C. Three standard hydrocarbons, n-heptadecane, heptadec-1-ene and anthracene are included for comparison purposes.

scale apparatus which permitted the efficient hydrogenation of less than 1 mg of material. The hydrogenated products were isolated in the normal way (filtration through alumina, evaporation, preparative thin-layer chromatography and weighing) and then examined by analytical thin-layer chromatography.

The alkanes were sieved (O'Connor *et al.*, 1962) using $\frac{1}{8}$ in. pellets of 5 Å molecular sieve activated at 375° C for 24 hr. Separation of n-alkanes from branched and cyclic alkanes could still be achieved with amounts of mixed alkanes of less than 1 mg when the apparatus was scaled down to an appropriate size.

Results

1. *Room-temperature solvent extraction of Green River Shale as control*

Powdered Green River Shale (20 g) was ultrasonically extracted with benzene/methanol (3 : 1) for 5 hr, with five changes of solvent. The total extract was then evaporated to give a gum (0·352 g, 1·76%) which was treated as summarised in Fig. 5. Infra-red examination of the alkenes indicated the presence of *trans* di-substituted and tri-substituted double bonds at 965 cm $^{-1}$ and 815 cm $^{-1}$, respectively. The thin-layer chromatographic behaviour of these alkenes is shown in Fig. 4.

Crude extract
(352 mg)

(i) Al_2O_3 column chromatography, eluted with 800 ml benzene.
(ii) Evaporated and weighed.

↓

Total hydrocarbon
(100 mg; 0·5% yield)

$AgNO_3/SiO_2$ preparative T.L.C. benzene/hexane (1:9).

↓

Total alkanes	*Total alkenes*	*aromatic hydrocarbons*
(43 mg; 0·22%)	(17 mg; 0·09%)	(13 mg; 0·07%)

FIGURE 5. — Flow diagram for the separation of the total alkane, alkene and aromatic fractions from a room-temperature solvent extract of Green River Shale.

The total alkane fraction was separated into a normal and branched/cyclic fraction by boiling a solution of the alkanes (1 part) in iso-octane with 5 Å molecular sieve (60 parts) for three days. This gave 12% n-alkanes and 78% branched and cyclic alkanes with 10% loss. The overall yield per 20 g shale was 0·02% n-alkanes and 0·17% branched and cyclic alkanes.

Gas chromatograms of the above hydrocarbon fractions are shown in Fig. 6. The branched/cyclic and normal hydrocarbons, specifically or generally indicated, have been identified by previous workers. In the present study, the identities of most of these peaks have been confirmed by combined gas chromatography – mass spectrometry and by co-injection gas chromatography.

2. *Pyrolysis of Green River Shale at 500°C*

Powdered Green River Shale from the same batch as above (1 g) was pyrolysed under vacuum (500° C, 0·3 mm Hg) for 3 hr and the pyrolysate (92 mg, 9·2% yield) was collected in a cold trap. This was then separated as shown in Fig. 7. An infra-red examination of the alkene fraction (i)

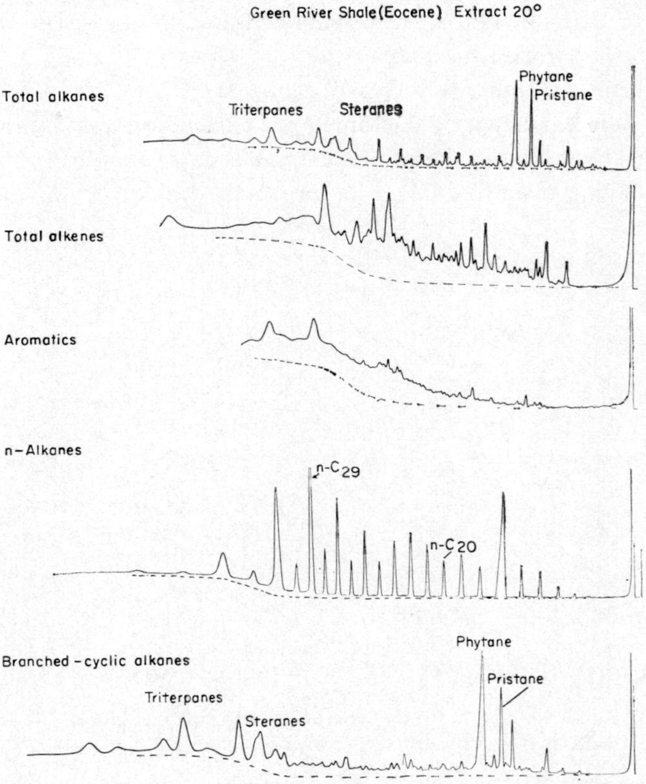

FIGURE 6. — Gas-chromatographic records of hydrocarbon fractions separated from a room-temperature solvent extract of Green River Shale. Column conditions: 10 ft×1/16 in., 2%–7-ring polyphenylether on Gas Chrom P; temp. programmed at 4°/min. from 100 to 300° C. The base-line signal for injection of solvent alone is also shown.

showed absorption due to *trans*-double bonds (965 cm^{-1}), with possibly some tri-substituted double bonds (weak absorption near 820 cm^{-1}), while fraction (ii) showed absorption due to vinyl double bonds (990 and 910 cm^{-1}) only. The total alkanes were then sieved as before, which gave the n-alkanes (~45% of the total alkanes; 0·60% overall) and the branched/cyclic alkanes (~55%; 0·70% overall). The normal and the branched/cyclic alkanes were then examined by combined gas chromatography–mass spectrometry; the C_{14}, C_{20} and C_{29} n-alkanes, pristane and phytane were identified (Fig. 8). The alkene fraction (i) was hydrogenated with 5% palladium charcoal as catalyst in ethyl acetate. However, incomplete hydrogenation occurred, which may have been due to the known difficulty in reducing sterically hindered, probably tri- and tetra-substituted, double bonds. Further work is in progress.

The alkene fraction (ii) was hydrogenated as above and the alkanes obtained were examined by combined gas chromatography-mass spectrometry. This showed that the prominent peaks corresponded to the normal alkanes ranging from C_{15}–C_{32}, but the small peaks were not identified.

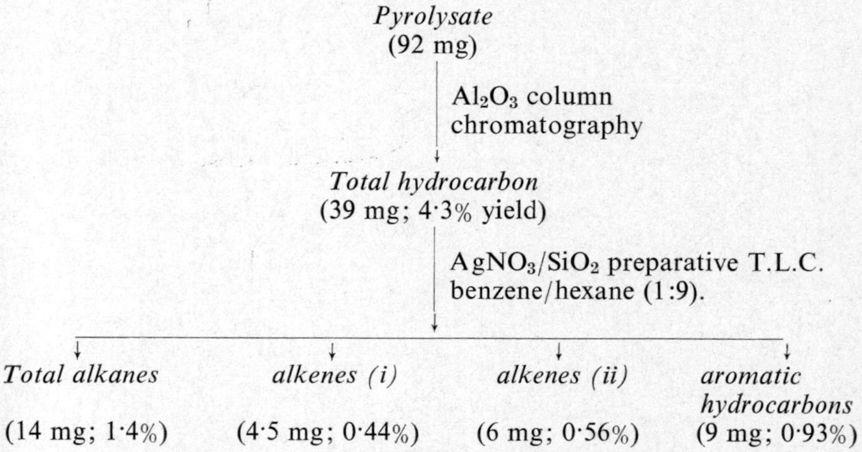

FIGURE 7. — Flow diagram for the separation of the alkane, alkene and aromatic fractions from the 500° C pyrolysate of Green River Shale.

3. *Prolonged thermal treatment of Green River Shale*

A pellet of the shale (2 g) was heated in a Parr bomb at 200° C for 10 days and then extracted. Virtually no difference was found between this extract and that from the room-temperature extract (Fig. 6), although there were slight indications of an increased content of n-alkanes of even-carbon number. Further studies at higher temperatures are in progress.

4. *Prolonged thermal treatment of mixtures of the alga Botryococcus braunii and bentonite*

The yields and some data on the composition of the alkane and alkene fractions are summarized in Table 1. A combined gas chromatographic–mass spectrometric study showed a complicated pattern of saturated hydrocarbons in the C_{34} region. The alkene fraction was similarly complicated.

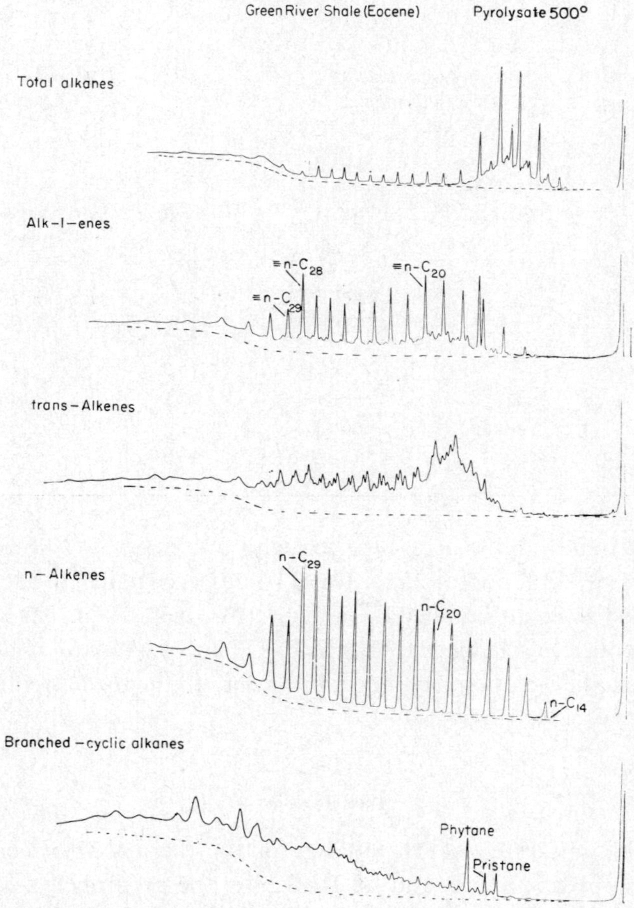

FIGURE 8. — Gas-chromatographic records of hydrocarbon fractions separated from the 500° C pyrolysate of Green River Shale. Column conditions as for Fig. 6.

5. *Pyrolysis of Scottish Torbanite*

The yields and composition of the alkane and alkene fractions are summarized in Table 1. Some data for the room-temperature solvent extract are given for comparison purposes (Maxwell, 1967). Pristane and phytane were present in the branched and cyclic alkane fraction. There were no alkenes apparent and the aromatic fraction has yet to be examined. After pyrolysis at 500° C the alkane fraction had increased to 2·7%, with a C.P.I. value of about 1·0. There were now 5·7% of alkenes present,

TABLE 1

	Botryococcus braunii (now living)		Torbanite ($\sim 3 \times 10^8$ years)	
Time heated	—	240 hr	—	6 hr
Temperature	20° C	200° C	20° C	500° C
Alkanes	0%	2% mainly C_{34} branched	0·26% n-C_{12} → n-C_{37} $\frac{\text{odd}}{\text{even}} \sim 1$	2% 3 : 1/normal: branched
Alkenes	80% C_{34} branched	25% mainly C_{34} branched	0%	6%

mainly alk-1-enes and *trans* n-alkenes, with a C.P.I. value of about 1·0, and an aromatic fraction of 22%. These results are similar to those found for a distilled shale oil supplied by industry (Maxwell 1967). The difference in the percentage of alkanes at 20° and 500° C is not likely to be due to the differences in the isolation technique, but rather to the formation of hydrocarbons, as discussed later.

Discussion

Although some results are available for the thermal treatment of the alga, *Botryococcus braunii*, and of Torbanite, the experiments conducted with Green River Shale have been given the most attention and provide the main background for the ensuing discussion.

The room-temperature solvent extraction of Green River Shale gave a total alkane fraction, the gas chromatogram of which is shown in Fig. 6,

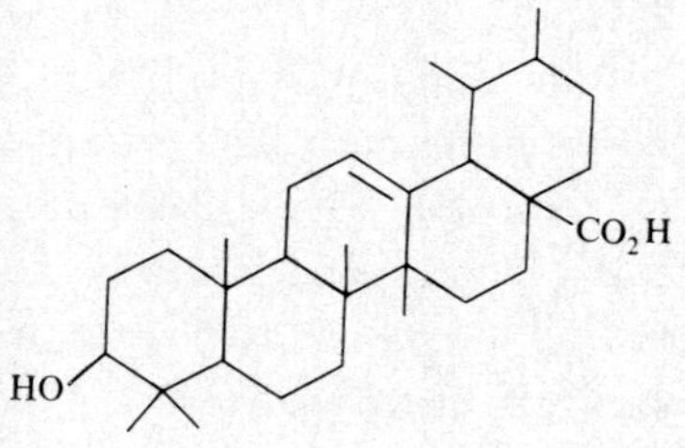

I

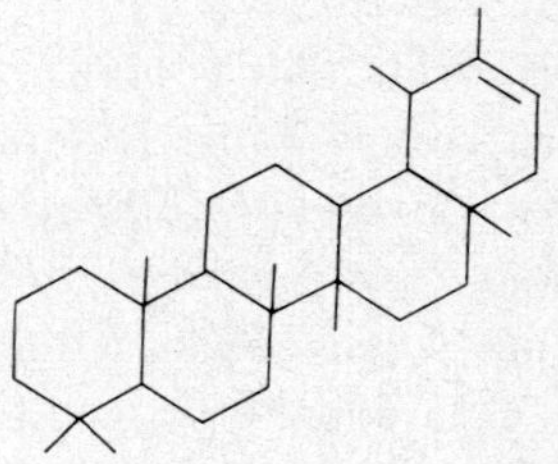

II

the results generally match those found by other workers. Unexpectedly, however, in addition to the alkanes, we also found alkenes in this extract, the yield being 0·085%, which is approximately one-third of the yield of alkanes. The gas chromatogram of these alkenes is shown in Fig. 6, and the pattern, although not analyzed as yet, suggests the presence of a limited number of alkenes in the C_{30} region: Examination of the infra-red spectrum of the mixture of alkenes showed not only the presence of *trans* di-substituted and vinyl double bonds, but also tri-substituted double bonds, and also a high percentage of methyl groups, some being present as *gem*-dimethyl groups. Since some of these alkenes evidently resist hydrogenation, it would seem reasonable to suspect that some of the higher molecular weight components of this fraction are unsaturated triterpenes. If this suggestion is confirmed, then other sediments and also crude oils should be examined for these compounds. Again it would be interesting to see what structural relationships exist between the oxygenated, unsaturated and saturated triterpenes in the Green River Shale. Two examples of triterpenes which are prominent in plants are the oxygenated, unsaturated triterpene ursolic acid (I) and the unsaturated triterpene lupene-1 (II). We are currently investigating this topic.

The yield of aromatic hydrocarbons (as with the alkenes this is based on a "cut" taken from a silver nitrate impregnated silica-gel coated plate and does not necessarily mean that all are aromatic hydrocarbons, or alkenes, since the observed R_f value depends on the functional groups, degree of substitution, hindrance due to alkyl substitution, etc.) is 0·067% which is approximately one quarter that of the alkanes. The peaks of long retention time in the gas chromatogram (Fig. 6), may be due to partially-aromatic polycyclic compounds of high molecular weight. The most likely origin of such compounds would be dehydrogenation of the triterpenoid or steroid system.

Comparison of the results obtained with the pyrolysis product (500° C) of Green River Shale with those from the room-temperature solvent extract reveals many differences. Firstly, pyrolysis (even at 300°) produces more volatile organic material than is obtained by extraction, and the percentage yield increases rapidly with temperature (up to 500°). Secondly, the percentage yields of alkanes, alkenes and aromatic hydrocarbons are greater by a factor of about ten for the 500° pyrolysates compared with the room-temperature extracts. When the gas-chromatographic patterns for the total alkanes derived from the room-temperature extraction (Fig. 6) and the 500° pyrolysis (Fig. 8), are compared, it is apparent that the proportion of n-alkanes is greater in the pyrolysate. This is confirmed on direct comparison of the gas-chromatographic data for the n-alkanes, obtained from the total alkanes by the molecular-sieve process. More even-numbered n-alkanes appear to have been generated, and the C.P.I. value for the pyrolysate is approximately 1·0 compared with 3·6 in the room-temperature extract. There also seems to have been generation of more n-alkanes around the C_{30} region, which could correspond to decarboxylation of the prominent long-chain fatty acids in the Green River Shale (Douglas *et al.*, 1966; Ramsay, 1966). Certain isoprenoids, e.g. phytane and pristane, are recognizable in the branched/cyclic fraction from the pyrolysate in which they were identified by combined gas chromatography–mass spectrometry. There are also high molecular weight cyclo-alkanes in the pyrolysate which could be triterpenoid, although there is a marked "background" due to many unresolved peaks in this region of the gas chromatogram. Although the overall amount of the branched/cyclic fraction of the pyrolysate has increased, it is not yet possible to state whether the concentrations of individual components have increased or decreased during pyrolysis.

In the 500° pyrolysis experiment, alkanes, alk-1-enes *trans*-alkenes and aromatic hydrocarbons were obtained (Fig. 8). Similar findings were reported by Iida *et al.* (1966) in a paper published after the above work was carried out. These authors reported the occurrence of n-alkanes, alk-1-enes and *trans*-alkenes in the range C_{13}–C_{18} in a "cut" from a Colorado Oil Shale distillate (b.p. 280–305°). Pyrolysis of Green River Shale at lower temperatures (Hoering and Abelson, 1963) gave results which showed that at 250° saturated and unsaturated hydrocarbo nswere obtained (in a ratio of about 9 : 1), and at 325° the authors state the major components of the pyrolysate were odd- and even-numbered n-paraffins, with appreciable amounts of branched and cycloparaffins.

The gas chromatograms of the two alkene fractions obtained by us from the pyrolysate are shown in Fig. 8. As with the alkane fractions, the proportion of straight-chain compounds is greatly increased in the pyrolysate compared with the room-temperature extract (Fig. 6). In the alkene fraction (ii), containing predominantly alk-1-enes, it is the even-numbered carbon chains, especially n-C_{20} and n-C_{28}, which predominate. This even-number dominance is significant in view of the odd-number dominance found for all biologically-originated alkane mixtures; thermally-induced generation of hydrocarbons in deeply-buried sediments might thus explain the disappearance of the odd-number dominance. The generation of the alk-1-enes described above might take place by one or more of the following pathways.

(i) Dehydration of even-numbered long-chain alcohols could give rise to terminal alkenes of even number. Alcohols vary in the ease with which they undergo dehydration to alkenes, the decreasing order of reactivity being tertiary > secondary > primary. However, the primary alcohols are known to dehydrate fairly readily under certain conditions; thus Roberti *et al.* (1941) found that cetyl alcohol (n-$C_{16}H_{33}OH$) gives a 65% conversion to hexadec-1-ene on passing the alcohol over alumina at 350° C.

$$CH_3(CH_2)_{13}CH_2CH_2OH \xrightarrow[350^\circ C]{Al_2O_3} CH_3(CH_2)_{13}CH{=}CH_2 + H_2O$$

(ii) A reaction scheme can be written, formally at least, whereby an even-numbered acid or ester would eliminate acetic acid to give an even-numbered alk-1-ene

$$R(CH_2)_2CH_2CO_2R^1 \rightarrow R{-}CH{=}CH_2 + CH_3CO_2R^1.$$

where R^1 = H *or* metal,
or Kerogen matrix,
or alkyl, e.g. long-chain wax esters.

An analogy for this reaction is provided by the classic rearrangement (McLafferty, 1963) undergone by carbonyl compounds (esters, acids, aldehydes, ketones, amides and carbonates) and other functional groups in the ionization chamber of a mass spectrometer.

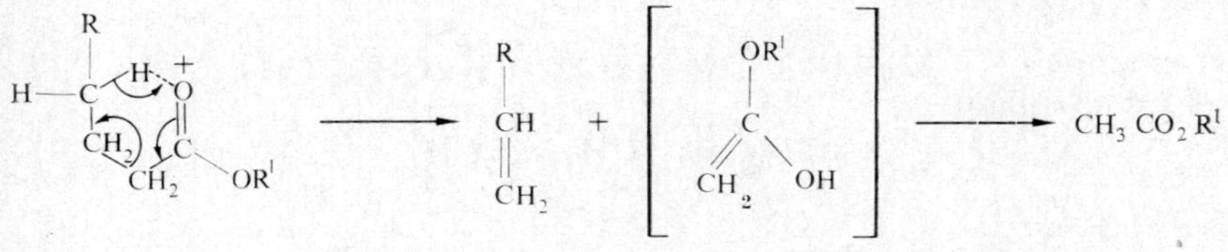

(iii) Decarboxylation of an even-numbered n-fatty acid furnishes principally the corresponding odd-numbered alkane, but Jurg and Eisma (1964, 1966) found that behenic acid (C_{22}) when heated with bentonite clay generates, in addition to the C_{21} alkane, a homologous series of n-alkanes and fatty acids with carbon chains both longer and shorter than that of the original acid. They suggest radical processes to account for the formation of these products.

(iv) Thermally-induced direct scission of carbon-carbon bonds in hydrocarbons furnishes alk-1-enes.

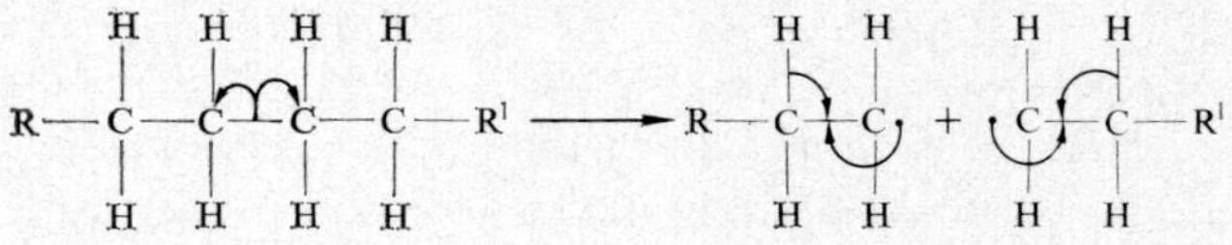

$$\longrightarrow R-CH=CH_2 + CH_2=CH-R^1 + 2H\cdot$$

The laboratory experiments of Holman *et al.* (1966) demonstrate that heating an n-alkane in helium at 600° C results in a smooth distribution of n-alkenes of steadily diminishing carbon number. Pyrolysis of a predominately straight-chain kerogen matrix might be expected to give similar results, and evidence for such a matrix is discussed by Forsman (1963), while oxidative studies of Green River Shale by Robinson *et al.* (1956) established the presence of large amounts of open-chain acids with no evidence of aromatic acids.

(v) Polymerization of ethylene, formed during the pyrolytic reaction from long-chain material, could give rise to even-numbered long-chain alk-1-enes. However, there are two objections to this scheme which are difficult to rationalize, (a) the presence of odd-numbered alk-1-enes and (b) the hydrocarbons formed by such a process are usually of very high molecular weight.

Polymerization (radical propagation) $CH_2=CH_2 \rightarrow CH_3(CH_2-CH_2)_n\,CH_2\cdot$

Termination (radical coupling) $CH_3(CH_2-CH_2)_nCH_2\cdot + \cdot CH_2(CH_2-CH_2)_nCH_3 \rightarrow$

$$\rightarrow CH_3(CH_2-CH_2)_{2n+1}CH_3$$

Termination (radical abstraction) $CH_3(CH_2-CH_2)_nCH_2\cdot + \cdot CH_2CH(H)(CH_2-CH_2)_nCH_3$

$$\longrightarrow CH_3(CH_2-CH_2)_nCH_3 + CH_2=CH(CH_2-CH_2)_nCH_3$$

(vi) The Fischer–Tropsch reaction (Lovelock, 1965; Meinschein 1963; Sharkey *et al.*, 1962) involving carbon monoxide and water at elevated temperatures and pressures in the presence of a catalyst, produces a smooth carbon-number pattern of alk-1-enes and n-alkanes, as the major products. One can conceive of some form of Fischer–Tropsch reaction taking place in the pyrolytic experiment, but this leaves unexplained the dominance of the even-numbered alk-1-enes.

At present the evidence seems insufficient for any decision to be reached concerning the reaction taking place during the pyrolytic generation of the alk-1-enes. However, it is likely that the *trans*-alkenes from the pyrolysate are mainly straight-chain, with the double bond varying in position over the length of the chain, and that their generation is probably related to that of the alk-1-enes. Isomerization of the latter is one possibility; another would involve generation by hydrogen abstraction processes,

e.g. $R\cdot + -CH_2-CH_2-CH_2-CH_2- \rightarrow RH + -CH_2-\cdot CH-CH_2-CH_2-$

$$R\cdot + -CH_2-\cdot CH-CH_2-CH_2- \rightarrow RH + -CH_2-\underset{trans}{CH=CH}-CH_2-$$

The alkanes, alkenes and aromatic hydrocarbons may have to be considered together as products of pyrolysis. For example, alkenes formed by pyrolysis might, in the presence of hydroaromatic systems, give rise to alkanes and aromatic hydrocarbons by hydrogen abstraction. The aromatic hydrocarbons could arise by cyclization of unsaturated intermediates (Badger *et al.*, 1966) and hydrogenolysis and by dehydrogenation of cyclo-alkanes; this latter process may be the more important. Radical processes are the obvious choice to explain these hydrogen transfers and bond breakages.

It is tempting to propose that the alkenes, generated infinitely more slowly under geological conditions, might suffer reduction by hydrogen transfer, thereby providing an explanation for the smooth carbon-number distributions of the n-alkanes in ancient sediments. If the present pyrolytic experiments are meaningful, and the results at least partially transferable to reactions at much lower temperatures over long periods of geological time, then alkanes, alkenes and aromatic hydrocarbons are in all probability geogenetically related. This is indicated by the observed trend of the C.P.I. value of the alkanes of Green River Shale from 3·6 to about 1·0 on pyrolysis, and the parallel appearance of the alk-1-enes with slight even-carbon-number dominance.

One intriguing point which urgently needs further study concerns the relatively great abundance of the straight-chain material in the pyrolysates of Green River Shale and Scottish Torbanite. The obvious interpretation is that this material represents pre-existing straight-chain material 'cracked' from the kerogen matrix. Welte (1965) found that pre-extracted samples of a non-marine Eocene oil shale, when heated, gave fresh extractable material, the C.P.I. value of which decreased with increasing temperature, becoming 1·0 at temperatures above 300°. It seems unlikely that the straight-chain hydrocarbons arise from pre-existing branched-chain materials (Holman *et al.*, 1966), since pyrolysis of n-paraffins yields a homologous series of n-olefins, while pyrolysis of branched paraffins yields mainly branched olefins. If this inference is correct then pyrolysis experiments will give useful information about the nature of kerogen. This type of experiment will be useful in the laboratory and, possibly, in automated planetary probes.

Acknowledgements

We thank the National Aeronautics and Space Administration (Grant NsG 101–61), Shell International and the Natural Environment and Science Research Councils for financial support (GR/3/0655 and B/SR/2398 respectively). We are much indebted to Dr. W. E. Robinson (Bureau of Mines, Laramie, Wyoming) for providing the Green River Shale and Dr. W. D. I. Rolfe (Hunterian Museum, Glasgow) for the sample of Torbanite. The freshwater alga *Botryococcus braunii* was collected from Oakmere, Cheshire by Mr. J. Osborne (warden of Rostherne N.N.R.). The authenticity and purity of the alga was confirmed by Dr. J. W. G. Lund, (Freshwater Biological Association, Windermere) and Dr. E. Conway (Botany Department, Glasgow University).

References

Badger, G. M., Donnelly, K., and Spotswood, T. M. (1966) The formation of aromatic hydrocarbons at high temperatures. XXVII. The pyrolysis of isoprene. *Aust. J. Chem.* **19,** 1023–1043.

Barghoorn, E. S., Meinschein, W. G. and Schopf, J. W. (1965) Paleobiology of a Precambrian shale. *Science*, **148,** 461–472.

Belcher, J. H. (1958) Physiological studies on the growth of freshwater algae in culture. Ph.D. Thesis, University College (London).

Blackburn, K. B. and Temperley, B. N. (1936) *Botryococcus* and the algal coals. *Trans. Roy. Soc. Edin.* **58,** 841–868.

Burlingame, A. L., Haug, P., Belsky, T. and Calvin M. (1965) Occurrence of biogenic steranes and pentacyclic triterpanes in an Eocene shale (52 million years) and in an early Precambrian shale (2.7 billion years); A preliminary report. *Proc. Nat. Acad. Sci.* **54,** (5), 1406–1412.

Cooper, J. E. and Bray, E. E. (1963) A postulated role of fatty acids in petroleum formation. *Geochim et Cosmochim. Acta*, **27,** 1113–1127.

Douglas, A. G., Douraghi-Zadeh, K., Eglinton, G., Maxwell, J. R. and Ramsay, J. N. (1966) This volume, pp. 315–334.

Douglas, A. G. and Eglinton, G. (1966) The distribution of alkanes. In *Comparative Phytochemistry*, T. Swain (Editor), pp. 57–71. Academic Press, London.

Dulhunty, J. A. (1943) Classification and origin of the New South Wales Torbanites. D. Sc. Thesis, University of Sydney, Australia.

Eglinton, G., Scott, P. M. Belsky, T., Burlingame, A. L., Richter, W. and Calvin, M. (1966) Isoprenoid alkanes in a Precambrian sediment. *In Advances in Organic Geochemistry*, 1964, G. D. Hobson and M. C. Louis, (Editors), pp. 41–74. Pergamon Press, London.

Eisma, E. and Gransch, J. A. (1966) This volume, pp. 407–426.

Forsman, J. P. (1963) Geochemistry of kerogen. In *Organic Geochemistry*, I. A. Breger (Editor), pp. 148–182. Pergamon Press, London.

Hanson, W. E. (1959) Some chemical aspects of petroleum genesis. In *Researches in Geochemistry*, P. H. Abelson (Editor), pp. 104–117. John Wiley and Sons, Inc., New York.

Hills, I. R. and Whitehead, E. V. (1966) This volume, pp. 89–110.

Hoering, T. C. and Abelson, P. H. (1963) Hydrocarbons from kerogen. *Carnegie Inst. Wash. Yr. Book*, **62,** 229–234.

Holman, R. T., Deubig, M. and Hayes, H. (1966) Pyrolysis chromatography of Lipids. 1. Mass spectrometric identification of pyrolysis products of hydrocarbons. *Lipids*, **1,** 247–253.

Iida, T., Yoshii, E. and Kitatsuji, E. (1966) Identification of normal paraffins, olefins, ketones and nitriles from Colorado oil shale. *Anal. Chem.* **38,** 1224–1227.

Jones, J. D. and Vallentyne, J. R. (1960) Biogeochemistry of organic matter. 1. Polypeptides and amino-acids in fossils and sediments in relation to geothermometry. *Geochim. et Cosmochim. Acta*, **21,** 1–34.

Jurg, J. W. and Eisma, E. (1964) Petroleum hydrocarbons: generation from fatty acids. *Science*, **144,** 1451–1452.

Jurg, J. W. and Eisma, E. (1966) This volume, pp. 367–368.

Leo. R. F. and Parker, P. L. (1966) Branched-chain fatty acids in sediments. *Science*, **152,** 649–650.

Lovelock, J. E. (1965) A physical basis for life detection experiments. *Nature*, **207,** 568–570.

Macgregor, M. (1938) Oil-shales and cannel coals of Scotland. In *Oil Shale and Cannel Coal*, pp. 6–17. Institute of Petroleum, London.

Maxwell, J. R. (1967) Studies in organic geochemistry. Ph.D. Thesis, Glasgow University.

Meinschein, W. G. (1963) Hydrocarbons in terrestrial samples and the Orgueil meteorite. *Space Sci. Rev.* **2,** 653–679.

Milton, C. and Eugster, H. P. (1959) Mineral assemblages of the Green River formation. In *Researches in Geochemistry*, P. H. Abelson (Editor), pp. 118–150. John Wiley and Sons, Inc., New York.

McLafferty, F. W. (1963) Decompositions and rearrangements of organic ions. In *Mass Spectrometry of Organic Ions*, F. W. McLafferty (Editor), pp. 309–342. Academic Press, London.

Morris, L. J. (1964) Specific separations by chromatography on impregnated adsorbents. In *New Biochemical Separations*, A. T. James and L. J. Morris (Editors), pp. 295–319. Van Nostrand Company Limited, London.

Nevers, N. de. (1966) Tar sands and oil shales. *Scientific American*, **214,** 21–29.

O'Connor, J. G., Burrow, F. H. and Norris, M. S. (1962) Determinations of normal paraffins in C_{20} to C_{32} paraffin waxes by molecular sieve adsorption. Molecular weight distribution by gas–liquid chromatography. *Anal. Chem.* **34,** 82–85.

Ramsay, J. N. (1966) Fatty acids in sediments. M. Sc. Thesis, Glasgow University.

Rea, D. G. (1966) Analytical methods for Landers. In *Biology and the Exploration of*

Mars, C. S. Pittendrigh, W. Vishnaic and J. P. T. Rearman (Editors), National Academy of Science, Washington.

Roberti, G., Minervini, C. and Berti, V. (1941) A method for the preparation of cetane. *Energia Termica*, 582–584; *C.A.* **36,** 5767^{c}.

Robinson, W. E., Cummins, J. J. and Stanfield, K. E. (1956) Constitution of organic acids prepared from Colorado oil shale, *Ind. Eng. Chem.* **48,** 1134–1138.

Robinson, W. E., Cummings, J. J. and Dineen, G. U. (1965) Changes in Green River oil shale paraffins with depth. *Geochim. et Cosmochim. Acta*, **29,** 249–258.

Sharkey, A. G., Shultz, J. L. and Friedel, R. A. (1962) Mass spectrometric determination of the ratio of branched to normal hydrocarbons up to C_{18} in Fischer–Tropsch products. *Anal. Chem.* **34,** 826–830.

Stewart, D. and Forbes, C. E. (1938) Retorting of oil shales in Scotland. In *Oil Shale and Cannel Coal*, pp. 96–114. Institute of Petroleum, London.

Teichmuller, M. (1962) Die Genese der Kohle. Compte Rendu du Quatrième Congrès, pour l'avancement des études de stratigraphie et de géologie du Carbonifère, Heerlen Septembre 1958, Tome III, 699–722.

Thomas, D. W. and Blumer, M. (1964) Porphyrin pigments of a Triassic sediment. *Geochim. et Cosmochim. Acta*, **28,** 1147–1154.

Welte, D. H. (1965) Kohlenwasserstoffgenese in Sediment gesteinen. *Sonderdruck aus der Geologischen Rundschau*, **55,** 131–144; *Bull. Am. Assoc. Petrol. Geol.*, **49** (12), 2246–2268.

INFLUENCE DE L'ORIGINE DES MATIÈRES ORGANIQUES ET DE LEUR DEGRÉ D'ÉVOLUTION SUR LES PRODUITS DE PYROLYSE DU KÉROGÈNE

par

M. BORDENAVE, A. COMBAZ et A. GIRAUD

Compagnie Française des Pétroles, Paris

I. Introduction

De nombreuses études ont été faites sur les bitumes extraits des roches. Elles ont apporté des résultats corrélables avec l'origine de ces roches ou les conditions physico-chimiques subies au cours de leur histoire géologique. On a ensuite cherché à rapprocher les résultats obtenus sur les bitumes extraits des roches, de ceux trouvés sur des pétroles bruts voisins.

Par contre, peu d'études ont été faites sur le kérogène qui représente tout de même près de 95 % du contenu organique des sédiments ayant subi un début de diagenèse.

Ce peu d'enthousiasme provenait essentiellement de la difficulté de caractériser un produit très complexe, insoluble dans les solvants organiques et lié aux argiles.

Différents chercheurs ont essayé de comprendre la structure des kérogènes, soit par l'étude de leur composition, soit par l'analyse de leur produit de dégradation obtenu en oxydant ou en hydrogénant le kérogène.

L'étude élémentaire des kérogènes a fait l'objet de nombreux travaux ; ceux de Breger et Brown[5] par exemple.

Citons en ce qui concerne la dégradation par oxydation les travaux de Bone et Himus[3], de Juettner[12], de Down et Himus[8] et surtout la méthode d'oxydation sélective de Khalifeh et Louis[17].

La dégradation par hydrogénation a été utilisée de façon intensive avec les charbons. Ceci dans un but industriel. Avec les kérogènes, par contre.

nous ne trouvons que peu de travaux, citons par exemple ceux de Kogerman et Kopwillem[14] puis de Forsman et Hunt[9].

La présente communication illustre l'influence de l'origine des matières organiques et de leur degré d'évolution sur les produits de pyrolyse du kérogène.

Une première étude[10] qui portait sur des échantillons bien définis par des données géochimiques et palynologiques nous avait permis de mettre au point une technique de pyrolyse et de chromatographie, et, de ce fait, de caractériser rapidement le kérogène d'un sédiment.

L'étude que nous présentons aujourd'hui est une application de cette technique à un niveau géologique déterminé: le Gothlandien du Nord Sahara, comprenant des formations transgressives à influence continentale et des formations plus franchement marines.

Après avoir rappelé quelques idées sur les mécanismes de genèse des hydrocarbures, nous dirons quelques mots sur les résultats que peut nous donner la pyrolyse dans l'étude des kérogènes. Enfin nous confronterons sur l'exemple choisi, les paramètres géologiques et les résultats d'analyses.

Les auteurs expriment leurs remerciements à la Direction de la Compagnie Française des Pétroles et à la Direction de la Compagnie Française des Pétroles (Algérie) pour avoir bien voulu autoriser la publication de cette communication.

Ils remercient MM. Pommier, Rouge et de Matharel de C.F.P.(A), MM. Byramjee et Ferrero de la C.F.P., des nombreux échanges d'idées qui ont facilité la réalisation de ce travail.

Ils sont très reconnaissants à M. Louis de l'Institut Français du Pétrole, aux Docteurs Long et Neglia de l'A.G.I.P., des discussions fructueuses qu'ils ont eues avec eux.

L'important travail d'analyses a été réalisé pour la plus grande part par MM. Drapeau et Merer (Laboratoire de la C.F.P. Levallois).

II. Rappel des mécanismes de genèse des hydrocarbures

Les progrès de l'océanographie et le nombre croissant d'analyses du contenu organique de l'eau de mer et des sédiments récents ou anciens, permettent de mieux comprendre les mécanismes de sédimentation des matières organiques.

La chute des cadavres sur le fond marin, et l'incorporation directe de leurs matières organiques dans les sédiments jouent, en général, assez peu. Exception faite cependant, de tests organiques de micro-fossiles tels les spores, pollens, Chitinozoaires . . . dont la substance est particulièrement resistante aux actions physico-chimiques ou biologiques.

À cette image s'oppose le caractère complexe des processus biologiques. L'ensemble des organismes marins forme un filtre efficace qui transforme la plus grande partie des substances organiques utilisables et ne laisse se déposer presque uniquement que des déchets sans valeur biologique.

Ces déchets sont de deux sortes : des bitumes provenant directement de la fraction lipide des organismes et des complexes plus ou moins condensés analogues aux acides humiques des sols.

Les substances de type humique, en solution dans l'eau de mer, ont tendance à être adsorbées par les argiles, ou à se condenser pour former des agrégats moléculaires complexes, comprenant souvent des hétérocycles avec du soufre, de l'azote et de l'oxygène[2, 4].

Ces substances se mélangent aux déchets marins non dégradés : tests organiques de micro-fossiles, spores . . . dont la proportion dans certains faciès peut devenir prépondérante.

À ces déchets d'origine marine s'ajoutent les apports continentaux essentiellement sous forme de solutions humiques provenant des sols et transportés par les rivières, de débris végétaux (cellulose, lignine. . .) et de micro-fossiles typiquement terrestres (spores, pollens. . .).

C'est cet ensemble de substances insolubles dans les solvants organiques que nous désignons sous le terme de *kérogène*.

La nature chimique des solutions de type humique d'origine marine ou provenant des sols se différencie par leur degré de condensation qui peut être estimé par le rapport C/H :[4]

$$8 < C/H < 10 \text{ pour les solutions marines}$$

$$12 < C/H < 22 \text{ pour les acides humiques des sols}$$

Ces différences se retrouvent dans les kérogènes correspondants. On peut donc s'attendre à ce que les influences continentales se traduisent par des différences de composition du kérogène[5].

Lorsque les sédiments sont enfouis à une profondeur suffisante l'action de la pression et surtout de la température provoque une dégradation du kérogène (rupture des chaînes latérales) et corrélativement une condensation des cycles. Cette action se traduit par une évolution, d'une part vers des structures de type graphitique et de l'autre, par la genèse d'hydrocarbures, qui vont se mélanger aux hydrocarbures du bitume initial.

Les hydrocarbures formés sont riches en hydrogène: le rapport C/H d'un pétrole brut varie en général peu autour d'une valeur moyenne égale à 7, celui d'un gaz est encore plus faible.

La genèse des hydrocarbures provoque un appauvrissement en hydrogène du kérogène restant dans le sédiment, et donc une augmentation du rapport C/H.

La capacité de produire des hydrocarbures, pour un kérogène donné, dépend en conséquence de son rapport C/H initial; on peut définir pour chaque type de kérogène un maximum théorique de productivité, correspondant à une évolution totale du kérogène considéré, vers une structure graphitique.

Ce maximum théorique sera d'autant plus faible que le rapport C/H initial sera grand, et d'après ce que nous avons vu précédemment, le maximum de productivité d'un kérogène d'origine au moins en partie continentale, est plus faible que celui d'un kérogène franchement marin.

Ce bref rappel sur les mécanismes de genèse des hydrocarbures montre l'intérêt que présente la connaissance du degré d'évolution d'un kérogène et de son aptitude à donner des hydrocarbures par pyrolyse à basse température. La technique de pyrochromatographie semble répondre à ces deux questions.

III. Pyrolyse du kérogène

III. 1. *Pyrolyse à basse température*

Nous ne reviendrons que de façon très sommaire sur la technique de pyrolyse et de chromatographie qui a fait l'objet d'une communication antérieure[10].

La figure 1 schématise l'ensemble de l'appareillage. On peut distinguer le circuit 1 de pyrolyse du circuit 2 de chromatographie.

Dans le circuit 1 nous avons successivement le dispositif de régulation de pression, l'ensemble de pyrolyse et enfin, avant la vanne régulatrice de débit, la vanne-piège qui permet de mettre le piège en série avec le circuit 1 ou le circuit 2.

Dans le circuit 2 nous avons un détendeur, le régulateur de débit du chromatogramme, la vanne-piège qui permet ici encore de mettre ou non le piège en série avec ce circuit et enfin, la colonne et le détecteur du chromatographe.

Ces deux circuits sont indépendants et peuvent recevoir des gaz vecteurs de nature et de débit différents.

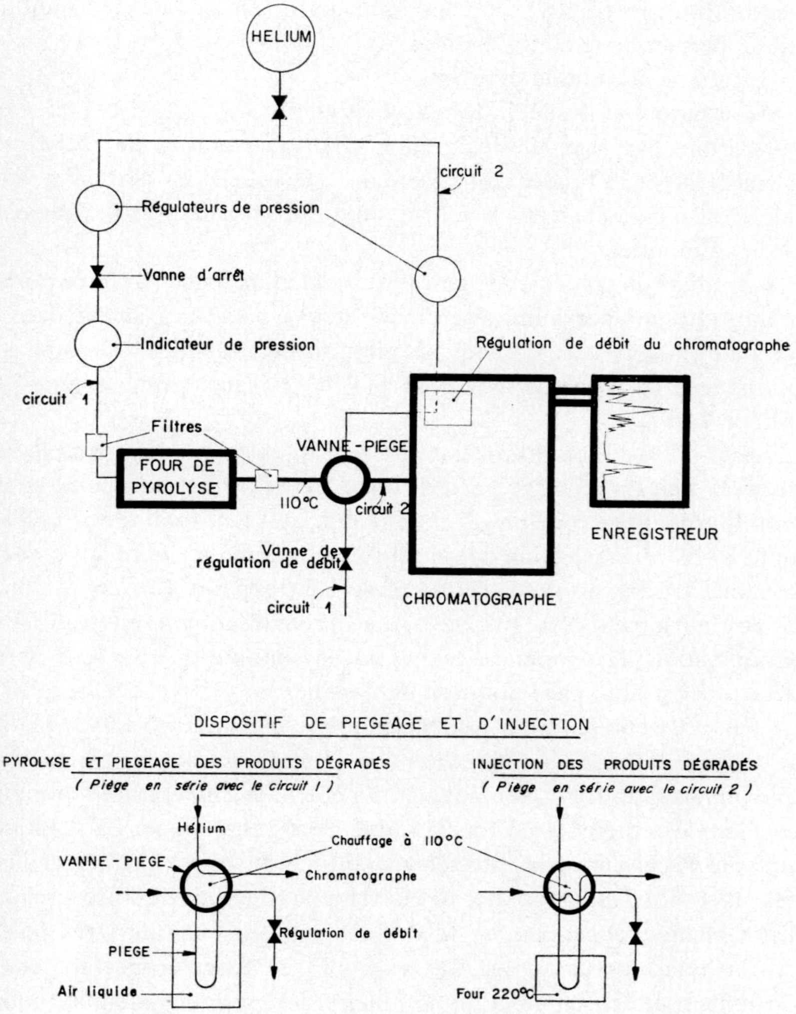

FIGURE 1.

De ce fait, il est possible de concilier les deux exigences suivantes qui sont contradictoires dans le cas d'un accouplement direct entre le pyrolyseur et le chromatographe :

— la durée d'émission des produits de pyrolyse doit être rapide d'où la nécessité d'avoir une température élevée.

— le cracking secondaire des produits dégradés doit être minimum.

L'utilisation du piégeage momentanée dans l'air liquide des produits dégradés permet donc d'être maître :

— de la température de pyrolyse,

— de la nature et du débit des gaz vecteurs.

Lors d'une pyrolyse, un échantillon (100 mg environ) de roche préalablement lavée et broyée, est pesé dans une nacelle en platine et cette dernière est introduite dans le four de pyrolyse chauffé à 280° C pendant environ 15 minutes.

Les produits dégradés sont entraînés instantanément par un courant d'hélium (10 cm^3 par minute) et sont retenus quantitativement dans le piège plongé dans l'air liquide. Le piège est un tube en U rempli sur une longueur de 5 cm de chromosorb P 80–100 mesh imprégné à 5 % d'apiézon L.

L'injection des produits de la pyrolyse dans le chromatographe est réalisée en plaçant le piège en série avec le circuit de chromatographie, ceci en tournant la vanne-piège, et en remplaçant l'air liquide par un four chauffé à 220° C. L'injection dans le chromatographe est de ce fait presque instantanée et les produits de pyrolyse sont séparés sur une colonne capillaire au squalane de 300 pieds. La chromatographie est réalisée en programmation de température (3,3° C par minute jusqu'à 120° C) et la détection est faite par ionisation de flamme.

La figure 2 représente un four de pyrolyse classique constitué par un tube de silice entouré d'un manchon chauffant. Ce four a une inertie thermique importante. Il est bien adapté aux études à température constante.

La figure 3 représente un four à faible inertie thermique. La résistance chauffante a été enroulée directement autour du tube de silice et l'isolement thermique est assuré par une enveloppe sous vide de silice argentée.

Cette technique permet, en travaillant sur des quantités très faibles de roche (cuttings de forage par exemple) de caractériser rapidement un sédiment. Le chromatogramme obtenu nous renseigne sur le potentiel en gaz, en composés aromatiques et en hydrocarbures paraffiniques.

Dans l'étude antérieure[10] nous avions vu que :

— une roche d'origine continentale donne, par pyrolyse, du méthane, de l'éthane et des composés aromatiques,

— une roche d'origine marine donne d'autant plus de dérivés benzéniques qu'elle est moins évoluée, mais des hydrocarbures paraffiniques sont toujours présents en plus ou moins grande quantité.

Dans la présente étude, une étape supplémentaire a été franchie en interprétant quantitativement les chromatogrammes. Nous pouvons

FIGURE 2.

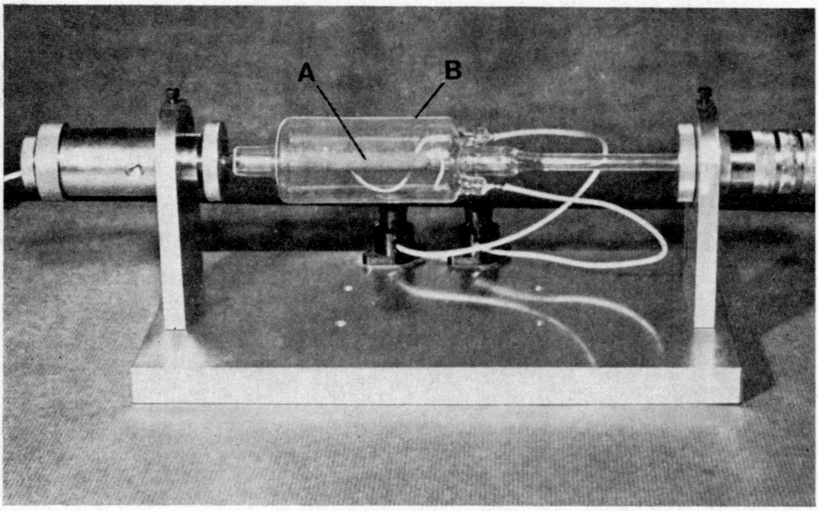

FIGURE 3.

ainsi estimer un potentiel en gaz en regroupant méthane, éthane et propane. De la même façon la somme des normal-paraffines comprises entre C_4 et C_{11} permet de chiffrer un potentiel en huile. Le potentiel en aromatiques enfin est obtenu en groupant benzène, toluène, xylène, cumène et xylénol.

Les nombres portés dans le Tableau I et la figure 12 et représentant les potentiels définis ci-dessus sont proportionnels à une masse de gaz, d'huile ou de composés aromatiques rapportée à une masse de carbone (kérogène initial exprimé en carbone) constante.

III. 2. *Pyrolyse à haute température*

Pour estimer le degré d'évolution actuel d'un type de kérogène matérialisé par son rapport C/H, nous aurions pu séparer ce kérogène de son support minéral, par attaque acide, et faire l'analyse élémentaire de la matière organique.

Une telle méthode a été appliquée à quelques échantillons, cependant sa mise en œuvre nécessite des quantités importantes de roche et beaucoup de travail.

Remarquons que la seule connaissance de ce rapport C/H n'est pas toujours suffisante. Il faut relier cette détermination aux résultats de la pyrolyse ou à la répartition des différentes familles de substances qui composent le kérogène. Quoi qu'il en soit, dans la présente étude nous avons résolu le problème complémentaire en faisant évoluer le kérogène artificiellement jusqu'à une limite maximum théorique.

C'est-à-dire que le kérogène associé à son support minéral est pyrolysé à 500° C, dans un four identique à celui utilisé dans la pyrolyse à basse température.

Les produits dégradés (ΔC) sont entraînés par l'hélium sur un oxydant qui transforme quantitativement le carbone en gaz carbonique. Ce dernier est finalement dosé par chromatographie en phase gazeuse.

Le rapport entre le carbone ainsi dosé et la teneur en carbone de l'échantillon avant pyrolyse, définit un taux de dégradation : $T_D = \dfrac{\Delta C}{C}$ qui est en liaison avec l'origine du kérogène, mais surtout avec son degré d'évolution, comme le confirme l'étude sur la carbonisation des spores et micro-fossiles organiques.

IV. Étude de deux niveaux argileux du Nord Sahara

Les échantillons dont nous parlerons ici appartiennent à deux niveaux argileux du Gothlandien correspondant à des types de sédimentation différente.

Parmi les puits sahariens auxquels notre méthode à été appliquée nous en avons sélectionné huit que nous désignerons par des numéros de I à VIII.

Sur les formations ordoviciennes, se sont déposés en discordance légère (phase taconique de l'orogénèse calédonienne) des remplissages irréguliers d'argile sablo-micacée, à passées microconglomératiques et à lentilles de grès ; ces formations se terminent en général par des grès quartzitiques.

Sur ces couches de passage, viennent transgresser des argiles et des marnes noires à aspect charbonneux, parfois silto-micacées, pyriteuses, avec des passées de calcaire détritique.

Cette transgression n'est pas partout synchrone, elle varie du Llandovery moyen-supérieur, dans le bassin de Polignac, au Taranon-Wenlock, dans le Nord Sahara.

La zone transgressive est caractérisée par un faciès très particulier que l'on a pu qualifier de «tourbière à algues», à cause de la grande abondance de Tasmanacées (fig. 4), elle contient aussi des Chitinozoaires et des Acritarches. Corrélativement, les teneurs en oligo-éléments y sont importantes. En particulier, la richesse en Uranium et Radium provoque une radio-activité facilement observable sur les diagraphies, d'où le nom de «Pic γ» donné au niveau correspondant.

Ce Pic γ suit la base de l'argile et bien que diachrone, correspond à un même type de sédimentation. Cette formation s'est déposée le long d'une côte basse et marécageuse bordant un continent arasé, mais probablement couvert de végétation[(6)].

Les fortes teneurs en oligo-éléments pourraient trouver une explication, à la fois dans des apports sous forme de complexes humiques provenant du lessivage du continent proche et dans un processus de fixation biologique par les algues.

Par la suite, la mer submerge les marécages et les conditions marines deviennent de plus en plus franches. Du point de vue lithologique les échantillons représentatifs de ce niveau ne diffèrent pas sensiblement du niveau précédent, il s'agit d'argiles noires, cependant plus silteuses.

Si les Tasmanacées demeurent, le plancton devient plus abondant et varié; la radio-activité est moins élevée, le lessivage continental continue mais la côte est plus éloignée.

En résumé, les deux niveaux de lithologie voisine sont caractérisés :
– l'un par un faciès transgressif littoral (zone γ)
– l'autre par un faciès plus franchement marin (désigné arbitrairement ici par zone C).

Nous devons estimer maintenant le degré d'évolution du kérogène. Il dépend essentiellement des températures et des pressions auxquelles les roches ont été soumises au cours de leur histoire.

Ces paramètres peuvent être évalués soit par la mesure de la cristallinité des illites (anchimétamorphisme)[15], soit par l'état de conservation des micro-fossiles organiques (spores végétales et micro-fossiles divers) contenus dans la roche[7].

«Les argiles sont parmi les minéraux des roches sédimentaires les plus sensibles aux transformations. En règle générale, il est admis qu'une diagenèse poussée suffit à faire disparaître Kaolinite et Montmorillonite et à les transformer en mica et chlorite»[19].

Le stade qui précède le métamorphisme ou anchimétamorphisme va, d'une part continuer à modifier les rapports entre les types d'argiles par transformation des espèces minérales et, d'autre part provoque un perfectionnement des réseaux cristallins. Plus les réseaux sont réguliers, plus les pics de diffraction aux rayons X sont aigus (travaux de Weaver et Kubler)[21]. Nous avons ainsi un moyen quantitatif de déterminer l'importance de l'anchimétamorphisme subi par la roche.

De son côté, par l'observation optique en lumière naturelle ou en fluorescence ultra violette, la palynologie apprécie l'état de conservation ou de dégradation des micro-fossiles. La couleur renseigne sur le degré de carbonisation, l'altération des formes renseigne sur les déformations subies du fait des pressions, et sur la destruction progressive des structures intimes par les actions thermiques, ou autres facteurs, tendant à la destruction des édifices moléculaires.

En général, ces deux méthodes donnent des résultats très corrélables, à des argiles bien cristallisées correspondent des spores très carbonisées.

Nous ne parlerons ici que de la deuxième méthode, plus sensible lorsque les températures et pressions n'ont pas atteint un certain seuil.

Au Sahara, l'état de conservation varie d'une façon régulière à l'échelle régionale ; depuis des spores de couleur jaune d'or montrant tous les détails de leur structure jusqu'à du matériel complètement carbonisé, noir et opaque, souvent difficilement déterminable (figs. 5 et 6).

Le degré de carbonisation a été chiffré de 0 à 4, 0 correspondant à un état de conservation parfait et 4 à une carbonisation poussée*.

* Le degré de carbonisation des spores contenues dans les roches paléozoïques ne semble pas dépendre du niveau stratigraphique auquel elles appartiennent. Il est, pour un forage déterminé, en première approximation, le même depuis l'Ordovicien jusqu'au Dévonien. Ce qui semble démontrer, dans l'exemple particulier de cette étude, que l'influence sur le degré de carbonisation de facteurs tels que conditions de dépôt ou action des substances radio-actives contenues dans la zone γ n'est pas importante.

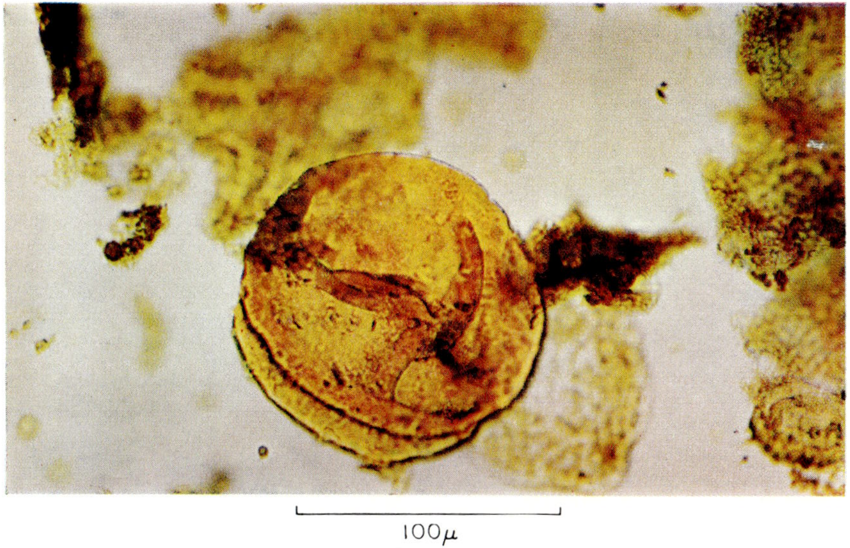

FIG. 4. Exemple de Tasmanacées

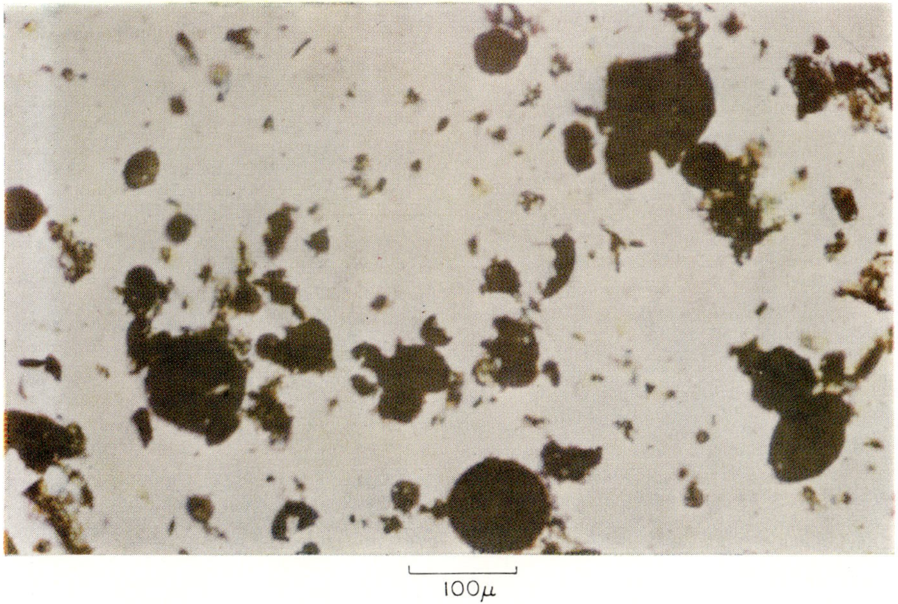

FIG. 5. Palynofaciès à Tasmanacées—environnement côtier—Carbonisation poussée—Forage II

FIG. 6. Palynofaciès à Tasmanacees—environnement côtier—Carbonisation insignifiante—Forage VIII

V. Interprétation des resultats

Dans le Tableau I, nous avons rassemblé, pour chaque niveau et pour chacun des puits, les diverses données géologiques et palynologiques ainsi que les résultats d'analyse. (5 à 10 échantillons ont été analysés par niveau et par puits.)

Nous donnons successivement pour chaque groupe d'échantillons étudié :

— le niveau géologique d'origine,

— le degré de carbonisation des spores,

— le taux moyen de dégradation du kérogène par pyrolyse à 500° C,

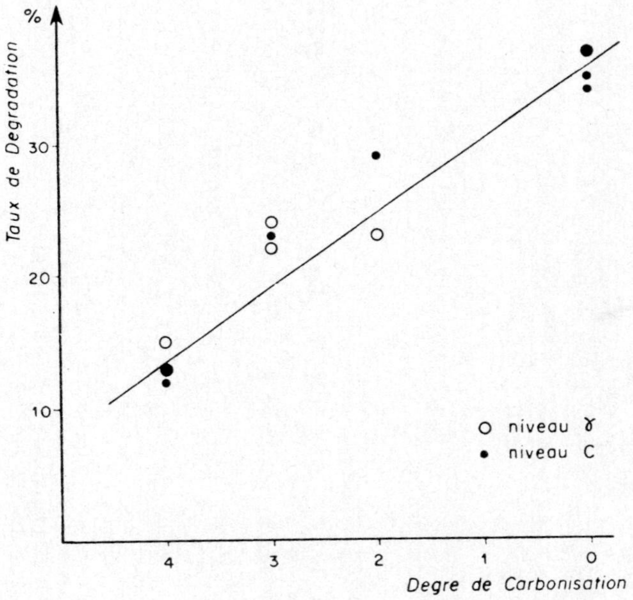

FIGURE 7.

et l'écart maximum à cette moyenne dans le niveau considéré.

— une valeur moyenne des potentiels en gaz, huile et composés aromatiques par pyrolyse à 280° C pendant 15 minutes. Ici encore nous donnons les écarts maxima aux moyennes considérées.

La corrélation entre l'état de carbonisation des spores et le taux de dégradation à haute température du kérogène est bonne (fig. 7).

Pour des échantillons isolés d'origine continentale, nous avions constaté la prédominance des hydrocarbures aromatiques sur ceux paraffiniques

TABLEAU I

Puits N°	I		II		III		IV		V		VI	VII	VIII	
Zone	γ	C	γ	C	γ	C	γ	C	γ	C	C	C	γ	C
Carbonis. (1)	4		4		3		3		2		0	0	0	
TD (2)	13±2	13±2	15±1	12±2	24±1	23±1	22±2	23±2	23±1	29±2	34±1	37±1	37±2	35±2
Pot. Gaz (3)	123±8	40±8	100±9	40±8	20±8	8±2	30±10	8±2	28±9	16±5	24±5	25±5	15±5	24±5
Pot. Huile (3)	8±1	10±1	9±1	8±1	2,5±1	25±10	3±1	28±10	3±1	7±5	5±2	4±2	6±2	4±1
Pot. Ar. (3)	9±1	5±1	11±1	6±1	2,5±1	8±5	2,5±1	8±5	3±1	2±1	1±0,5	0,5±0,1	8±2	0,5±0,1

(1) Degré de carbonisation.
(2) Taux de dégradation par pyrolyse à 500° C.
(3) Potentiel en gaz, huile, composés aromatiques, par pyrolyse à 280° C.

dans les produits de la pyrolyse alors que les roches marines donnaient une plus grande quantité d'hydrocarbures paraffiniques[10]. Nous retrouvons des résultats analogues, le rapport hydrocarbures aromatiques/hydrocarbures paraffiniques est d'autant plus élevé que les échantillons considérés sont plus proches de la transgression, donc plus littoraux.

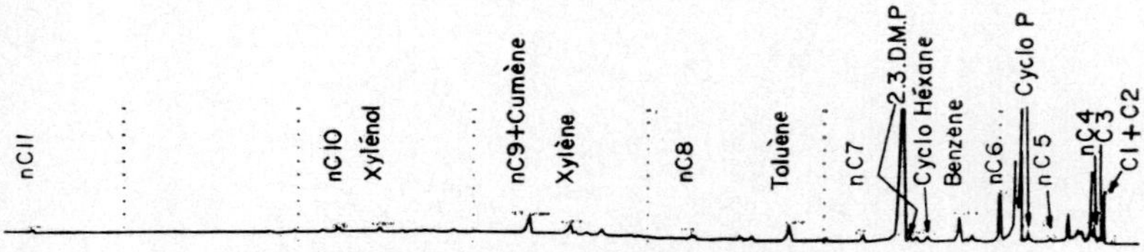

Fig.8 - Puits I, Zone γ. Carbonisation 4

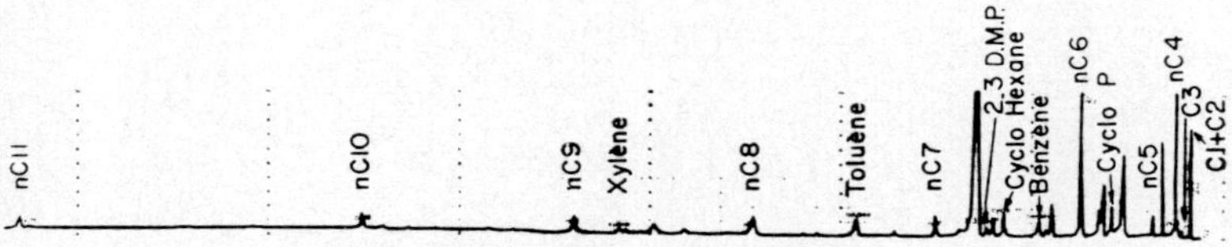

Fig.9 - Puits II, Zone C. Carbonisation 4

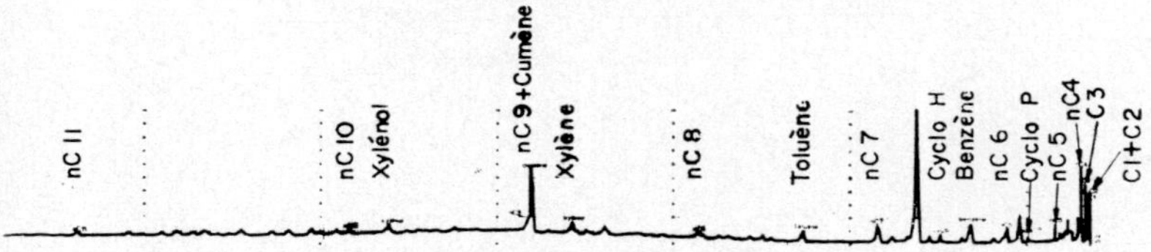

Fig.10 - Puits III, Zone γ. Carbonisation 3

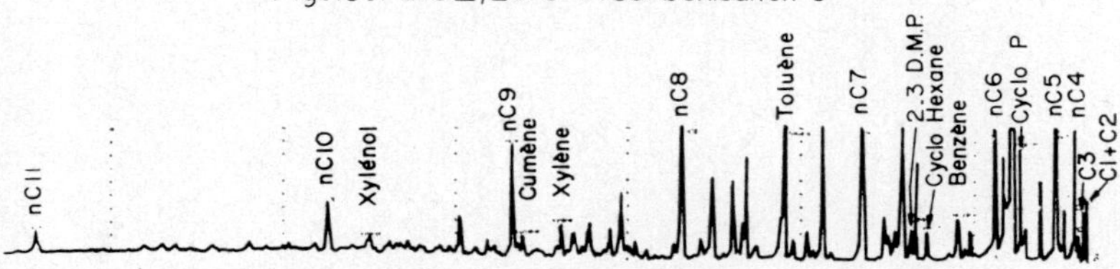

FIG. 11. Puits III. Zone C. Carbonisation 3.

Inversement, pour le niveau C, plus franchement marin, les hydrocarbures paraffiniques sont plus abondants que les aromatiques dans les produits de la pyrolyse. Nous donnons, figures 8, 9, 10 et 11 quelques exemples de chromatogrammes qui permettent de juger des différences obtenues.

La carbonisation et le taux de dégradation permettent de définir trois provinces qui présentent des caractéristiques tranchées :

— La première province comprend les forages I et II. Elle est caractérisée par un taux de dégradation faible, et une carbonisation poussée (T_D moyen : 13, carbonisation : 4).

La réponse à la pyrochromatographie donne un potentiel en huile faible, mais un potentiel en gaz important, surtout pour le niveau γ.

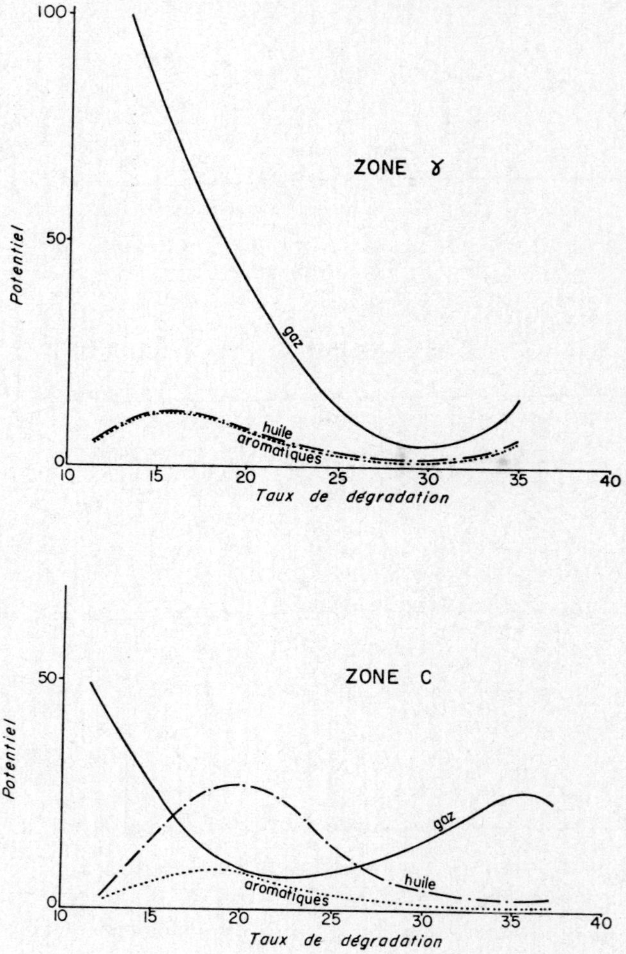

FIGURE 12.

On peut en conclure que le kérogène de cette province est très évolué ($18 < C/H < 22$) ; son aptitude actuelle à donner des hydrocarbures est faible, et encore il ne s'agit presqu'exclusivement que de gaz.

Il semble que nous ayons affaire ici à une roche mère en fin d'évolution.

— Dans la deuxième province qui comprend les forages III, IV et V, le taux de dégradation est voisin de 23 et la carbonisation est intermédiaire (2 à 3) ; le rapport C/H du kérogène se situe entre 13 et 19.

La pyrochromatographie donne des potentiels en gaz variables, les potentiels en huile sont faibles pour les échantillons prélevés dans la zone γ. Les potentiels en huile du niveau C sont moyens à forts pour les forages IV et V.

Nous avons affaire à un kérogène évolué, c'est-à-dire que la roche a probablement fonctionné comme roche-mère, son potentiel actuel peut être encore excellent ; il suffirait que les conditions de température et de pression augmentent légèrement dans la roche pour que le kérogène donne de l'huile et un peu de gaz.

— Le kérogène de la troisième province est peu évolué (taux de dégradation moyen = 36, $11 < C/H < 15$), ce qui correspond à un matériel palynologique parfaitement conservé.

La réponse à la pyrochromatographie est faible à moyenne donnant de l'huile et du gaz. La roche n'a pas été soumise à des pressions et températures très fortes. Elle a pu fournir des hydrocarbures mais en quantités faibles.

D'une manière générale, les potentiels en gaz, huile et produits aromatiques varient régulièrement en fonction du taux de dégradation TD.

Pour le niveau C, lorsque le kérogène évolue, c'est-à-dire lorsque le taux de dégradation diminue, le potentiel en gaz, d'abord assez fort, passe par un minimum, puis croît rapidement ; le potentiel en huile par contre, passe par un maximum très net, correspondant sensiblement au minimum de gaz ; le potentiel en aromatiques suit le potentiel en huile tout en restant nettement plus faible (fig. 12).

Cette répartition est probablement due au fait que les kérogènes au début de leur évolution engendrent du gaz et des huiles très lourdes qui n'apparaissent pas dans les chromatogrammes, ils donnent naissance ensuite à des huiles moyennes, puis légères, et enfin à du gaz.

Pour le niveau γ, le potentiel en gaz d'abord faible augmente très vite. Les potentiels en huile et en produits aromatiques, sensiblement égaux, sont généralement très faibles ; ils passent par un maximum pour un taux de dégradation TD = 15 (fig. 12).

Pour les deux niveaux γ et C il faut aussi noter une simplification des chromatogrammes, c'est-à-dire une diminution des produits ramifiés par rapport aux normal-paraffines, au fur et à mesure de la diminution du taux de dégradation.

Conclusion

L'étude que nous venons de présenter constitue une première tentative en vue de préciser l'origine et le degré d'évolution d'un kérogène.

La structure du kérogène d'une roche actuelle, dépend de son origine et de son évolution. Cette évolution est fonction de l'historique des conditions physico-chimiques, et en particulier, des températures et des pressions subies au cours des temps géologiques. Ces conditions peuvent être évaluées par certains critères: cristallinité des argiles ou degré de carbonisation des micro-fossiles organiques par exemple.

Nous avons vu que la connaissance du degré d'évolution d'un kérogène était primordiale, puisque en effet une telle évolution s'accompagne de la genèse d'hydrocarbures.

Il faut donc préciser dans la structure actuelle de ce kérogène quelle est la part qui revient à l'origine de celle qui revient à l'évolution.

Dans l'étude que nous venons de présenter, le problème était simplifié : nous avons travaillé sur deux faciès différents, mais chacun était à peu près constant sur toute l'étendue de la zone considérée. La technique de pyrolyse et de chromatographie nous a permis d'une part de définir un taux de dégradation qui est en liaison, dans les deux niveaux considérés, avec le degré d'évolution du kérogène et d'autre part, d'estimer des potentiels en huile, gaz, composés aromatiques qui dépendent de l'origine des kérogènes et de leur degré d'évolution et renseignent sur la nature des produits engendrés au cours de la diagenèse.

La prochaine étape consiste à mieux connaître la part de l'origine. Pour cela, outre une meilleure connaissance de sédimentologie organique actuelle, nous devons, avant de rendre la méthode opérationnelle, étendre l'application de notre technique à des faciès différents et relier la caractérisation d'un kérogène à la répartition des différentes familles de substances qui le composent :

— le kérogène sous forme de matière organique amorphe et provenant essentiellement des substances de type humique ;
— les débris divers ;
— les spores et micro-fossiles.

Nous devons, de la même façon, continuer nos études sur la caractérisation du kérogène séparé de son support minéral.

Nous essaierons enfin de corréler l'ensemble des résultats obtenus avec les kérogènes, de ceux trouvés sur la matière organique soluble.

Bibliographie

1. P. H. Abelson. Organic geochemistry and the formation of petroleum, Proc. 6th World Petroleum Congress. Frankfurt/Main, 1964.
2. R. G. Bader *et al.* Recovery of dissolved organic matter in sea water and organic sorption by particulate material. *Geochemica et Cosmochemica Acta*, **19,** 236–243, 1960.
3. W. A. Bone et G. W. Himus. *Coal, Its Constitution and Uses.* London, 1936.
4. O. K. Bordovskiy. Accumulation and transformation of organic substances in marine sediments (traduit du Russe). *Marine Geology*, **3,** n° 1 et 2, 1965.
5. I. A. Breger et A. Brown. Kerogen in the Chattanooga shale. *Science* (*Washington*), **137,** 221/24, 1962.
6. A. Combaz. Remarques sur les niveaux à Tasmanacées du Paléozoïque Saharien (sous presse). *The Paleobotanist*. Lucknow, Inde, 1966.
7. A. Combaz. Les Palynofaciès. *Revue de Micropaléontologie*, **7,** n° 3, 205–218, 1966.
8. A. L. Down et G. W. Himus. A preliminary study of the chemical constitution of kerogen. *J. Inst. Petrol.* **27,** 426–45, 1941.
9. J. P. Forsman et J. M. Hunt. Insoluble organic (kerogen) in sedimentary rock of marin origin. In *Habitat of Oil*, pp. 747–778. AAPG, Tulsa, 1958.
10. A. Giraud et D. Taisne. Application de la pyrolyse et de la chromatographie pour la caractérisation du kérogène associé à une roche sédimentaire. *Recherche des roches-mères* (à paraître), 1966.
11. J. M. Hunt. Geochemical data on organic matter in sediments. International Scientific Oil Conference, Budapest, 1962.
12. B. Juettner. Mellitic acid from coal, cokes and graphites. *J. Amer. Chem. Soc.*, **59,** 208–213, 1937.
13. A. V. Kochenov, V. V. Zinev'yev et S. A. Lovaleva. Some features of the accumulation of Uranium in peat bogs. Transl. in *Geochemistry International*, **2,** n° 1, 1965.
14. P. N. Kogerman et J. Kopwillem. Hydrogenation of Estonian Oil shale and Shale oil. *J. Inst. Petrol. Technologists*, **18,** 1932.
15. B. Kubler. Les argiles indicateurs de métamorphisme. *Revue Institut Français du Pétrole*, **19,** n° 10, 1964.
16. M. Louis. Études géochimiques sur les «Schistes cartons» du Toarcien du Bassin de Paris. In *Advances in Organic Geochemistry*, Pergamon Press, 1964.
17. Y. Khalifeh et M. Louis. Contribution à la reconnaissance des roches-mères de pétrole. *Revue Inst. Franç. Pétrol.*, **10,** 340, 1955.
18. J. A. Martin et P. Bessot. Sur les acides humiques extraits d'une tourbe acide: étude de leurs propriétés, et de leurs aptitudes à fixer et à transporter l'Uranium. In *Advances in Geochemistry*, Pergamon Press, 1964.
19. G. Millot. *Géologie des argiles*. Masson, Paris, 1964.
20. L. R. Moore. The microbiology, mineralogy and genesis of a Tonstein. *Proc. Yorkshire Geological Society*, **34,** part 3, n° 12, 1964.
21. C. E. Weaver et B. Kubler. Possible uses of clay minerals in search for oil. *Proc. VIIIth Conference of Clays and Clay Minerals*, pp. 214–227, 1960.
22. D. H. Welte. Relation between petroleum and source rock. *Bull. Amer. Ass. Petrol. Geol.*, **49,** n° 12, 2246–2268, 1965.

CHARACTERIZATION OF THE INSOLUBLE ORGANIC MATTER OF SEDIMENTS BY PYROLYSIS

by

J. A. GRANSCH and E. EISMA

Koninklijke/Shell
Exploratie en Produktie Laboratorium
Rijswijk, the Netherlands

Introduction

Sedimentary rocks may contain autochthonous organic matter in concentrations ranging from traces to nearly 100%. Its origin may be terrestrial as in the case of peats, lignite and coal seams, aquatic, or a mixture of both, as in the case of many oil shales and presumed oil-source rocks.

In characterising this autochthonous organic matter attention must be paid to three important aspects: its quantity, its type and the degree of diagenesis to which it has been subjected.

One way of establishing the quantity and type and degree of diagenesis of the organic matter in a rock would be to separate it completely from the rock and subject it to normal chemical analysis. Generally, a small part of the organic matter is soluble in benzene, ether and chloroform and it can easily be isolated. A much larger part is insoluble, however, and this presents difficulties. Forsman and Hunt[(1)] and King, Goodspeed and Montgomery[(2)] concentrated the insoluble organic matter or separated it from the inorganic matter by dissolving and destroying the latter with hydrochloric and hydrofluoric acid. This procedure presents two difficulties:

1. It is hard to obtain the organic matter free of inorganic constituents, which is important since these constituents interfere with the hydrogen and oxygen determinations.

2. It is hard to ensure that all the organic matter has been obtained. If only part of the organic matter is isolated, this may not be representative of the bulk of the organic matter. The alternative, then, is to determine the quantity and type and degree of diagenesis of the insoluble organic matter in the sediment, without first isolating it (or after removing the carbonates only).

For an estimate of the quantity of the organic matter in a sediment, wet oxidation and pyrolysis are among the methods that have been used.[3, 4] Generally, the organic carbon content is determined by combustion and this figure is then used for further calculations since 70–90% of the organic matter usually consists of organic carbon.

Khalifeh and Louis[5] used a combination of wet partial oxidation and determination of organic carbon, both carried out on the sediment. The ratio "pouvoir réducteur"/organic carbon is used to characterize the sedimentary organic matter.

We, too, have tried to characterize sedimentary organic matter without isolating it from the sediment. Since in coal chemistry the determination of the fixed carbon content by pyrolysis has been found most useful in characterizing coals, we have applied a similar method. Unfortunately, the simple method as applied in coal chemistry would give large errors if applied to sediments with large amounts of inorganic constituents. The method had therefore to be modified slightly. This will be discussed later, and a number of results will be given. First, however, we will discuss the significance of the content of volatile matter and fixed carbon as it is used in coal chemistry and the type of information that can be obtained from it.

The significance of volatile matter and fixed carbon contents of coals

In coal chemistry, the amount of volatile matter is defined as the loss in weight (corrected for moisture and ash) of a powdered coal when heated under standard conditions to about 900° C. The percentage residue resulting from the pyrolysis is the percentage fixed carbon: % fixed carbon = 100%—% volatile matter. The fixed carbon content of a coal depends on its temperature history and on the proportions in which the various macerals appear in its composition. Figure 1 shows the fixed carbon content, determined by M. Teichmüller,[6] of samples of coal seams penetrated in the boring Münsterland 1, Germany. The figure illustrates well the influence of depth and, since temperature increases with depth, that of temper-

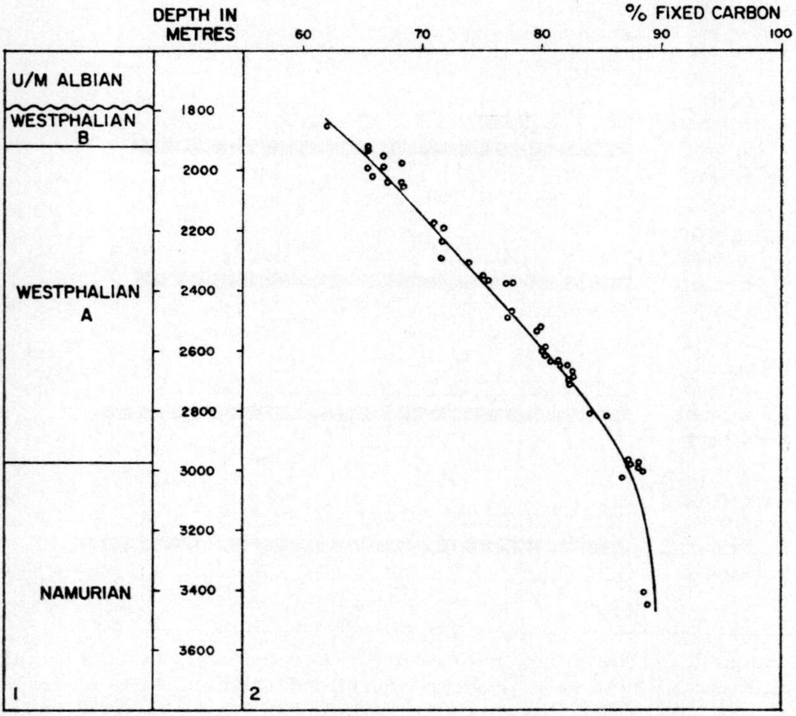

FIGURE 1. — Hilt's rule in boring Münsterland 1, Germany. 1. According to Riechwien, Schuster, R. Teichmüller and Wolburg. *Fortschr. Geol. Rheinld. Westf.* **II**, 16/1–14 (1963). 2. According to M. Teichmüller *Fortschr. Geol. Rheinld. Westf.* **II**, 132–5 (1963).

ature (Hilt's rule). Figure 2 shows the fixed carbon content of macerals isolated by C. Kröger *et al.*(7) from four German coals differing in rank. Macerals are microscopic constituents of coals more or less comparable with the minerals of a rock. The maceral vitrinite is related to the remains of wood, exinite to the exines of pollen and spores and cuticles of plant leaves and micrinite to dark opaque matter of unknown origin. The fixed carbon content of the different macerals, particularly in coals of low rank such as those of Flöz R and Zollverein, differs widely. In coals of high rank, such as that from Flöz Wilhelm, these differences have disappeared. Since these macerals are present in different proportions in different coals it can be inferred that the composition of the coals, at least those of low rank, influences the fixed carbon content.

It has been widely assumed that the fixed carbon content indicates the

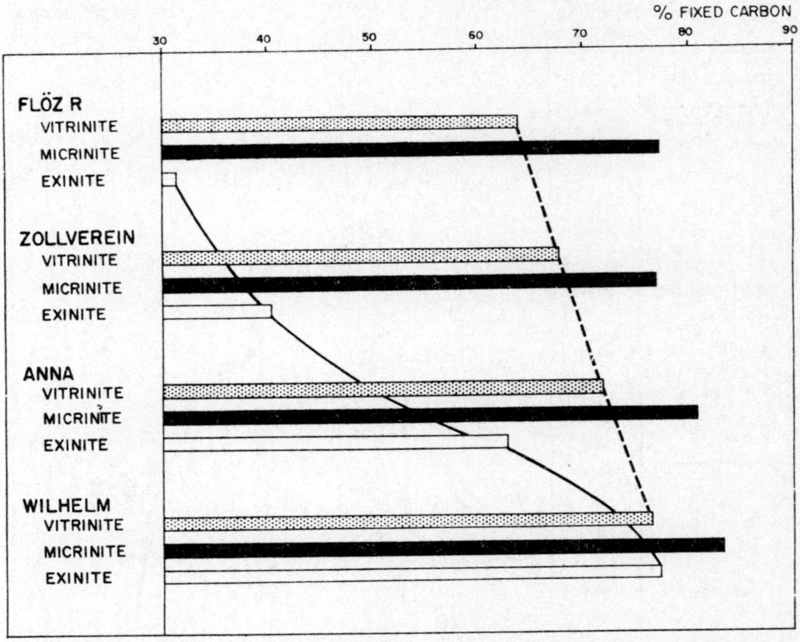

FIGURE 2. — Fixed carbon in macerals from four German coals, according to C. Kröger, A. Pohl, M. Kuthe, H. Havestadt, and H. Bürger, *Brennstoffchemie*, **33**, 1957.

temperature history of a coal. Since Kröger's results show that the composition also has a distinct influence on the fixed carbon content, however, the fixed carbon content should be measured in the vitritic part of the coals, which largely consists of the maceral vitrinite.

Method

Since both organic and inorganic matter will lose weight appreciably during pyrolysis, the loss of weight of a sediment sample on pyrolysis, unlike that of coals, is not solely characteristic of the organic matter. We have, therefore, modified the pyrolysis method used for determining fixed carbon in such a way that we measure the loss of weight not of the sample but of the organic carbon.

Figure 3 illustrates the analytical procedure followed. After removal of soluble organic matter and inorganic carbon, the total organic carbon in part of the sample (C_T) is determined while the residual organic carbon (C_R) in another part is determined after pyrolysis at 900° C for $1\frac{1}{2}$ hours

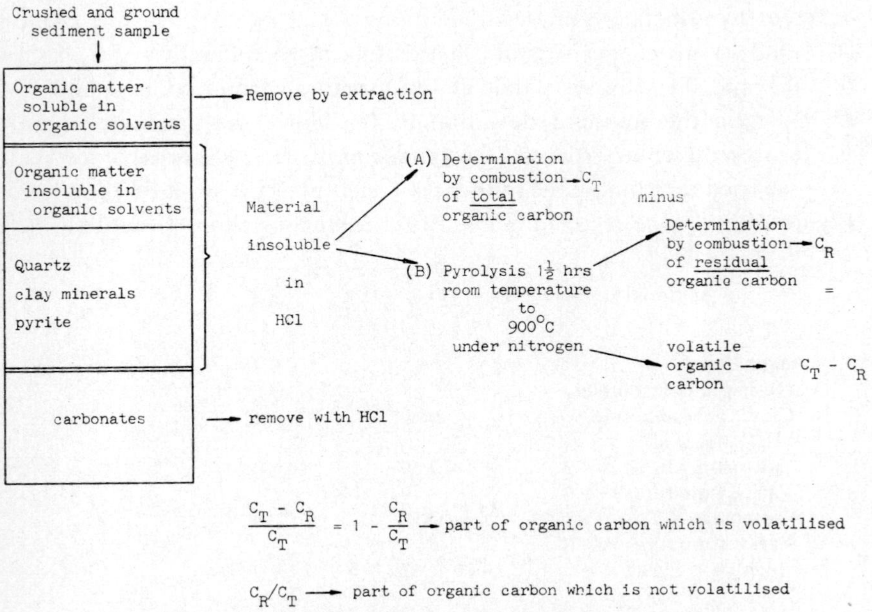

FIGURE 3. — Procedure for determination of C_R/C_T ratio.

in a current of nitrogen. The ratio $(C_T - C_R)/C_T$ represents the part of the organic carbon that can be volatilized. We use the ratio C_R/C_T to characterize the insoluble organic matter.

Figure 4 gives an idea of the accuracy of the C_R/C_T ratio for eight

% C_T	Standard deviation C_T[1)]	% C_R	Standard deviation C_R[1)]	C_R/C_T	Standard deviation C_R/C_T[2)]
5	0·062	4	0·163	0·8	0·038
5	0·062	2·5	0·111	0·5	0·027
5	0·062	1	0·050	0·2	0·011
2	0·050	1·6	0·078	0·8	0·039
2	0·050	1·0	0·050	0·5	0·025
2	0·050	0·4	0·020	0·2	0·010
0·8	0·025	0·64	0·032	0·8	0·032
0·8	0·025	0·40	0·020	0·5	0·020

1) Calculated from 229 C_T and 238 C_R measurements.
2) Calculated according to O. L. Davies (1947): Statistical Methods in Research and Production. London, Edinburgh.

FIGURE 4. — Standard deviation of C_T, C_R and C_R/C_T of some hypothetical samples.

different hypothetical samples, containing 0·8, 2 and 5% organic carbon (C_T) and of which the organic matter has three different C_R/C_T ratios 0·2, 0·5 and 0·8. It shows that at low organic carbon content and low C_R/C_T ratio the standard deviation is 4%. Since the insoluble organic matter in sedimentary rocks is always accompanied by inorganic minerals we have tried to estimate the influence of their presence on the C_R/C_T ratio. Figure 5 shows the results obtained for various mixtures. In all cases the

	Material	% C_T	C_R/C_T	
A	Asphaltenes (n-heptane insoluble)	79·1	0·48	
B	Clay, carbonate-free, H_2O_2-treated	0·06	n.d.	
C	Various mixtures A+B	1·59–7·59	0·47–0·51	$\Delta C_R/C_T = +0{\cdot}06$
	Various mixtures A+B compressed 7·5 ton/cm²	1·59–7·59	0·52–0·54	
	Various mixtures A+B plus 10% Pyrite	4·16–7·59	0·53–0·54	
D	Seat earth, coal layer, carbonate-free	9·34	0·76	$\Delta C_R/C_T = +0{\cdot}04$
B	Clay, see above	0·06	n.d.	
E	Various mixtures B+D	0·98–3·18	0·76–0·80	
F	Bituminous rock, carbonate-free	8·32	0·39	$\Delta C_R/C_T = +0{\cdot}05$
G	F, H_2O_2-treated	0·12	~0·6	
H	Various mixtures F+G	1·04–4·79	0·40–0·44	

$\Delta C_R/C_T$ = largest difference in C_R/C_T ratio found.

FIGURE 5. — Influence of inorganic matter on C_R/C_T ratio.

C_R/C_T ratio after addition of inorganic matter increases by about 0·05, indicating that a slight influence of inorganic matter on the C_R/C_T ratio should be taken into account.

C_R/C_T ratios of peats, lignite and humic coals

Figure 6 illustrates the results obtained with two Dutch peats, a German lignite, and New Zealand and Dutch coals of various ranks. The results are arranged according to increasing organic carbon and fixed carbon content and they show that the C_R/C_T ratios of the Dutch coal samples and the New Zealand coal sample HI 2669, which have the rank of high-volatile bituminous A coal or higher, increase regularly with increasing organic carbon and fixed carbon content. The peats and lignite and the other New Zealand coal samples which are of lower rank than that of a high-volatile bituminous A coal have C_R/C_T ratios that do not regularly increase with organic carbon or fixed carbon content.

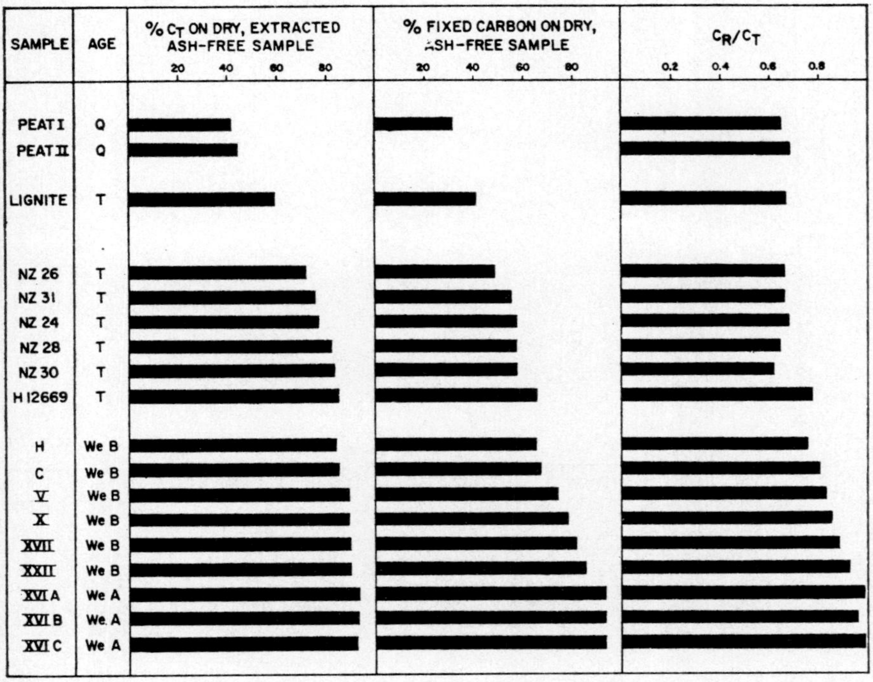

FIGURE 6. — C_R/C_T ratios of peats and New Zealand and Dutch coals. Peat I. De Peel, Netherlands. Peat II. Broek in Waterland, Netherlands. Lignite Herzogenrath, "Maria Theresia", Germany. NZ 26—H12669. New Zealand coals. H—XXII. Dutch Staatsmijn Maurits. XVI. Dutch Staatsmijn Wilhelmina.

Figure 7, in which the C_R/C_T ratios have been plotted against the percentages of fixed carbon, shows the same feature. The C_R/C_T ratios of coals of high-volatile bituminous A rank or higher are linearly related to their fixed carbon content, whereas those of the peats, German lignite and low-rank New Zealand coals do not show this relation with their fixed carbon content. Since the fixed carbon content of coals having the rank of high-volatile bituminous A coal or higher reflects the temperature history of the coals, this also holds for the C_R/C_T ratios of coals in the same diagenetic range. In the range below high-volatile bituminous A coals the fixed carbon (and C_R/C_T) is unsuitable for the characterization of coal. A second point of interest is that humic coals at an early diagenetic stage already have C_R/C_T ratios above 0·6.

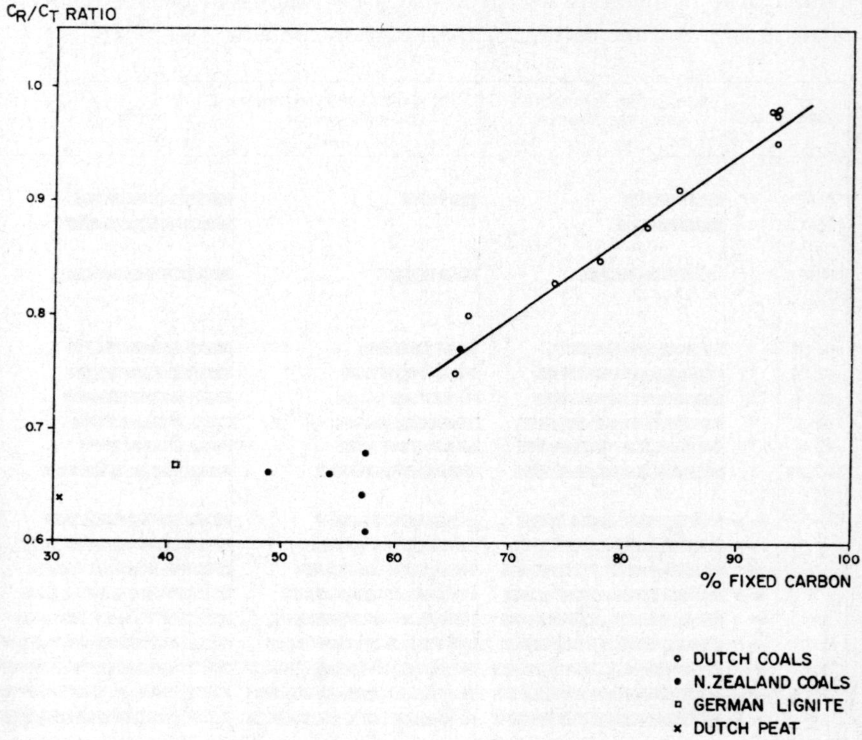

FIGURE 7. — C_R/C_T ratios versus % fixed carbon of various coal samples.

C_R/C_T ratios of shales associated with coals (Westphalian B) from the Netherlands

Figure 8 represents some analytical results for samples of seat earths and roof layers and of coal seams with which these sediments are associated. The Dutch coals are the same as those used in Fig. 6. The organic carbon content of the shale samples ranges from 0·9 to 20·2%. With a single exception the C_R/C_T ratios of the sediments are the same as or very near to the C_R/C_T ratios of the samples from the corresponding coal seams.

As the sediments associated with each of the coal seams lie immediately above or below the seams, these sediments must have been subjected to the

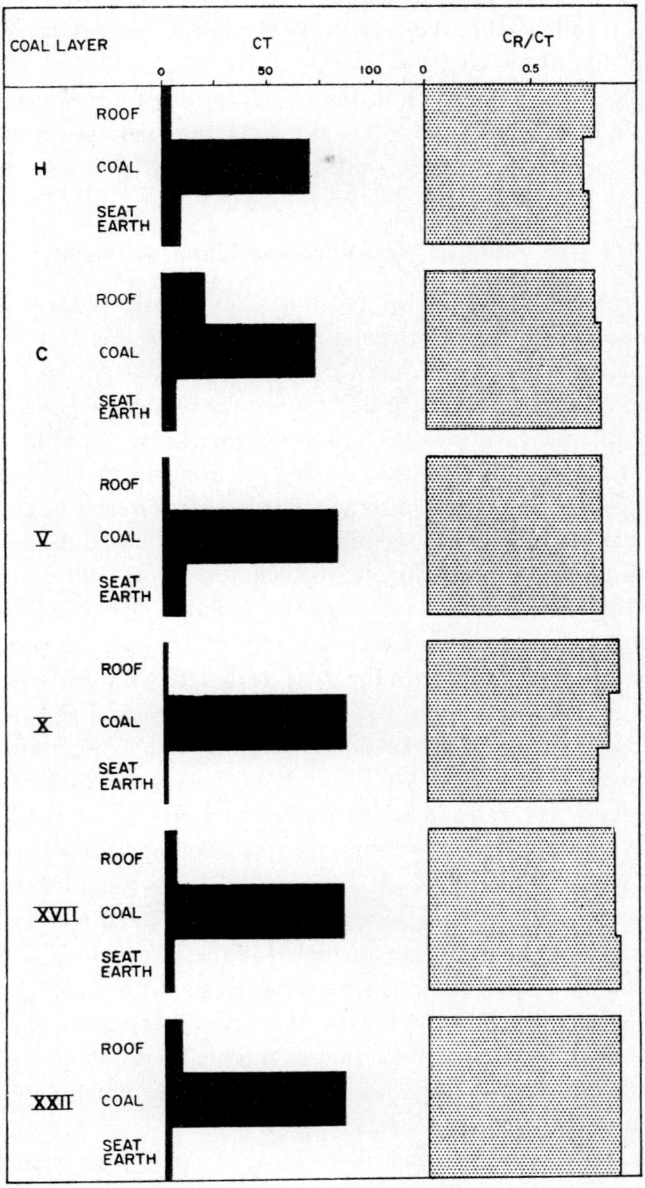

FIGURE 8. — C_R/C_T ratios of coals from Westphalian B, Netherlands, and associated shales.

same temperature as the coals with which they are associated. Since the C_R/C_T ratios of the sediments are equal to, or very nearly equal to, those of the coals it follows that the organic matter in this respect does not differ much from that of the coals.

The results obtained show that the C_R/C_T ratios of humic coaly matter embedded in inorganic matter follow the same trend in relation to the temperature history as do the C_R/C_T ratios of the coals.

C_R/C_T ratios of West European Liassic sediments

We determined C_R/C_T ratios of some 250 Liassic sediment samples from various locations in Western Europe. Some of the locations are in the Lower Saxony Basin (Germany).

Khalifeh, Louis and Roucache determined the reduction-power (pouvoir réducteur)/ organic carbon (PR/C) ratio of other Liassic samples from this area and M. Teichmüller measured the reflection of vitrinite particles in these samples.[8] On the basis of these reflection measurements, an approximate rank can be assigned to the organic matter contained in the samples. Our samples, like those investigated by Khalifeh *et al.*, are derived from Posidonienschiefer layers and although the locations of the samples partly differ, it may be assumed that they are very much alike. It was possible to compare the C_R/C_T ratio with the PR/C ratio and with the rank.

Figure 9 represents the results of this comparison and shows that the C_R/C_T ratio increases with increasing rank in the same manner as the PR/C ratio decreases. Hence the C_R/C_T and PR/C ratios correlate. The figure shows that the C_R/C_T ratio increases in the order given from 0·2 for Bethel to 0·9 for Vehrte, and that the samples from Bethel, Haverlahwiese and Hemmelte West 17 all have C_R/C_T ratios well below 0·6. The diagenetic range of the organic matter in these samples, as estimated by Teichmüller, is approximately subbituminous to high volatile bituminous B/C and in this range humic coaly matter, as we have seen, has a C_R/C_T ratio well above 0·6. This shows that the Liassic sediments investigated have C_R/C_T ratios that are much lower than that of humic coals of the same rank and this indicates that the organic matter in these sediments differs considerably from that of humic coals.

The Lias samples of Suderbruch 49 have C_R/C_T ratios near those of low-rank humic coals and therefore we cannot conclude from their C_R/C_T ratio whether or not their organic matter is similar to that of humic coals.

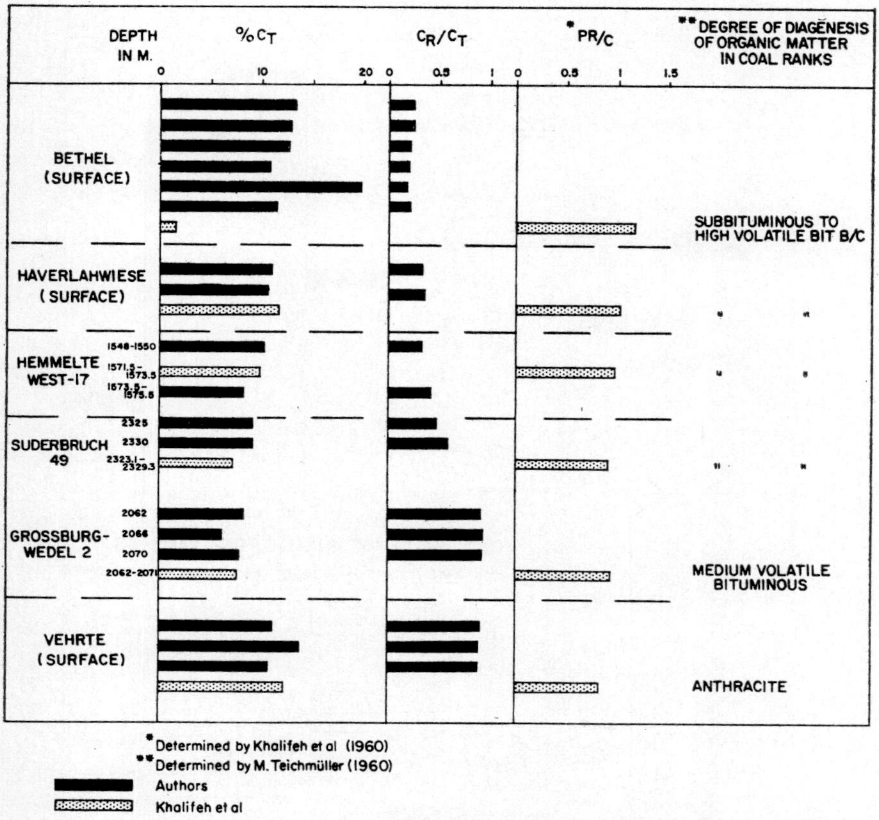

FIGURE 9. — C_R/C_T, PR/C ratios and degrees of organic diagenesis of samples from Posidonienschiefer, Lower Saxony Basin, Germany.

The samples of Grossburgwedel 2 and from Vehrte have C_R/C_T ratios of about 0·9. The rank of the organic matter in these samples is high and it is possible that the high C_R/C_T ratios are due to coalification of bituminous matter as well as to coalification of humic coaly matter.

We have analysed various other Liassic sections and these sections have in common that they generally are comparatively thin (less than 100 metres).

Figure 10 gives the C_R/C_T ratio of Liassic samples having less than 2% organic carbon as well as those having more than 2% organic carbon. For comparison we have given the C_R/C_T and fixed carbon/C_T ratios—which are better comparable with the C_R/C_T ratios than the % fixed carbon—for various coals. The FC/C_T ratios have been calculated from

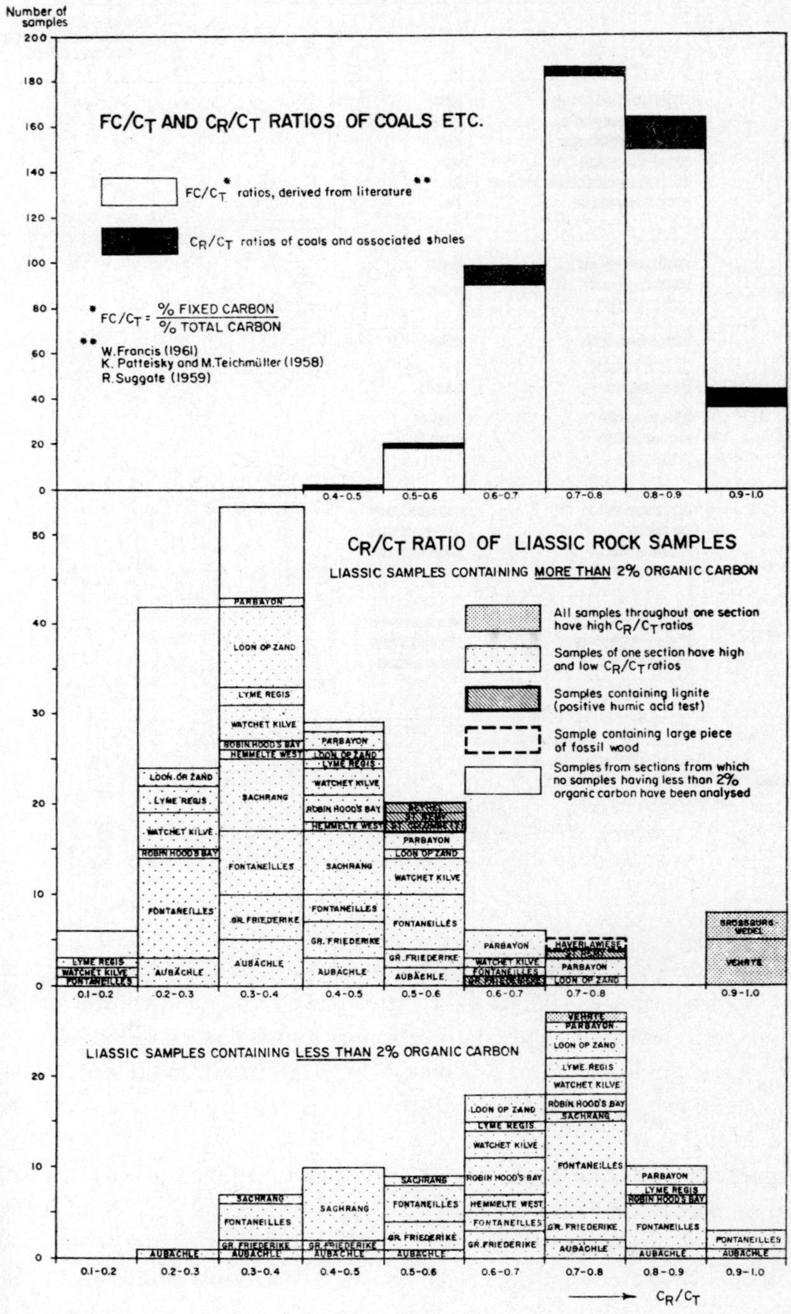
Number of samples
FC/CT AND CR/CT RATIOS OF COALS ETC.
FC/CT* ratios, derived from literature**
CR/CT ratios of coals and associated shales
* FC/CT = % FIXED CARBON / % TOTAL CARBON
** W. Francis (1961)
K. Patteisky and M. Teichmüller (1958)
R. Suggate (1959)
0.4-0.5
0.5-0.6
0.6-0.7
0.7-0.8
0.8-0.9
0.9-1.0
CR/CT RATIO OF LIASSIC ROCK SAMPLES
LIASSIC SAMPLES CONTAINING MORE THAN 2% ORGANIC CARBON
All samples throughout one section have high CR/CT ratios
Samples of one section have high and low CR/CT ratios
Samples containing lignite (positive humic acid test)
Sample containing large piece of fossil wood
Samples from sections from which no samples having less than 2% organic carbon have been analysed
PARBAYON
LOON OP ZAND
LYME REGIS
WATCHET KILVE
ROBIN HOOD'S BAY
HEMMELTE WEST
SACHRANG
FONTANEILLES
GR. FRIEDERIKE
AUBACHLE
HAVERLAHWIESE
GROSSBURGWEDEL
VEHNTE
0.1-0.2
0.2-0.3
0.3-0.4
LIASSIC SAMPLES CONTAINING LESS THAN 2% ORGANIC CARBON
CR/CT

FIGURE 10.

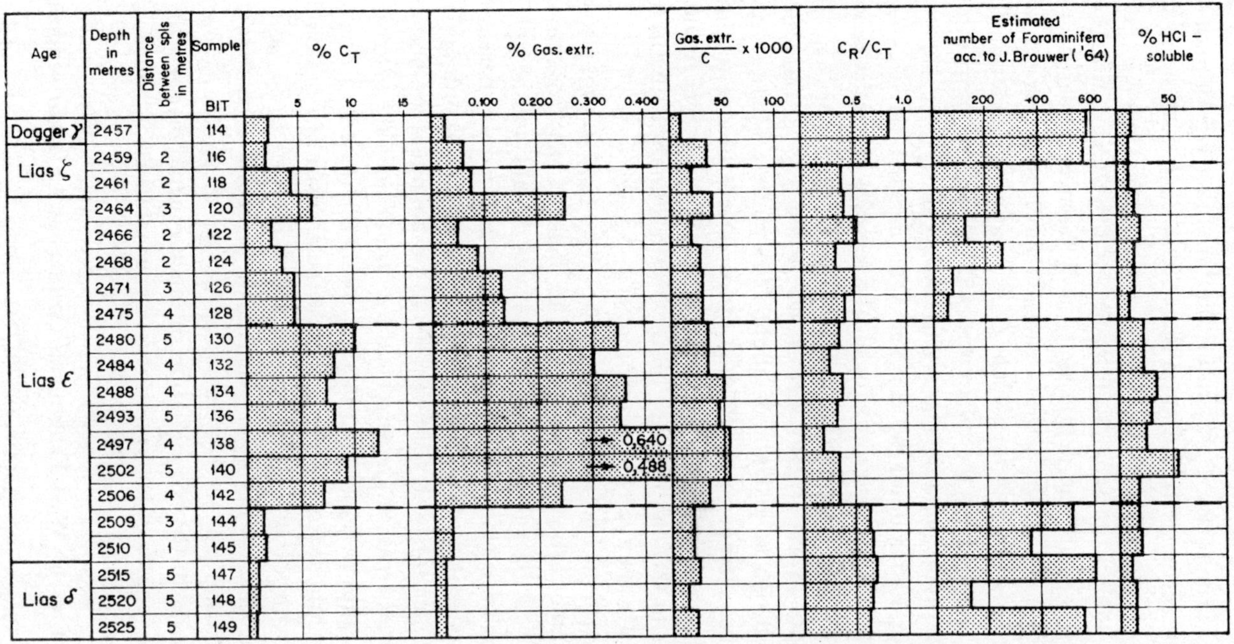

FIGURE 11. — Analytical results of samples from the boring Loon op Zand 1 (Netherlands), plotted in stratigraphic order.

literature data. The figure shows that most of the samples having less than 2% organic carbon have C_R/C_T ratios that are like those of humic coals, whereas most of those having more than 2% organic carbon have much lower C_R/C_T ratios.

Figure 11 shows some analytical results for the Dutch boring Loon op Zand 1. In this figure the depths and the differences in depths between the layers from which the various samples have been derived are also given. The total thickness for which C_R/C_T ratios are determined is 68 metres. The influence of the difference in depth on the temperature history or rank of the organic matter therefore can only be slight. Thus, if the organic matter in all the samples were the same the C_R/C_T ratios would be more or less the same for all the samples or, perhaps, increase slightly downwards. In contrast, there are two jumps in the C_R/C_T ratio, one at a depth of about 2461 metres, where the C_R/C_T ratio suddenly decreases and one at a depth of about 2509 metres, where it suddenly increases. These jumps in the C_R/C_T ratios cannot be due to the slight increase in depth. Instead it is more likely that the differences found in the C_R/C_T ratios in this and in other Liassic sections analyzed by us are due to differences in the type of organic matter.

The interval in the Lias epsilon having the lowest C_R/C_T ratios also has the highest organic-carbon content and the highest gasoline-extract per carbon ratios. Although the gasoline extracts do not consist entirely of hydrocarbons they generally are very rich in them and we therefore have taken them as estimates for hydrocarbons. Moreover, according to a personal communication by J. Brouwer (1964), Foraminifera are absent from this interval whereas in the layers above and below they are present. As far as is known at present, Liassic Foraminifera seem to have been exclusively benthonic and we assume that the absence of Foraminifera in the interval with the highest C_T contents and lowest C_R/C_T ratios indicates an anaerobic environment, whereas the presence of benthonic Foraminifera indicates an aerobic environment. An anaerobic environment apparently leads to the accumulation of much organic matter with a low C_R/C_T ratio, which is favourable for the generation of hydrocarbons.

C_R/C_T ratios of West Venezuelan Cretaceous La Luna sediment samples

It has been generally assumed that the West Venezuelan Cretaceous La Luna formation and to a lesser extent, the Cretaceous Cogollo contain the source beds which have produced a substantial part of the West Venezuelan Lake Basin crude oils.[9-13] Many samples of the La Luna contain porphyrin/vanadium complexes as do the majority of the Lake Basin crudes and the porphyrin investigation has confirmed the opinion that the La Luna contains outstanding oil source rocks.[14] These oil source beds, which consist of the La Luna limestones and marls, were deposited in an environment unfavourable to benthonic life, but favourable to the accumulation of planktonic remains on the sea floor, and over a large area are homogeneous in facies. If we assume that the similarity of the facies throughout large parts of the La Luna source beds indicates that the

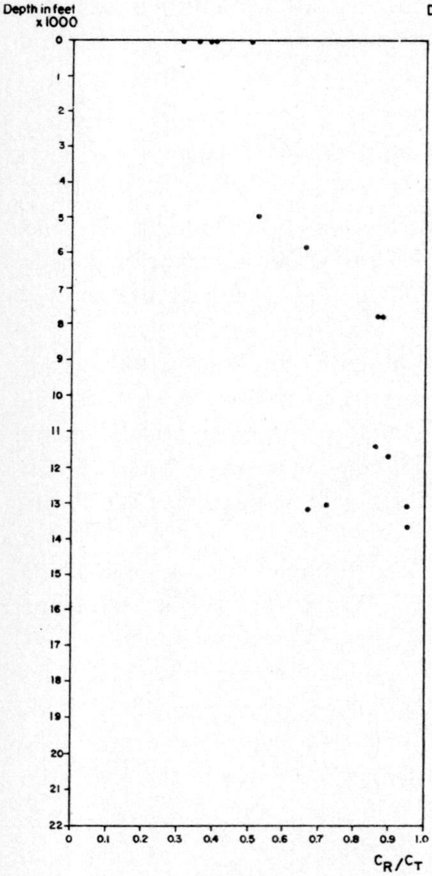

FIGURE 12. — Relation between C_R/C_T ratio and depth of La Luna samples.

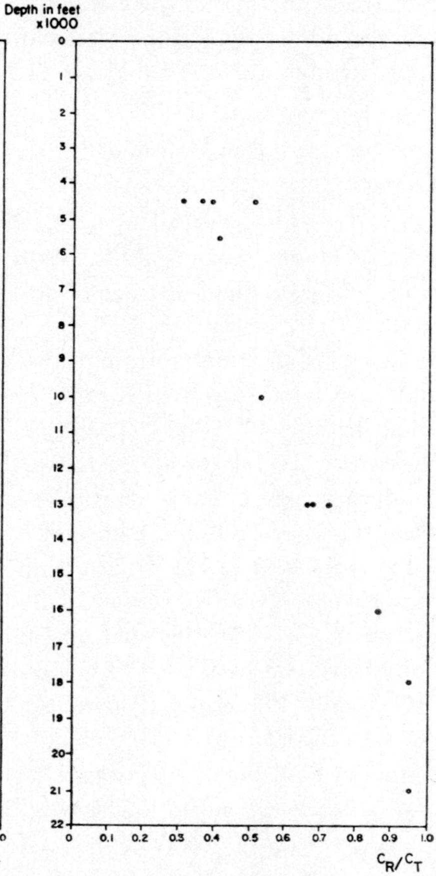

FIGURE 13. — Relationship between C_R/C_T ratio of La Luna samples and estimated maximum possible overburden.

organic matter in these parts is similar in nature, the differences in the C_R/C_T ratio should be ascribable to differences in temperature history.

Figure 12 shows that there is some tendency for the C_R/C_T ratio to increase with depth. For part of the La Luna, Dr. J. K. A. Habicht (personal communication, 1957) estimated the maximum possible overburden.

Figure 13 shows that there is a good correlation between the C_R/C_T ratios and this overburden, suggesting that the C_R/C_T ratio of the organic matter increases as a result of increasing temperature. Ultimately therefore, the C_R/C_T ratio of both coaly and non-coaly organic matter tends to approach unity on continued diagenesis. Humic coaly matter, however, starts thermal diagenesis with a C_R/C_T ratio of approximately 0·6; whereas non-coaly organic matter probably starts with a C_R/C_T ratio of approximately 0·2 or even less.

Comparison of C_R/C_T ratios with other data

Extracts

The percentages of ether extract for the Dutch coal sample *H*, which has the lowest rank, is 0·139. With increasing C_R/C_T ratio and hence rank, the percentage extract of the Dutch coals regularly decreases to 0·002%.

The Liassic samples from Grossburgwedel and Vehrte, the organic matter of which has a high rank, have extract/carbon ratios × 100 that range between 0·1 and 0·2. Liassic samples containing organic matter having a low rank, however, have ratios ranging between 2 and 23. Evidently the extract/carbon ratio decreases with increasing C_R/C_T ratio when this ratio is due to high rank.

In sections like that of Loon op Zand, samples with a high C_R/C_T ratio have low, whereas those with low C_R/C_T ratios have higher extract per carbon ratios, indicating that organic matter unfavourable to the generation of hydrocarbons even of low rank also has low extract/carbon ratios.

Generally, therefore, a high C_R/C_T ratio is accompanied by a low extract/carbon ratio and the high C_R/C_T ratio may be due to organic matter of high rank or to organic matter unfavourable to the generation of hydrocarbons and of low rank.

Atomic H/C ratios

In order to assess the value of the atomic *H/C* ratio in combination with the C_R/C_T ratio, we determined the atomic *H/C* ratios of several

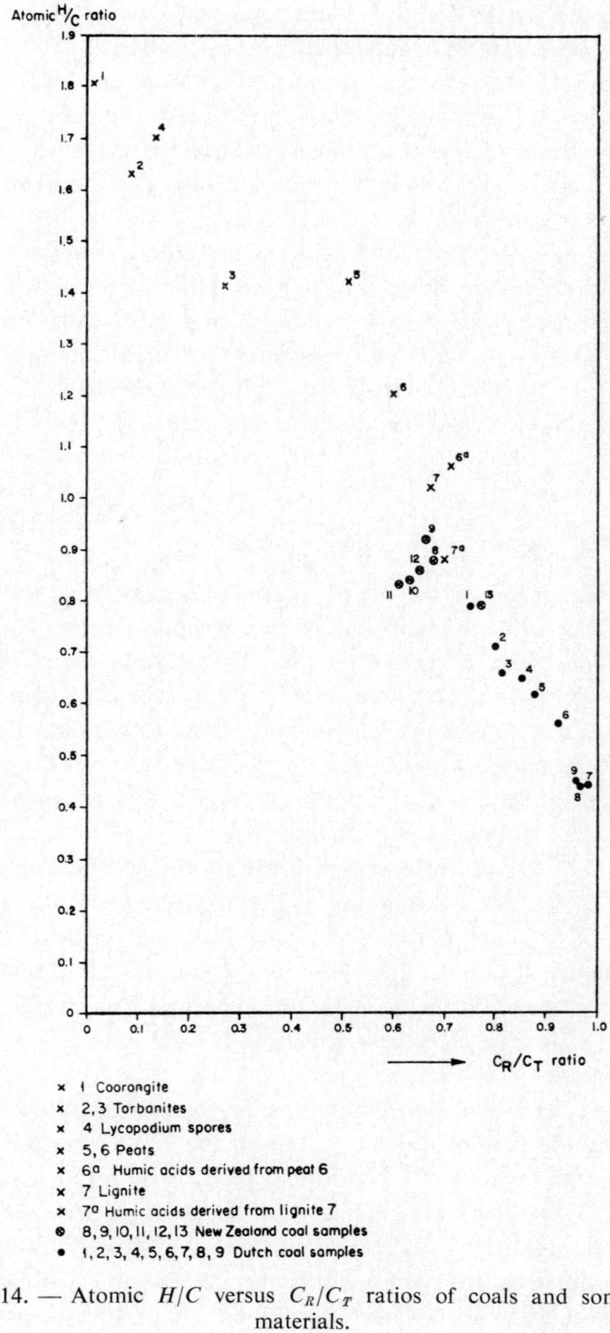

FIGURE 14. — Atomic H/C versus C_R/C_T ratios of coals and some other materials.

types of organic matter (see Fig. 14). Humic acids, peat, lignite and humic coals have C_R/C_T ratios ranging from 0·6 to about 1 and H/C ratios ranging from 1·4 to 0·4. The samples of coorongite, torbanite (which represents alginite) and lycopodium spores (which represent exinite) have C_R/C_T ratios ranging between 0·05 and 0·27 and H/C ratios ranging from 1·8 to 1·4. These results show that with increasing C_R/C_T ratio the H/C ratio decreases and that algal coaly matter and exinite of low rank are much richer in hydrogen than humic coals. Apart from this, however, the H/C ratio gives no more information about the samples than does the C_R/C_T ratio. It is possible, though, that more data on H/C and C_R/C_T ratios of, say, samples of alginite, etc., of higher rank may give additional information, but this will only be possible if a satisfactory method can be developed for the determination of organic hydrogen in sediments.

Final remarks

The determination of the C_R/C_T ratio is an empirical method. It is possible that it is slightly influenced by inorganic matter. However, the correspondence between the C_R/C_T ratios of coaly sediments and the coal seams with which they are associated, the correspondence between C_R/C_T ratio and reflectance of vitrinite particles of some Liassic sediments and the relation between C_R/C_T ratios and environment of Liassic sediments all indicate that the C_R/C_T ratios reflect properties that are inherent to the insoluble organic matter itself.

The C_R/C_T ratio is concerned with the total insoluble organic matter in contrast to rank determinations such as translucency of pollen[(15)] and reflectance of vitrinite particles.[(16)] Like the volatile matter and fixed carbon contents of coals it can only be used to indicate the temperature history if the insoluble organic matter represents one type, i.e. either favourable (with respect to the generation of hydrocarbons) or unfavourable (e.g. humic coaly) matter.

The C_R/C_T ratio, however, has in many cases been used to indicate differences in the type of organic matter. If the C_R/C_T ratio is below 0·6 it indicates organic matter favourable to the generation of hydrocarbons. If samples of various sediment layers of one section are compared, the unfavourable organic matter (C_R/C_T ratio 0·6–0·8) can be distinguished from favourable, as was demonstrated by the example for the Loon op Zand section. Generally, however, when the C_R/C_T ratio is above 0·6, rank determinations are necessary for the distinction between favourable and unfavourable organic matter.

In conclusion, for the interpretation of the C_R/C_T ratio three cases have to be distinguished:

1. The organic matter occurring in different sediments has attained the *same degree of diagenesis*: differences in the C_R/C_T ratios then represent *different types* of organic matter.
2. The organic matter occurring in different sediments is of the *same type*: differences in the C_R/C_T ratio then represent *different degrees of diagenesis*.
3. The organic matter occurring in different sediments represents *different types* and *different degrees of diagenesis*. Other parameters have to be used to establish the degree of diagenesis. When sporomorphs are present translucency measurements can be applied, or when vitrinite particles can be detected their reflection can be used.

Acknowledgement

The authors wish to thank the management of Shell Research N.V. for permission to publish this paper.

References

1. J. P. Forsman, and J. M. Hunt, Insoluble organic matter (kerogens) in sedimentary rocks of marine origin. *Habitat of Oil*, p. 747, 1958.
2. L. H. King, F. E. Goodspeed, and D. S. Montgomery, A study of sedimented organic matter and its natural derivatives. Department of Mines and Technical Surveys. Ottawa Mines Branch Research Report R 114, 1963.
3. P. D. Trask, and H. W. Patnode, *Source Beds of Petroleum*. Tulsa, 1942.
4. H. P. Mojen, Zum chemischen Verhalten der Bitumina. In *Zur Paläogeographie und Bitumenführung des Posidonienschifers im deutschen Lias* (B. Brockamp.) *Archiv für Lagerstättenforschung*, Heft 72, Berlin, 1944.
5. Y. Khalifeh, and M. Louis, Contribution à la reconnaissance des roches-mères du pétrole. *Revue de l'Institut Français du Pétrole*, **10**, 340, 1955.
6. M. Teichmüller, Die Kohlenflöze der Bohrung Münsterland 1. (Inkohlung, Petrographie, Verkokungsverhalten.) *Fortschr. Geol. Rheinld., Westf.*, **11**, 129–78, 1963.
7. C. Kröger, A. Pohl, M. Kuthe, H. Havestadt, and H. Bürger, Die physikalischen und chemischen Eigenschaften der Steinkohlengefügebestandteilen. *Brennstoff-chemie*, **33**, 1957.
8. Y. Khalifeh, M. Louis, J. Roucache, and M. Teichmüller, Die Umwandlung organischer Substanz von bituminösen Schiefern bei der Metamorphose. Ein Beispiel aus dem Posidonienschiefer Niedersachsens. *Erdöl u Kohle*, **13**, 725, 1960.

9. G. T. Philippi, Identification of oil source beds by chemical means. Papers Congreso Geologico Internacional 20th Sesión Ciudad de Méxicao, pp. 25–38, 1956.
10. H. D. Hedberg, Cretaceous limestone as petroleum source rock in N. W. Venezuela. *B.A.A.P.G.* **15,** 229, 1931.
11. H. D. Hedberg, Discussions of article by C. G. de Juana and L. R. Ponte. Fundamental geological characteristics of the Venezuelan oil basin. *Proc. 3rd W.P.C.* Section 1, 1951.
12. H. D. Hedberg, Geologic aspects of origin of petroleum. *B.A.A.P.G.* **48,** 1755–1803, 1964.
13. J. Dufour, Some oil-geological characteristics of Venezuela. *Proc. 4th W.P.C.*, Rome, 1955.
14. J. A. Gransch, and E. Eisma, Geochemical aspects of the occurrence of porphyrins in West Venezuelan mineral oils and rocks. Paper presented at 3rd International Meeting on Organic Geochemistry, London, 1966, this volume, pp. 69–86.
15. C. C. M. Gutjahr, Carbonisation measurements of pollen grains and spores and their application. Thesis, Leiden, 1966.
16. M. Teichmüller, Métamorphose du charbon et prospection du pétrole. *Revue de l'Industrie Minérale*, **40,** 100–111, 1958.

Other Literature (Fig. 10)

1. W. Francis, *Coal.* Edward Arnold Ltd., London, 1961.
2. K. Patteisky, and M. Teichmüller, Examen des possibilités d'emploi de diverse échelles pour la mesure du rang des charbons et propositions pour la délimination des principaux stades de houillification. *Revue Ind. Minérale, numéro special*, pp. 121–37, 1958.
3. R. Suggate, New Zealand coals. Their geological setting and its influence on their properties. *New Zealand Department of Scientific and Industrial Research, Bulletin* 134, 1959.

A LASER-MICROPYROLYSIS SYSTEM FOR THE STUDY OF ORGANIC SEDIMENTS AND INCLUSIONS

by

F. J. Vastola, P. H. Given, R. R. Dutcher and A. J. Pirone

Departments of Fuel Science and Geology,
Pennsylvania State University, University Park, Pa. U.S.A.

Abstract

An instrument has been devised and assembled as follows. A solid sample having a polished or otherwise prepared surface up to 15×3 mm is mounted in the ionization chamber of a time-of-flight mass spectrometer. The surface is examined with a microscope that looks through a window above the sample, and a micro-area selected for study. A low-energy laser beam is projected through the microscope objective and pyrolyzes the selected area of the sample; craters as small as 10 microns diameter can be made with present optics. The pyrolysis products are released into the electron beam, and are analysed. A well-resolved mass spectrum is displayed on an oscilloscope and photographed.

Preliminary runs have been made on coals, various pure minerals, and organic high polymers. Particles of mass number in the range up to 300 are prevalent, and are thus large enough to represent significant fragments of the original solid structure. The mass numbers found indicate that highly unstable primary products are formed and analysed. The technique has been used to study the composition of various macerals in place in coal.

EXPERIMENTAL DIAGENESIS OF THE NAUTILUS SHELL

by

CH. GRÉGOIRE

Department of Biochemistry, University of Liège, Belgium

THE structure of mother-of-pearl, which forms the inner shell layer in various molluscs, can be compared with that of a brick wall.[6, 30] The bricks correspond to the tabular aragonite crystals, the mortar to the organic matrix or nacreous conchiolin.

In this matrix, freed of this layer by decalcification, details of structure so far undetected with the conventional microscope were observed by means of the electron microscope.[10, 14, 15, 18, 19] The organic matrix appears in the form of lace-like reticulated sheets, with irregularly cylindrical trabeculae, studded with hemispherical protuberances, delimiting fenestrations. Three general patterns of structure (nautiloid (Fig. 1), gastropod and pelecypod), differing in the size, the shape of the trabeculae and distribution of fenestrations, have been provisionally recognized in these sheets.

In further studies,[9, 11–13, 16] remains of nacreous conchiolin, variously altered in their structure, though still biuret-positive, were consistently found with the electron microscope in the shells of more than 150 species of Cenozoic, Mesozoic and Palaeozoic (including Ordovician) molluscs. The results are based on the examination of more than seven thousand micrographs. In some of these organic remains, the nacreous conchiolin pattern was remarkably well preversed, especially in several Pennsylvanian nautiloids buried in the asphaltic sandstones of Oklahoma (Buckhorn asphalt).[13, 20]

Most of these shells had retained their original aragonite and architecture, with moderate recrystallization into calcite in some samples or parts of samples.[9, 31] Preserved systems of interlamellar, and intercrystalline organic matrices were detected in Buckhorn asphalt specimens of *Pseudorthoceras knoxense* (nautiloid) and of *Eoasianites* sp. (ammonoid).[16] Fragments of relatively well-preserved matrices were also found, although infrequently, in shells in which aragonite had been recrystallized into calcite or replaced by other minerals (e.g. *Isorthoceras sociale*, Ordovician, Maquoketa formation).[16]

FIGURE 1. — *Nautilus pompilius* L. (modern). Fragment of lace-like reticulated sheet of nacreous conchiolin left by decalcification of the nacreous layer in the living chamber. Sturdy, irregularly cylindrical trabeculae or cords, studded with hemispherical protuberances of various sizes, and separating elongate openings, characterize the nautiloid pattern. Compare these normal structures with their experimental alterations in Figs. 2–4, 7, 8, 10–12. Shadowcast with palladium. Direct print. ×42,000.

In weathered fragments of the Buckhorn asphalt material (e.g. exfoliated, outermost portions of the inner nacreous layers in nautiloid shells)[16, 20] and in many other fossil shells from all ages and formations, the nacreous conchiolin matrices were variously modified in their structure. The alterations recorded in the trabeculae of the reticulated sheets consisted most frequently of flattening, broadening, thinning, curling, shrinkage, disappearance of the tuberosities, coalescence into perforated membranes, ribbons, strands or blades, and, in more advanced stages of degradation, of fragmentation into twisted segments, resulting finally in the formation of lenticular or spheroidal, pebble-shaped corpuscles and discs of various sizes. Bloating of the trabecular fragments and possibly swelling of the protuberances were also observed. These alterations occurred in shells (e.g. many Cretaceous specimens) in which aragonite had been preserved and in shells recrystallized into calcite. In the latter, traces of the original nacreous brickwall architecture were seen with the electron microscope scattered among mosaics of coarse calcitic crystals in specimens as old as Ordovician,[16] in agreement with the findings of original architecture in fossils with the conventional microscope.[4, 5, 23, 24]

In decalcified fossil samples, it is not always possible to identify with certainty the strongly altered organic matrices as remnants of original organic shell fabrics. However, detection of transitional aspects between trabeculae and pebbles in limited areas of modern and of still intact fossil materials, and preservation of the specific protidic constitution in Holocene, Oligocene and Eocene shells,[7] in which identical structural alterations had been previously detected by electron microscopy,[11, 12] suggested that these structures were actually altered conchiolin.

In order to obtain additional evidence for the identification of these remains as conchiolin, to decrease the risk of confusion with foreign contaminants and to study the relations between modifications on the ultrastructure level and changes in chemical composition, nacreous fragments of *Nautilus macromphalus* Sowerby and of *Nautilus pompilius* L. were exposed, in investigations still in progress, to some of the factors involved in diagenesis, namely increase of temperature and pressure. The results, summarized below, are based on the examination of more than 2700 micrographs.

In the first group of experiments, nacreous fragments from the inner layer of the living chamber, placed in open vessels, were heated dry in an oven for various lengths of time (a few minutes to 21 days) to temperatures ranging from 150° C to 800° C. Preliminary results have been reported previously.[17]

The effects of dry heat on mother-of-pearl consist of a progressive increase in brittleness of the originally hard substance, of multiple cleavage into mineral sheets and of transformation at 500° C and above into a snow-white powdery substance. The nacreous samples kept at 200° C and 300° C for 5 hours closely resembled the brittle and strongly variegated nacreous substance of many Cretaceous tests, such as those of the Ripley and Coon Creek formations, for example. X-ray powder diffraction analysis[9] revealed transformation of the original aragonite into calcite in the samples heated to 400° C and 500° C, a result in agreement with previous data.[22, 26, 29]

In all the samples, including those heated to 800° C, the biuret reaction, performed on structures remaining after decalcification of the samples, showed under the conventional microscope pink and violet flakes, corresponding to shreds of nacreous conchiolin. In the samples heated to 800° C, however, the colours were very faint.

FIG. 2

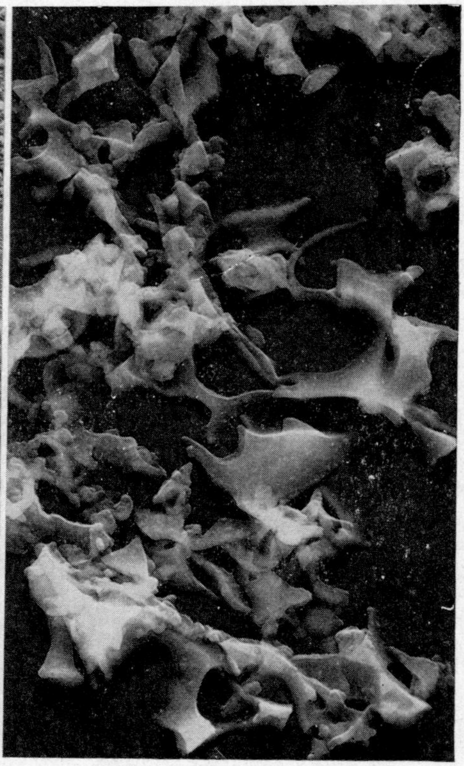

FIG. 3

FIGURE 2.—*Nautilus pompilius* L. (modern). Mother-of-pearl from the inner layer of the shell wall in the living chamber, heated in open air in a dry oven to a temperature of 400° C for 5 hours. Decalcification of the brittle and cleaved calcitic substance resulting from this treatment left residues consisting of spheroidal, pebble-shaped and granular bodies, frequently clustered by desiccation on the supporting film. Shadowcast with palladium. Direct print. × 42,000.

FIGURE 3.—*Nautilus pompilius* L. (modern). Fragment of inner nacreous layer of the shell wall in the living chamber, heated under vacuum in a sealed quartz tube to a temperature of 400° C for 5 hours. The remnants of decalcification of the calcitic material, shown in this picture (fragments of membranes and of contorted ribbons, angulated blades, wedge-shaped strands) differ thoroughly in their structure from those illustrated in Fig. 2. Shadowcast with platinum. Reversed print. × 42,000.

The alterations of the remains after decalcification, examined in the electron microscope, consisted predominantly in the samples heated above 225° C for 5 hours, of the dislocation of trabeculae of reticulated sheets into rounded pebble-shaped corpuscles (Figs. 2 and 4), closely resembling those recorded among the decalcification remnants of many nacreous samples of Palaeozoic and of Mesozoic mollusc shells (Fig. 5).

In a second group, nacreous fragments placed under vacuum in sealed quartz tubes, were exposed to conditions of time and temperature identical with those of the first group.

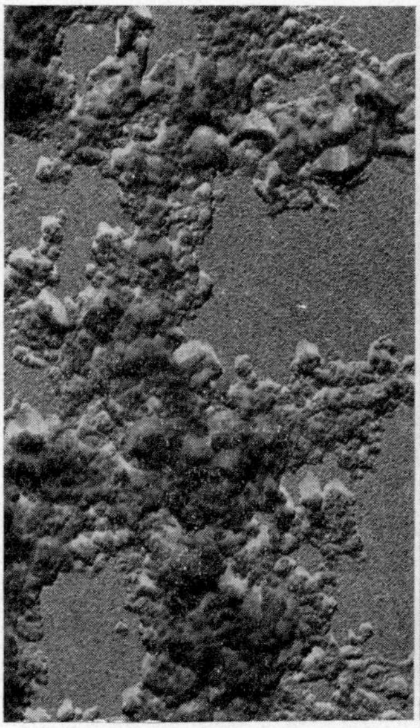

FIG. 4

FIG. 5

FIGURE 4. — *Nautilus pompilius* L. Mother-of-pearl from the inner layer of the shell wall in the living chamber, heated in open air in a dry oven to a temperature of 600° C for 5 hours. Decalcification of the cleaved, lustreless, calcitic, grey-white substance resulting from this treatment left predominantly organic residues consisting of spheroidal corpuscles of various sizes, resembling those recorded in the organic residues of fossil nacreous substance prepared by the same procedures (see Fig. 5). Direct print ×42,000.

FIGURE 5. — Nacreous fragment from *Sphenodiscus* sp. (ammonoid), Cretaceous, Bluffs of Owl Creek formation, 3 miles NE. of Ripley, Tippah C°, Mississippi, U.S.A., supplied by U.S. Geological Survey (20 843, 707 alpha; our files: 389-2), Dr. Reeside, Dr. Sohl and Dr. G. Arthur Cooper. Biuret-positive remnants of decalcification of this material appear predominantly in this area in the form of irregularly spheroidal corpuscles, clustered by desiccation on the supporting film. Note the close resemblance with the structures shown in Fig. 4, produced artificially. Direct print. × 42,000.

On breaking open the tubes, a pungent smell of ammonia, H_2S, and occasionally of tar and of petrol, was noticeable. Compared with the samples heated in open air, the fragments heated to the same temperature in sealed tubes appeared to be less brittle and had kept a variegated or black coloration at higher temperatures (500° C and above). The samples heated to 900° C for 5 hours were transformed into brittle snow-white slabs.

In the decalcification residues, the trabeculae of the reticulated sheets were less and differently altered from those in the samples of the first group heated to the same temperature for the same or for shorter periods of time (compare Fig. 2 with Fig. 3, and Fig. 8 with Figs. 11 and 12).

The remnants of decalcification exhibit a great diversity of structures, resulting from the dislocation and flattening of the conchiolin trabeculae into polygonal, wedge-shaped blades, contorted, angulated strands, flat discs (Figs. 3 and 7). Coalescence and flattening transformed the reticulated sheets into loose and plane networks, appearing as perforated membranes, with broadened fenestrations (Fig. 12) and foldings in the form of rings and pseudo-fibres (Fig. 7). The original outlines of the nautiloid pattern are recognizable in Fig. 11. Similar changes in conchiolin structures were observed in Palaeozoic and Mesozoic remnants of nacreous organic matrices (Fig. 6 and ref. 9, plates III and VII). Pebble-shaped corpuscles, characterizing the alterations in organic matrices of conchiolin in the first group, were absent or less frequent in the material heated in sealed tubes.

In samples of this group heated to 900° C for 5 hours, demineralization of the calcitic components left abundant biuret-positive elements. A part of this material (not shown) consisted of strongly folded and pitted membranes, of other flattened membranes perforated by broad, oval or rounded holes. In other complex structures, resembling the crystalline aragonitic flaggings of the nacreous lamellae viewed tangentially, polygonal areas were delimited by broad, electron-opaque strands. Layers of substance, filling these areas, appeared to be made up predominantly of clusters of rounded particles and vesicles, 65 Å in diameter, frequently assembled in contorted pitted ribbons and opaque patches of substance. Similar particles were clustered in hexagonal figures of considerable size (e.g. 6 microns in length), still exhibiting the outlines of the original nacreous tabular crystals of aragonite.

In a third group of experiments, nacreous fragments, dry or soaked in wet sea mud or in sea water, were compressed in the ranges of 2 (in autoclave) to 60,000 kg per cm² (in steel cylinders), at various temperatures (20° C to 650° C). Because of frequent explosions, the experiments were not of long duration (a few seconds to 10 min).

Comparison, in preliminary experiments in collaboration with Mr. R. Lorent, with the results of the first group indicates preservation and pos-

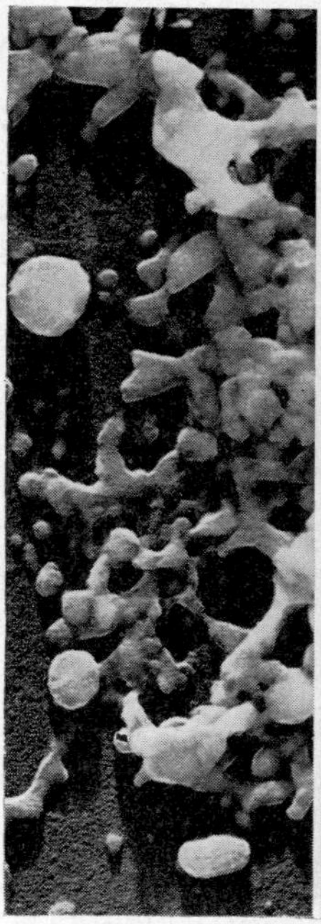

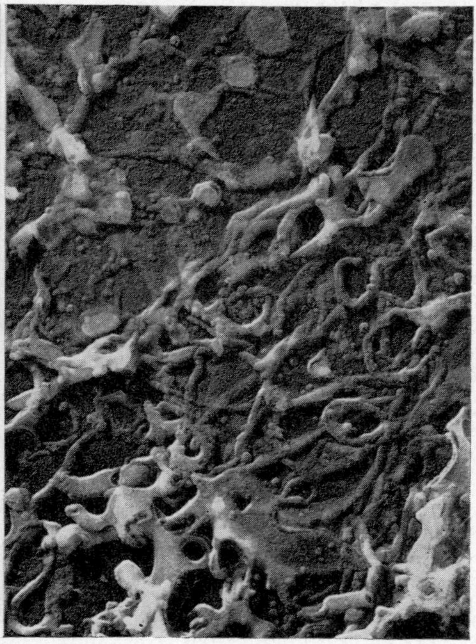

FIG. 6 FIG. 7

FIGURE 6. — *Domatoceras* or *Stearoceras* sp. (nautiloid), Permian, San Andres limestone, Rio Penasco River, east of Alamogordo, New Mexico, U.S.A., supplied by Dr. Rousseau H. Flower. The biuret-positive structures consistently detected in all the residues of decalcification of this specimen were spheroidal or lenticular corpuscles and debris of networks composed of twisted, ribbon-like, partly flattened fragments of trabeculae. Note the resemblance of these fragments of trabeculae to similar alterations produced artificially (Fig. 7, bottom left) Shadowcast with palladium. Reversed print. × 42,000.

FIGURE 7. — *Nautilus pompilius* L. Fragment of the inner nacreous layer of the shell wall in the living chamber, heated under vacuum in a sealed quartz tube to a temperature of 600° C for 5 hours. A great diversity of structures, resulting from alterations in conchiolin reticulated sheets, appears in these remnants of decalcification: perforated membranes, twisted or contorted ribbons with dumpy expansions, cords disposed in rings and hooks, discs, small spheroidal bodies. Several fibre-shaped structures are probably folds in the membrane lying on the background. Platinum shadowcast. Reversed print. × 42,000.

sibly increase in hardness of compressed nacreous substance heated to temperatures which induced, in open air, considerable brittleness of the mother-of-pearl and its transformation into powdery material.

FIG. 8 FIG. 9 FIG. 10

FIGURE 8. – *Nautilus pompilius* L. Mother-of-pearl from the inner layer of the shell wall in the living chamber, heated for 5 hours in the open air to a temperature of 800° C. The nacreous material, transformed into snow-white, cleaved, powdery, brittle, mineral sheets (CaO and traces of calcite), left after decalcification faintly biuret-positive shreds of substance appearing in the electron microscope under various shapes: spheroidal corpuscles, contorted ribbons, fibres, frequently agglutinated into fragments of membranes. Compare with alterations in sealed tubes at the same temperature: Figs. 11 and 12. Palladium shadow cast. Direct print. × 42,000.

FIGURE 9. — Shell fragment of a large nautiloid (*Rutoceratidae* sp.). Upper Middle Devonian (Givetian), Sötenicher Mulde, Sötenich, Eifel, Germany (685-2). Loose networks of relatively flattened membraneous strands, delimiting a broad polygonal fenestration, and a few clusters of small granular corpuscles are shown. Compare with similar alterations in trabeculae produced artificially: Fig. 10. Palladium shadowcast. Direct print. × 42,000.

FIGURE 10. — *Nautilus pompilius* L. Fragment of mother-of-pearl from the inner layer of the shell wall in the living chamber, soaked in sea water and heated to a temperature of 20° C to 500° C in 2 min, 30 sec and of 500° C for 5 min under a pressure of 60,000 kg am per cm² (Grégoire and Lorent, unpublished). Flattening and coalescence of trabeculae into membraneous strands with local bloatings were the predominant alterations in this preparation. Compare with similar changes in fossil material, shown at a slightly higher magnification in Fig. 9. Palladium shadowcast. Direct print. × 25,000.

In the decalcification remnants of these samples, conversion of reticulated sheets by flattening or coalescence of their trabeculae into stout, strand-like structures and perforated membranes, was predominant over changes into pebble-shaped corpuscles, as in samples heated under vacuum, and, as in the latter, the alterations differed greatly from those observed in open air.

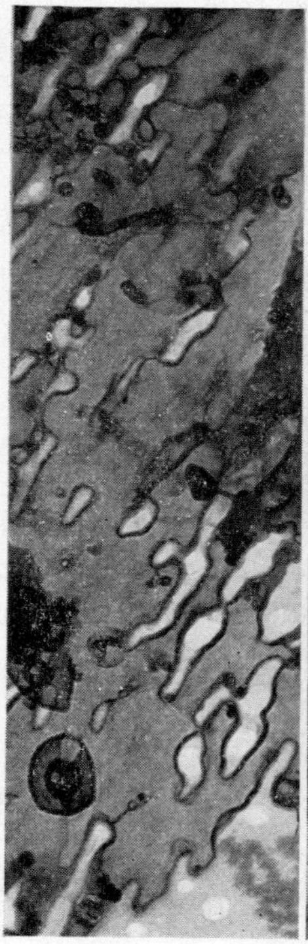

FIG. 11

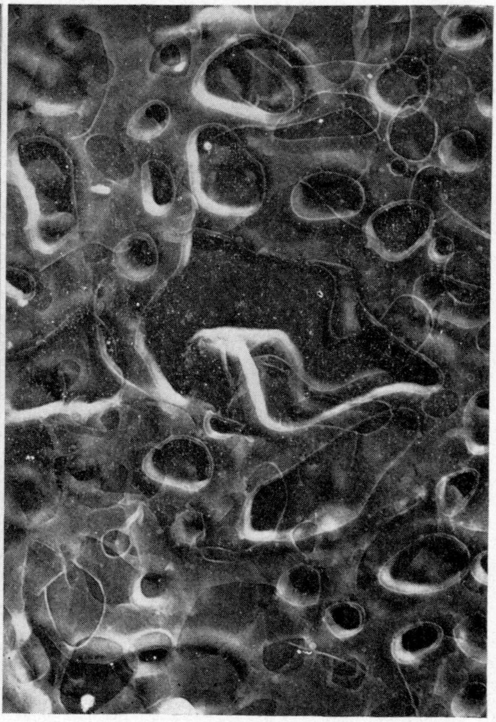

FIG. 12

FIGS. 11 and 12. — *Nautilus pompilius* L. Fragments of mother-of-pearl heated under vacuum in a sealed quartz tube to a temperature of 800° C for 5 hours. Compare with Fig. 8. Substantial remnants of decalcification appear to be composed of perforated membranes, resulting from flattening and spreading of trabeculae of conchiolin reticulated sheets. Faint outlines of the nautiloid pattern are recognizable in Fig. 11 (compare with Fig. 1). In Fig. 12, two or three extremely thin, transparent, perforated membranes are superimposed on the apparently swollen, similarly perforated membrane lying on the background. Fig. 11: unshadowed preparation. Direct print. × 42,000. Fig. 12: platinum shadowcast. Reversed print. × 42,000.

Comments and conclusions

1. Several alterations so far observed in nacreous layers of fossil shells from various ages and formations[9, 11–13, 16] were reproduced artificially in the corresponding layers of the modern *Nautilus* shell by exposure of this material in dry and wet environments to heat and pressure, in open air and under vacuum. The results suggest the possibility of combined electron microscope and biochemical stepwise studies of the mechanisms of action of these diagenetic factors. These studies might also furnish supplementary controls for the identification of remnants of conchiolin structures in which the original pattern is no more recognizable.

Several of the conditions of temperature and of pressure realized in these investigations probably did not reflect actual diagenetic circumstances, and, therefore, do not permit direct extrapolation to the palaeontological record.

2. Fossil conchiolin remains and heated modern nacreous organic matrices are fragile substances. The procedure used here for giving this material an adequate degree of electron permeability required for observation in the electron microscope (decalcification, thinning of too dense materials by means of the ultrasonic vibrations, for example) are obviously responsible in part for a certain degree of mechanical disintegration and dispersion of structures. However, all the samples included in this study were submitted to identical conditions of preparation and are thus comparable. On the other hand, by different methods of preparation, such as replication of conchiolin remains freed on polished nacreous surfaces by etching, alterations identical with those illustrated in Figs. 1–12 were also observed, together with slightly different modifications, which will be described elsewhere, in altered organic matrices left undisturbed on the mineral background or still partly embedded in it.

3. Constant findings of biuret-positive structures among the remnants of decalcification of fossil shells and of modern nacreous substance altered experimentally, are consistent with biochemical data concerning preservation, in fossil shells, of polypeptides[1, 3], of assemblages, after elimination of the free amino acids, of several amino acids characteristic of conchiolin scleroproteins, in Eocene and Oligocene mother-of-pearl[7], and in shell wall and cameral deposits of Pennsylvanian nautiloids.[21]

Detection of biuret-positive shreds in samples heated to temperatures as high as 500–900° C was first considered with suspicion. However, careful search for artifacts and for possible sources of contamination by foreign structures, especially during collection of the heated materials and preparation of the decalcification remnants, has so far been unsuccessful. Under the conventional microscope, the biuret positive shreds left by decalcification of all the samples, including those heated to 900° C, appeared in the form of clusters of pink-violet, polygonal flakes, like those of the true conchiolin fragments. In addition, the presence, in the samples contained in heated sealed tubes, of substantial membraneous structures obviously of organic nacreous origin (see Figs. 11 and 12), precludes any confusion with foreign material. On the other hand, it has been shown that amino acids dry heated to 200° C for a few hours polymerize to polypeptides,[8] but data concerning the action of higher temperatures could not be found.

4. The obvious differences in preservation of conchiolin remnants between nacreous fragments heated in open air on one hand, and those heated under vacuum and under pressure on the other hand, probably reflect differences in biochemical preservation: it has been shown that amino acid degradation is greatly enhanced in the presence of oxygen.[2] The biochemical composition differs in shells heated in open vessels and under vacuum in sealed tubes.[25]

Volatile constituents have been reported to develop in heated evacuated and sealed tubes containing Pleistocene shell samples.[25] Pressure exerted by these gases are possibly responsible for considerable degree of flattening presented by organic matrices of modern mother-of-pearl heated under vaccuum to 800° C (Figs. 11 and 12). Pressure has, however, not been measured in the present material.

5. Constant findings of organic nacreous matrices in fossil shells from various ages and formations and the presence of substantial organic remains in samples of modern mother-of-pearl heated to high temperatures,suggest that conchiolin is either highly resistant by itself to disintegration factors or is efficiently protected from the influence of these factors by its close association with mineral matter and its sheltering between alternating mineral layers in the architecture of mother-of-pearl. The latter possibility is in agreement with a recent suggestion[27] that amino acids (and possibly their polymerization compounds) might exhibit a higher degree of stabilization when associated with minerals.

A detailed, illustrated description of results, including modifications of the mineral architecture of mother-of-pearl, especially in transitionae stages in which aragonite is recrystallized into calcite, will be reported elsewhere.

Acknowledgements

I thank Dr. Rousseau H. Flower, who supplied specimens of *Domatoceras*, U.S. Geological Survey (Dr. Reeside and Dr. N. F. Sohl) and Dr. G. Arthur Cooper, who supplied a specimen of *Sphenodiscus*, Prof. Dr. H. Lambot (Diamant Boart), for generous hospitality in his laboratories, Dr. G. Vanderschrick (Diamant Boart), for suggesting adequate modifications of the high pressure equipment, Dr. R. Lorent, who performed compaction preparations, Mr. D. Coutsouradis, for authorization to use groups of ovens of the C.N.R.M., Dr. A. Davin and Mr. G. Casier, who prepared the sealed tubes and performed the heating, Dr. A. Lutts and Mr. J. Grandjean, who recorded and analysed X-ray diffraction patterns.

References

1. P. H. Abelson, Annual Report of the Director of the Geophysical Laboratory, 1953–1954: *Carnegie Inst. Wash. Yearbook*, **53,** 97–101, 1954; *Paleobiochemistry, Scientific American*, **195,** 83, 1956.
2. P. H. Abelson, Some aspects of Paleobiochemistry, *Ann. N.Y. Acad. Sciences*, **69,** 276–285, 1957.
3. P. H. Abelson, Organic constituents of fossils. *Treatise on Marine Ecology and Paleoecology*, 2. Geol. Soc. Amer. Memoir 67, 87–92, 1957.
4. R. G. C. Bathurst, The replacement of aragonite by calcite in the molluscan shell wall. In *Approaches to Paleoecology* (John Imbrie and Norman D. Newell, Ed.), pp. 357–376, Wiley, New York, 1964.
5. O. B. Bøggild, The shell structure of the mollusks. *Kong. Danske Vidensk. Selsk. Skrifter. Natur-vidensk. Math. afd.*, 9 Raekke, II, 2, 233–325, 1930.
6. E. Ehrenbaum, Untersuchungen über die Struktur und Bildung der Schale der in der Kieler Bucht häufig vorkommenden Muscheln. *Zeitschr. Wissenschaft. Zool.* **41,** 1–47, 1885.
7. M. Florkin, Ch. Grégoire, S. Bricteux–Grégoire, and E. Schoffeniels, Conchiolines de nacres fossiles. *C. R. Acad. Sc., Paris*, **252,** 440–442, 1961.
8. S. W. Fox, and M. Middlebrook, Anhydrocopolymerization of amino acids under the influence of hypothetically primitive terrestrial conditions. *Feder. Proc.* **13,** 211, 1954.
9. J. Grandjean, Ch. Grégoire, and A. Lutts, On the mineral components and the remnants of organic structures in shells of fossil molluscs. *Bull. Acad. Roy. Belgique, Cl. Sciences*, 5e série, Tome L, 562–595, 1964.
10. Ch. Grégoire, Topography of the organic components in mother-of-pearl. *J. Biophys. Biochim. Cytol.* **3,** 797–808, 1957.
11. Ch. Grégoire, Essai de détection au microscope électronique des dentelles organiques dans les nacres fossiles, (ammonites, nautilides, gastéropodes et pélécypodes). *Arch. Internat. Physiol. Bioch.* **66,** 674–676, 1958.
12. Ch. Grégoire, A study on the remains of organic components in fossil mother-of-pearl. *Bull. Inst. Roy. Sc. Natur. Belgique*, **35,** n° 13, 1–14, 1959.
13. Ch. Grégoire, Conchiolin remnants in mother-of-pearl from fossil Cephalopoda. *Nature*, **184,** 1157–1158, 1959.

14. Ch. Grégoire. Further studies on structure of the organic components in mother-of-pearl, especially in Pelecypods. Part I. *Bull. Inst. Roy. Sc. natur. Belgique*, **36,** n° 23, 1–22, 1960.
15. Ch. Grégoire, On submicroscopic structure of the *Nautilus* shell. *Bull. Inst. Roy. Sc. Natur. Belgique*, **38,** n° 49, 1–71, 1962.
16. Ch. Grégoire, On organic remains in shells of Paleozoic and of Mesozoic cephalopods (nautiloids and ammonoids), *Bull. Inst. Roy. Sc. Natur. Belg.*, **42**, n° 39, 1–36, 1966.
17. Ch. Grégoire, Thermal changes in the *Nautilus* shell, *Nature*, **203,** 868–869, 1964.
18. Ch. Grégoire, Gh. Duchâteau, and M. Florkin. Structure, étudiée au microscope électronique, de nacres décalcifiées de mollusques (Gastéropodes, Lamellibranches et Céphalopodes). *Arch. Internat. Physiol.* **58,** 117–120, 1950.
19. Ch. Grégoire, Gh. Duchâteau, and M. Florkin, La trame protidique des nacres et des perles. *Ann. Inst. Océanogr.* **31,** 1–36, 1955.
20. Ch. Grégoire and C. Teichert, Conchiolin membranes in shell and cameral deposits of Pennsylvanian cephalopods, Oklahoma. *Oklahoma Geology Notes*, **25,** 175–201, 1965.
21. P. E. Hare, Quoted by Lowenstam.(28)
22. C. Hintze, *Handbuch der Mineralogie*, **1,** 3, 2982. Walter de Gruyter Co., Berlin und Leipzig, 1930.
23. H. Hölder. Ueber Gehäusebau, insbesondere Hohlkiel jurassischer Ammoniten. *Palaeontographica*, Abteilung A, **102,** 18–48, 1952.
24. J. D. Hudson, Pseudo-pleochroic calcite in recrystallized shell-limestones. *Geol. Mag.* **99,** 492–500, 1962.
25. J. D. Jones, and J. R. Vallentyne, Biochemistry of organic matter. I. Polypeptides and amino acids in fossils and sediments in relation to geothermometry. *Geochim. & Cosmochim, Acta*, **21,** 1–34, 1960.
26. C. Klein. *Nachr. d. Ges. d. Wiss., Göttingen*, 1883. Quoted by Mügge.(29)
27. H. Kroepelin, Some organic compounds extracted from Posidonia shale. *Proc. Intern. Meet. Org. Geochem.*, Milan, 1962.
28. Heinz A. Lowenstam, Biologic problems relating to the composition and diagenesis of sediments. *The Earth Sciences*, (Thomas W. Donnelly Ed.), pp. 137–195, Rice University, Semicentennial Publ., Univ. Chicago Press, 1963.
29. O. Mügge, Krystallographische Untersuchungen über die Umlagerungen und die Structur einiger mimetischer Krystalle. *Neues Jahrb. Miner. Beilage* Bd XIV, 246–318, 1901.
30. W. J. Schmidt, Bau und Bildung der Perlmuttermasse. *Zool. Jahrb. Anat. Abt.* **45,** 1–148, 1923.
31. F. G. Stehli, Shell mineralogy in Paleozoic invertebrates, *Science*, **123,** 1031–1032, 1956.

INDICATIONS OF HIGH-TEMPERATURE PROCESSES IN ORGANIC GEOCHEMISTRY

by
G. MUELLER

Institute of Molecular Evolution, University of Miami, Coral Gables, Florida, U.S.A., and University of Concepcion, Chile.

1. Introduction

The "inorganic geological thermometers" hitherto utilized indicate, as a general rule, a single point in the temperature-pressure scale of the geological history of the formation in question. The main aim of the present paper is an illustrative discussion of the utilization of the "carbonaceous complex" as an indicator of the thermal history (mainly the maximum temperature) and other conditions which prevailed within a given locality of the Earth's crust. The considerable sensitivity of the carbonaceous complex throughout the temperature range of 50° C to about 800° C, *a priori* renders it a most useful geological indicator. Other advantages when compared with the inorganic indicators are as follows:

A. Its continuous and, under most geological conditions, irreversible loss of volatiles with increase of temperature (or decrease of pressure) is a unique advantage for the study of thermal gradients.

B. In certain cases, more than one property of the complex can be proved to change with the physical conditions (loss of volatiles and consistency, for example) which enables us to estimate both the maximum temperature and pressure, which were reached within the given geological setting.

In the course of research, from 1948 onwards, the very instructive bitumen occurrences within the North Derbyshire ore field were studied from an organic geochemical and genetic point of view.[1–4] The present paper aims at the utilization for the purposes of geological thermometry and manometry of the results gained through that experimental work and field observation. Data from recent experiments, especially devised for the purpose in question, are also interpreted.

It appears from these studies that processes of mobilization of carbonaceous sediments, through more or less localized agencies of mineralization, igneous activity, and possibly also faulting and folding, are quite widerspread within the Earth's crust. The number of processes involving elevated temperatures in Organic Geochemistry seems to be greater in extent and in variety, than has hitherto been supposed.

2. The Outlines of the Geology of the North Derbyshire ore field

The ore field is in a Lower Carboniferous Limestone dome, about 16 miles from north to south and 8 miles wide, overlain along most of its flanks by the Lower Carboniferous "Edale" shales.[5, 6] The limestone contains numerous vents, sills, dikes and flows of basalt.[7, 8] The ore field has well-defined zoning, which was previously interpreted as due to variations in the temperature of mineralization.[9, 10, 11] The presumed highest-temperature, fluorite, blende and galenite–bearing zone is situated along the eastern limestone–shale boundary, and it is, therefore, supposed that the mineralizing solutions emanated mainly from a concealed source to the east of the ore field. Farther to the west, the veins contain a gangue in which barite reaches a maximum, and towards the central and western portions of the dome, within the lowest temperature zone, calcite is the main hydrothermal mineral, with subordinate galenite and occasional chalcopyrite.

It should be stressed that the above outlined thermal zoning applies only for the major lodes of the area. The side veins within the higher temperature zones tend to change their mineral composition towards the lower temperature assemblages, within a few yards or tens of yards from the main deposits.

3. The reconstruction of the genetic history of bitumens from Derbyshire

The bitumens of Derbyshire are invariably associated with hydrothermal veins or basaltic intrusions. The following two theories have been suggested in relation to their genesis.

A. Condensation of magmatic, juvenile carbonaceous complexes.

B. Distillation of biogenic sediments.

Sylvester- Bradley and King[13] keep an open mind regarding both of the above modes of formation of the Derbyshire bitumens. The writer[4, 12] prefers hypothesis B on the following grounds:

A. There appears to be a good statistical interrelation between the percentage of bitumens in the veins, and the percentage of carbonaceous sediments within the country rock.

B. There is also a well-defined relationship between the type of primary organic sediments, and the type of bitumen assemblages within veins of differing fluorite and ore percentages. The increase of such percentages was interpreted as indicative of increase of temperature of mineralization (see 2, above).

Within areas in which the shale sediments contain low grade kerogens (Ecton Mine, Tissington, Staffordshire, and Mill Close Mine) the calcitic veins contain oil of asphaltic base. With the increase of the percentage of fluorite and sulphide ore the bitumen within the veins becomes inspissiated, in turn, to pittasphalt, asphalt, asphaltite, graphitite and graphite, whereas a "fractionation series" of lower boiling bitumens appears within progressively broader zones of the surrounding country rock. The above described localities were termed "undifferentiated" deposits by the writer.

In areas with a bituminous coal type sediment (Eyam, Bakewell, Wirksworth and Ashover) the bitumen in the low temperature veins is of albertite type. The high fluorspar and ore percentage veins contain, beside albertite, a black variety of bernalite (which was termed "olefinite" in the writer's earlier works), black elaterite and a dark brown ozocerite. In these "imperfectly differentiated" deposits it appears that the original hydrocarbon mixture separated into immiscible liquids, as demonstrated in a more spectacular manner in the "well differentiated" deposits, described below.

The primary carbonaceous sediment within areas with "well differentiated" bitumen assemblages (Castleton, Bradwell and Matlock) has proved to resemble anthracitic coals. The mainly calcitic veins contain the brittle variety of carbonite of mainly aromatic composition. In the high-temperature veins, the resinous variety of carbonite contains essentially aromatic molecules. The bulk of olefinic molecules is concentrated in the remarkable extruded rods of brown and green varieties of bernalite (see 4.4) and the paraffins into oil with ozocerite concretions. The "fractionation series" outside the veins contains a fusible variety of carbonite, a lower molecular, open-chain elaterite, and foxite, which is a soft to hard resin composed mainly of high O/H ratio molecules.

It appears that within areas with the "undifferentiated deposits" the original kerogen-type shale sediments were distilled and, in the case of mineralization at relatively elevated temperature, the distillate became

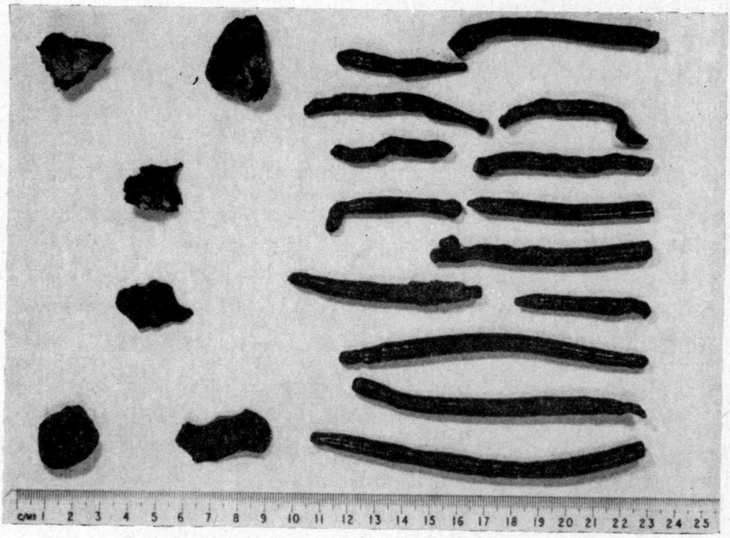

FIGURE 1. — Suite of the differentiated bitumens. Left, from top downwards: foxite, elaterite, vaseline, carbonite; Right: extruded forms of bernalite. All the illustrated specimens are from the "bernalite flat". Windy Knoll, Castleton.

FIGURE 2. — Solid section of extruded droplets of bernalite. Note carbonite in the extrusion channel of double droplet at centre, also small carbonite lined veinlet at extreme right. The gangue of the vein is mainly calcite, with some barite, fluorite and marcasite. L., $\times \frac{1}{2}$.

fractionated along the thermal gradient which surrounded the vein. The essentially coal-type carbonaceous sediments of the districts with "imperfectly differentiated" and "well-differentiated" type deposits, yield progressively more aromatic coal tars on distillation, which polymerize into albertite and carbonite-type substances in the mainly calcitic veins.

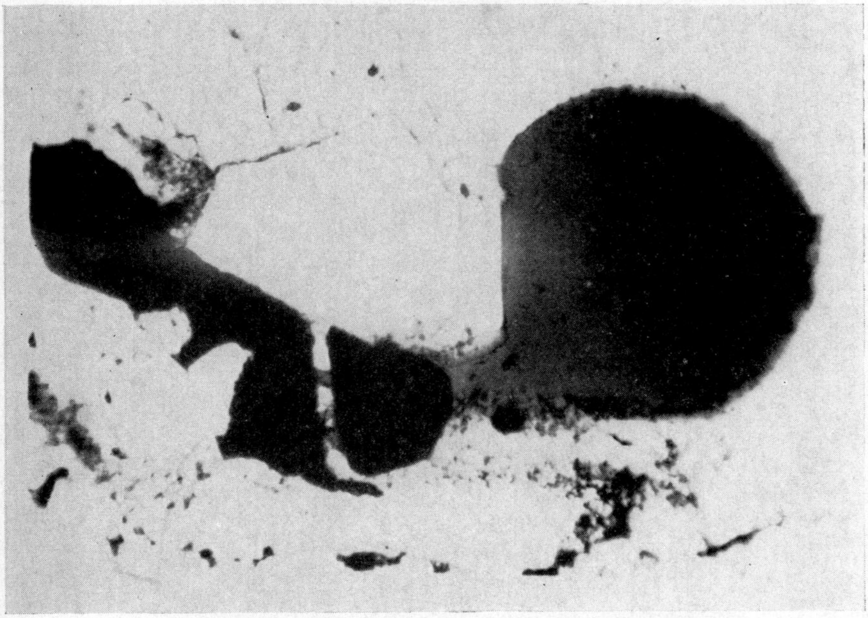

FIGURE 3. — Extruded globule of opaque carbonite followed with globule of bernalite in which minute globules of carbonite occur as inclusions. The country rock is at the left bottom corner of the section. The calcite crystal at upper centre apparently exerted pressure on the extrusion channel in the course of its growth, which caused the snapping of the already solid carbonite, and the expansion of two threads of the still plastic bernalite, in the crack of the former, O.L., × 15.

In order to account for the increase in the H/C ratio of the bitumen assemblage of the high-temperature veins, a process of hydrogenation of the original aromatic distillates has to be postulated (see 4.4). The hydrogenated and highly polymerized products tend to separate into immiscible liquids (see Figs. 2, 3 and 4) producing the "differentiated" deposits, which were observed for the first time in the course of the writer's field work within the area.

For the main ends of geological thermometry and manometry, which is the purpose of the discussion, the detailed reconstruction of the genetic history and chemical descriptions of bitumens are only secondary requirements, because the work is mainly based on the bitumens' physical properties. For more details of studies of their geochemistry and genesis

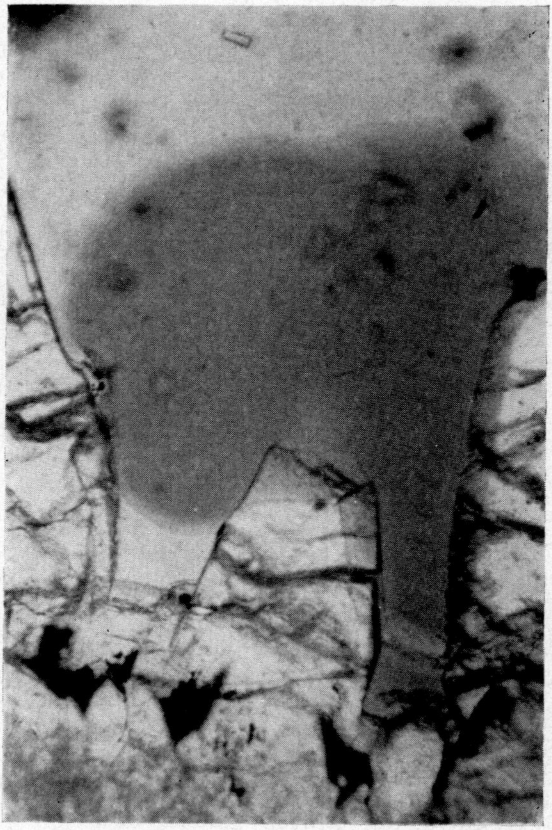

FIGURE 4. — Droplet of dark brown bernalite extruded into a larger droplet of elaterite. The extrusion channel traverses the calcite gangue from the country rock (limestone) at the bottom of the microphotograph. Funnel-shaped scales of carbonite are seen at the margin of the country rock, and sharp phase boundaries between the three differentiates are evident. O.L., × 40.

the reader is referred to the previous works of the writer.[2, 3] An up-to-date chemical study of the remarkable "differentiated" bitumen deposits is at present in progress.

4. Organic geochemical thermometers and manometers in the North Derbyshire ore field

1. *Thermometry based on the melting points of bitumens*

Most of the types of bitumens within the Derbyshire deposits entirely fill the available space between crystals of the gangue minerals, and occasionally reveal "solidified puddles". These types of structures clearly indicate that the bitumen was injected as a thin liquid ("liquid injection forms") into the vein. The presence of globules indicates "viscous injection", and the presence of rods indicates entry of the material into the vein in a plastic state which will be discussed later. Each bitumen appears to have entered the vein in one or two of the characteristic ways which can be summarized as follows:

Bitumens injected as liquids: (Oil), pittasphalt, asphalt, asphaltite, graphitite, carbonite, elaterite, foxite, ozocerite.

Bitumens injected as viscous liquids: Albertite, black bernalite, brown bernalite, alyphite, elaterite.

Plastic injection: Green bernalite, brown bernalite.

It appears that, in most cases at least, the injection of the bitumen was simultaneous with the actual process of mineralization, as indicated through the following structures. (A) In specimens from the Odin Mine and Windy Knoll (Castleton District) solidified puddles of carbonite rest on an older generation of calcite crystals, and they are overgrown, in turn by a younger generation of marcasite and sphalerite. (B) Inclusions of oil, bernalite, carbonite, etc., appear in gangue and ore minerals from numerous localities. (C) It was often observed that within veins which are filled with bitumens the crystals of gangue minerals develop unusually freely grown rounded or globular habits, indicating the presence of a relatively viscous medium, in the course of their actual growth.

It appears self-evident that the temperature during the entry of a given bitumen into the vein, must have exceeded its melting point (or temperature of plastification, as the case may be). The following two sources of error should, however, be taken into consideration, before utilizing the organic phase as a geological thermometer: (1) Increased pressure may lower the melting points of those bitumens, which at atmospheric pressure lose volatiles at temperatures below the melting point, or partially melt with marked decomposition. This factor is non-existent in the paraffinites, and is relatively small even in the case of the bernalites as shown in Graph 3 following. (2) Post-mineralization polymerization, oxidation, and other processes in the bitumens may tend to increase the melting point; it would, however, be expected that such processes would be accompanied by shrinkage. It appears from the study of shrinkage cracks that post-mineralization contraction was well below 2% by volume in the case of the asphalts, asphaltites, mutabilites, elaterites and bernalites, which were utilized for the purposes of thermometry. The carbonites show, as a rule, 5% to 12% contraction in volume, which seems to explain the polymerization to their present infusible or partially fusible state.

It appears from the foregoing considerations that the present-day melting points of the "injected" bitumens may be only slightly higher than those of the original substances, which entered the vein in the course of mineralization. This indicates the possibility of utilizing the bitumens as indicators of the "minimum temperature" which prevailed in the course of mineralization. In order to check this possibility, bitumens were collected for the determination of melting points, from a series of veins of increasing fluorite and sulphide ore percentages, since these are generally interpreted (see 2, above) as indicative of increasing temperature of mineralization. The mean percentages of sulphide ores and fluorspar (calcite with some barite being the other gangues) were determined by means of the standard sampling procedures in veins exposed in working mines, disused mines and quarries. In the case of inaccessible mines, relatively less reliable data were taken from old mining records, or by sampling the tips, and the latter are shown in brackets in Table 1.

The melting point of ozocerite is well defined, that of the rest of the bitumens is somewhat more gradual, involving softening at first, followed by the formation of a liquid which may occasionally still contain flakes of semi-solid material. In order to form the "injection structures" the essential condition is that the bitumen should form a liquid with a well-defined, even meniscus. For the experimental work, 0·2 g samples of coarsely powdered material were heated at a rate of 5° per minute in a 3 mm diameter test tube placed in a sulphuric acid bath. The formation of a clear surfaced liquid could be determined in all cases to the nearest 10°, as shown in Table 1. It was also found in the case of control experiments on bernalite from the Glebe Mine, Eyam, that the melting point reveals no detectable change through changing the heating rate to 1° per minute, or making the experiment in a smaller capillary tube, although in the latter case it is more difficult to assess the temperature of formation of the liquid, because occasionally a semi-solid flake may reach the dimensions of the drooplet. It was found that on very rapid heating, or burning, bernalite fragmentates at temperatures considerably above its melting point on slower heating, due to an initial very highly polymerized state.

TABLE 1

Fluorite + Ore %	Locality	Type of bitumen	Melting point
0%	Waterhouses, Quarry, (Recryst. fossils)	Pittasphalt	40°
0%	Tissington, Quarry, small veins.	Pittasphalt	65°
1%	Wirksworth, Road cutting, small veins.	Albertite	90° partial
2%	Ecton Mine, side veins.	Pittasphalt	30°
(5%)	Ecton Mine, tips.	Asphalt	90°
6%	High Rake Mine, Matlock, side veins.	Elaterite	120°
11%	Bakewell, Chert Quarry, Veins.	Ozocerite	120°
12,5%	Glebe lead Mine, Eyam. Min. Breccia.	Bernalite	240°
(20%)	Ecton Mine, tips.	Asphaltite	180° partial
22%	High Rake mine, Matlock. Lode.	Bernalite	330°
(25%)	Odin Mine, Castleton. Lode and tips.	Bernalite	320°
29%	Winnatts Head, Castleton, Quarry, vein.	Bernalite	350°
35%	Pindale, Castleton, Quarry, Veinlets.	Bernalite	320°
45%	Windy Knoll, Castleton, Veinlet.	Bernalite	310°
71%	Windy Knoll, Castleton, Quarry, Flat.	Bernalite	370°

In the case of the "differentiated type deposits", the bitumen with the highest melting point from each vein is included in Table 1. It is rather unfortunate that the low fluorite and ore proportion veins of the "well-differentiated" type deposits, and the high fluorite and ore proportion veins of the "undifferentiated" deposits contain no fusible bitumens. However, all the data which could be collected so far clearly demonstrate the presence of relatively higher melting bitumens in the veins for which a relatively higher temperature of formation is postulated through quite independent considerations of the thermal zoning of the inorganic minerals of the hydrothermal deposits.

2. The "fractionation series" surrounding major mineral veins

It was found, as a general rule, as indicated earlier, that the "initial boiling point" (arbitrarily taken at 1% loss at a heating rate of 10° C per minute of vacuum dessicator dried material) decreases with the distance from the centre of any given hydrothermal lode. As the volatilization of bitumens depends greatly on pressure, which cannot be determined from I.B.P. data, the latter serve primarily for the reconstruction of thermal gradients on a relative scale. With the utilization of a geological manometer (see 3, following) absolute values of temperature and pressure can be obtained in certain vein systems.

The two vein systems which proved to be exceptionally well exposed were (a) a mineralized flat, the "bernalite flat" of some 8 m length and situated some 2 m beneath the surface, in the disused limestone quarry at Windy Knoll, which is 2 miles to the west of Castleton. (b) A lode, presumably a major side vein at High Rake, Matlock Bath, which was exposed in a new test pit some 10 m beneath the land surface. Both deposits contain relatively fresh bitumens, little affected by surface oxidation and escape of volatiles.

The "fractionation series" of the "bernalite flat" is rather unsymmetrical. Covering the flat is a lightly mineralized breccia mainly with calcite, which contains with increasing distance from the central mineralized zone, a maximum in percentage of a yellow oil, brown elaterite and foxite in turn; beneath the flat there are a few 1–3-cm thick calcitic veinlets in compact reef limestone. These contain mainly foxite, although traces of elaterite and oil are also present. The spectacular decrease of the I.B.P.s of the solid bitumens with distance from the main deposit are well shown in Graph 1. The liquid and semi-solid bitumens (oil, ozocerite and elaterite are shown by a broken line in the graph) reveal a rather less pronounced curve. This may be due to the diffusion of these bitumens after mineralization. It is, therefore, possible that the more extreme value for the thermal gradient surrounding the vein, given by the solid differentiates and semi-solid differentiates would be closer to the actual conditions which prevailed in the course of mineralization. The coincidence of the I.B.P.s of bernalite and carbonite indicate that the bitumens were formed under identical conditions.

Essentially the same conclusions can be drawn from the "fractionation series", illustrated in Graph 2, for the vertical lode at High Rake. In this case, hollow calcite and oil lined cavities appear in the veinlets of the compact grey limestone both to the east and the west of the lode. These seem to indicate (see Graph 2) escape of still more volatile constituents.

Both the Windy Knoll and High Rake deposits are of "high fluorspar and ore percentage", 71% and 22%, respectively. There are indications of the trend of reduction of I.B.P. of bitumens, with increasing distance

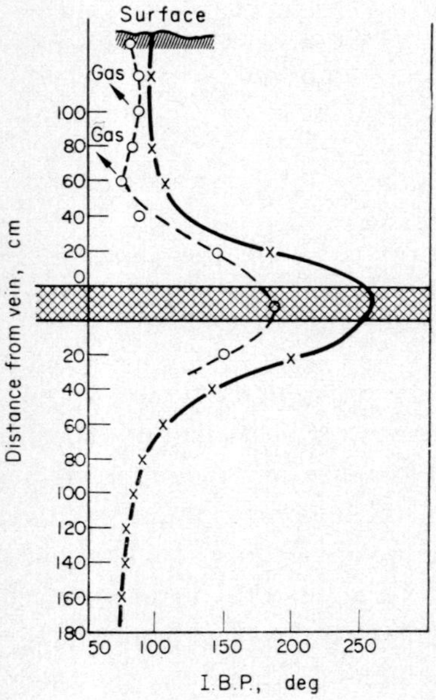

GRAPHIC 1

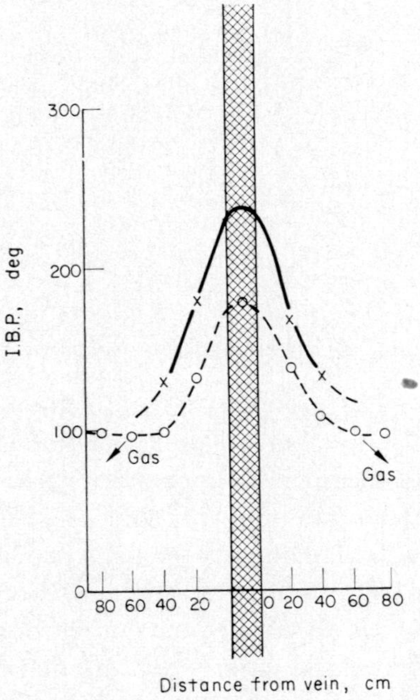

GRAPHIC 2

from the vein, in a considerable number of other deposits of varying mineral composition. Exact curves could not be plotted for these, however, because of the sporadic distribution of bitumens, and/or poor exposures. The data are summarized in Table 2.

It appears from the data given in Table 2 that the thermal gradients which surrounded the veins increase with increasing fluorspar and ore proportion, as expected theoretically. The sharp thermal gradients which are indicated from the interpretation of the "organic thermometers" seem to point to a close-to-surface mineralization with hydrothermal solutions with a fast rate of flow. Such indications, which cannot be obtained from any inorganic geological thermometers, are of considerable interest in the study of hydrothermal mineral deposits in general, which is outside the scope of the present paper.

TABLE 2

Fluorite + Ore %	Locality, type of deposit	Initial boiling point range in	
		Main vein	Side vein
2%	Windy Knoll, Castleton "Carbonite lode"	120–145°	85–120°
(5%)	Speedwell Mine, Castleton. Lode	240°	110°
(20%)	Ecton Mine, Staffordshire. Lodes	140–170°	100–120°
(25%)	Odin Mine, Castleton, Lode	230–280°	85–125°
80%	Treak Cliff, Castleton, Irregular veins	220°	100°

3. The utilization of Bernalite for the determination of the temperature and pressure which prevailed during mineralization

The presence of a "fractionation series" surrounding the "bernalite flat" of Windy Knoll (and other veins) seems to indicate that in the course of the highest temperature phase of the mineralization, the bitumens were in equilibrium with their volatiles, within the apparently quite open vein system. Although the longer rods of bernalite are at present in a fragmented state within the oily matrix of the vein (Fig. 1) the "extrusion structures" revealed by the shorter rods and droplets of the same material (Figs. 2, 3 and 4) clearly indicate that the bernalite and other bitumens entered the vein in a plastic state through narrow "extrusion channels" between the crystals of growing gangue minerals. These considerations point to the injection of the bernalite by the pressure of more volatile organic molecules, in a *plastic state*, and *in equilibrium with its volatiles*. At atmospheric pressure the I.B.P. of bernalite proved to be 100° below its incipient plastification. This indicates, as expected, that the mineralization occurred at elevated pressures. The coincidence of the plasticity range with the I.B.P. would indicate the likely pressure–temperature conditions, with minor sources of errors due to polymerization, etc., as discussed above.

In the course of experimental work, distillations of bernalite under reduced pressure were carried out using standard methods, and between 1 and 20 atm in an autoclave in a steam atmosphere. In the case of the high pressure experiments the I.B.P.s could not be observed visually, and a series of experiments were conducted with 10° increases of temperatures (heating for 10 min) until 1% loss of volatiles was reached, at a given pressure. The "plasticity range" was determined through microscopic inspection of the course powder after each successive heating to 10° higher temperatures. The definition of the "states" of the powder were as follows;

Solid state: no visible rounding of the edges of the fragmented pieces under the microscope, little or no cohesion of the particles of the 40-mesh powder.

Plastic state: rounding of edges of powder, which sinters into a coherent cake.

Liquid state: well-defined meniscus forms on the surface, irregular gaps between the original grains disappear, although bubbles persist as the material forms a viscous liquid.

It is seen from Graph 3, that the pressure at 20 atm, which could be reached with the autoclave used, was inadequate to produce coincidence between I.B.P. and the range of plastification. A rather short extrapolation, however, gives us for coincidence the range of *330° to 350° and 35 to 80 atm*. It is interesting to note that this is according to data on water

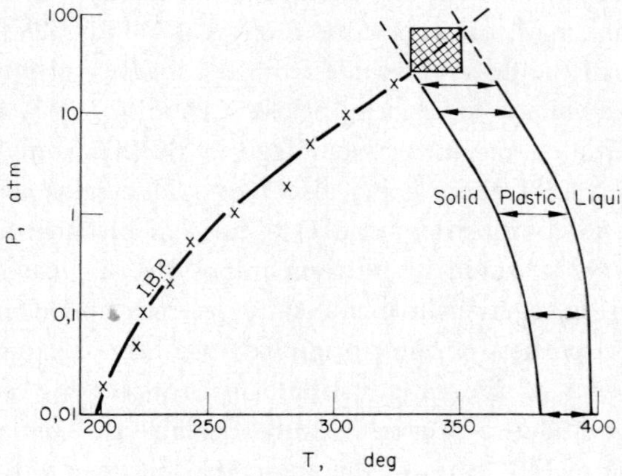

GRAPHIC 3

(see for example, F. G. Keyes, *Crit. tables* (1928), vol. 3, p. 233.) within the vapour phase. This indicates that, at least during the highest temperature phase of mineralization, the conditions must have been "pneumatolitic", in contrast to the "hydrothermal conditions" which are attributed, without any concrete proof, in the literature, for this type of fluorspar–galena–blende–pyrite deposit.

4. Chemical processes, and their interpretation

What we may term "unit processes", in organic chemistry, usually produce varying yields through a very considerable range of pressures and temperatures. The evidences of the occurrence of a given general reaction within the geological history of a given bitumen deposit, can therefore give only very approximate values for the likely conditions which prevailed in its geological history, particularly when considering the unknown time scale available for the reaction. For this reason the indications for certain reactions within the Derbyshire deposits, are only briefly discussed below.

(A) HYDROGENATION. The general increase of the mean H/C ratio of the bitumen assemblages with increase of fluorite and ore percentage of the veins within the "well-differentiated type" deposits seems to indicate strongly that hydrogen must have been added to the original fractions at relatively higher temperatures.

The most spectacular location which supports the hypothesis of "hydrogenation" is the small, old limestone quarry at Windy Knoll, Castleton. The H/C ratio of the de-ashed shale from the locality proved to be 0·51, that of the large calcitic lode with "brittle carbonite" 0·87, and the mean for the differentiated bitumen assemblage in the flat (only some 15 m. from the former) 1·79, there being also four minor veins of intermediate inorganic mineral composition, and H/C ratio of bitumen assemblage in this small quarry. It would be outright impossible to visualise a process which would bring together in such a small space bitumens of such divergent primary properties, be their origin sedimentary or magnetic. A gradual hydrogenation of aromatic distillates with increasing temperature of mineralization would, however, readily explain the remarkably well-defined increase of H/C ratio of the associated bitumen with the increase of fluorite and ore percentage of the vein.

The hydrogen could originate, according to the views of the writer, directly from coal distillation, and also through the industrially utilized reaction; $H_2O + CH_4 = 3H_2 + CO$.

Subsequently the hydrogen would be added to the aromatic coal-tar-type primary distillates, and in this connection it is interesting to note that one of the successful catalysts which are used in the Bergius process of coal hydrogenation is ZnS, a substance which is present in quantities of from 0·1% to 5% in the veins in question. The Bergius process reaches an optimum at 450° and 200 atm. These temperatures (and pressures) are not quite reached within the highest temperature "bernalite flat" at Windy Knoll, as indicated above. The fact that the temperatures of formation of the other deposits within the quarry of Windy Knoll are even farther from the optimum of the Bergius process, explains the progressively lower values for the H/C ratios of the bitumen assemblages. Another possible factor, beside that of temperature, in the lowering of the H/C ratio in the calcitic veins, may be the lower proportion of ZnS for catalyzing the reaction. In this connection it is interesting to note that the relatively sulphide-poor and fluorite-rich deposits between Windy Knoll and Castleton (Winnatts Head and Treak Cliff) contain less of the low O/H ratio oil and elaterite, and more of the hard varieties of high O/H ratio mutabilite.

(B) UNMIXING (DIFFERENTIATION). The process of unmixing of the main types of molecules in the case of the "differentiated type" deposits, seems to indicate polymerization under relatively high-temperature conditions, although the processes of condensation of organic molecules are so complex, that no likely range of temperature can be delineated for the occurrence of the phenomenon.

(C) ELIMINATION OF O AND N. The bitumens in the "fractionation series" and the low-temperature, calcitic vein of Windy Knoll contain an average of approximately 6·5% oxygen and 0·9% nitrogen. The bitumen assemblage in the mineralized zone of the presumably highest temperature "bernalite flat" contains 0·2% oxygen and below 0·1% nitrogen. The elimination of oxygen and nitrogen from the high boiling fractions within the central portions of the vein system, is quite in accord with general experience of crude oil distillation and cracking, and other distillation processes in industry. It should be noted in this connection that all the bitumen assemblages of Windy Knoll, and the whole North Derbyshire ore field in general, contain an approximately constant amount of 1·2% sulphur, and some industrial experience indicates that the sulphur is not eliminated from the lighter fractions under the usual conditions of distillation.

(D) ABSORPTION OF URANIUM. It was found by the writer[3] that within the Windy Knoll and adjacent bitumen deposits the U-equivalent of the radioactivity of bitumens: (A) is much lower in the shale bitumen than in the distilled ones; (B) it increases somewhat with the temperature of the vein; (C) it increases with the O/H ratio of differentiates from the same location; (D) in the case of the hitherto investigated bernalite rods, it is about five times greater in the 0·5 mm surface crusts than in the interior. All these, and several other evidences discussed by the writer seem to indicate than the bulk of the uranium became absorbed into the bitumens during mineralization. The absorption of uranium in marine sediments is well known, and it is not therefore an indication of elevated temperature. The slight increase of uranium in bitumens from higher temperature veins may be due to the fact that the mineralizing solutions in question were richer in the metal.

(E) CARBOXILATION. The significance of metabitumite must be briefly discussed. This bitumen is found in the form of small droplets within limestone, which surrounds the "Speedwell vent", Castleton, and some other intrusions. The droplets were found to be restricted to within 20 yd distance from the igneous rocks, and the field evidence indicates that they must be condensed distillation products of the intrusion. The substance has a high density range, between 1·52 and 1·68. It shows stronger graphitic spacings than the carbonites in X-ray powder diagrams, yields only water on distillation, it has less than 10 ppm uranium, and it is insoluble and not fusible. Its non-fusibility precludes its utilization for the purposes of physical thermometry. The quite anomalously high O/C ratio of 0·294 may, however, prove to be significant. It is conceivable that in the course of the traversing of the limestone by the intrusive, much CO_2 would form, and that this would carboxylate the original, mainly aromatic, coal-tar-type distillates. In this connection it is interesting to note that similar essentially non-radioactive substances from the proximity of intrusions from Karelia, U.S.S.R., namely the shungites, also have an anomalously high O/C and O/H ratio. Carboxylation has been used so far in industry and in the laboratory for the introduction of the acid group into well-defined molecules, and not carbonaceous complexes. The data available indicate that the optimum of the reaction in general would lie in the high-pressure and medium-temperature field, which would be expected in the case of contact effects of the rather shallow intrusions in question, when the pressure would be increased through the liberation of CO_2.

5. Reconstruction of the geological history of the Bernalite flat, Windy Knoll

In the course of describing the examples given so far, attempts have been made to reconstruct the highest temperature stage in the geological history of the location in question. This is because in the same way as the inorganic geological thermometers, the carbonaceous complex tends to "freeze in" the equilibrium or semi-equilibrium reached at the highest temperature. In certain instances, however, detailed study of what may be termed regressive metamorphism of the carbonaceous complex may unveil further details in the geological history of the deposit in question. The bitumen assemblage within the "bernalite flat" of Windy Knoll appears to reveal several rather striking features, and in what follows an attempt has been made to reconstruct the history of this locality. Many of the features exist within other deposits also, and for this reason the work is intended primarily as an illustration in the exploration of possibilities.

The following phases can be distinguished in the mineralization and post-mineralization history of the deposit:

PHASE 1. Deposition of the first generation of small calcite crystals with marcasite inclusions. It is seen within a smaller pocket to the north of the main flat that the deposition of this generation was simultaneous with the injection of the differentiates (see Figs. 2, 3 and 4). In the main flat it appears that the extruded olefinite was partially dislodged from the original extrusion channels. It had been partially expelled through the extrusion of other differentiates from the same channel (as shown in Figs. 2 and 3), as is indicated by the presence of rods which show fused endings on both extremities (Fig. 1).

PHASE 2. Considerable mechanical agitation is indicated following the extrusion of bernalite, when the latter was still in a plastic state. Dilation of the vein explains, for instance, the 7–8 mm diameter rods ending in 0·5 mm diameter and up to 5 cm long threads.

PHASE 3. All the first generation bernalites, injected during phase 1 and 2, became covered with highly carbonized crusts (Fig. 5). This must have been caused by an event of brief duration, such as increase of temperature or release of pressure.

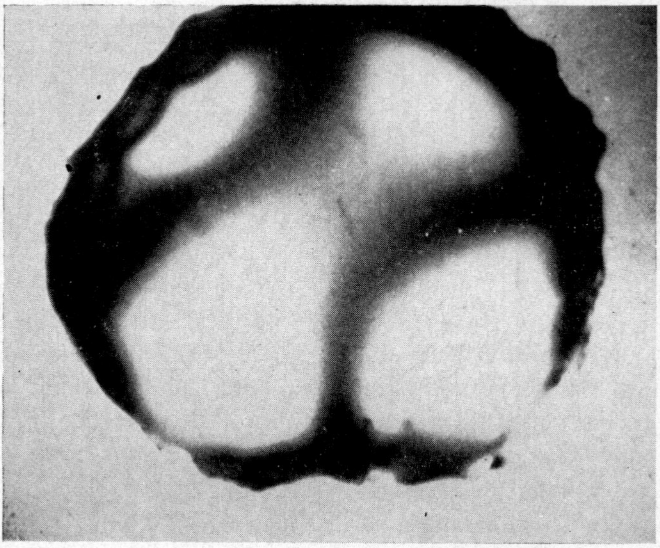

FIGURE 5. — Transverse section of a bernalite rod, which shows carbonized surface zone and intensive strain anomaly. Polarized light, × 18.

PHASE 4. A second generation of bernalite entered the vein, the rods and droplets of which do not have the black surface crust.

PHASE 5. A second generation of inorganic minerals namely marcasite, blende and galenite, deposited on the bernalite rods, fragmented pieces and solidified puddles of carbonite. On the surface of some snapped rods this crust of 1·5 mm diameter is considerably thinner, indicating that their fragmentation occurred in the course of this marcasite phase of mineralization.

PHASE 6. Mechanical disturbances fracture further carbonite and bernalite. These younger fractures are not overgrown with marcasite, but are covered with 0·1–0·5 mm films of a differentiate of higher H/C ratio than that of the fragmented surface below. Thus fragments and even vesicles within carbonite are lined with bernalite films, which often reveal "tack pattern" as if the surfaces covered with tacky bernalite had been forced asunder (Fig. 6). On the broken bernalite rods alyphite, a soft light yellow resin of H/C = 1·76 appears, which shows "bubble patterns" evidently indicating the escape of gases (Fig. 7). It appears that they were

FIGURE 6. — "Tack patterns" of bernalite on a fragmented surface of resinous carbonite. Reflected light, ×30.

FIGURE 7. — "Bubble pattern" of a film of alyphite on a snapped rod of bernalite. Reflected light, × 40.

separated from the bitumen beneath them through some distillation process which was more gradual than the brisk heating which caused the surface carbonization of phase 3. The higher H/C ratio of the exuded film may indicate the continuation of hydrogenation.

PHASE 7. A third generation of large calcite, fluorite and barite crystals formed in which the bernalite rods, fragments of carbonite and oil droplets became included (Fig. 8). The remarkably freely grown habits of

FIGURE 8. — Freely grown calcite crystal with overgrown, marcasite covered rods of bernalite, inclusions of oil and fragmented carbonite. Reflected light, $\frac{3}{4}$ nat. size.

these crystals indicate the presence of the viscous oil, at the phase in question, presumably in emulsified form.

PHASE 8. Recent oxidation of the marcasite caused absorption of SO_2 by the olefines of the elaterite and oil, and this is released on distillation.

6. Search for thermal effects of mineralization and faulting within the Lota Coalfield of Chile

Some studies were carried out on the bituminous coalfield of Eocene age, with the main object of determining the effects of faulting, which cannot be readily done within the Derbyshire area, which contains no coal.

It was found that an 8–cm broad purely calcitic vein, which traverses the coal, contains no discrete bitumens, although some brownish coloration of the crystals of calcite may be due to diffused organic matter. The total volatiles of the coal bordering the vein proved to be 35·02%, well within the range of average coal from the coal measures in question. This seems to indicate, that the deposition of this calcitic vein occurred at temperatures below those for the distillation of the coal, perhaps below 150°, since the heating curve of the coal at 1 atm is as shown in Table 3.

TABLE 3

Temperature	Loss	Total loss	Type of volatiles
105°	1·84%	1·84%	Water
200°	0·91%	2·75%	Water
300°	0·83%	3·58%	Water
350°	3·59%	7·17%	CH_4, etc.
400°	4·06%	11·23%	

An exposure of a fault of some 3 m throw, which displaces the third coal seam of the Schwager mine, was studied in some detail. Five samples from different distances from the fault were heated to 800° for half an hour, with the results shown (Table 4).

It appears from Table 4 that the volatiles in the coal have not been detectably affected by the fault, indicating the essentially negative result, that in the course of the faulting, the temperature did not exceed, at least in the case of the fault of small throw in question, the temperature of initial devolatilization of the coal.

TABLE 4

Distance from fault	Ash	Total volatiles, ash-free basis
0–1 cm	26·03%	35·30%
10–12 cm	17·76%	34·89%
50–55 cm	1·09%	35·25%
Approx. 2 m	2·07%	35·01%
Approx. 100	1·00%	35·15%

7. The possibilities of assessing temperature at different phases of mineralization of the same vein

Preliminary work on geological thermometry on a considerable range of bitumen deposits is being carried out and the results of this are going to be published in the future. At present, only some findings on inclusions of bitumens in quartz crystals from a vein deposit of the Precambrian basement of SSW. Africa will be briefly mentioned, because these specimens reveal yet another potentiality of the method.

The material originates from veins which traverse granite. There is, therefore, a strong probability that the inclusions of bitumens are abiogenic magmatic condensates. The inclusions (up to 3 mm long) contain a vaselinous oil, which produces crystals with paraffinic optical properties when frozen, some dark brown resinous flakes (possibly olefinic differentiates), and black globules, which appear to be probably aromatic differentiates; almost all the twenty or so inclusions contain gas bubbles. It was found that the bubbles immobilize, indicating the solidification of the main paraffinic phase, at different temperatures within different growth zones within the crystal. A dark brown, orange fluorescing liquid inclusion "freezes" at $+11°$, and this is within the deepest growth zone of the crystal; a greenish-blue fluorescing inclusion, which is the closest to the surface, freezes at $-23°$, and inclusions at intermediate levels freeze between the above two extremes of temperature. The interrelations between depth and melting point can be explained when supposing that the crystal formed under generally cooling conditions of the mineralizing solutions. The I.B.P.s of the diverse inclusions may give the actual temperatures at 1 atm, and the pressure may be assessed from the gases present in the inclusions. This work will be carried out in the future.

5. Some theoretical considerations

According to the results of field observations and experimental work described in this paper, it appears that within certain localities of the Earth's crust, carbonaceous complexes are chemically altered and redistributed at elevated temperatures and pressures. The study of metamorphism of the carbonaceous complex was restricted so far to readily recognisable localized agencies, namely mineral veins, intrusives and faults. It is indicated by the field studies that bitumens associated with such structures are not rare, particularly so far as hydrothermal veins are considered. All the same, it appears that if carbonaceous complexes with elevated temperatures of origin would be restricted to such structures, then the volume of these complexes within the Earth's crust would be restricted to "museum curiosity level".

In considering the likely importance of high-temperature organic geochemical processes, it should be stressed, however, that not all more or less localized rises of temperature above the geothermal gradient are necessarily marked by the well-recognizable features of mineral veins, intrusives or faults. Localized flows of heat which were not accompanied by substantial quantities of transported matter or displacements are indicated by zones of metamorphism of inorganic sediments within innumerable locations of the Earth's crust. It is generally estimated that the inorganic minerals show detectable thermal change only at temperatures exceeding about 500° C. Fundamental organic chemical changes do occur, however, at considerably lower ranges of temperatures, particularly in the "geological time scale", say between 200° and 300°, under which the localized heating event cannot be any longer detected through the alteration of clay minerals, etc. There must be in addition to the numerous areas of "inorganic metamorphism", a very considerably greater number and volume of "organically metamorphosed" localities within the Earth's crust.

It seems to the writer that the sporadic distribution of oil fields within a huge volume of sapropelic sediments, could be fully or partially accounted for, by the hypothesis of production of petroleum by localized metamorphic agencies. It appears from the results of field investigations in Derbyshire that petroleum can be produced either through low-temperature hydrothermal metamorphism of kerogen-type sediments, or through more intensive alteration of coal-type sediments. Trace element distribution studies may, possibly in the future, reveal locations of metamorphism within the range of alteration of the carbonaceous complex. As an alternative to this, criteria could be eventually worked out, which would enable us to distinguish between the alteration products of distinct types of sediments at different temperature (and pressure) ranges. Besides its theoretical interest such knowledge may increase our power of selectivity in the prospecting for petroleum and other economically important bitumens. It appears to the writer that at first, systematic studies of the molecular (and elemental) composition of those carbonaceous complexes for which a definite elevated temperature origin is indicated should be undertaken through the utilization of the "organic geochemical thermometers" on the lines of the present research, and the properties of the primary carbonaceous sediments should be carefully determined. Next, carbonaceous sediments, etc., should be studied, in the geological setting of which, the possibility of localized metamorphic effects are least indicated. In case consistent differences should be found between the presumably high- and low-temperature carbonaceous complexes then the results of the work could be applied for the determination of temperature of origin of those deposits which would have less certain geological clues as to the thermal history. It appears from the results of the present work, that oils of asphaltic base are of lower temperature of origin than oils of paraffinic base. The presence of olefinic molecules and the phenomenon of "differentiation" are characteristic of high-temperature conditions of formation, similarly to high M.P.s and high I.B.P.s of the bitumens. The above are, of course, only preliminary observations. A great number of other possibilities are open in the future in the determination of more detailed molecular composition through chromatography and mass spectroscopy, determination of even-odd preference of diverse paraffins and alkenes, the analyses for functional groups through I.R. and U.V. spectra, etc.

Finally, some thought may be given to the consideration of the mode of formation of amino-acids and polypeptides, which are the prerequisites for the emergence of life. From the experimental results of S. W. Fox[14] the conclusion may be drawn that conceivable processes at elevated temperatures do lead to the formation of polypeptides, whereas this is not the case for low temperatures. The present work seems to accentuate the role of elevated temperature processes in organic geochemistry in general, indicating that there must have been abundant opportunities on the surface of our planet for such high-temperature amino-acid polymerizations.

I would like to express my thanks for a stimulating conversation on the subject with Prof. S. W. Fox of the University of Miami, and Prof. J. D.

Bernal of Birkbeck College, London. This work was carried out with the financial support of NASA grant R 7493R, the Nuffield Foundation, and the University of Concepcion, Chile. It is contribution No. 117 of the Institute of Molecular Evolution, University of Miami, Coral Gables, Florida, U.S.A.

References

1. G. Mueller, A genetical and geochemical survey of Derbyshire mineral deposits. Ph.D. Thesis, University College, University of London, 1951.
2. G. Mueller, The theory of genesis of oil through hydrothermal alternation of coal type substances within certain Lower Carboniferous strata of the British Isles. *Compt. Rend. XIXe Session, Cong. Geol. Internacional*, Alger, Sect. 12, Fasc. 12, p. 279, 1954.
3. G. Mueller, Distribution of Uranium in naturally fractionated organic phases. *Proc. Int. Geol. Congress*, XXIth Session, Norden, Part 15, p. 123, 1960.
4. G. Mueller, Evidence of abiogenic hydrocarbons. *Nature*, **198,** 731, 1963.
5. J. Shirley, and E. L. Horsefield, The Carboniferous Limestone of the Castleton–Bradwell area, North Derbyshire. *Quart. J. Geol. Soc.* **96,** 271, 1940.
6. D. Parkinson, The stratigraphy of the Dovedale area, Derbyshire and Staffordshire. *Quart. J. Geol. Soc.* **105,** 242, 1950.
7. S. I. Tomkeieff, On the volcanic complex of Calton Hill. *Quart. J. Geol. Soc.* **85,** 703, 1928.
8. F. Wolverton Cope, A Thoilitte dike near Buxton, Derbyshire. *Geol. Mag.* **70,** 414, 1933.
9. C. B. Wedd, and C. Cooper Drabble, The fluorspar deposits of Derbyshire. *Trans. Inst. Min. Eng.* **35,** 301, 1908.
10. G. W. Wilson, and co-workers, Barite and Whitherite, *Spec. Rep. Min. Res. Great Britain*, p. 64, 1922.
11. G. Mueller, The distribution of coloured varieties of fluorites within the thermal zones of Derbyshire mineral deposits. *Compt. Rend. XIXe Session Cong. Geol. Internacional*, Alger, Sect. 13, p. 523, 1954.
12. G. Mueller, The properties of extraterrestrial hydrocarbons and theory of their genesis. *Proc. VIth World Petroleum Congress*, Frankfurt-am Main, Sect. 1, Paper 29, PDl, p. 1, 1963.
13. P. C. Sylvester-Bradley, and R. J. King, Evidence for abiogenic hydrocarbons. *Nature*, **198,** 728, 1963.
14. S. W. Fox. A theory of macromolecular and cellular origins. *Nature* **205,** 328, 1965.

POSSIBLE TAXONOMIC INTERPRETATION OF SOME PALAEOZOIC AND PRECAMBRIAN CARBOHYDRATE RESIDUES

by

F. M. Swain, G. V. Pakalns and J. G. Bratt

Department of Geology and Geophysics
University of Minnesota, Minneapolis

Introduction

Although carbohydrate residues have been reported from Palaeozoic rocks (Swain, 1958; Palacas, 1959; Palacas *et al.*, 1960; Rogers, 1965; Swain, 1966; Swain and Rogers, 1966; Swain *et al.*, 1967) and possibly from Precambrian rocks (Swain *et al.*, 1958), little attempt has been made to identify the residues with specific taxonomic groups of organisms. A comprehensive review of the geochemistry of carbohydrates up to 1962 was provided by Vallentyne (1963).

It would be of interest to know whether the carbohydrate residues in Palaeozoic and older geological materials came from marine or non-marine sources, whether from plants or animals and what phylum or lower rank taxon is represented in individual fossiliferous specimens. Although the difficulties in such determinations are numerous it is believed that a few suggestions can be made based on present as well as some earlier studies.

Methods

Recent and fossil samples were obtained from several localities, the details of which are available from the authors. The recent samples were stored in deep freeze; the marine algae from Dillon Beach were freeze-dried. The rock samples were cleaned, dipped in chromic acid solution to degrade possible surface organic contaminants, washed in distilled water, crushed and powdered and stored in glass or polyethylene containers.

Total carbohydrate determinations were made in triplicate using the phenol-sulfuric colorimetric method of Dubois *et al.* (1956). Absorption spectra were measure of the colored products obtained by the reaction of phenol and furfurals to learn whether hexoses (absorption maximum 490 mμ) or pentoses (absorption maximum 480 mμ) were predominant. One to two-gram samples of Recent specimens and 10–20-g samples of fossil and rock samples were analyzed for total carbohydrates.

Individual monosaccharides were analyzed by treatment of 1–2 g Recent specimens and 500–1000-g fossil or rock samples with 0·5 N H_2SO_4 at the boiling temperature of the mixture for 8 to 10 hours. Carbonaceous material in a few rock and fossil samples was concentrated by adding kerosene to an aqueous suspension of the ground rock and stirring with a magnetic stirrer. The carbonaceous residue which separated into the kerosene phase, amounting to 10–20 g, was washed with acetone and with distilled water and treated with 0·5 N H_2SO_4. The sulfuric acid hydrolyzates were neutralized with $BaCO_3$, desalted on ion exchange columns and separated by paper chromatography as described earlier (Swain and Rogers, 1966). Butanol: acetic acid: water (4:1:5 v/v) and aniline: phthalic acid anhydride: water-saturated butanol (0·9 g: 1·6 g: 100 ml) were the developing solution and spray reagent, respectively. Some of the samples were separated on thin-layer sheets (Eastman Type K 301 R, Silica Gel), using butanol: acetone: water (4:5:1 v/v) or n-propanol: ethyl acetate: water: 25% ammonia solution (5:1:3:1 v/v) as developing solutions and the same spray reagent as above. The unknown chromatograms and known co-chromatographed mixtures were scanned in a recording densitometer equipped with an integrator.

Enzymatic determinations of D-glucose and D-galactose were obtained with Worthington glucose-oxidase and galactose-oxidase products. The photodensity of the resulting colored solutions was read at 420 mμ in a spectrophotometer and the values compared with those of standards for quantitative determinations.

Several of the acid hydrolyzates of fossils that showed little or no carbohydrate material by the other methods of analysis were separated, as their chloromethyldimethyl silyl derivatives, on gas chromatographic columns provided with an electron-capture detector.

Amino sugars were detected on amino acid chromatograms of HCl extracts of sediment samples run on a Phoenix amino acid analyzer.

In a recent paper Myers and McCready (1966) discussed the penetration of several kinds of rock by the bacteria *Serratia marcescens*. Under controlled conditions, at atmospheric pressure, p^{32}-labeled *marcescens* was able to penetrate dry Berea Sandstone having 100–300 m.d. permeability to a depth of at least 14·25 in. Samples in which the bacteria could penetrate at least 3 in. included Mississippian limestone of less than 0·1 m.d. permeability, Early Devonian limestone of 1·7 m.d. permeability and Late Mesozoic sandstone of less than 0·1 m.d. permeability. Their study was made primarily to demonstrate that micro-organisms reported from ancient rocks could be of recent origin (Lipman, 1928; Gruner,1922; Barghoorn and Schopf, 1965), but merits consideration also by the organic geochemist. Although care was taken in the present study to select unweathered impervious shales and limestones, many of which released H_2S gas on being broken, there remains the possibility of such contamination as that discussed by Myers and McCready. Dr. Samuel Kirkwood and associates of the University of Minnesota Biochemistry Department are now repeating the bacterial experiments on a Jurassic core from 11,000 ft in East Texas supplied by the authors. If bacterial penetration proves negligible into the core, it will be studied for carbohydrate contents.

Carbohydrate components of some living materials of geological importance

To provide a comparison of fossil and Recent carbohydrate components, analyses of Recent specimens made previously in this laboratory have been assembled (Rogers, 1962, 1965a; Swain, 1966) and additional analyses were made. The following groups of analyses are presented below: (1) Freshwater algae from Minnesota, (2) Marine algae from California, (3) Freshwater and terrestrial higher plants from Minnesota and California, (4) Marine invertebrates from California.

Freshwater algae from Minnesota

Total carbohydrate contents of freshwater algae were given by Rogers (1965), Fogg (1956) and Hough *et al.* (1952). The monosaccharide components of algae from two lakes in central Minnesota determined by Rogers (1965a) are shown in Fig. 1, plotted as per cent of total acid-hydrolyzable carbohydrates. Glucose and galactose predominate over arabinose and xylose in the lake-algae samples. Of the four principal algal polysaccharides: agar, carageenan, alginic acid and laminaran (Pigman *et al.*, 1957, p. 686) agar and carageenan contain important quantities of galactose, and the carageenan of a marine algae *Ascophyllum* contains

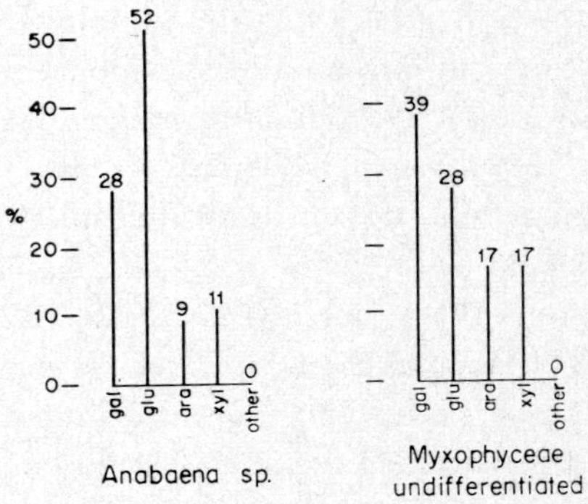

FIGURE 1. — Monosaccharide components of freshwater algae from Minnesota.

xylose (Conway and Young, 1966). The arabinose content of the algal polysaccharides is not known, but is small in the marine algae analyzed here, as noted below. Pectins may be the source of the arabinose in the freshwater algal samples from Minnesota.

Marine algae from California

The total carbohydrates contents of some marine algae are given in Table 1; other analyses are given by Bidwell *et al.* (1952). The monosaccharides separated chromatographically from acid hydrolyzates in several marine algae are shown in Figs. 2a and 2b in per cent of acid-hydrolyzable sugars. The carbohydrates of the chlorophyte *Ulva lobata*, unlike the other marine algae studied, contain a large percentage of rhamnose and about twice as much glucose as galactose. Another chlorophyte, *Codium* sp., however, contained galactose and glucose in 3:1 proportions, and no other sugars. The phaeophytes *Pelvetiopsis* sp., *Laminaria* sp. and *Egregia menziesii* contain relatively a larger percentage of ribose than the other marine algae and variable amounts of other sugars. The *Rhodophyta* are characterized by general predominance of glucose and galactose over other monosaccharides each of which, except for the ribose of *Gracilaria sjoestedii*, are less than 10%. Fucose was not determined separately in our analysis; it may have chromatogrammed with xylose.

TABLE 1. — TOTAL CARBOHYDRATE MOISTURE AND ASH CONTENTS OF LIVING MARINE PLANTS AND ANIMALS FROM CALIFORNIA (SAMPLES COLLECTED FROM DILLON BEACH, MARIN COUNTY, AND MUSSEL BEACH, SONOMA COUNTY)

Species	Moisture %	Ash %	Glucose equivalent mg/g, dry wt., ash free
Plants			
Pelvetiopsis sp.	64	25·3	275
Laminaria sp.	76	31·3	143
Ulva lobata (Kutzing)	69	33·5	192
Coralline red alga	62	56·0	273
Red alga	54	24·5	830
Prionitis sp.	76	18·9	795
Codium sp.	53	21·4	467
Gigartina sp.	50	46·0	
Zostera marina Linn.	70	22·5	488
Animals			
Mytilus californianus Conrad	20	13·0	95
Mitella polymeris (Sowerby)	82	19·7	53
Anthopleura xanthogrammica (Brandt)	78	21·5	179
Cribrina sp.	82	24·1	712
Tegula funebralis (Adams)	80	22·2	325
Pisaster sp.	79	22·2	51

Aquatic higher plants from Minnesota and California

Table 1 gives the total carbohydrates contents of *Zostera marina* specimens, and Bidwell *et al.* (1952), McNair (1952), Rogers (1965) and Swain (1967) gave analyses of non-marine species. The carbohydrate components of some higher plants, both marine and non-marine forms, are shown in Figs. 3(a)–(c). Xylose, together with glucose and galactose are the predominant monosaccharides in the plants studied. Rhamnose is also an important constituent of the marine pondweed, *Zostera marina*, but is absent or nearly so in the freshwater pondweeds. Arabinose on the other hand is an important element in the freshwater pond weeds but not in *Zostera*. The polysaccharides of the higher plant specimens were not studied.

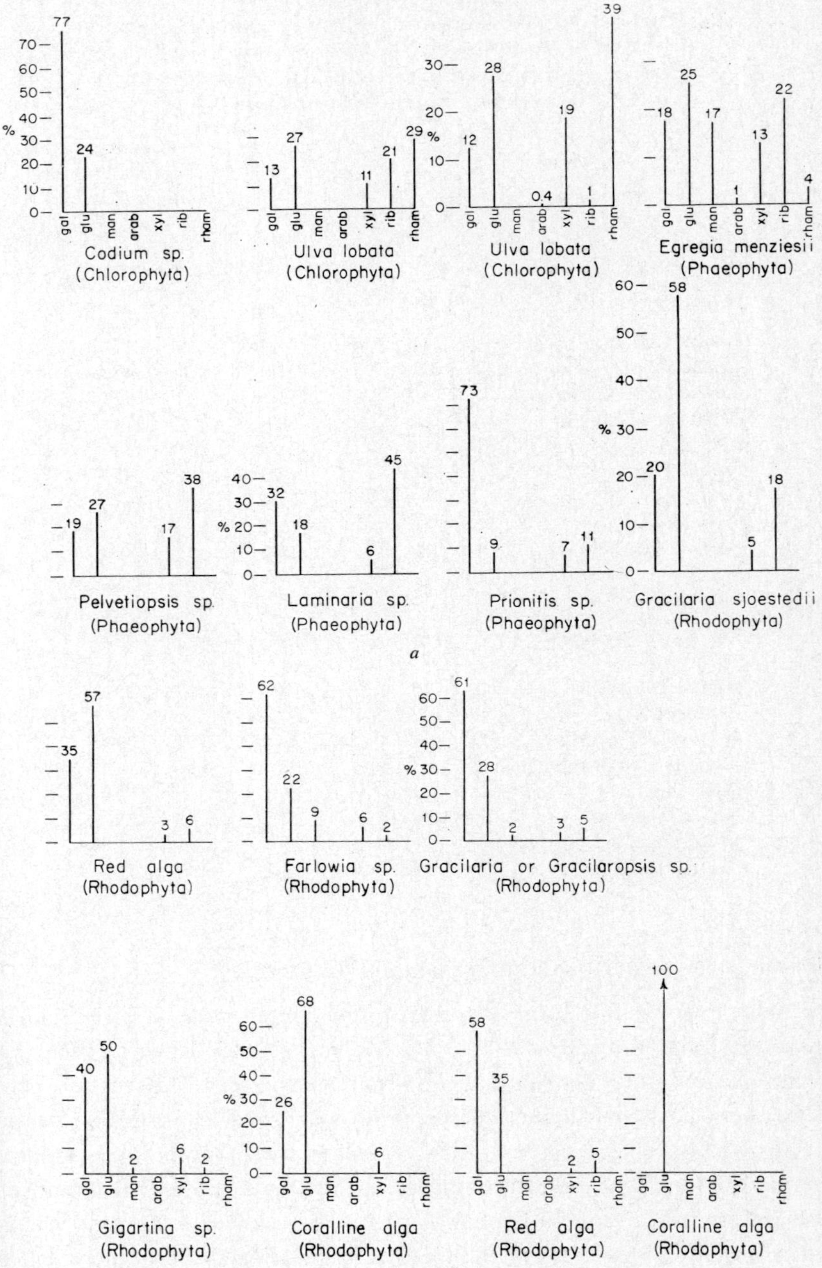

FIGURE 2. *a* and *b* — Monosaccharide components of marine algae from California; the values for xylose may include some L-fucose.

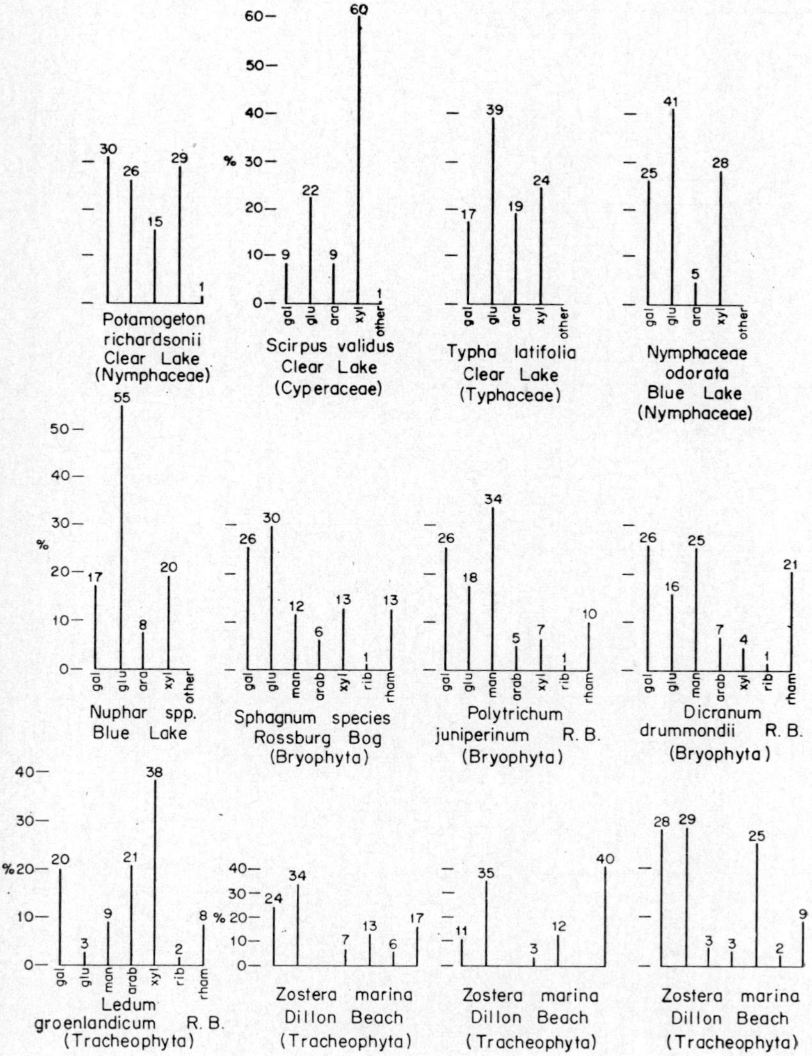

a

FIGURE 3. *a* — Monosaccharide components of higher plants from Minnesota and California.

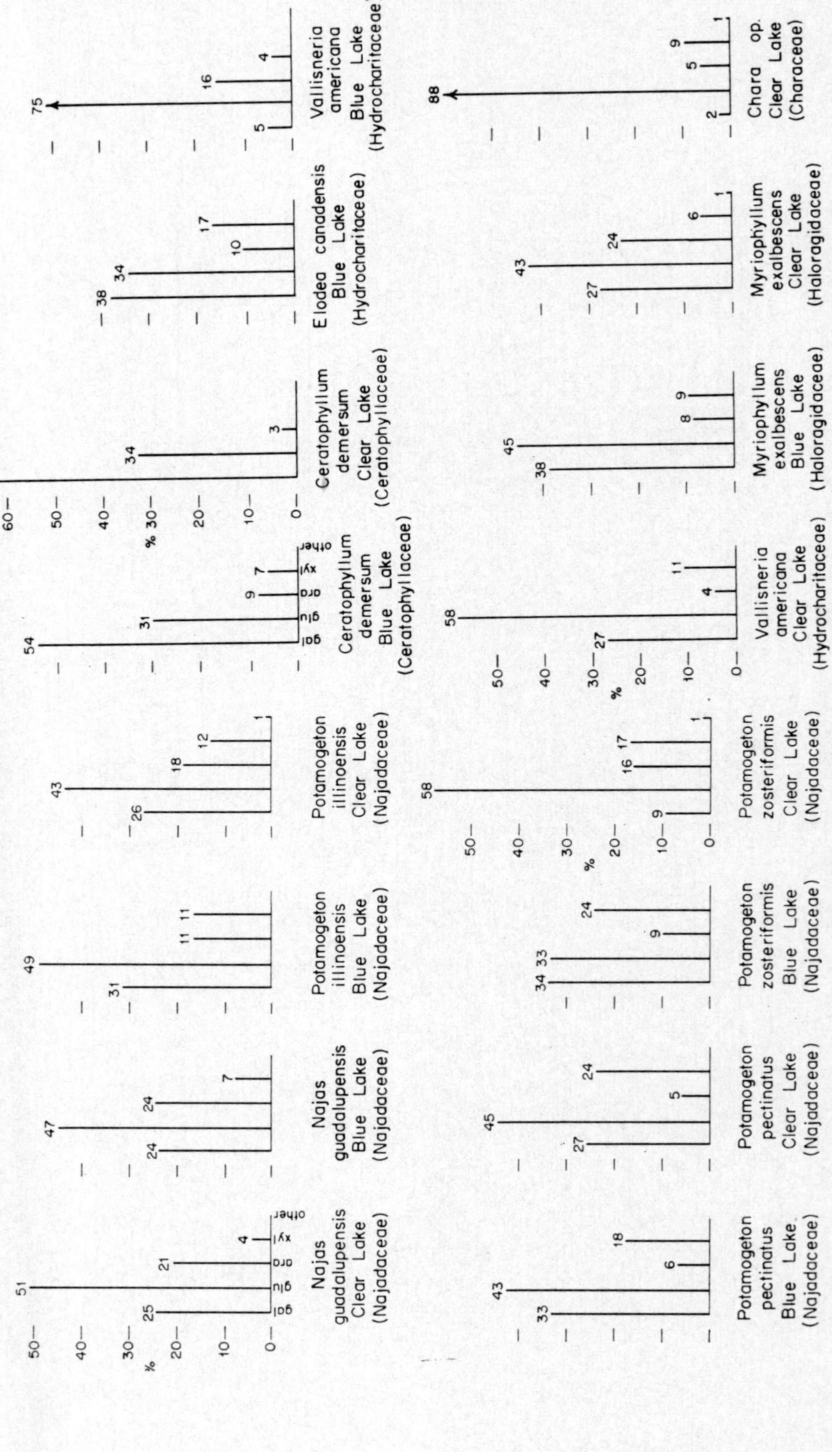

FIGURE 3. *b*, *c*, — Monosaccharide components of higher plants from Minnesota and California.

Marine invertebrate animals from California

The total carbohydrate content of several marine invertebrates is given in Table 1. The carbohydrate components of several marine invertebrates are shown in Fig. 4. Glucose predominates over the other carbohydrates

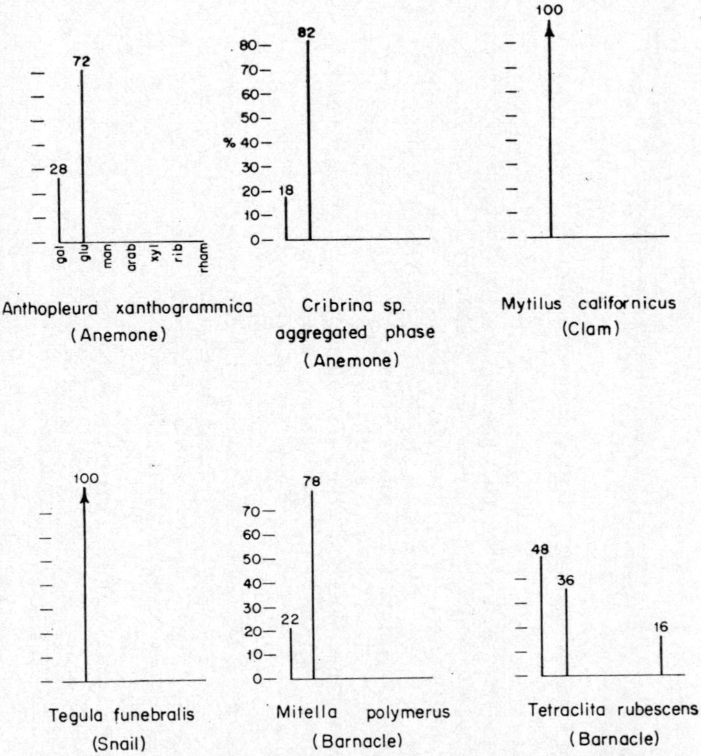

FIGURE 4. — Monosaccharide components of marine invertebrates from California.

or is the only constituent except in one of the barnacles where galactose is the predominant sugar. Ribose also occurs prominently in the barnacle. The polysaccharides of the animal specimens were not investigated.

Total carbohydrate residues in Precambrian rocks

The total carbohydrate contents of several samples of Precambrian carbonaceous rocks from Minnesota, Ontario and Victoria Island, N.W.T. are shown in Table 2. The samples range in age from early Precambrian

TABLE 2. — TOTAL CARBOHYDRATE CONTENTS OF PRECAMBRIAN CARBONACEOUS ROCK SAMPLES FROM NORTHERN NORTH AMERICA

Formation and age	Locality	Description and no. of samples	Total carboh. ppm	Abs. max. mμ
Coutchiching, Early Precambrian	Near Coutchiching, Ont.	Dark green quartz-chlorite schist (3)	6–14	488
Soudan, Early Precambrian	Lower Minnesota	Graphite schist (2)	7·5–8	488
Thomson, Middle Precambrian	Carlton, Minnesota	Dark gray argillite (1)	25[1]	486
Cuyuna, Middle Precambrian	Crosby, Minnesota	Greenish-gray argillite (1)	16	486
Biwabik, Intermediate Slate Member, Middle Precambrian	Babbit, Minnesota	Dark gray argillite (2)	0-trace	488
Biwabik, Middle Precambrian	Virginia, Minnesota	Algal structures in ferruginous chert (1)	trace	490
Rove, Middle Precambrian	Gunflint Lake, Minnesota	Dark gray argillite (2)	0-trace[1]	490
Rove, Middle Precambrian	Thunder Cape, Ont.	Carbonaceous argillite, near basic igneous intrusion (2)	15	484
Rove, Middle Precambrian	Pigeon Point, Minnesota	Evenkite (solid hydrocarbon $C_{21}H_{44}$) masses in algal structure chert (2)	0-trace	490
Wynniatt, Late Precambrian or Cambrian	Victoria Island, N. W.T., Canada	Dark gray argillite (1)	9	488
Killian, Late Precambrian or Cambrian	Victoria Island, N.W.T., Canada	Dark gray argillite (2)	0	—

[1] Presence of D-glucose verified by enzymatic test.

(Coutchiching and Soudan) to late Precambrian or Cambrian (Victoria Island samples).

No coarsely-particulate carbonaceous matter was seen in any of the samples, and most of the organic matter is apparently either in the form of humic acids or occurs as finely divided particulate material.

Several of the Precambrian samples were tested for monosaccharides by paper chromatography and enzymatic reactions as discussed below.

Total carbohydrate contents of some palaeozoic fossils and carbonaceous rocks

The total carbohydrate contents of a variety of rock and fossil specimens were reported previously (Swain, 1966; Swain and Rogers, 1966; Swain, Rogers, Evans and Wolfe, 1967).

Examples of the total carbohydrate contents of several Middle Ordovician and Upper Silurian samples from central Pennsylvania, not previously published, are given in Table 3. The samples were selected as representative of nearshore mud flat sedimentary environments, and contain finely divided organic debris, presumably consisting of detrital shore vegetation. As noted previously (Swain and Rogers, 1966) for unmetamorphosed Devonian dark shale and limestone samples, a range of 50–100 ppm of total carbohydrates is typical.

The amount of material extractable with 4% NaOH that would react as carbohydrates in the phenol-sulphuric acid test is shown in the "hemicellulose" column in Table 3. This varies from about a third to more than a half of the total carbohydrate contents of the samples. The presence of polysaccharides other than cellulose, starch and glycogen is suggested by the hemicellulose content (Vallentyne, 1963, p. 461). The occurrence of glucuronic acid, a component of hemicellulose, has been noted in several Devonian samples by Rogers (1965b). Hemicellulose fractions obtained by 5% NaOH extraction of Miocene silicified wood have, on hydrolysis, yielded several monosaccharides and polysaccharides (Jones and Merler, 1956; Brasch and Jones, 1959). It appears that further carbohydrate study should be made of alkali-extracts of geological materials in spite of the instability of monosaccharides in alkaline solution in laboratory preparations.

Samples of *Sigillaria approximata* Fontaine and White from the basal Greene Formation, Lower Permian of southwestern Pennsylvania were tested for total carbohydrates with the results shown in Table 4. The

TABLE 3. — TOTAL CARBOHYDRATES AND HEMICELLULOSE CONTENTS OF EARLY PALAEOZOIC ROCKS FROM APPALACHIAN REGION

Formation	Geological Age	Location	Description	Total Carbohydrates ppm	Hemicellulose ppm
Oranda	Middle Ordovician	Plainfield, Pa.	Black dense limestone	57·5	30·0
Wills Creek, 44 ft below top	Upper Silurian	Mt. Union, Pa.	Dark green-gray shale	56·8	32·5
Wills Creek, 39 ft below top	Upper Silurian	Mt. Union, Pa.	Light gray limestone	90·0	34·9
Tonoloway, 30 ft above base	Upper Silurian	Mt. Union, Pa.	Dark gray, lithographic limestone	46·5	24·9
Tonoloway, 42ft above base	Upper Silurian	Mt. Union, Pa.	Dark gray shale	58·6	32·2
Tonoloway, 83 ft above	Upper Silurian	Mt. Union, Pa.	Dark gray, shaly, laminated limestone	75·5	29·7
Tonoloway, 108 ft above base	Upper Silurian	Mt. Union, Pa.	Dark gray, silty, laminated limestone	57·5	34·7
Keyser	Upper Silurian	3·5 miles SW. of McVeytown, Pa.	Dark gray silty limestone, *Gypidula prognostica*	72·1	27·4
Keyser, 94·5 ft below top	Upper Silurian	2·5 miles SW. of McVeytown, Pa.	Dark gray laminated limestone, H_2S odor	56·5	32·4
Keyser, 9·4 ft below top	Upper Silurian	2·5 miles SW. of McVeytown, Pa.	Dark gray silty massive limestone	42·0	37·4
Keyser, 0–1 ft below top	Upper Silurian	2·5 miles SW. of McVeytown, Pa.	Medium gray, lithographic limestone, H_2S odor	62·5	34·9

values are in the general range of other fossil specimens studies, and the absorption maxima indicate the presence of both hexoses and pentoses, the former predominating.

TABLE 4. — TOTAL CARBOHYDRATE CONTENTS OF SPECIMENS OF *Sigillaria approximata* FROM BASAL GREENE FORMATION, LOWER PERMIAN, VANCE, PENNSYLVANIA

Specimen no.	Trunk	Max. mμ.	Leaf	Max. mμ
1	50	482	19	484
2	n.d.	n.d.	29·5	488
3	26	488	25	488
4	38	488	25	482
5	48	488	22	488
6	29	484	50	488
7	82	488	60	488
8	153	482	48	484
9	n.d.	n.d.	70	488

The total carbohydrate and hemicellulose contents of several Pennsylvanian plants and the associated matrix from central Pennsylvania are shown in Table 5. The total carbohydrate values of the fern leaves is higher than most of the other fossils studied, that of the dark shale matrix and of the woody stem and trunk material is about average for that type of material elsewhere.

Monosaccharide components of some recent sediments

Studies of carbohydrate components in various types of modern lake sediments from Minnesota (Rogers, 1965a; Swain, 1967) showed that glucose, galactose, xylose, arabinose and to a lesser extent mannose, ribose, rhamnose, and glucuronic acid occur residually in lake and bog deposits; traces of maltose, lactose, and sucrose were also found. The structural polysaccharides, cellulose (glucose only), mannan (mannose only), xylan (xylose + arabinose), hemicellulose (glucuronic acid + xylose) and bacterial polysaccharides (glucose, galactose, glucuronic acid and glucosamine) may be reasonably assumed to have contributed to the monosaccharide residues of the lake sediment samples studied. It is doubtful whether nutrient polysaccharides (starch, inulin, glycogen)

TABLE 5. — TOTAL CARBOHYDRATE AND "HEMICELLULOSE" CONTENTS OF FOSSIL PLANTS AND ASSOCIATED MATRIX OF PENNSYLVANIAN ROCKS FROM CENTRAL PENNSYLVANIA (IN PPM OF GLUCOSE EQUIVALENT, DETERMINED BY PHENOL-SULFURIC ACID METHOD)

Species	Geol. horizon and age	Locality	Tot. carb. fossil	"Hemi-cell". fossil	Tot. carb. matrix	"Hemi-cell". matrix
Neuropteris cf. *fimbriata* Lesq. (leaf)	Allegheny Group, Middle Pennsylvanian	Dudley, Bedford County, Pa.	125	70	62	40
Neuropteris cf. *vermicularis* Lesq. (leaf)	Allegheny Group, Middle Pennsylvanian	Finleyville, Bedford County, Pa.	230	14	70	55
Neuropteris sp. (stem)	Allegheny Group, Middle Pennsylvanian	Beccaria, Cambria County, Pa.	83	80	50	45
Pecopteris cf. *vestita* (leaf)	Allegheny Group, Middle Pennsylvanian	Broad Top City, Bedford County, Pa.	128	110	72	53
Pecopteris cf. *vestita* Lesq. (leaf)	Allegheny Group, Middle Pennsylvanian	Finleyville, Bedford County, Pa.	270	125	70	55
Odontopteris cf. *Schlotheimi* Bt. (leaf)	Allegheny Group, Middle Pennsylvanian	Finleyville, Bedford County, Pa.	260	135	70	55
Dendrophycus ? sp. (stems)	Allegheny Group, Middle Pennsylvanian	Wood, Bedford County, Pa.	64	52	n.d.	n.d.
Cordaites ? sp. (stem)	Allegheny Group, Middle Pennsylvanian	Van Ormer, Cambria County, Pa.	74	n.d.	65	45

exist in the sediments, as they probably would have been utilized as food before burial in the sediments could take place.

Ribose was relatively uncommon in living aquatic plants associated with the sediments (Rogers, 1965a) and in *Sphagnum* moss (Swain, 1967), but was fairly common in copropelic lake sediments. These sediments are comprised of settled algal plankton and of other organic debris worked over by mud-eating organisms and bacteria. Such material is believed to be the source of the ribose in the lake sediments.

Carbohydrate components of Recent sediments from the Pacific Ocean near Santa Barbara, California, at a water depth of 582 metres were studied by Prashnowsky, Degens, Emery, and Pimenta (1961). The total carbohydrates in a 4-metre core ranged from 0·002 to 0·63 mg/g of the dry sediment and attain a maximum concentration 90–100 cm below the sediment surface. Individual sugars in the sediments include galactose, glucose, mannose, rhamnose, ribose, xylose, arabinose and traces of fucose. The authors suggested that marine rather than terrestrial sources were mainly responsible for the carbohydrates. Marine algae, including seaweeds and diatoms, and microorganisms were suggested as possible sources. The ribose content of the sediments possibly was derived from nucleic acids or from vitamins or enzymes. Galactose, glucose, and rhamnose showed maximum concentration in the upper metre of sediment. Xylose and arabinose had their maxima at 1 metre or shallower and a second maximum at 2·6 metres. Mannose and ribose showed variable concentration, but generally were higher in the upper metre than elsewhere. The variations in the individual sugars are ascribed to diagenetic changes in *Eh* that resulted in uneven preservation of sugars, and to source differences.

Amino sugars, including glucosamine and possibly galactosamine, were found in copropel samples from Rossburg Bog, Minnesota whilst being analyzed for amino acids. Glucosamine presumably originated from bacterial polysaccharides or in the chitin from Cladocerans or other arthropod carapaces which occur in the sediment. Amino sugars have not yet been found in rock samples.

Monosaccharides in Precambrian rocks

Glucose-oxidose tests on acid hydrolyzates of several of the samples showed the presence of D-glucose (Table 6). The occurrence of the D enantiomorph of glucose is of interest in strongly suggesting a biogenic

TABLE 6. — MONOSACCHARIDE COMPONENTS OF PRECAMBRIAN ROCKS IN PPM
(a, CHROMATOGRAPHIC ANALYSIS; b, ENZYMATIC ANALYSIS)

Formation and age	Locality	gal	glu	ara	unkn	Σ
Coutchiching schist, Early Precambrian	Near Coutchiching, Ontario	—	—	—	(a)	—
Thomson argillite, Middle Precambrian	Carlton, Minnesota	0·19(a)	0·19(a)	0·09(a)	—	0·47(a)
Thomson argillite	same	—	0·41(b)	—	—	0·41(b)
Rove argillite, Middle Precambrian	Gunflint Lake, Minnesota	—	0·29(b)	—	—	0·29(b)
Rove Formation evenkite, Middle Precambrian	Silver Isle, Lake Superior, Minnesota	—	0·87(a)	—	—	0·87(a)
Rove Formation evenkite	same	—	0·30(b)	—	—	0·30(b)

nature of part of the carbonaceous matter. The structural nature of the rest of the carbohydrates is not known.

The small, restricted suites of Precambrian monosaccharides, if at all representative of the original organic matter, indicate that only *Thallophyta* or *Protista* contributed to the organic material. Although the lack of fossilized remains of higher animals in the rocks is not direct evidence, it seems probable that this form of life did not contribute to the carbohydrate residues studied here.

Monosaccharides in Palaeozoic rocks

In addition to common glucose, galactose and arabinose which plausibly are from thallophytic plant sources, ribose was noted in many samples of black Devonian shale by Rogers (1965b). These shales formed under conditions of somewhat stagnant waters in which the settled organic debris was of partly marine and partly terrestrial origin (Swain and Rogers, 1966). Algal plankton that had been affected by the activity of bottom-dwelling bacteria and other organisms could have been a likely source material for ribose in the geological samples.

Mannose was also found by Rogers (1965b) in several Devonian samples, both black shales and gray shales. In some samples mannose was the predominant monosaccharide. The importance of mannose as a constituent of yeast and its relative scarcity in other materials that would be likely to occur in geological samples suggests to the writers that yeasts provided the source of mannose in the Recent as well as the Devonian sediment samples. The presence of purine residues in the Devonian rocks reported previously (Swain, 1964) also suggests yeasts as a source of organic material in these rocks. Marine yeasts in nearshore environments have been shown (van Uden and Taysi, 1964) to be in much larger concentrations in or near bays.

Glucuronic acid which, with xylose, was found in several of the Devonian rock samples studied (Rogers, 1965b), presumably was in the form of hemicellulose residues in the rocks. The source of the hemicelluloses in Devonian time presumably would have been in the trunks of scale trees and large ferns. Xylose has been separated from a *Lepidodendron* of Mississippian age (Swain *et al.*, 1967), but larger Devonian plants have not yet been analyzed for xylose.

Palacas (1959) did not find the association xylose–glucuronic acid in pre-Devonian samples from the Appalachian region, which suggests that hemicellulose-bearing plants, if they existed, did not supply measurable quantities of residues to the pre-Devonian sediments. As noted above, however, base-extractable material, possibly of hemicellulose nature, was found in some of the Ordovician and Silurian samples studied here.

Palacas (1959) found glycerol in a number of Ordovician rock samples where it was associated with glucose and a few other monosaccharides. In many other Palaeozoic rock samples glycerol was not found, although glucose and other sugars were present. The glycerol may have originated in sorbitol, a polyol that occurs in some seaweeds as well as in higher plants, but its origin remains uncertain.

Monosaccharides of Palaeozoic fossils

The monosaccharide components of some Palaeozoic fossil specimens are being investigated, but only a few positive results have been obtained so far (Table 7). The small quantity of fossil residue available for study has been a limiting factor.

The brachiopod-shell monosaccharide residues are glucose and a trace of galactose, as would be expected for lower forms of animals. The Devonian cordaite *Callixylon* contains no glucose, but has a moderate amount of galactose, possibly indicating an agar type of polysaccharide. The *Cordaites* sp. in the Pennsylvanian coal ball yielded very little monosaccharides, but arabinose and xylose are present in addition to glucose and galactose. The Devonian psilopsid *Zosterophyllum* and the Devonian *Lepidophloios* and the Permian *Sigillaria* (lycopsids) have yielded only glucose and galactose by enzymatic tests in the samples so far studied, but other specimens are being analyzed.

The Mississippian lycopsid *Lepidodendron* contained small amounts of both pentoses, including xylose, and hexoses as would be expected of woody-tissue plants. Other individual Palaeozoic fossils are being studied for carbohydrates by the writers with the purpose of finding possible taxonomic variations in sugar components.

Oligosaccharide and polysaccharide content of Palaeozoic rocks

The polymeric components of the carbohydrate residues were not studied separately here. Rogers (1965b, 1967) using enzymatic techniques on water extracts of Devonian dark shales identified β-1-3-linked glucans (from laminarians) and possibly β-1-4-linked glucans (from cellulose).

TABLE 7. — CARBOHYDRATE COMPONENTS IN SULFURIC ACID EXTRACTS OF PALAEOZOIC FOSSILS, IN PPM (a, SEPARATED BY PAPER CHROMATOGRAPHY; b, IDENTIFIED BY ENZYMATIC TESTS)

Species	Geol. horizon and age	Locality	gal	glu	ara	xyl	rham	unkn	Σ
Echinocoelia ambocoeloides	Mahantango Fm., Middle Devonian	Huntingdon, Pa.	tr(b)	0·31 (a, b)	—	—	—	(a)	0·31
Zosterophyllum sp.	Lower Devonian	Seal Rock Ldg., Gaspe Bay, Canada	6·26(b)	20·0(b)	—	—	—	—	—
Callixylon sp.	Upper Devonian	Escuminac Bay, Quebec	26·22(b)	0(b)	—	—	—	—	—
Lepidodendron cf. *gaspianum*	Pocono Fm., Lower Mississippian	Trough Creek Park, Pa.	0·12(a)	0·47 (a, b)	—	0·95(a)	0·21(a)	—	1·63
Lepidophloios sp.	Lower Pennsylvanian, No.1 Coal	Globe Mine Jackson, Ohio	1·17(b)	1·34(b)	—	—	—	—	—
Cordaites sp.	Cherokee Gr., Middle Pennsylvanian	Pittsburg & Midway mine No. 19	0·07(a)	0·09(a)	0·01(a)	0·05(a)			0·22(a)
		Cherokee Co., Kansas	0·005(b)	0·335(b)					
Sigillaria cf. *approximata*	Greene Fm., Lower Permian	Vance, Pa.	0–0·2(b)	0–3·50(b)					

No conclusive results on polymeric components by paper chromatography of acid and base extracts of Palaeozoic rocks were obtained by Palacas (1959), Palacas *et al.* (1960), Rogers (1965) or in the present work. Several low Rf, presumably polymeric, substances were detected by those writers and by us on paper chromatograms, but have not been resolved further.

Summary and conclusions

1. Glucose and galactose are predominant over other monosaccharides in specimens of two freshwater algae, but arabinose and xylose are also important constituents. As arabinose is not a common component of glycerol polysaccharides, pectins may be the source in the freshwater samples.

2. Among marine algae studied, the chlorphyte *Ulva* contains relatively large amounts of rhamnose and the phaeophytes are relatively rich in ribose. The carbohydrates of the rhodophytes are predominantly glucose and galactose. Galactose occurs in polysaccharides agar and carageenan of marine algae, one or both of which may be the source in the present samples.

3. The higher aquatic plants studied, both marine such as *Zostera*, and freshwater types, have significant amounts of xylose and/or arabinose in addition to glucose and galactose.

4. Although most of the Precambrian samples tested had little or no measurable carbohydrates the more bituminous samples yielded small quantities of total carbohydrates and of glucose; the latter was determined both chromatographically and by enzymatic tests. In addition possible traces of galactose and arabinose in one sample were detected chromatographically. The source of the Precambrian carbohydrate residues was most likely *Protista* or *Thallophyta*, although a source in metazoan animals can not be ruled out on the basis of these residues. There are no structures of any kind that suggest metazoans in the Precambrian rocks studied, but Protista structures have been reported in the sequence (Barghoorn and Tyler, 1965). The Precambrian rocks studied are, to the best of knowledge, of marine origin, but whether the carbohydrate residues were derived from marine or freshwater sources is not known.

5. The total carbohydrate residues of Palaeozoic rocks and fossil specimens range from traces to over 400 ppm. The richer samples are not fossil shells or larger plant fossils, but shales and some limestones of predomi-

nantly nearshore origin, or where offshore plankton accumulations occurred. The total carbohydrates of several fossil ferns from the Broad Top and Allegheny Mountain coal fields of central Pennsylvania were relatively higher than most of the other plant fossils studied. It is suggested that the degree of humification accounts for the observed difference, although thc amount of pre-fossil decomposition and of subsequent rock metamorpism may also account for some of the variation.

6. Varying amounts of base-extractable carbohydrate-reacting material occur in some of the Palaeozoic rocks and fossils studied. The exact chemical nature of this "hemicellulose" fraction is unknown, but the fraction is sufficiently high to warrant further study.

7. Although glucosamine and possibly galactosamine occur among the carbohydrate components of Recent sediments, amino sugars have not been found in the Palaeozoic specimens studied.

8. The monosaccharide residues of Precambrian rocks so far studied are limited to small quantities of glucose and galactose and a trace of arabinose, which suggest a thallophytic or protistid source. Although a possible contribution by metazoans can-not be ruled out, lack of any other fossil evidence of metazoans in the rocks studied argues against such a possibility. In one Precambrian sample from the Rove Formation the enzymatically determined D-glucose is considerably lower than that measured chromatographically. The mildly metamorphic state of the sample offers the possibility that part of the carbohydrate materials has undergone racemization, but this, of cours, has not been verified by chemical or optical tests.

9. The ribose content of Devonian dark shales may have been contributed either by phaeophyte algae or by microorganisms acting on settled algal plankton or both. The mannose of some of the same dark shales as well as other Devonian black and gray shale samples could have been supplied by marine yeasts. Glucuronic acid–xylose associations in some Devonian rock samples may have originated in the hemicelluloses of larger land plants. This combination of sugars has not yet been found in the pre-Devonian samples studied, although some have given gross tests for hemicelluloses. The glycerol, associated with glucose and a few other monosaccharides in Ordovician rock samples, possibly originates in the polyhydroxy alcohol sorbitol which occurs in some modern seaweeds.

10. Monosaccharides in only small amounts have been extracted so far from individual Palaeozoic plant and animal species. Glucose has been identified chromatographically in Devonian branchiopods, Mississippian

Lepidodendron and Pennsylvanian Cordaites. These three fossils also gave positive glucose-oxidase tests for D-glucose. Small amounts of galactose have also been detected enzymatically in Devonian, Mississippian and Permian plants. Other chromatographically determined monosaccharides in the *Lepidodendron* include xylose and rhamnose, as would be expected in woody-tissue plants.

11. Except for several unresolved possibly oligosaccharide compounds having low *Rf* values on paper chromatograms, the only polymeric carbohydrates that have been found in the Palaeozoic samples studied are those characterized by M. A. Rogers as β-1-3 glucans and possibly β-1-4 glucans. The former, if indigenous, may have been derived from laminaran, a marine algal polysaccharide, while the latter may have come from cellulose.

Acknowledgements

The work has been supported by Grant No. NGR-24-005-054 of the National Aeronautics and Space Administration. Dr. G. A. Sellers ran the amino acid chromatograms to test for amino sugars. Vernon de Ruyter, Gregory Gohn, David Rowlands and Fred J. Gunther assisted with the laboratory and field work. Drs. Samuel Kirkwood and M. A. Rogers provided helpful advice. The writers express sincere thanks for this support and assistance.

References

Barghoorn, E. S. and Schopf, J. W. (1965) Microorganisms from the late Precambrian of Central Australia. *Science*, **150**, 337–339.

Barghoorn, E. S. and Tyler, S. A. (1965) Microorganisms from the Gunflint chert. *Science*, **147**, 563–577.

Bidwell, R. G., S., Krotkov, G. and Reed, G. B. (1952) Paper chromatography of sugars in plants. *Can. J. Bot.* **30**, 291–305.

Brasch, J. D. and Jones, J. K. N. (1959) Investigation of some ancient weeds. *Tappi*, **42**, 913.

Conway, E. and Yound, E. E. (1966) Rewiev of 5th International Seaweed Symposium. *Science*, **151**, 358–359.

DuBois, M., Gilles, K. A., Hamilton, J. K., Rebers, P. A., and Smith, F. (1956) Colorimetric method for determination of sugars and related substances. *Anal. Chem.* **28**, 350–356.

Fogg, G. E. (1956) The comparative physiology and biochemistry of the blue-green algae. *Bact. Rev.* **20**, 148–165.

Gruner, J. W. (1922) The origin of sedimentary iron formations: the Biwabik formation of the Mosabi range. *Econ. Geol.* **17**, 407–460.

Hough, L., Jones, J. K. N., and Wadman, W. H. (1952) An investigation of the polysaccharide components of certain freshwater algae. *J. Chem. Soc.* 3393–3399.

Jones, J. K. N. and Merler, E. (1956) The hemicelluloses of the fossilized wood of *Cedrus penhallowii*. *Canad. J. Chem.* **34**, 840.

Lipman, C. B. (1928) The discovery of libing micro-organisms in ancient rocks. *Science*, **68**, 272–273.

McNair, J. B. (1932) Some properties of plant substances in relation to climate of habitat: volatile oils, raponins, cyanogenetic glucosides and carbohydrates. *Am. Jour. Bot.* **19**, 168–193.

Myers, G. E. and McCready, R. G. L. (1966) Bacteria can penetrate rock. *Canad. J. Microbiology*, **12**, 477–484.

Palacas, J. G. (1959) Geochemistry of carbohydrates. Unpublished Ph.D. dissertation, University of Minnesota.

Palacas, J. G., Swain, F. M. and Smith, F. (1960) Presence of carbohydrates and other organic compounds in ancient sedimentary rocks. *Nature*, **185**, 234.

Pigman, W. ed. (1957) *The Carbohydrates*. Academic Press, New York, 902 p.

Prashnowsky, A., Degens, E. T., Emery, K. O., and Pimenta, J. (1961) Organic materials in recent and ancient sediments. Part 1. Sugars in marine sediment of Santa Barbara Basin, California. *Neues Jahrb. Geol. Paleont. Min.* **8**, 400–413.

Rogers, M. A. (1962) Carbohydrates in plants and sediments from two Minnesota lakes. Unpublished M.S. thesis, University of Minnesota.

Rogers, M. A. (1965a) Carbohydrates in aquatic plants and associated sediments from two Minnesota lakes. *Geochim. et Cosmochim. Acta*, **29**, 183–200.

Rogers, M. A. (1965b) Organic geochemistry of some Devonian black shales from eastern North America. Carbohydrates. Unpublished Ph.D. dissertation, University of Minnesota.

Swain, F. M. (1958) Organic materials of early Middle Devonian, Mt. Union area, Pennsylvania. *Amer. Assoc. Petroleum Geol. Bulletin*, **42**, 2858–2891.

Swain, F. M. (1964) Residual organic substances in Paleozoic rocks of central Pennsylvania. Program of Amer. Conf. on Coal Sci., Pennsylvania State University, p. 3.

Swain, F. M. (1966) Distribution of some organic substances in Paleozoic rocks of central Pennsylvania. American Chemical Society, Advances in Chemistry Series, *Coal Science* volume, Ser. 55, Washington, pp. 1–21.

Swain, F. M. (1967) Stratigraphy and biochemical paleontology of Rossburg Peat (Recent), North-central Minnesota: R. C. Moore commemorative vol., Teichert and Yochelson (eds.) University of Kansas Press, Lawrence, Kans., pp. 445–475.

Swain, F. M., Blumentals, A. and Prokopovich, N. (1958) Bituminous and other organic substances in Precambrian of Minnesota. *Amer. Assoc. Petroleum Geol. Bulletin*, **42**, 173–189.

Swain, F. M. and Rogers, M. A. (1966) Stratigraphic distribution of carbohydrate residues in Middle Devonian Onondaga beds of Pennsylvania and western New York. *Geochim. et Cosmochim. Acta*, **30**, 497–509.

Swain, F. M., Rogers, M. A., Evans, R. D. and Wolfe, R. W. (1967) Distribution of carbohydrate residues in some fossil specimens and associated sedimentary matrix and other geologic samples. *J. Sed. Petrology*, **37**, 12–24.

Vallentyne, J. R. (1963) Geochemistry of carbohydrates. In Breger, I. A., ed., Organic Geochemistry, pp. 456–503. Pergamon Press, Oxford.

van Uden, N. and Taysi, I. (1964) Occurrence and population densities of yeast species in an estuarine-marine area. *Limnol. Oceanogr.* **9**, 42–45.

ALIPHATIC HYDROCARBONS FROM THE PRE-CAMBRIAN OF NORTH AMERICA AND SOUTH AFRICA

by

J. ORÓ and D. W. NOONER

Department of Chemistry
University of Houston
Houston, Texas 77004

Introduction

To help explain the origin of life, it would be useful to know how far back in time life processes existed. Chemical evidence rather than fossil organisms, would be best to establish the existence of life in the remote past because extremely ancient life may not have left fossils or the fossils might have disappeared through geological change. This has led to a search for biological markers. One class of compounds which meet one of the important requirements for a marker, i.e. structural specificity, is the saturated isoprenoid hydrocarbons. Consequently, searches have been made for these compounds in ancient sediments, and when found the implication is that they are remnants of early life.[1–6] Some of this work is briefly summarized below.

Alkanes of high molecular weight, including the saturated isoprenoid hydrocarbons pristane and phytane, have been found in shale 1 billion years old from the Nonesuch formation of Michigan[1, 2] and chert 2 billion years old from the Gunflint iron formation of Ontario, Canada.[3] Work carried out on rocks from the Soudan iron formation (2·7 billion years old) of Minnesota at the same time, but independently, of our work has been reported.[4, 5] In this paper we report our work on (i) rocks from the Soudan iron formation and (ii) sedimentary rocks (3·2 billion years old) from the Fig Tree series of the Swaziland system, eastern Transvaal, South Africa.

Experimental

Samples

Two types of rock from the Soudan iron formation were analyzed. These were a carbonaceous lens (sample a) from 1800 feet below ground in the Soudan mine, Soudan, Minn., and a carbonaceous silicate rock (sample b) from an outcrop on a hill north of the high school at Tower, Minnesota.

Several rocks from the Fig Tree series were analyzed. These were black chert (sample No. 1) and a carbonaceous graphitic-like silicate rock (sample No. 2) from the Sheba gold mine near Barberton, South Africa, and black chert (sample No. 3) from an outcrop near the Daylight mine, 28 km east-northeast of Barberton. In addition, an extract made in the Geophysical Laboratory, Carnegie Institute, Washington, D.C., of a core specimen of shale from near Barberton was designated as sample No. 4.

Procedure

The techniques used in analyzing the above described samples have been presented elsewhere[6-9] and only brief summaries will be made here. The procedure used with Soudan samples a and b and Fig Tree samples 1, 2, and 3 was to pulverize the rocks (after appropriate steps had been taken to remove surface contamination; see below) in a Carver press test cyclinder, extract the organic compounds with benzene-methanol (3 : 1) in an all-glass Soxhlet-type extractor, recover the organic material, fractionate it by means of silica gel chromatography, and analyze the alkane fraction by gas chromatography and gas chromatography-mass spectrometry.[10] The extract (sample No. 4) supplied by Dr. T. C. Hoering was obtained from the Fig Tree shale by the procedure briefly described below. Three to four hundred grams of the shale (core sample) were broken in to small pieces and then ground in a ball mill to approximately 200 mesh. Solvent extraction (n-pentane was used in this case) was facilitated by means of an ultrasonic generator and was carried out for 4 hours. During this period, the solvent temperature increased from room temperature to the boiling point of the solvent (n-pentane). The suspended material was removed by centrifugation and the solution was then concentrated almost to dryness under a stream of purified nitrogen. Colloidal copper was added to remove sulfur; however, removal of sulfur was not quantitative. The extract was then fractionated on silica gel. The procedure of

Meinschein and Kenny[8] was followed, except that the aliphatic hydrocarbons were eluted with n-pentane. By carefully blowing this low-boiling solvent to a very small volume, many of the relatively low-molecular-weight hydrocarbons, ordinarily lost by evaporation to dryness when n-heptane is used, were recovered also. The n-pentane eluate was analyzed by gas chromatography and gas chromatography-mass spectrometry.

Stainless steel capillaries coated with Polysev† or Carbowax 20M terminated with *p*-phthalic acid† were used in the gas chromatographic separations.

Results

Soudan Rocks

Inside pieces of a carbonaceous lens (sample a) and a carbon-rich silicate rock (sample b) were analyzed. The outside, 0·5 cm and above, was removed with a carborundum saw and the inside pieces to be analyzed were thoroughly washed with extracting solvent. Figure 1A shows a chromatogram of the alkanes from the carbonaceous lens obtained using a polyphenyl ether (Polysev) as the liquid phase in a stainless steel capillary. Normal alkanes from C_{16} through C_{19} were detected. Normal pentadecane was not present, but had probably been lost during the evaporation procedure. Another analysis of the lens[6] showed this hydrocarbon to be present. More significantly the saturated isoprenoids, norpristane, pristane and phytane (peaks a and b, respectively) were also found. Similar results were also obtained using a capillary column with Carbowax 20M terminated with *p*-phthalic acid as the liquid phase. In addition to these tentative identifications based on retention time, mass spectral evidence for the presence of pristane and phytane was obtained using the LKB type 9000 gas chromatograph-mass spectrometer combination and the aforementioned capillary coated with Polysev.

The chromatographic pattern of the paraffinic hydrocarbons from the carbon-rich silicate rock is shown as Fig. 1B. Chromatographic evidence for the presence of norpristane, pristane, and phytane in the carbon-rich silicate rock was obtained. However, due to technical difficulty only the spectrum of norpristane was obtained when the sample was analyzed on the LKB type 9000 instrument. The hydrocarbons pattern of the two

† Polysev [*m*-bis *m*-(phenoxyphenoxy)-phenoxy benzene] was obtained from Applied Science Laboratories, Inc., State College, Pa; Carbowax 20M (polyethylene glycol) terminated with *p*-phthalic acid was obtained from Wilkens Instrument and Research, Inc., Walnut Creek, Calif.

samples from the Soudan were thus found qualitatively to be quite similar. Quantitatively, there are significant differences, the carbonaceous lens containing approximately 0·54 ppm paraffinic hydrocarbons versus 0·02 ppm for the silicate rock.

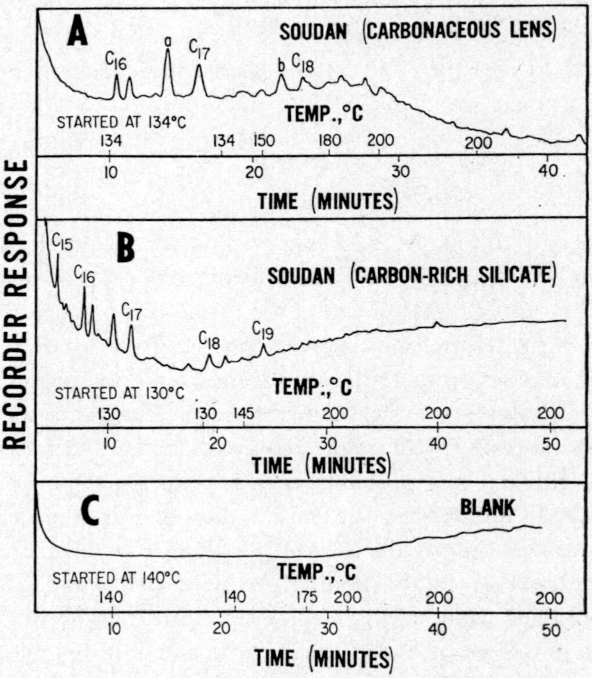

FIGURE 1. — Gas chromatographic separation of alkanes from the Soudan rocks with a capillary column (Polysev). (A) Stainless steel tubing, 75 m long ×0·076 cm I.D., coated with Polysev. Barber–Colman series 5000 gas chromatograph equipped with a flame-ionization detector. Nitrogen pressure, 700 g/cm^2. Small split. Carbonaceous lens (sample from Cloud) extracted, 10·7 g. About 1/10 of the n-heptane eluate was injected. Attenuation, 30. Isothermal at 134° C for 18 min, then programmed at approximately 6·0° C per min to 200° C. (B) Stainless steel tubing, 75 m long×0·076 cm I.D., coated with Polysev. F and M model 810 gas chromatograph equipped with a flame-ionization detector. Nitrogen pressure 700 g/cm^2. No split. Carbon-rich silicate (graphitic rock obtained from Cloud) extracted, 32·4 g. About 1/3 of the n-heptane eluate was injected. Range, 10^2; attenuation, 1. Isothermal at 130° C for 18 min, then programmed at approximately 6·0° C per min to 200° C. (C) Stainless steel tubing, 91·5 m long ×0·076 cm I.D., coasted with Polysev. F and M model 810 gas chromatograph equipped with a flame ionization detector. Nitrogen pressure, 1015 g/cm^2. No split. Blank. About 1/2 of the n-heptane eluate was injected. Range, 10^2; attenuation, 1. Isothermal at 140° C for 22 min, then programmed at approximately 6·0° C per min to 200° C.

One of the blanks, that for the carbon-rich silicate rock, is shown in Fig. 1C. The other blank also did not contain significant amounts of hydrocarbons. The amounts of hydrocarbons in each blank were less than 1% of that found in the corresponding sample.

Fig Tree Rocks

Figure 2 shows the chromatograms obtained using Polysev as the liquid phase in stainless steel capillary tubes. The hydrocarbons shown in Fig. 2A were obtained from a chert sample (sample no. 1) treated as follows: At least 0·5 cm of the outside surface material was sawed away with a carborundum saw. The inside pieces of rock (three) were then treated with hot chromic acid solution (40 g sodium dichromate, 90 ml water and 700 ml concentrated sulfuric acid) for 5 min, washed with distilled water, dried, rinsed with benzene-methanol (3:1) solvent, and then crushed in a Carver press test cylinder[(3)]. Normal alkanes ranging from C_{15} through C_{26} were found to be present. The peaks designated as a and b correspond in their chromatographic position to pristane and phytane, respectively. The peak immediately following n–C_{16} contains Δ–C_{16}, but probably also contains norpristane. The shape of the peak clearly shows that it represents at least two compounds. The peaks immediately following n–C_{18}, n–C_{20}, and n–C_{22} are normal C_{18}, C_{20} and C_{22} monoenes, respectively. The retention times do not correspond to the respective n-alkylcyclohexanes. The identifications indicated here were made on the basis of mass spectrometric analysis and retention times of known hydrocarbons. The gas chromatographic analysis of this sample made on the LKB type 9000 gas chromatograph-mass spectrometer combination is shown in Fig. 2E.

A second analytical determination of alkanes was carried out with smaller pieces (inside and outside) of the same specimen of chert. The removal of possible surface contaminants was performed this time, not by treatment with chromic acid, but by carefully extracting the rock surface with organic solvent. For this, about six pieces of chert weighing approximately 100 g were placed in an all-glass Soxhlet extractor and extracted for 13 hours with benzene-methanol (3:1). The pieces were then dried, crushed, extracted and the extract analyzed.

The chromatographic pattern of the paraffinic hydrocarbons recovered from the sample is shown in Fig. 2C. As shown, the chromatographic pattern is the same as that of Fig. 2A.

Approximately the same level of hydrocarbons (0·02 ppm) was found in sample No. 1 (black chert) by the two separate and slightly different analytical determinations. The mass spectrometric results of the second determination were essentially as obtained for the first. Furthermore, in

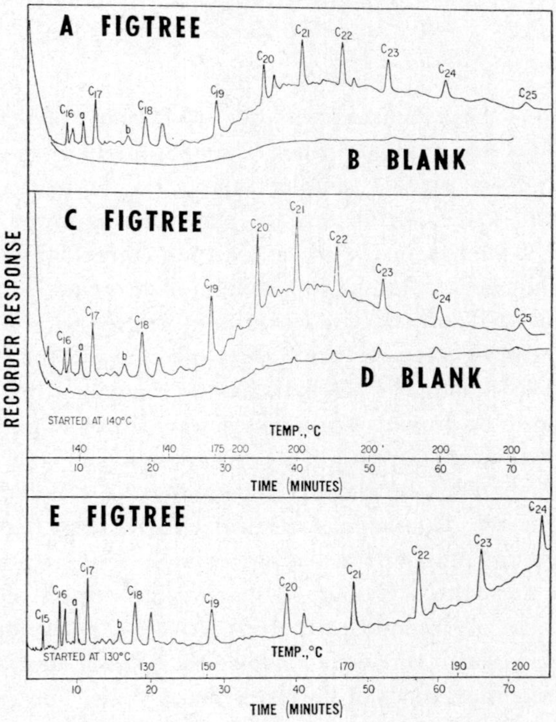

FIGURE 2. — Gas chromatographic separation of alkanes from the Fig Tree chert with capillary columns (Polysev). (A) Stainless steel tubing 91·5 m long ×0·076 cm I.D., coated with Polysev. F and M model 810 gas chromatograph equipped with a flame ionization detector. Nitrogen pressure, 1050 g/cm². No split. Chert extracted (sample from Hoering; pulverized inside piece), 100·5. About 1/2 of the n-heptane eluate was injected. Range 10; attenuation, 4. Isothermal at 140° C for 22 min, then programmed at approximately 6° C per min to 200° C. (B) Blank (pulverized glass substituted for sample), treated in the same way as the sample. (C) Analytical equipment and conditions same as in (A). Chert extracted (sample from Hoering; pulverized piece which included some solvent-cleaned outside material), 97·9. About 1/2 of the n-heptane eluate was injected. (D) Blank (pulverized glass substituted for sample), treated in the same way as the sample. (E) Stainless steel tubing, 75 m long×0·076 cm I.D., coated with Polysev. LKB type 9000 gas chromatograph-mass spectrometer combination. Helium flow, 24 ml/min. No split. Same sample and eluate as in (A). About 1/3 of the eluate was injected. Gain, 6. Isothermal at 130° C for 18 min, then programmed at approximately 1·5° C per min to 200° C.

the second analysis, two mass spectra were taken of the peak immediately following n—C_{16}, one on the ascending side of the peak before inflection and the other on the descending side. The components of this peak were found in fact to be norpristane and the n–C_{16} monoene as indicated by the first analysis. The mass spectra of the isoprenoid hydrocarbons of this sample are presented in Fig. 3. They are essentially identical with those

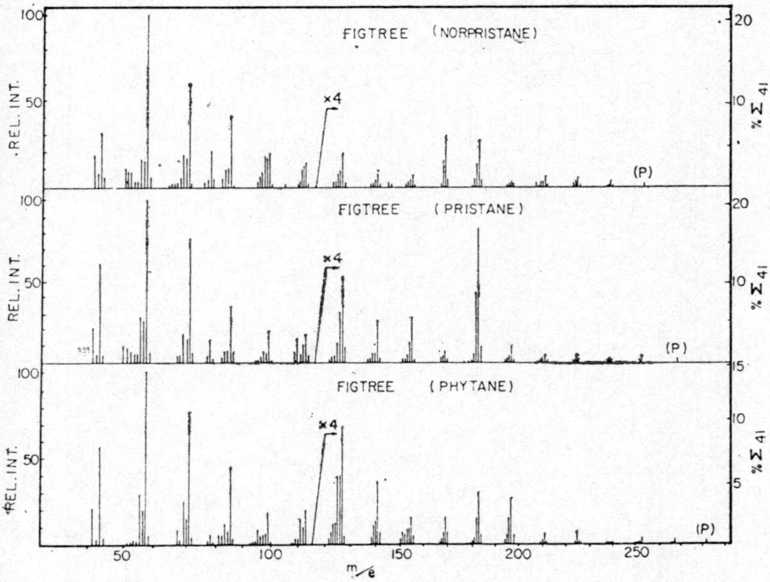

FIGURE 3. — Mass spectra of saturated isoprenoid hydrocarbons from the Fig. Tree chert sample extracted in Houston. Obtained using an LKB type 9000 gas chromatograph-mass spectrometer.

given by standards. The controls or procedural blanks (glass was carried through all analytical steps, including pulverization in the Carver press, associated with the particular samples) are shown in Figs. 2B and 2D. As seen the alkane contribution of the blanks amounts to less than 1% and 5%, respectively, of the amounts found in the actual samples.

Sample No. 2 (graphitic-like silicate obtained from Hoering) was sawn to give several 8–10 g sized pieces, at least 0·5 cm removed from the surface. These inside pieces were then boiled for 1 hour in each of three portions of extraction solvent (benzene-methanol, 3:1) to remove inadvertent surface contamination. Sample No. 3 (chert obtained from Barghoorn) was boiled for 1 hour in each of three portions of benzene-methanol (3:1) to remove surface contamination. In the case of the chert, the outside surface was not sawn away. The n-heptane eluates from these samples were analyzed on a 305 m long × 0·076 cm inside diameter stainless steel capillary coated with Polysev. Extremely low levels of hydrocarbons (approximately 0·003 ppm) were obtained. Small peaks representing hydrocarbons in the C_{12}–C_{19} range were obtained. Pristane and phytane were detected (chromatographically) in the graphitic-like silicate, but not in the chert.

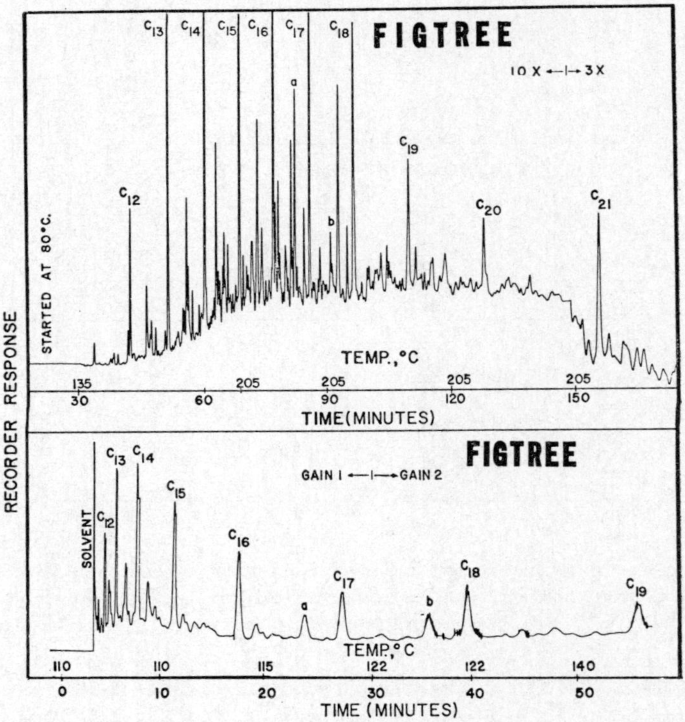

FIGURE 4. — Gas chromatographic separation of alkanes from the Fig Tree shale with capillary columns (Polysev).

(Upper) Stainless steel tubing, 305 m long×0·076 cm I.D., coated with Polysev. Barber-Colman series 5000 gas chromatograph equipped with a flame ionization detector. Helium pressure, 2100 g/cm². Small split. Shale extracted at Washington, D.C. by Hoering (see text), about 400 g. About 1/10 of the n-pentane eluate was injected. Attenuation, indicated on figure. Programmed from 80° C to 205° C at approximately 2·0° C per min.

(Lower) Stainless steel tubing, 75 m long×0·076 cm I.D., coated with Polysev. LKB type 9000 gas chromatograph-mass spectrometer combination. Helium flow, 24 ml/min. No split. Same sample and eluate as above. About 1/20 of the n-pentane eluate was injected. Gain, indicated on figure.

The above work was carried out in the University of Houston laboratories during September 1965.[11] In order to obtain additional data on Fig Tree Precambrian sediments, a core of the Fig Tree shale was extracted (sample No. 4) at the Geophysical Laboratory of the Carnegie Institute of Washington by Dr. T. C. Hoering (see above). The hydrocarbons obtained were analyzed (Fig. 4) at Houston during and after December 1965. The analysis on a Polysev capillary column, 75 m long × 0·076 cm inside diameter, gave results (Fig. 4, lower) similar to those of Fig. 2, except that more of the low-molecular-weight alkanes and less of the higher-molecular-weight paraffins were detected. This is, indeed, what would be expected by the use of n-pentane as a solvent instead of n-heptane. Furthermore, this sample may also be richer in lighter hydrocarbons than the other samples.

A good mass spectrum was obtained for pristane (Fig. 5) on the LKB type 9000 gas chromatograph-mass spectrometer, and recognizable mass spectra, but not as clean as those shown in Fig. 3, were also obtained for norpristane and phytane.

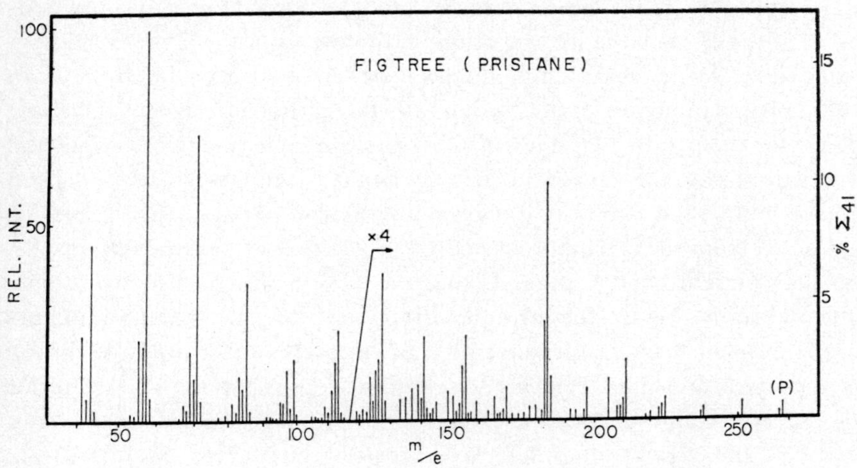

FIGURE 5. — Mass spectrum of pristane from the Fig Tree shale sample extracted in Washington. Obtained using an LKB type 9000 gas chromatograph-mass spectrometer.

The analysis of the sample on the long (305 m) Polysev capillary column gave extremely good resolution and indicated that some of the peaks on the earlier chromatogram obtained with the shorter column (75 m) represented more than one compound (Fig. 4, upper). Normal hydrocarbons ranging from C_{12} through C_{23} were observed. Norpristane and phytane are present in smaller amounts than pristane and are near compounds which have retention times quite close to those of these two isoprenoids. Therefore, it is not surprising that clean mass spectra of norpristane and phytane were not obtained on LKB type 9000 instrument, using the shorter column. The core sample of shale extracted at Washington also contained more hydrocarbons (approximately 0·15 ppm) than the chert and graphitic-like rock samples extracted at Houston (0·02 ppm).

The results obtained at the two laboratories are in fairly good agreement considering that two different extraction procedures were used and that a certain amount of heterogeneity in the distribution of compounds in the rocks must be allowed for as briefly discussed later.

Discussion

Soudan Rocks

Fossils found in the Nonsuch shale[1] and the Gunflint chert[12] give added confidence that the hydrocarbons extracted from them are of biological origin. While fossils have not been observed directly in the Soudan rock, observations on pyrite found with the carbonaceous lens gave evidence for the presence of microstructures of possible biological origin and indications that the carbon in the lens might be derived from blue-green algae.[13] Also the similarity of chromatograms of paraffinic hydrocarbons extracted from the Soudan rocks (Fig. 1)[6] to that of chromatograms run on the Gunflint chert[3] suggests that the carbon was formed from blue-green algae However, definite conclusions regarding the origin of alkanes in the Soudan formation cannot be made at present as a result of carbon isotope measurements. Hoering's method of comparing the $^{13}C/^{12}C$ ratios of the extractable and non-extractable carbon was applied to the Soudan[4, 5] and significant differences were obtained. This could indicate that the hydrocarbons entered the rock after deposition. Also, it should be pointed out that the presence of pristane and phytane is not necessarily, per se, evidence of biogenic origin. Preliminary evidence has recently been obtained that pristane and phytane may be synthesized abiotically,[14, 15] and thus these isoprenoids must be examined more critically to establish their chirality before they can be accepted as unequivocal evidence of biological activity.

Fig Tree Rocks

The analyses of samples 2 and 3 given above demonstrate the extremely low level of hydrocarbons present in some samples of the Fig Tree series rocks. Because of the small amount of hydrocarbons in these rocks, much larger samples (at least 350 g versus the approximately 35 g used in each case) would be required for accurate quantitative determination. These results also show that significant variations in the hydrocarbon content of these rocks exist. Variations in hydrocarbon content can be expected from Precambrian sediments and indeed similar variations have been observed in our Laboratory in rocks from the Gunflint iron formation of southern Ontario. In a sense the very existence of such hydrocarbon-deficient rocks makes it possible to run ideal analytical controls for samples from the same formation which may be richer in hydrocarbons. The heterogeneity of distribution of hydrocarbons may be the result of differences in composition or in thermal metamorphism. This heterogeneity may be taken as an indication that the hydrocarbons were not incorporated in the rocks by a general process of petroleum migration in more recent times.

Paraffinic hydrocarbons are present in the Fig Tree rocks at extremely low levels (0·003–0·15 ppm). Sufficient data are not available to ascertain whether or not the hydrocarbons date from deposition. However, the treatment of the samples precludes the hydrocarbons being merely surface contaminants. This conclusion is in line with the fact that the $^{13}C/^{12}C$ ratios of the extractable and insoluble organic matter of the Fig Tree rocks are essentially identical.[9] Entities have been found in the Fig Tree shale that have the morphology of fossil bacteria and possibly algae,[16] and these hydrocarbons may be remains from this early life.

The occurrence of olefins in some samples of Fig Tree chert is somewhat unexpected. Although olefins have been observed in Pennsylvania crude,[17] in general they are relatively rare in petroleum crudes and sediments. Their presence in the cherty Precambrian sediment is not easily understood unless one assumes that the olefins can be preserved in the chert matrix for long periods of time or that they may be intermediate products in the thermal formation of aromatic hydrocarbons. The latter compounds, particularly benzene and toluene, have been observed to be released during pyrolysis of the Fig Tree rocks.[9] The order of distribution shown by the predominance of even carbon-numbered olefins can be explained by either a biological or an abiological origin. Thus, they could result from the dehydration of biological even carbon-numbered alcohols or they could have been synthesized by prebiotic or postbiotic surface catalytic processes involving the condensation of ethylene.[15]

The relatively large amounts of pristane found in the sediment may be taken as an indication that the isoprenoid hydrocarbons are probably biogenic, although as pointed out earlier there are indications that small amounts of isoprenoid hydrocarbons can be synthesized by Fischer-Tropsch processes, and, therefore, they could also be considered of prebiotic origin.[14, 15] However, more work needs to be done on this type of synthesis before advancing any speculations.

It should be added that after this work was done[11] isoprenoid and paraffinic hydrocarbons were detected in samples of the Fig Tree chert which had apparently undergone a minimum amount of thermal metamorphism[18] and in similar rocks which presumably were more metamorphosed.[9] The organic geochemistry of Precambrian sediments and the problems involved in assessing the role of contamination and metamorphism, and in determining whether the hydrocarbons date from sediment deposition or are the result of later infiltrations into the rock have been recently discussed in some detail by Hoering.[9]

Because of the compact nature of some of these rocks, especially the chert, and the fact that internal samples were analyzed, it is probable that the hydrocarbons are indigenous to the rock. The fact that microfossils have been found in some of these rocks would suggest a biochemical origin for these hydrocarbons about $3{\cdot}2 \times 10^9$ years ago.

Conclusions

The saturated isoprenoids, norpristane, pristane, phytane, and normal alkanes have been found in rocks from the Soudan and Fig Tree formations. In addition monoalkenes were found in the Fig Tree rocks. The following conclusions regarding these compounds can be made.

(i) The hydrocarbons are in the rocks and are not merely surface contaminants.

(ii) The hydrocarbons, or their precursors, found in the cherty Fig Tree rocks may have been deposited with the sediments, as indicated by the identity of the $^{13}C/^{12}C$ carbon isotope ratios of extractable and non-extractable organics, and by gross correlation of black color, microfossils, and hydrocarbon content. These observations apply equally well to the Gunflint chert, and only partially to the Soudan rocks which show different carbon isotope ratios.

(iii) The similarity of the patterns indicate that the same or related organisms may have produced the hydrocarbons extracted from the Soudan and Fig Tree rocks.

(iv) Although it is believed the hydrocarbons in the rocks may be residues of ancient life, additional research will have to be carried out to decide more definitely whether these hydrocarbons date from deposition and whether they are of biological or abiological origin.

Acknowledgments

We thank T. C. Hoering for Fig Tree samples and valuable collaboration; E. S. Barghoorn, P. Cloud, Jr., and W. G. Meinschein for other samples and valuable comments; E. Gelpi and H. B. Skewes for assistance in obtaining mass spectra; and A. Zlatkis for aid in utilizing high resolution gas chromatography. This work was presented in part at the 4th meeting of the Group for the Analysis of Carbon Compounds in Carbonaceous Chondrites, Washington, D. C., April 21, 1966, and was supported in part by NASA Research Grants (NsG-257 and NGR-44-005-020) from the National Aeronautics and Space Administration.

References

1. W. G. Meinschein, E. S. Barghoorn and J. W. Schopf. *Science*, **145,** 262 (1964); E. S. Barghoorn, W. G. Meinschein and J. W. Schopf. *Science*, **148,** 461 (1965).
2. G. Eglinton, P. M. Scott, T. Belsky, A. L. Burlingame and M. Calvin. *Science*, **145,** 263 (1965); G. Eglinton, P. M. Scott, T. Belsky, A. L. Burlingame, W. Richter, and M. Calvin. Space Sciences Laboratory, University of California, Berkeley, Technical Report Series No. 6, Issue No. 9, January 1965.
3. J. Oro', D. W. Nooner, A. Zlatkis, S. A. Wikström and E. S. Barghoorn. *Science*, **148,** 77 (1965).
4. T. Belsky, R. B. Johns, E. D. McCarthy, A. L. Burlingame, W. Richter and M. Calvin. *Nature*, **206,** 446 (1965).
5. W. G. Meinschein. *Science* **150,** 601 (1965).
6. J. Oro', D. W. Nooner, A. Zlatkis and S. A. Wikström. *Life Sc. Space Res.* **4,** 63 (1966).
7. J. Oro', D. W. Nooner and S. A. Wikström. *Science*, **147,** 870 (1965).
8. W. G. Meinschein and G. S. Kenny. *Anal. Chem.* **29,** 1153 (1957).
9. T. C. Hoering. In *Researches in Geochemistry*, vol. II, edit. P H. Abelson. John Wiley, N. Y., in press.
10. R. Ryhage. *Anal. Chem.* **36,** 759 (1964); J. T. Watson and K. Biemann. *Anal. Chem.* **36,** 1135 (1964); R. Ryhage, S. Wikström and G. R. Waller. *Anal. Chem.* **37,** 435 (1965).

11. M. Calvo-Hernando. In *YA*, Brief News report (Madrid). February 5, 1966.
12. S. A. Tyler and E. S. Barghoorn. *Science*, **119,** 606 (1954); E. S. Barghoorn and S. A. Tyler. *Ann. N.Y. Acad. Sci.* **108,** 451 (1963); *Science*, **147,** 563 (1965).
13. P. E. Cloud, Jr., J. W. Gruner and H. Hagen. *Science*, **148,** 1713 (1965).
14. M. H. Studier, R. Hayatsu and E. Anders. Work presented at the 4th meeting for the Analysis of Carbon Compounds in Carbonaceous Chondrites, April 21, 1966, Washington, D.C.
15. J. Oro', J. Han and D. W. Nooner. *Ibid.*
16. E. S. Barghoorn and J. W. Schopf. *Science*, **152,** 758 (1966).
17. M. Fred and R. Putscher. *Anal. Chem.* **21,** 900 (1949); F. A. Haak and K. Van Nes. *J. Inst. Petrol.* **37,** 245 (1951).
18. W. G. Meinschein. Personal communication; see also ref. 16.

HYDROCARBONS IN METEORITES

by

J. ORO', D. W. NOONER and R. J. OLSON

Department of Chemistry
University of Houston
Houston, Texas 77004

Introduction

Since the beginning of the last century when the Alais meteorite was examined,[1] it has been known that carbonaceous meteorites (chondritic and achondritic meteorites known as carbonaceous chondrites) contained carbon compounds. Many early investigators reported indications of hydrocarbons in these meteorites.[2] A review of the nineteenth-century studies on meteorites has been written by Cohen.[3]

Interest in the carbon compounds in meteorites waned following the nineteenth century until Mueller[4] in 1953 reported carbonaceous material containing chlorine, hydrogen, nitrogen, oxygen, and sulfur in the Cold Bokkeveld meteorite. This revival of interest has resulted in investigations in which traces of organic compounds[5] and particularly hydrocarbons[6–8] were found in carbonaceous and other meteorites.

The origin of these hydrocarbons and other organic compounds in meteorites is still unknown, although proposals have been made to explain their existence.[5, 6, 9, 10] Our work is intended (i) to provide a comparative study of the distribution of hydrocarbons in the three major types of carbonaceous chondrites and other meteorites, (ii) to determine if hydrocarbons having the same molecular distribution as those found in meteorites are found in sediments and other terrestrial samples, and (iii) to determine to what extent the hydrocarbons found in meteorites are due to contamination.

Methods

Twenty-nine meteorites were analyzed for hydrocarbons according to the following procedure:

The meteorites (0·2 to 8·6 g) were pulverized to approximately 100-mesh, placed in an all-glass Soxhlet-type extractor[11] and extracted with 50 ml of a benzene-methanol (3 : 1) mixture for 8 hours. Colloidal copper was added to the system to remove any free sulfur extracted from the meteorites. The extract was transferred to a beaker and the solvent removed by evaporation at 40° C in a stream of purified nitrogen. The contents of the beaker were quantitatively transferred to the top of a heptane-washed silica gel column (1×16 cm) by means of three 3-ml aliquots of n-heptane. The silica gel had previously been activated by heating in an electric furnace at about 425° C for at least 8 hours. The column was then successively eluated with 15 ml each of n-heptane, carbon tetrachloride, benzene, and methanol. The n-heptane eluate for each extract was evaporated to dryness as was done for the original extract. Most of the benzene eluates were also taken to dryness; the other eluates were retained for analysis. The carbon tetrachloride elution step was later omitted because of the possibility that the benzene and methanol eluates would be analyzed by electron capture techniques.[12]

The residues from the n-heptane and benzene eluates, which usually contained less than 0·1 mg hydrocarbon, were dissolved in benzene and suitable aliquots taken for analyses. The analysis of the paraffinic hydrocarbons from the Orgueil, Boriskino, Mighei, Murray, Santa Cruz, Mokoia, and Odessa (graphite nodule) was carried out by gas chromatography and by a combination of gas chromatography and mass spectrometry. The other meteorites were analyzed by gas chromatography only. A Barber–Colman series 5000, a Barber–Colman series 5320, and/or an F and M model 810 gas chromatograph equipped with flame ionization detectors were used for these analyses. Capillaries coated with Apiezon L,† Polysev,† Igepal CO-880† and SF–96,† Paraplex G–60,† and Carbowax 20 M terminated with *p*-phthalic acid† were used with these instruments. Identification of compounds separated was made on the basis of retention

† Apiezon L (a high temperature grease), SE–30-cyclohexanedimethanol succinate, Polysev [*m*-bis *m*-(phenoxyphenoxy)-phenoxy benzene] and Gas Chrom P (a diatomaceous earth support) were obtained from Applied Science Laboratories, Inc., State College, Pa. The other materials were obtained from the following sources: Carbowax 20M (polyethylene glycol) terminated with *p*-phthalic acid from Wilkens Instrument and Research, Inc., Walnut Creek, Cal.; Igepal CO–880 [nonylphenoxypoly (ethyleneoxy)-ethanol] from General Aniline and Film Corporation, New York, N.Y.; Paraplex G–60 (soya bean oil epoxide), from Rohm and Haas Company, Philadelphia, Pa.; SE–30 (a silicone rubber gum) and SF–96 (methyl silicone fluid) from the General Electric Company, Waterford, N.Y.; and F–60 silicone oil from Dow Corning Corporation, Midland, Mich.

times of known hydrocarbons either individually or as mixtures. The combined gas chromatographic-mass spectrometric analyses of the Orgueil, Murray, and Mokoia meteorites were carried out on a chromatograph of the type developed by Haahti(13) attached to an Atlas CH4 mass spectrometer by means of two molecular separators.(14) Columns packed with SE–30,† SE–30-cyclohexanedimethanol succinate† or F–60,† on acid-washed silanized Gas Chrom P†, were used with this system. Gas chromatographic-mass spectrometric analysis of the Boriskino, Mighei, Murray, Santa Cruz, Mokoia, and Odessa meteorites were made on an LKB type 9000 gas chromatograph-mass spectrometer combination. A stainless steel capillary coated with Polysev was used with this system.(15) In addition the Mighei meteorite was analyzed on an Hitachi Perkin-Elmer RMU–60 mass spectrometer-Perkin-Elmer 801 gas chromatograph combination equipped with Biemann(16) molecular separators. A stainless steel capillary coated with Polysev was used with this instrument combination. Identification was afforded, when using the gas chromatographic-mass spectrometric combination systems, by taking the mass spectra of the individual components as they were eluted from the column.

The analyses of the benzene eluates were carried out on a high-resolution capillary column [1000 ft of stainless steel capillary, specially cleaned(17) and coated with Polysev] installed in a Barber–Colman series 5000 gas chromatograph equipped with a flame-ionization detector. Tentative identification of some of the compounds eluted was made by comparing their retention times with those of known aromatic hydrocarbons. Ultraviolet spectroscopic examinations were performed on some of the samples in cyclohexane solution using a Cary model 11 spectrophotometer.

Results

The meteorites examined were in four groups: i.e. Wiik(18) types I, II, III, and a fourth heterogeneous group. The classification of meteorites in the fourth group is according to Mason.(19) The analyses for compounds from these meteorites in the n-heptane and benzene eluates of the silica gel chromatography separations are considered separately below. In addition our use of high-resolution gas chromatography and the identification of isoprenoids are also given.

Heptane Eluates

This fraction contains primarily paraffinic hydrocarbons.[11] Analyses of these eluates from the different meteorites will be reported in detail elsewhere (20).

CARBONACEOUS CHONDRITES

Type I. The Alais, Ivuna, and Orgueil meteorites were examined from this group. No marked similarity was observed in the distribution of paraffinic hydrocarbons among these meteorites. The paraffinic hydrocarbons of the Alais ranged from C_{16} through C_{28}; the Ivuna from C_{16} through C_{28}; and the Orgueil from C_{16} through C_{28}. The hydrocarbon present in the largest amounts for the Alais, Ivuna, and Orgueil meteorites was n–C_{23}, n–C_{17}, and n–C_{19}, respectively.

The Alais (one sample) contained 5·9 ppm paraffins; the Ivuna (one sample) 22·6 ppm; and those in the Orgueil (8 samples) ranged from $<$ 0·1 ppm (inside piece) to 25·8 ppm.

Isoprenoid hydrocarbons (through the C_{20} isoprenoid) were detected chromatographically in the meteorites although all the samples of Orgueil did not contain these hydrocarbons. No predominance of odd over even carbon-number alkanes in the range analyzed was observed except for Orgueil, and Orgueil did not show it for all samples. Analyses of inside samples and of samples from different sources showed the distribution of aliphatic hydrocarbons in Orgueil to be quasi-homogeneous qualitatively and heterogeneous quantitatively. The alkanes in Ivuna, and to a lesser extent Alais, appear to have bimodal distributions, the second maximum in both cases occurring at n–C_{23}.

Type II. Nine meteorites from this group were analyzed. These meteorites together with (i) the total amount of isoprenoids; (ii) the total amount of n-alkanes; (iii) the range of alkanes; and (iv) the paraffinic hydrocarbon present in largest amounts, respectively, for each are listed as follows: Al Rais (11·0 ppm; 38·5 ppm; C_{15}–C_{24}; and n–C_{17}), Bells (2·4 ppm; 9·8 ppm; C_{16}–C_{21}; and n–C_{16}), Boriskino (18·5 ppm; 54·3 ppm; C_{16}–C_{24}; and n–C_{18}), Cold Bokkeveld (6·8 ppm; 39·2 ppm; C_{15}–C_{24}; and n–C_{18}), Mighei (24·2 ppm; 103·5 ppm; C_{15}–C_{25}; and n–C_{18}), Murray (4·6 ppm; 30·7 ppm; C_{15}–C_{29}; and n–C_{21}), Nogoya (9·8 ppm; 82·6 ppm; C_{16}–C_{24}; and n–C_{20}), Renazzo (0·2 ppm; 8·7 ppm; C_{17}–C_{25}; and n–C_{19}), and Santa Cruz (35·4 ppm; 114·6 ppm; C_{16}–C_{25}; and n–C_{19}).

The saturated isoprenoid hydrocarbons, norpristane, pristane, and phytane, were identified chromatographically in all these meteorites. Mass spectral confirmation of the presence of isoprenoids was obtained for the Boriskino, Mighei, Murray, and Santa Cruz meteorites on the LKB type 9000 gas chromatograph-mass spectrometer combination. These were the only meteorites of this group analyzed by mass spectrometry. There is little doubt but that the other meteorites would have given the same results.

The meteorites in this group gave hydrocarbon patterns that are similar qualitatively in the C_{16}–C_{21} range. However, as indicated above, there is considerable variation in the amounts of hydrocarbon present. The analyses listed above are for the first sample obtained of the meteorites. In the case of the Boriskino, Cold Bokkeveld, Mighei, Murray, Renazzo, and Santa Cruz meteorites second samples were obtained. Only the Boriskino, Mighei, and Santa Cruz showed significant differences in the amounts of hydrocarbon present in the samples (the second samples contained 1/2 to 1/5 the amounts of hydrocarbon present in the first samples).

Type III. The meteorites together with the analytical results in the same order as given for the type II meteorites are listed below:

Felix (3·2 ppm; 124·8 ppm; C_{15}–C_{25}; and n–C_{19}), Grosnaja (3·0 ppm; 412·0 ppm; C_{17}–C_{29}; and n–C_{23}), Kaba (0·4 ppm; 11·5 ppm; C_{17}–C_{25}; and n–C_{19}), Lancé (14·1 ppm; 148·9 ppm; C_{16}–C_{25}; and n–C_{19}), Mokoia (7·8 ppm; 21·5 ppm; C_{15}–C_{24}; and n–C_{17}), Ornans (0·0 ppm; 8·3 ppm; C_{19}–C_{25}; and n–C_{21}), Vigarano (16·6 ppm; 206·4 ppm; C_{15}–C_{25}; and n-C_{18}), and Warrenton (0·1 ppm; 8·1 ppm; C_{16}–C_{25}; and n–C_{19}). The meteorites in this group do not show qualitative or quantitative homogeneity in the paraffinic hydrocarbons present. Four of the type III meteorites, i.e. Felix, Lancé, Mokoia, and Vigarano, have hydrocarbon patterns similar to those of the type II meteorites.

Except for Ornans, isoprenoid hydrocarbons were detected chromatographically in all these meteorites. The presence of norpristane, pristane, and phytane in the Mokoia meteorite was established unequivocally with the LKB type 9000 gas chromatograph-mass spectrometer combination.

The paraffinic hydrocarbons in the Warrenton meteorite appeared to have a bimodal distribution, the second maximum being at n–C_{23}. The Grosnaja meteorite contained only the smallest amounts of pristane and phytane, but large amounts of normal paraffinic hydrocarbons having a symmetrical distribution peaking at n–C_{23} as they are found in industrial waxes. This may indicate that the Grosnaja sample is contaminated.

HETEROGENEOUS GROUP OF METEORITES. These are meteorites we have analyzed that do not belong, or have not been assigned, to the three types of Wiik. These include the "non"-carbonaceous olivine-pigeonites Karoonda and Chainpur that on the basis of elemental analyses have been classified as types III-A and III-B, respectively, by Schmitt;[21] Indarch, an enstatite chondrite; Bruderheim, an olivine-hypersthene chondrite; St. Caprais*, apparently a carbonaceous chondrite but mislabeled; and graphite and troilite nodules from iron meteorites.

The Bruderheim, Chainpur, and Karoonda meteorites contain only traces of isoprenoid hydrocarbons and small amounts of alkanes (0·8, 2·8, and 3·2 ppm, respectively). The normal hydrocarbons in these meteorites range from C_{15} or C_{16} through C_{27}. The maximum peaks for Bruderheim and Chainpur are n-C_{22}, and for Karoonda n–C_{24}. Indarch contained only the slightest trace of hydrocarbons and resembled some inside samples of Orgueil.[8] The hydrocarbons of St. Caprais have a distribution resembling those for type II and certain type III meteorites. St. Caprais contains 26·1 ppm isoprenoids and 146·9 ppm alkanes. The range is C_{15} through C_{23}, with the maximum at n–C_{17}.

Analyses of paraffinic hydrocarbons from both outside and inside graphite nodules (Canyon Diablo and Odessa iron meteorites) and from a troilite nodule (Toluca iron meteorite) show that these samples contain hydrocarbons with qualitative distributions very similar to those of type II meteorites. The outside of the graphite nodules contained about 15 ppm paraffins, whereas, the inside samples contained only about 3 ppm. The outside of the troilite nodule contained less than 0·2 ppm hydrocarbons and the inside essentially zero (equivalent to a procedural blank). Chromatographic evidence for the saturated isoprenoid hydrocarbons norpristane, pristane, and phytane were obtained for all nodules. Mass spectrometric evidence for pristane and phytane in the outside of the Odessa nodule was established with the LKB type 9000 gas chromatograph-mass spectrometer combination.

Benzene Eluates

Meinschein[7] has shown that the aromatic hydrocarbons of carbonaceous chondrites are concentrated in the benzene eluates from elution

* See Nooner and Oró[20] for a discussion of the meteorite sample herein designated as St. Caprais.

chromatographic separation over silica gel. Since other non-aromatic-hydrocarbon compounds may also be present in the benzene eluates, quotation marks are placed around the term aromatic hydrocarbons when referring to the components of the meteoritic benzene eluates.

Compounds (probably predominantly aromatic hydrocarbons) of the benzene eluates were separated by gas chromatography for the following meteorites:

Type I: Alais, Ivuna, and Orgueil. Type II: Al Rais, Bells, Boriskino, Cold Bokkeveld, Mighei, Murray, Nogoya, Renazzo, and Santa Cruz. Type III: Felix, Indarch, Lancé, Mokoia, Ornans, Vigarano, and Warrenton; and fourth heterogeneous group: Canyon Diablo (graphite nodule), Chainpur, Karoonda, Odessa (graphite nodule), and St. Caprais. With the exception of the Al Rais, Bells, Mokoia, Murray, Orgueil, and Santa Cruz meteorites, which contained 4, 16, 4 to 83, 1 to 8, <1 to 172 and <1 to 10 ppm, respectively, of "aromatic hydrocarbons" separable by gas chromatography, the meteorites contained less than 1 ppm of these compounds. The graphite nodules in particular contained essentially no ($<0{\cdot}1$ ppm) "aromatic hydrocarbons". Since three-ring polyaromatic compounds are about the maximum (in number of rings) eluated by our gas chromatographic procedure, these quantitative estimates may be low. Ultraviolet absorption spectra were run on the benzene eluates of the Orgueil and Mokoia meteorites. Absorption spectra were obtained at 334·0, 343·0, 358·0 and 380·0 mμ for Orgueil, which suggests the presence of three-ring and angular four-ring polyaromatic hydrocarbons. The similar analysis of Mokoia showed very little ultraviolet absorption in the 330–390 mμ region.

The chromatograms of some of the eluates were very complex, showing 60 to 140 components. Tentative identifications, based on retention time, indicated that the following compounds may be included with those present in the eluates: naphthalene, 2-methylnaphthalene, biphenyl, acenaphthalene, acenapthene, fluorene, phenanthrene, and anthracene. The analyses of the benzene eluates of the meteorites will be reported in detail elsewhere[22].

High Resolution Gas Chromatography

Gas chromatography using extremely long stainless steel capillaries (700 to 1000 ft.), which have been specially cleaned and coated[17] to give maximum separation and efficiency, were used to (i) explain the

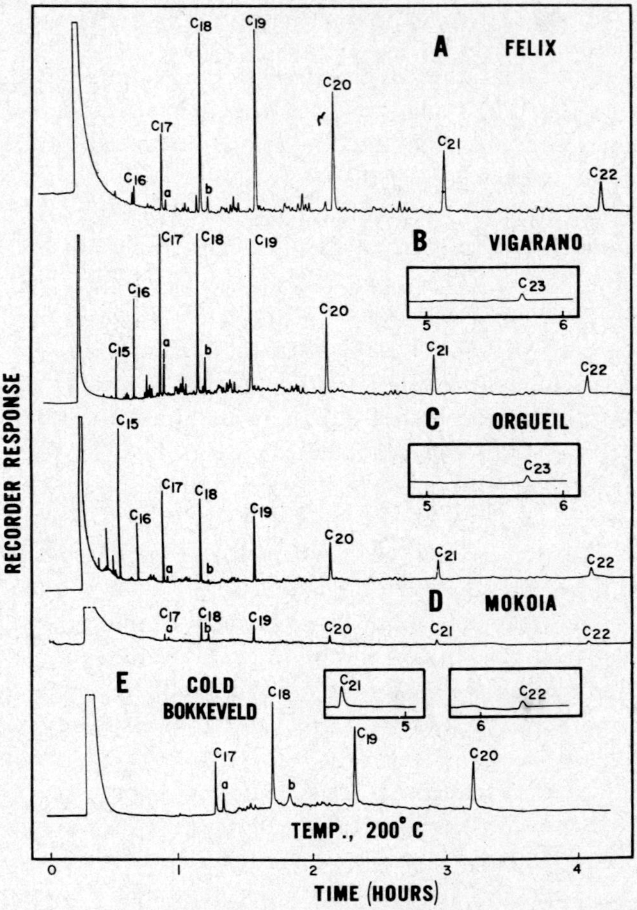

FIGURE 1. — Gas chromatographic separation of alkanes from the Felix (type III), Vigarano (type III), Orgueil (type I), Mokoia (type III), and Cold Bokkeveld (type II) meteorites with a long capillary high resolution column (Igepal CO–880 and SF–96). Stainless steel tubing, 210 m long × 0·05 cm I.D., cleaned, and coated with Igepal CO–880 and SF–96. Barber-Colman series 5000 gas chromatogram equipped with a flame ionization detector. (A) Helium pressure, 1760 g/cm^2. Small split. Felix meteorite extracted (sample from Olsen), 1·3 g. About 1/10 of the n-heptane eluate was injected. Attenuation, 30. Isothermal at 200° C. (B) Analytical equipment and conditions same as in (A). Vigarano meteorite extracted (sample from Olsen), 0·8 g. About 1/16 of the n-heptane eluate was injected. (C) Analytical equipment and conditions (except attenuation) same as in (A). Orgueil meteorite extracted. (No. 1 sample from Cavaillé), 2·1 g. About 1/8 of the n-heptane elute was injected. Attenuation, 10 until just after C_{15} was eluted, then 30 for the remainder of the run. (D) Analytical equipment and conditions same as in (A). Mokoia meteorite extracted (sample from Mueller), 0·5 g. About 1/20 of the n-heptane eluate was injected. (E) Column same as in (A). F and M model 810 gas chromatogram equipped with a flame ionization detector. Helium pressure, 1760 g/cm^2. No split. Cold Bokkeveld meteorite extracted (sample from Mason), 0·4 g. About 1/30 of the n-heptane eluate was injected. Range, 10. Attentuation, 2. Isothermal at 200° C.

"baseline hump" observed when alkanes from some meteorites are analyzed, and (ii) separate the extremely large number of compounds (60 to 140) found in some benzene eluates.

A large hump obscuring the baseline occurs when aliphatic hydrocarbons from meteorites and other samples, e.g. ancient sediments, are analyzed under conditions which give relatively narrow separations of the normal alkanes[8]. Programmed runs made on SE-30 packed columns and Apiezon L capillaries sometimes give especially large humps. Analyses of blanks showed the humps could not be attributed to column bleed alone. The relative size of the humps could be somewhat reduced by dilution and adjusting sample size. Since the hump was most likely caused by unresolved hydrocarbons, we decided to analyze hydrocarbons, which ordinarily gave significant baseline humps, from the Orgueil, Cold Bokkeveld, Felix, Mokoia, and Vigarano meteorites on one of the high-resolving columns (700-ft stainless steel capillary coated with SF-96 and Igepal CO-880). The results obtained are shown in Fig. 1. No significant baseline humps are observed, and up to 20 compounds can be found between some of the adjacent major peaks.

Some of the benzene eluates, e.g. those from the Mokoia, Murray and Orgueil meteorites give similar results (large humps with poor separation) when analyzed on short fast columns. By using high-resolution columns, e.g. a 1000-ft stainless steel capillary coated with Polysev, baseline separation of up to 140 components was obtained.

Identification of Isoprenoids

We have previously obtained chromatographic (separation on three liquid phases, Apiezon L, Carbowax 20M terminated with *p*-phthalic acid, and Polysev) and to some extent mass spectral evidence[8] for the presence of pristane (2, 6, 10, 14-tetramethylpentadecane) and phytane (2, 6, 10, 14-tetramethylhexadecane) in extracts of meteorites. Additional chromatographic evidence on an additional phase is given for these compounds in the Orgueil, Cold Bokkeveld, Felix, Mokoia, and Vigarano meteorites in Fig. 1.

We have now obtained unequivocal evidence for the presence of pristane and phytane in the Boriskino, Mighei, Murray, Santa Cruz, Mokoia, and Odessa (graphite nodule) meteorites. With the exception of Mighei, Santa Cruz, and Odessa, norpristane (2, 6, 10-trimethylpentadecane) spectra were also obtained. A typical gas chromatographic trace and spectra

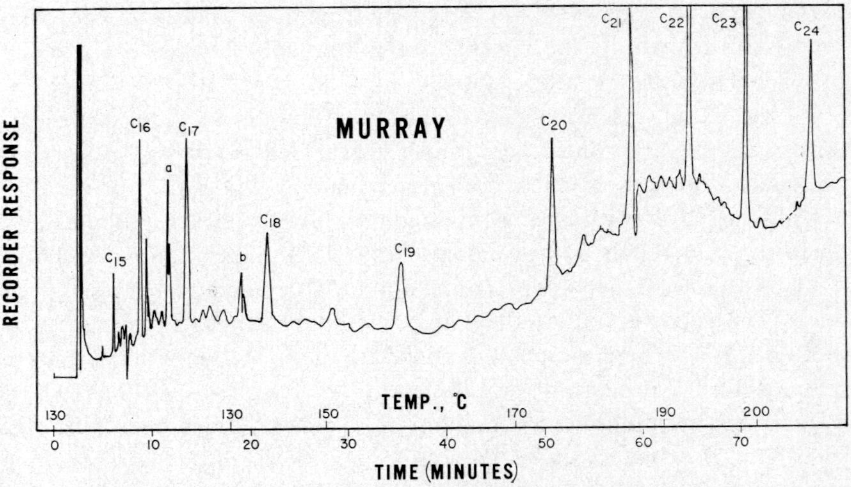

FIGURE 2. — Gas chromatographic separation of alkanes from the Murray meteorite (type II) with a capillary column (Polysev). Stainless steel tubing, 75 m long×0·076 cm I.D., coated with Polysev and installed in an LKB type 9000 gas chromatograph-mass spectrometer combination. Helium flow, 24 ml per min. No split. Meteorite extracted (sample from Mueller), 0·4 g. About 1/2 of the n-heptane eluate was injected. Gain, 1. Temperature programming shown on figure.

(those of the Murray) are shown in Figs. 2 and 3, respectively. The mass spectrometric analysis, of course, cannot tell whether or not these isoprenoid hydrocarbons are of biogenic origin. Pristane and phytane are designated by a and b, respectively.

Discussion

By carefully cleaning equipment and checking quality of solvents we demonstrated with numerous procedural blanks that the hydrocarbons observed could not be contaminants introduced during the meteorite analyses. Based on the information now available, the source of the hydrocarbons in meteorites cannot be unequivocally assigned. Perhaps the answer is near at hand from the analysis of returned lunar samples. Analysis of a freshly fallen and carefully handled meteorite might also indicate the more probable sources of the hydrocarbons.

There are several possible explanations of the hydrocarbons in meteorites:

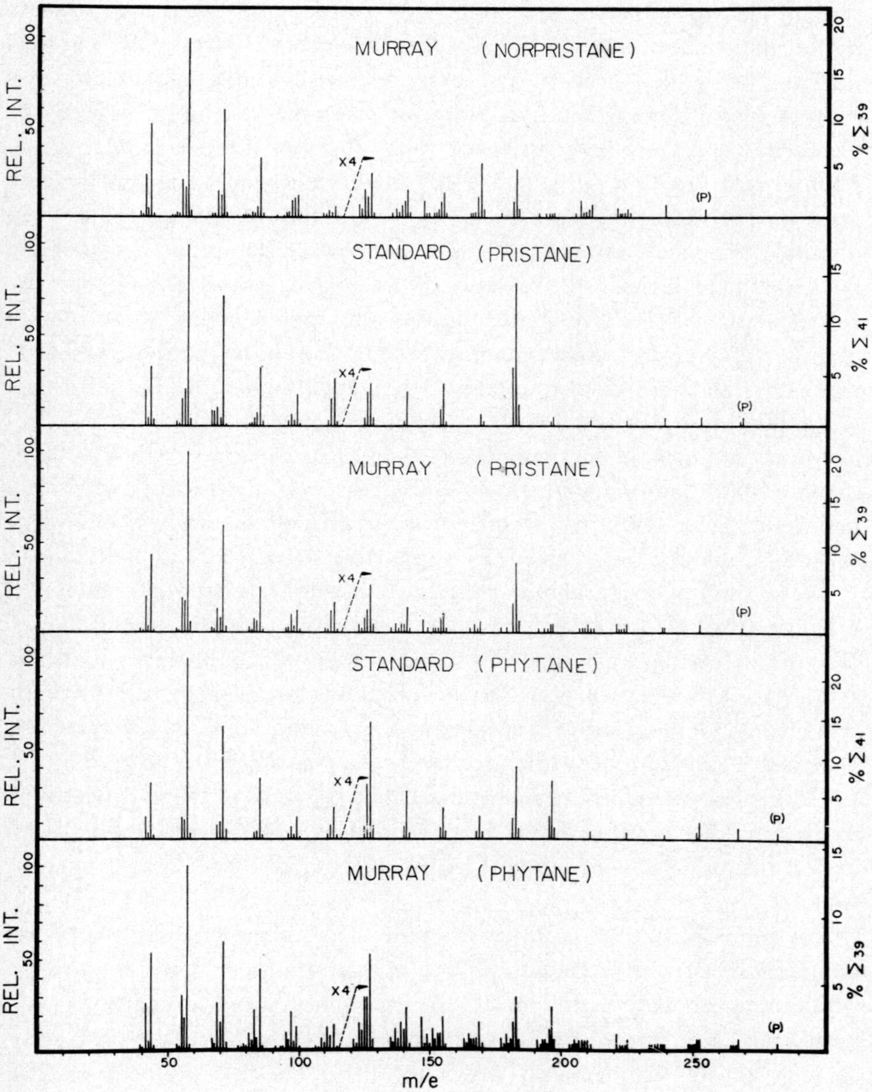

FIGURE 3. — Mass spectra of isoprenoids from the Murray meteorite (type II). Mass spectra were obtained using an LKB type 9000 gas chromatograph-mass spectrometer combination.

(i) they may be remnants of extraterrestrial life; (ii) they may be fossil terrestrial hydrocarbons introduced after arrival of the meteorites on earth; (iii) they may be remnants of life that existed on the earth-moon system billions of years ago and were subsequently returned to earth by lunar meteoritic impacts;[23] (iv) they may be terrestrial hydrocarbons that are being produced at the present time by microorganisms; and (v) they may be abiogenic hydrocarbons produced during genesis of the meteorites and their flight in space. The relatively large amounts of isoprenoid hydrocarbons in some of the meteorites tend to support a biogenic origin of the hydrocarbons,[8, 24] even though evidence has been obtained that small amounts of isoprenoid hydrocarbons can be produced by abiological means.[25] We have found hydrocarbons giving the type II meteoritic hydrocarbon pattern in Precambrian sediments;[8] petroleum crudes, varnish, and cobalt naphthenate;[26] water containing industrial wastes;[27] and terrestrial graphites.[26] Except for possibly the terrestrial graphites, these are fossil hydrocarbons; the hydrocarbons in the Precambrian sediments being indigenous, and those in varnish, cobalt naphthenate, and contaminated water having their ultimate origin in crude petroleum. The similarity of meteoritic hydrocarbons and fossil hydrocarbons could be interpreted as supporting both (ii) and (iii) above.

The presence of paraffinic hydrocarbons in terrestrial graphite and graphite from iron meteorites having distributions similar to the hydrocarbons in certain carbonaceous chondrites is evidence against abiotic synthesis. The hydrocarbons in the graphites are almost certainly not indigenous. Geological evidence about the origin of one of the graphites analyzed[28] and the heat histories of iron meteorites[29] indicate that the hydrocarbons must have been acquired. The presence of the hydrocarbons in the graphites could possibly be explained by contamination with fossil hydrocarbons or by deposition of hydrocarbons produced by microorganisms.

The compositional complexity of some of the benzene eluates supports (i) abiogenesis of at least part of the hydrocarbons in meteorites or (ii) contamination with highly metamorphized biogenic hydrocarbons, e.g. petroleum. No support for extraterrestrial life as a source of the hydrocarbons can be justified until all of the other possible explanations can be excluded. This is probably not possible with the samples now available.

Conclusions

Paraffinic hydrocarbons were found in varying amounts ($< 0{\cdot}1$–415 ppm) in all of the meteorites. The origin of these hydrocarbons is not known; however, one of the meteorites (the Grosnaja) may have been contaminated with paraffin wax during storage. A study of the chromatograms and mass spectrograms from the analysis of the 29 meteorites permits the following conclusions.

1. Chromatographic evidence of pristane and phytane was obtained for almost all the meteorites. In most of the type II and in some of the type III meteorites evidence for norpristane was also obtained. Mass spectrometric identification of isoprenoids was obtained for the Boriskino, Mighei, Murray, Santa Cruz, Mokoia and Odessa (graphite nodule) meteorites.

2. Analysis of several inside pieces and samples from different sources of Orgueil show the hydrocarbons in this meteorite to be heterogeneous quantitatively and quasi-homogeneous qualitatively. Only Orgueil showed any odd over even C-number preference in the alkanes, and it did not show it for all the analyses. Possible bimodal hydrocarbon distributions were observed for the Ivuna, Alais, and Warrenton meteorites.

3. Type II meteorites were the only group in which members have consistently similar chromatographic patterns for their alkane content. Between-group comparisons showed Mokoia, Lancé, St. Caprais, and graphite nodules from the Canyon Diablo and Odessa meteorites to have hydrocarbon patterns similar to the type II meteorites.

4. Many of the meteoritic hydrocarbon patterns were similar to the patterns given by fossil hydrocarbons (hydrocarbons in Precambrian sediments, petroleum crudes, and products derived from petroleum).

5. There was no correlation between the amount of paraffinic hydrocarbons extracted from a meteorite and the amount of material in the corresponding benzene eluate. The amount of material in the benzene eluates was usually small ($<$ 0·1–172 ppm). Some meteorites, especially Orgueil and Mokoia, showed quantitatively heterogeneous distributions of the "aromatic hydrocarbons".

Acknowledgements

We thank the donors of the meteorite samples, E. Anders, Enrico Fermi Institute for Nuclear Studies, University of Chicago, Chicago, Ill.; A. Cavaillé, Museum of Natural History, Montauban, France; R. E. Folinsbee, Department of Geology, University of Alberta, Edmonton, Alberta, Canada; K. Fredricksson, Smithsonian Institution, U.S. National Museum, Washington D. C; G. Kurat, Mineralogy-Petrography, Division, Museum of Natural History, Vienna, Austria; E. P. Henderson,

Smithsonian Institution, U.S. National Museum, Washington, D. C.; G. W. Hodgson, Research Council of Alberta, Edmonton, Alberta, Canada; G. I Huss, American Meteorite Laboratory, Denver, Col.; B. Mason, American Museum of Natural History, New York, N. Y.; C. B. Moore, Arizona State University and Nininger Meteorite Collection, Tempe, Arizona; G. Mueller, Department of Crystallography, Birkbeck College, University of London, London, England; B. Nagy, Department of Chemistry, University of California (San Diego), La Jolla, Cal.; E. J. Olsen, Department of Geology, Chicago Natural History Museum, Chicago, Illinois; and R. A. Schmitt, General Atomic Division of General Dynamics Corporation, San Diego, Cal., whose generous response to our request for samples made this investigation possible. We also thank E. Gelpi and H. B. Skewes, University of Houston, Houston, Texas; A. Struck, Perkin-Elmer Corp., Norwalk, Conn., and S. A. Wikström, Baylor University College of Medicine, Houston, Texas for assistance in obtaining mass spectra. We are also grateful to V. J. Coates, Perkin-Elmer Corporation, Norwalk, Conn. and E. C. Horning, Baylor University College of Medicine, Houston, Texas for making the gas chromatography-mass spectrometry combinations of their respective laboratories available to us and to A. Zlatkis and P. Simmonds, University of Houston, Houston, Texas for advice and assistance in carrying out the gas chromatographic analyses.

References

1. L. A. Pages and Dhombres–Firmas. *J. phys. chim. d'hist. nat.* **62,** 440 (1806); G. Monge, A. F. Fourcroy, D. L. Berthollet and L. N. Vauquelin. *Ann. chim. et phys.* **59,** 34 (1806); L. J. Thenard, *Ann. chim. et phys.* **59,** 103 (1806); *J. phys. chim. d'hist. nat.* **63,** 56 (1806).
2. F. Wöhler. *S. B. Akad. Wiss. Wien,* **33,** 207 (1858); *ibid.* **34,** 7 (1859); *ibid.* **35,** 6 (1859); S. Meunier. *Comp. rend.* **109,** 977 (1889); H. Roscoe. *Phil. Mag.* **25,** 319 (1863); G. Tschermak. *S. B. Akad. Wiss. Wien., Abt.* **2, 62,** 857 (1870); *ibid.* 860; C. Friedheim. *Sitzung. konigl, preuss. Akad. Wissensch., Berlin,* 364 (1888); P. Berthelot. *Comp. rend.* **67,** 849 (1868).
3. E. Cohen. *Meteoritenkunde.* E. Schweizerbart'sche Verlagshandlung, Stuttgart, 1894.
4. G. Mueller. *Geochim. Cosmochim. Acta,* **4,** 1 (1953).
5. I. R. Kaplan, E. T. Degens, and J. H. Reuter. *Geochim. Cosmochim. Acta,* **27,** 805 (1963); M. Calvin and S. K. Vaughn. *Space Research,* **1,** 1171 (1960); M. H. Briggs. *Nature,* **191,** 1137 (1961); K. I. Sztrokay, V. Tolnay, and M. Faldervari–Vogl. *Acta Geol. Acad. Sci. Hungarica,* **7,** 57 (1961).
6. B. Nagy, W. G. Meinschein, and D. J. Hennessy. *Ann. N.Y. Acad. Sci.* **93,** 25 (1961).

7. W. G. Meinschein, B. Nagy, and D. J. Hennessy. *Ann. N.Y. Acad. Sci.* **108,** 553 (1963).
8. J. Oró, D. W. Nooner, A. Zlatkis and S. A. Wikström. *Life Sci. Space Res.* **4,** 63 (1966).
9. M. H. Briggs and G. Mamikunian. *Space Science Rewiews,* **1,** 647, (1963).
10. M. H. Studier, R. Hayatsu, and E. Anders. *Science,* **149,** 1455 (1965).
11. G. W. Meinschein. Esso Research and Engineering Co. Contract No. NASw-508 Quarterly Reports, April 1, 1964.
12. J. E. Lovelock and S. R. Lipsky. *J. Amer. Chem. Soc.* **82,** 431 (1960).
13. E. O. A. Haahti. *Scand. J. Clin. and Lab. Invest.* **13,** Supplement 59 (1961).
14. R. Ryhage. *Anal. Chem.* **36,** 759 (1964).
15. R. Ryhage, S. Wikström, and G. R. Waller. *Anal. Chem.* **37,** 435 (1965).
16. J. T. Watson and K. Biemann. *Anal. Chem.* **36,** 1135 (1964).
17. R. Teranishi. Communication to A. Zlatkis, 1966.
18. H. B. Wiik. *Geochim. Cosmochim. Acta,* **9,** 279 (1956).
19. B. Mason. *Meteorites.* Wiley, New York, 1962.
20. D. W. Nooner and J. Oró. *Geochim. Cosmochim. Acta.* **31,** 1359 (1967).
21. R. A. Schmitt. General Atomic (Division of General Dynamics), Contract No. NASw-843, Quarterly Progress Report Ga-6414, March, April, May, 1965.
22. R. J. Olson, J. Oró, and A. Zlatkis. *Geochim. Cosmochim. Acta.* **31,** 1935 (1967).
23. H. C. Urey. *Science,* **147,** 1262 (1965); *Nature,* **193,** 1119 (1962).
24. W. G. Meinschein, E. S. Barghoorn, and J. W. Schopf. *Science,* **145,** 262 (1964); E. S. Barghoorn, W. G. Meinschein, and J. W. Schopf. *Science,* **148,** 46 (1965); G. Eglinton, P. M. Scott, T. Belsky, A. L. Burlingame and M. Calvin. *Science,* **145,** 262 (1964); G. Eglinton, P. M. Scott, T. Belsky, A. L. Burlingame, W. Richter, and M. Calvin–Space Sciences Laboratory, University of California, Technical Report Series No. 6, Issue No. 9, January, 1965; J. Oró, D. W. Nooner, A. Zlatkis, S. A. Wikström, and E. S. Barghoorn. *Science,* **148,** 77 (1965); T. Belsky, R. B. Johns, E. D. McCarthy, A. L. Burlingame, W. Richter, and M. Calvin. *Nature,* **206,** 446 (1965); W. G. Meinschein. *Science,* **150,** 601 (1965).
25. M. H. Studier, R. Hayatsu, and E. Anders. Presented at the 4th meeting for the Analysis of Carbon Compounds in Carbonaceous Chondrites, April 21, 1966, Washington, D. C.; J. Oró, J. Han, and D. W. Nooner. *Ibid.*
26. J. Oró and D. W. Nooner. To be submitted to *Geochim. Cosmochim. Acta.*
27. J. Oró and D. W. Nooner. Unpublished experiments.
28. M. F. Carman. Private communication, 1966.
29. J. A. Wood. *Icarus,* **3,** 429 (1964).

PRIMITIVE PHOTOSYNTHETIC ORGANISMS

by
PATRICK ECHLIN

The Botany School, University of Cambridge

PHOTOSYNTHESIS provides an energy source for biological processes by converting a fraction of the radiant solar energy into stored chemical energy in organic molecules. This process is the ultimate energy source for all biological activity. Man depends on photosynthesis for food and his store of fossil fuels. Every year about a hundred billion (10^{11}) tons of carbon dioxide are converted to carbohydrate by photosynthesis and in terms of energy conversion this is many times more than all energy man currently produces from fossil fuels. Yet in terms of the total solar radiation of all wavelengths reaching the earth, the overall efficiency of photosynthesis in nature is less than 2%.

The overall reaction of photosynthesis in green plants may be summarized in the following equation: $CO_2 + H_2O + \text{Light} \xrightarrow{\text{chlorophyll}} (CH_2O) + O_2$. Photosynthesis involves conversion of carbon dioxide and water into carbohydrate with the liberation of oxygen, using solar radiation as an energy source and a natural pigment (chlorophyll a) as a photocatalyst. The energy of the light falling on the chloroplasts is stored mostly as oxidation-reduction energy by the transfer of electrons to a higher energy state. A small fraction of the light energy is involved in the formation of energy rich phosphate compounds, and these too ultimately become part of the oxidation-reduction energy used in the dark fixation of CO_2 to carbohydrate.

The chloroplast is the organelle in higher plants where this energy conversion takes place and may be considered the photosynthetic counterpart of the mitochondrion, both organelles being more or less autonomous within the cell. Both are major sites of energy conversion, with the additional photosynthetic function of the chloroplast to convert carbon dioxide to carbohydrate. The chloroplasts of higher plants and some algae are seen in the electron microscope to be enclosed by a double membrane and to contain dense areas or grana. The grana consist of closely stacked layers of disks each of which is made up of a pair of membranes (lamellae) about 70 Å thick enclosing a space 100 Å thick. The grana lamellae are connected to less dense lamellar structures, the stroma lamellae. This lamellar structure is also found in the blue-green algae and in certain of the photosynthetic bacteria. But in these organisms the membranes of the photosynthetic apparatus are not contained within a discrete structure but ramify through the whole cell. Other photosynthetic bacteria do not appear to have lamellae, the photosynthetic machinery being confined to vesicular chromatophores dispersed around the cell periphery.

The developmental sequences which lead to the formation of the chloroplast in the modern living plant have been established with some degree of certainty. The greening that occurs when etiolated plants are exposed to light provides a spectacular example of the development of an organelle in response to an external stimulus. The synthesis of chlorophyll is accompanied by profound morphological changes as proplastids differentiate into mature chloroplasts. This phenomenon of the development and differentiation of the photosynthetic apparatus has been particularly well investigated in the alga *Euglena*, where the whole process occurs upon exposure of dark-grown organisms to light in the presence of suitable inorganic salts. Development of the mature chloroplast from pro-plastids (small bodies in the cytoplasm of dark-grown cells) involves the synthesis of photosynthetic pigments, enzymes and co-factors used in the photosynthetic process. Changes also occur in lipids and proteins, all at the expense of intracellular material as seen in the considerable turnover of ribonucleic acid and protein. The formation of the complete chloroplast is under genetic control. A similar process is seen in the development of the pigment system in photosynthetic bacteria and is also considerably influenced by the environment, particularly by the light intensity.

Although much is known about the structure, mode of action and development of the photosynthetic machinery in a wide range of organisms, very little is known with certainty when in geological time photosynthesis began, or the nature of the organisms in which this process may have occurred. It is obvious from the vast deposits of fossil fuels that photosynthesis is an ancient biochemical mechanism and according to Berkner and Marshall probably started about 3×10^9 years ago, but did not reach its maximum until the development of a land flora about 4–450 million years ago[5, 6, 7, 8]. The indications are that with slight variation during various geological ages, this maximum has been maintained up to modern times. The earth is about 4·5 billion years old, with the oldest rocks about 3 billion years old and although an element of scepticism has traditionally prevailed among geologists and biologists over the existence and nature of pre-Cambrian life, there is now good evidence, which suggests that there was widespread biological activity as far back as half the usually accepted age of the earth. The Earth's history may be divided into two parts, the pre-Cambrian which has a sparse fossil record, and the Phanerozoic beginning with the Cambrian 600 million years ago, which is rich in fossil records. For many years the only fossils which were known to be widespread and abundant in the pre-Cambrian were the stromatolites which are generally attributed to the blue-green algae. These fossils appear in most pre-Cambrian limestones and include those from the African Bulawayan dated at $2{\cdot}7\times10^9$ years old.[27] The animal record is practically non-existent, in the pre-Cambrian, and it is only at the base of the Cambrian period itself that there is an appearance of abundant animal life. No attempt will be made here to review the literature pertaining to pre-Cambrian fossils, and the reader is referred to the comprehensive reviews by Glaessner[25] and Rutten.[37] It is, however, of significance to consider some of the more recent discoveries of microfossils in the pre-Cambrian and the geochemical evidence of life during this period.

Barghoorn and Schopf[3] found collections of structurally and organically preserved microfossils from upper pre-Cambrian limestones of Central Australia dated at between 700 and 900 million years old. The microorganisms which exhibit distinct histological preservation are interpreted as consisting of photosynthetic green and blue-green algae made up of filaments and aggregates of single cells. Barghoorn and Schopf consider that these fossils may be representative of a transitional period in the evolution of pre-Cambrian plant life during which the simpler green algae were emerging and diversifying.

Barghoorn, Meinschein and Schopf[2] and Meinschein, Barghoorn and Schopf[28] found some rather poorly preserved algal and possibly fungal remains from shales of the billion year old Nonesuch formation of northern Michigan. These workers and Eglington *et al.*[20] were able to show that the same microfossils are associated with small amounts of crude oil, and biologically significant organic molecules such as porphyrins and the isoprenoid hydrocarbons phytane and pristane.

Phytane and pristane are biologically interesting because they are degradation products of the photosynthetic pigment chlorophyll. Phytane arises by reduction and dehydration of the branched chain alcohol phytol, and one of the carboxy groups of chlorophyll is esterified with this phytol alcohol. Pristane which has one carbon less than phytane is also derived from phytol, and both pristane and phytane are related to phytol which in turn is considered to be characteristic of the presence of chlorophyll. These two hydrocarbons are also found in certain animal sources, but barring contamination it is more likely that the chlorophyll and the products of chlorophyllous photosynthesis were the source of the oil found associated with the Nonesuch shale.

Tyler and Barghoorn[39] and more recently Barhoorn and his associates[1] have discovered structurally preserved fossils in rocks along the northern shore of Lake Superior, which morphologically resemble modern blue-green algae. On the basis of Ar^{40}/K^{40} and Rb^{87}/Sr^{87} dating the rocks are estimated to be about of $2{\cdot}0 \times 10^9$ years old. The most abundant organism was a filamentous assemblage which closely resembles *Oscillatoria* and *Lyngbya*, two modern blue-green algae, and these were associated with spherical bodies which could have been uni-cellular blue-green algae or the endospores of filamentous forms. It should be noted however, that although most of the organisms found in this particular fossil bed closely resembled blue-green algae, certain affinities were shown with other groups of microorganisms such as bacteria and fungi, and in some cases no living counterpart is believed to be in existence.

Schopf and his associates have also been investigating these fossil-bearing rocks and have found what appears to be the oldest definite occurrence of bacteria in fossil records.[38] Examination of the material in the electron microscope revealed rods and cocci which in shape and size resemble certain modern iron bacteria. Chemical studies of the Gunflint Chert deposits, as these fossil beds are called, indicate that the organic matter was probably produced from photosynthesis, but it is not known whether the palaeoenvironment at the time was reducing or oxidizing[1, 34] Chromatographic and mass spectroscopic studies indicate that the rocks containing the fossils, also contain phytane and pristane, both of which are the degradation products of the phytol alcohol of chlorophyll.[34] Pristane and phytane, are found not only as derivatives of green plant chlorophylls, but also in bacteria, and rather widely among animals as well, so the possibility is not excluded that the Gunflint microflora may have contained bacteria. It also cannot be entirely ruled out that these and other isoprene hydrocarbons could have been produced abiotically.

Barghoorn and Schopf are justifiably hesitant to draw any conclusions about the nature of the gaseous composition of the earth's atmosphere during Gunflint times, but Barghoorn speculates that this era may represent an intermediate stage in the evolution of the highly aerobic atmosphere which must have prevailed at the beginning of the Cambrian era. Cloud and his associates in Minneapolis have recently made a similar study of rocks from the Soudan Iron formation in N. Minnesota which are estimated to be $2{\cdot}7 \times 10^9$ years old.[(13)] These rocks contain pyrite balls possessing microstructures which may be of biological origin, although the morphological evidence is inconclusive. The C^{13}/C^{12} and S^{34}/S^{32} ratios and the finding of pristane and phytane are both consistent with a biological origin, though it is not certain whether the hydrocarbons are endemic to the rocks. Cloud has also discussed at some length the significance of the pre-Cambrian flora described by Barghoorn,[(12)] and believes that the microstructures represent the remains of autotrophic organisms which were in the line of evolution towards green plant photosynthesis.

In a more recent investigation, the Minneapolis groups have undertaken an electron-microscopic investigation of the microflora from the Gunflint deposits.[(11)] Elaborate precautions were taken both to exclude and/or recognize vital contaminants and artefacts of non-vital origin, which could mimic the morphology of organisms. They conclude that the Gunflint microflora consists of an association of early and simple organisms in which both bacterial and blue-green algal stages of development may be present. Earlier this summer[(4)] Barghoorn and Schopf reported finding microfossils from the pre-Cambrian of South Africa. The minute organisms, which have been placed in the new genus Eobacterium, were found organically and structurally preserved in the black chert of the Fig Tree series and dated by the Rb/Sr method at $3{\cdot}1 \times 10^9$ years old. Optical and electron microscopy of thin sections, macerations and surface replicas revealed a number of uni-cellular isolated rods about half a micron long and a quarter of a micron across. Their studies also revealed filamentous organic structures, which may be of biological origin, and complex alkanes which appear to contain small amounts of the isoprenoids phytane and pristane. These organic remnants are the oldest known evidence of biological organization in the geological record, and would indicate that life on earth must have originated in the preceding third of the earth's history.

Barghoorn and Schopf point out that because of insufficient information regarding the palaeobiochemistry of these deposits, no valid conclusions can be drawn with regard to the physiology of these microfossils and the environment in which they may have lived.

The universe in which we live is believed to be composed largely of hydrogen and helium and this factor together with terrestrial geochemical evidence strongly suggests that at an early stage in its history the terrestrial atmosphere was strongly reducing and lacked oxygen. Oparin,[32, 33] Calvin,[10] Wald[41] and many others have put forward ideas and speculations concerning terrestrial biogenesis. The majority of workers in this field visualize the evolution of the simple cell through various stages.[23] Colloidal coacervates with associating symbiotic macromolecules gradually formed a limiting membrane, which in turn leads to the possibility of different concentrations, or species of material being able to exist inside and outside this primitive "cell". The first true organisms were probably anaerobic heterotrophs which obtained their energy from the partial oxidation of the abundant store of geothermally produced organic substances, such as some of the abiogenic hydrocarbons. These forms gave rise to organisms in which energy from absorbed light added to the anaerobic metabolism, and these in turn later developed photosynthetic systems capable of carbon assimilation. These primitive anaerobic photosynthetic forms are believed to have given rise to organisms capable of the photolysis of water and the consequent release of molecular oxygen. Aerobes which then used this oxygen to respire more efficiently established the metabolic foundations of the organisms found in the modern biosphere.

In a study of biochemical and biological evolution, Commoner has pointed out that those enzymes which are specifically associated with aerobic metabolism are closely associated with complex intracellular structures.[14] The relevant enzymes are integrated into mitochondria and chloroplasts, and their electron transport systems are tightly bound to their respective carriers. In contrast, those electron carriers involved in anaerobic metabolism are not integrated into subcellular organelles and are relatively loosely bound. Such a relationship supports the contention that the origin of highly organized structures may have occurred in time somewhere between the development of anaerobic metabolic systems and systems which use molecular oxygen. It is of some significance that bacteria and blue-green algae, which appear close to the start of the sequence of biological evolution, possess relatively simple intracellular structures, and even in the forms which are aerobic, the enzymes associated with oxidative metabolism are located on the cell membrane or on simple elaborations of this structure.

There have been several suggestions as to both the mode and time of origin of oxygen in the atmosphere. Berkner and Marshall have calculated that the atmosphere began to contain oxygen approximately $3{\cdot}0 \times 10^9$ years ago. Free oxygen must be derived chiefly from the break up of water into hydrogen and oxygen, and can either occur by the photodissociation of water vapour by ultraviolet light or by the process which occurs during green plant photosynthesis. Urey[40] has shown that oxygen production by photodissociation becomes self-regulated, because the production of oxygen shades the water vapour from solar radiation and only a very small amount of oxygen would build up in the atmosphere. The rise of oxygen in the atmosphere could only have occurred through a process of photosynthesis. Only when the rate of production of oxygen exceeded the balance of photodissociation to O and O_3 by UV radiation and the consequent loss, could the equilibrium be upset and oxygen begin to accumulate.

Following the generation of photosynthetic oxygen and the consequent increase to 1% of present atmosphere level, an oxidative metabolism could evolve, which would provide organisms with many times more energy per molecule, than the more primitive fermentation processes. Berkner and Marshall consider that there were two critical phases in the development of an oxygen-containing atmosphere. In the first phase, which is believed to have occurred at the beginning of the Palaeozoic era 600×10^6 years ago, the oxygen level reached 1% of present day levels, and would thus permit aerobic processes to occur. But due to high UV radiation levels, any living organisms would be confined to the water bodies. It has been calculated[8] that at least thirty feet of water would be required to protect cells from the lethal effects of ultra-violet light. The second critical phase, which is thought to have occurred 420 million years ago at the beginning of the Silurian period, was when the oxygen level reached 10% of present atmospheric levels. This would allow the development of a land flora and fauna, since at such oxygen concentrations the UV-induced ozone production would have built up to sufficient levels to become self-regulatory and shield the earth from lethal radiation. There is no evidence of advanced forms of life until the Cambrian period 600 million years ago, and it is assumed that there was insufficient oxygen in the atmosphere to support the development of organisms containing specialized structures. It is significant to note that no multicellular obligate anaerobic organisms are known at present.

Having considered the palaeo-environment during the time when photosynthesis began to develop, let us briefly consider the organisms which are thought to have been living during this era. The fossil evidence points to the existence of the blue-green algae and the bacteria. Considering the modern counterparts of these two groups of organisms, although containing sophisticated biochemical machinery, they are structurally simpler than the higher plants and animals.[15] For convenience it is possible to group the blue-green algae and bacteria into the Pro-caryota. (All other plants and animals are grouped into the Eucaryota.) Organisms placed in the Pro-caryota do not contain the complex membranous organelles associated with oxidative phosphorylation and energy production, do not possess discrete nuclei and lack the sophisticated genetic divisional mechanism of higher organisms. There is also some biochemical evidence which supports including the bacteria and blue-green algae in the Pro-caryota. All available evidence suggest that sterols, which are widely dispersed in the Eucaryota, are absent in the Pro-caryota. A common feature of the Pro-caryotic cell wall is the presence of the amino acid, diamino-pimelic acid, which has not been found in the cell wall of the Eucaryotes. A reflection of this may be seen in the increased sensitivity of blue-green algae and bacteria to penicillin. This morphological simplicity, coupled with a complex biochemistry, would permit a radiation-induced diversity of asexual genetic expression, and is presumably the type of structure which would be likely to be evolved during terrestrial biogenesis.

It is of interest to note that with the exception of a few obligate anaerobes, porphyrins and tetra-pyrroles are found in all living cells. Tetra-pyrroles in the form of the chlorophylls and cytochromes are strong visible light absorbers, and the former have a vital function in the transformation of radiant energy into a form available for the metabolism of photosynthetic forms of life. Geochemical evidence suggests that such compounds were found in the earliest organisms where presumably they were selected in the initial stages of the growth of an oxygenic atmosphere because of the protection they afforded to the highly-reduced nucleic-acid genome of the primitive cellular systems. They were subsequently associated with the photoproduction of ATP. It would thus appear that in the evolution of photosynthesis the only fundamental biochemical changes have been the sources of hydrogen donors and acceptors, for all modern biochemical evidence indicates that with a very few exceptions[21] the process of light-activated carbon-fixation is the same in all organisms. The first photosynthetic systems were in all probability heterotrophic anaerobes which were able to oxidize carbohydrates and make ATP via tetra-pyrroles. Such forms evolved via mutation and selection pressures to aerobic phototrophs which were able to use the hydrogen from water instead of gaseous hydrogen as an electron donor for the photosynthetic pigments.

Although the utilization of solar energy probably preceded the origin of life, the first autonomous organism to use the sun's energy was probably similar in many respects to modern photosynthetic bacteria.

This particular group of micro-organisms are capable of carrying out a photosynthetic metabolism which differs from green plants and blue-green algae in that it fails to use water as a reductant, and hence does not proceed with the evolution of oxygen, but depends on the presence of extraneous oxidizable compounds which are dehydrogenated with the simultaneous reduction of carbon dioxide. A variety of simple substances can be used as oxidizable substrates, such as hydrogen sulphide, reduced sulphur compounds, molecular hydrogen and several organic acids and alcohols. All can be grown in strictly anaerobic cultures when illuminated, and those members which can be grown in the presence of air, can also be grown in the dark. It would thus appear that members of this group which are photosynthetic are also obligate anaerobes.

The structures associated with photosynthesis in the photosynthetic bacteria are usually in the form of peripherally located vesicular chromatophores or short lamellae, and are quite distinct from the circumferential arrangement of lamellae seen in the blue-green algae. There is, however, a common membranous structure associated with *all* photosynthetic organisms, a membrane bound space, the space being in the shape of a round vesicle in some of the photosynthetic bacteria, and a thylakoid or closed pouch in all other photosynthetic organisms, including blue-green algae. There is a wide variety in the arrangement of the thylakoids in photosynthetic organisms: some are in groups of three or four, others are differentiated into grana and stroma. The work of molecular biologists suggests that the photosynthetic activity is confined to small particles which themselves are located either on the membrane or within the closed pouch.[35, 31] These primary structures associated with photosynthesis appear in all the photosynthetic organisms examined so far, and it would appear unlikely that such a structural unit evolved independently several times giving rise to the various groups of photosynthetic organisms which are now found on earth. Although organisms similar to modern photosynthetic bacteria could have been the first photosynthetically active cells in the anoxic primaeval atmosphere, they would make no contribution to the increase in the oxygen content of the air, and like their modern counterparts would probably only make a small contribution to the amount of organic matter.

The blue-green algae which are the other abundant group of organisms in the fossil flora of pre-Cambrian rocks are most likely to have been the first photosynthetic organisms to produce oxygen. The modern blue-green algae are similar in many respects to the present day bacteria, but show a significant difference in that they produce oxygen during photosynthesis, and possess a photosynthetic apparatus consisting of peripherally arranged lamellae, rather than the vesicular chromatophores commonly found in photosynthetic bacteria.[18]

Gaffron[24] has shown that green plant chloroplasts and possibly blue-green algae evidently contain the components of a thermodynamically simpler synthetic process, very similar to that of anaerobic photosynthetic bacteria. The selective effect of metabolic poisons has shown that part of the photosynthetic mechanism of green plants corresponds to that in anaerobic purple bacteria. The other part of the process, the evolution of O_2 which is not affected by the inhibitors may be thought of as a later evolutionary adaption, and possession of it would permit primitive photosynthetic organisms to evolve O_2, Gaffron postulated that more than 2×10^9 years ago a new version appeared of the then already existing photochemical reaction which presumably was occurring in primitive photosynthetic bacteria, and the overall redox potential gradually shifted to a point where the two mechanisms barely overlapped. The two pigment systems combined and formed an electron transport chain which enabled water to be photochemically decomposed with the consequent evolution of oxygen.

It would seem probable that blue-green algae and bacteria arose from a common ancestor, since it is unlikely that so many features common to the cells of both groups arose independently more than once.

The blue-green algae may have evolved via anaerobic forms which used the reducing gases in the primordial atmosphere to make carbohydrates. Such cells were presumably photo-autotrophs and in the initial anaerobic stages the presence of oxygen may have been an objectional by-product and might have either been accepted by the ferrous ions which are thought to have been abundant in water at that time or would have quickly reacted with partially oxidized rocks.[7] Because the reaction rates of both O and O_3 are many orders of magnitude greater than the reaction rates of O_2 with respect to surface materials, surface oxidation rates even in low concentrations of atmospheric oxygen could exceed present values. There would have been a selection of organisms which were able to transfer hydrogen to free oxygen via tetra-pyrroles (cytochromes) instead of to nitrogen or other inorganic compounds. Gradually free oxygen would have become a tolerable part of the primaeval environment until the gas would have accumulated in the atmosphere on a large scale. A few modern blue-green algae which live on or even in strongly reducing muds are certainly micro-aerophillic, if not completely anaerobic. These forms are colourless and consequently are non-photosynthetic and are probably closely related to bacteria. One species of blue-green algae, an Oscillatoria, has been reported to use H_2S as a hydrogen donor in photosynthesis, which is essence is the photosynthetic mechanism of sulphur bacteria. Such anaerobic forms are admittedly atypical and constitute a group of colourless blue-green algae which share characteristics with both bacteria and the more typical blue-green algae. These colourless blue-green algae are invariably found in bottom muds which are strongly reducing. Certain species of algae including a few which are normally able to produce oxygen during photosynthesis are capable of highly specialized anaerobic respiration for short periods of time.

If the same wide range of photosynthetic and accessory pigments present in modern forms were found in either the hypothesized ancestor of blue-green algae or in the first blue-green algae themselves, this would be an advantage in the development of the photosynthetic machinery. The presence of blue, green, orange and red pigments within the same photosynthetic organism would cover the whole spectral range of white light, and allow the enhancement of photosynthetic reactions by superimposition of colours. Certain modern blue-green algae and bacteria are able to fix atmospheric nitrogen, and possession of such a character would presumably have an adaptive advantage in photosynthetic organisms evolving in a nitrogenous reducing atmosphere.

Taking the ougrp as a whole, modern blue-green algae are more resistant to UV radiation than most other organisms. Extrapolating back to primitive forms, possession of such a characteristic would endow a certain evolutionary advantage in an environment of increased UV radiation flux. Blue-green algal colonies of various kinds cover themselves with mineral particles—either by entrapping and binding extranous grains or by secreting $CaCO_3$ and silica. Such mineralization, which is also seen in the pre-Cambrian stromatolites, may have served to shade the cells and colonies from all radiation wavelengths, including ultra-violet.

The presence of a thick mucilaginous sheath, which would be an advantage in resisting desiccation and ultra-violet radiation, and the ability of some forms to live at high temperatures, as in hot geothermal springs, would be further support for the contention that blue-green algae were among the first photosynthetic organisms. Warm ponds associated with hot springs rich in nutrient mineral and geothermally produced organic compounds and bathed in ultra-violet light for the photochemical synthesis of organic substrates, seem prime candidates for the origin of life, and subsequently, photosynthesis. It has been suggested that life and life processes could not have originated in small ponds because such ponds dry out. But many blue-green algae and some bacteria are able to withstand alternate hydration and dehydration, and by extrapolation back from modern micro-organisms this is likely to have been a characteristic shared with ancient forms. Berkner and Marshall envisage such pools as needing convection currents gentle enough to supply nutrients from the bottom and organic material synthesized photochemically at the surface, but not so strong as to circulate the organisms near the lethal surface. The ancient Bulawayan stromatolites of Southern Rhodesia, which are generally attributed to blue-green algae, are $2 \cdot 7 \times 10^9$ years old, and could well have been the base of such ponds. The stromatolites show no macro-cellular detail and their interpretation as fossils rests on the following three lines of evidence. No known inorganic processes form such structures, blue-green algae produce such forms today, and few well preserved stromatolites do reveal microscopic cellular detail.

Many evolutionary trees have been constructed in an attempt to establish the phylogeny of the higher plants from primitive ancestors. Both bacteria and blue-green algae may have evolved to forms which are now extinct and which in turn may have given rise to the higher plants. The unicellular bacteria may themselves have given rise to filamentous forms similar to modern Ascomycetes, which in turn may have been the progenitors of the fungi. Similarly the unicellular blue-green algae may have developed filamentous, parenchymatous and heterotrichous habits, followed by compartmentation of the specific functional sites within the cell and may in turn have given rise to the other algae and subsequently higher plants.

On both morphological and biochemical grounds the blue-green algae share characteristics with the higher photosynthetic organisms. Bloch[(9)] has pointed out that the universal presence of the lipid α-linolenic in the photosynthetic apparatus of blue-green algae and higher plants and its absence in photosynthetic bacteria, may mark the beginning of separate membrane-bound units of cell function. Similarly, galactosyl glycerides, the carriers for α-linolenate, are only found in photosynthetic organisms which evolve O_2. It would appear that the ability to photolyse water may go hand in hand with the biosynthesis of lipids, which in turn may have paved the way for the specialization and differentiation of cells. It is unlikely, therefore, that the photosynthetic bacteria gave rise to the higher photosynthetic organisms, because of the relative morphological simplicity of the photoactive chromatophores, the absence of oxygen production during photosynthesis, and the differences in both the primary and accessory photosynthetic pigments.

A totally different view on the origin of the photosynthetic apparatus in higher plants is that the blue-green algae through modification and adaptation occasioned by parasitism and symbiosis are the progenitors of the chloroplasts in higher plants. Such an hypothesis would argue that chloroplasts as distinct plant organelles are relics of earlier autonomous structures rather than being created by the association of specific proteins formed in the cell. This hypothesis is not entirely new and was first considered by Mereschkowsky[33] and Famintzin[22] at the beginning of this century, and was briefly alluded to a few years ago by Ris and Plaut[36] and more recently by Kistev, Swift and Bogorad.[26] Recent advances in geology, biochemistry and knowledge of the ultrastructure of cells, indicate that endosymbiosis may again be considered as a possible evolutionary step in the origin of complex cell systems, and the author has recently examined in detail the evidence from the various scientific disciplines.[16, 17, 19] By extrapolating back from what we know about the biology of modern micro-organisms, and using what little evidence there is about the palaeobiology of pre-Cambrian micro-fossils, speculations have been presented concerning the origins of photosynthesis, photosynthetic organisms and possible chloroplasts.

References

1. E. S. Barghoorn and S. A. Tyler, Microorganisms from the Gun Flint Chert. *Science*, **147,** 563–577, 1965.
2. E. S. Barghoorn, W. G. Meinschein and J. W. Schopf, Paleobiology of a pre-Cambrian shale. *Science*, **148,** 461, 1965.
3. E. S. Barghoorn and J. W. Schopf, Microorganisms from the late pre-Cambrian of Central Australia. *Science*, **150,** 337–9, 1965.
4. E. S. Barghoorn and J. W. Schopf, Microorganisms three billion years old from the pre-Cambrian of South Africa. *Science*, **152,** 758–63, 1966.

5. L. V. Berkner and L. C. Marshall, The history of the growth of oxygen in the earth's atmosphere. *Progress in Radio Science*, vol. III. *The Ionosphere*. Elsevier Publishing Company, New York, 1965.
6. L. V. Berkner, and L. C. Marshall, The history of the oxygenic concentration of the earth's atmosphere. *Discussions of Faraday Society*, **37**, 122–41, 1964.
7. L. V. Berkner and L. C. Marshall. On the origin and rise of oxygen concentration in the earth's atmosphere. *Journal of the Atmospheric Sciences* **22**, 225–61, 1965.
8. L. V. Berkner and L. C. Marshall, History of major atmospheric components. *Proc. Nat. Acad. Sci.* **53**, 1169–1226, 1965.
9. K. Bloch, Lipid Patterns in the evolution of Organisms, in *Evolving Genes and Proteins*, ed. by V. Bryson and H. J. Vogal. Academic Press, New York, 1965.
10. M. Calvin. Chemical evolution. Xth int. Bot. Congress General Congress Lecture 4, Edinburgh, 1964.
11. P. E. Cloud and H. Hagen, Electron microscopy of the Gun Flint Microflora. Preliminary results. *Proc. Nat. Acad. Sci.* **54**, 1–8, 1965.
12. P. E. Cloud, Significance of the Gun Flint (pre-Cambrian) microflora. *Science* **148**, 27–35, 1965.
13. P. E. Cloud, J. W. Gruner, and H. Hagen, Carbonaceous rocks of the Soudan Iron Formation. *Science*, **148**, 1713–16, 1965.
14. B. Commoner, Biochemical, biological and atmospheric evolution. *Proc. Nat. Acad. Sci.* **53**, 1183–94, 1965.
15. P. Echlin and I. Morris, The relationship between the blue-green algae and the bacteria. *Biological Rewiews*, **40**, 143–87, 1965.
16. P. Echlin, The cyanophytic origin of higher plant chloroplasts. Proceedings 14th Annual Meeting British Phycological Society London, 1966.
17. P. Echlin, The origins of photosynthesis. *Science Journal*, April 1966.
18. P. Echlin, The blue-green algae. *Scientific American*, June 1966.
19. P. Echlin, The fine structure of Glaucocystis nostochinearum in relation to the cyanophytic origin of chloroplasts. Proceedings 6th International Congress of Electron Microscopy, Kyoto Japan, 1966.
20. G. Eglinton, P. M. Scott, T. Belisky, A. L. Burlingame and M. Calvin, Hydrocarbons of biological origin from a one billion year old sediment. *Science*, **145**, 263–4, 1964.
21. M. C. W. Evans, B. B. Buchanan and O. I. Arnon, A new ferredoxin-dependent carbon reduction cycle in a photosynthetic bacterium. *Proc. Nat. Acad. Sci.* **55**, 928–34, 1966.
22. A. Famintzin, Die Symbiose als Mittel der Synthese von Organismen. *Biologisches Centralblatt* **27**, 353–64, 1907.
23. S. W. Fox. (ed.) *The Origins of Pre-biological Systems*. Academic Press, New York, 1965.
24. H. Gaffron, The role of light in evolution: The transition from a one quantum to a two quantum mechanism. In *The Origins of Pre-biological Systems*, ed. S. W. Fox. Academic Press, New York, 1965.
25. M. F. Glaessner, Pre-Cambrian fossils. *Biol. Reviews*, **37**, 467–94, 1962.
26. N. Kislev, H. Swift, and L. Bogorad, Nucleic acids of chloroplasts and mitochondria in Swiss chard. *J. Cell Biol.* **25**, 327–44, 1965.
27. A. M. Macgregor. A pre-Cambrian algal limestone in S. Rhodesia. *Trans. Proc. Geol. Soc. S. Africa*, **43**, 9–16, 1940.
28. W. G. Meinschein, E. S. Barghoorn and J. W. Schopf, Biological remnants in a pre-Cambrian sediment. *Science*, **145**, 262–3, 1964.
29. W. G. Meinschein, Soudan formation: Organic extracts of early pre-Cambrian rocks. *Science*, **150**, 601–5, 1965.

30. C. Mereschkowsky, Über Natur und Ursprung Chromatophoren im Pflanzenreiche. *Biologisches Centralblatt* **25,** 593–604, 1905.
31. K. Muhlethaler, H. Moor and J. W. Szarkowski, Ultrastructure of the chloroplast lamellae. *Planta* **67,** 305–23, 1965.
32. A. I. Oparin, *The Origin of Life* 3rd ed. New York Academic, Press, 1957.
33. A. I. Oparin, (ed.) *Origin of Life on Earth.* London, Pergamon Press, 1959.
34. J. Oro, D. W. Nooner, A. Zlatkis, S. A. Wikström and E. S. Barghoorn, Hydrocarbons of biological origin in sediments about 2 billion years old. *Science*, **148,** 77–79, 1965.
35. R. B. Park, Substructure of chloroplast lamellae. *J. C. B.* **27,** 151–61, 1965.
36. H. Ris and W. Plaut, Ultrastructure of DNA containing areas in the chloroplasts of chlamydomonas. *J. Cell Biology*, **13,** 383–91, 1962.
37. M. G. Rutten, *The Geological Aspects of the Origin of Life on Earth.* Elsevier, Amsterdam, 1962.
38. J. W. Schopf, E. S. Barghoorn, M. D. Maser, and R. O. Gordon, Electron microscopy of fossil bacteria two billion years old. *Science*, **149,** 1365–7, 1965.
39. S. A. Tyler and E. S. Barghoorn. Occurrence of structurally preserved plants in pre-Cambrian rocks of the Canadian Shield. *Science*, **119,** 606–8, 1954.
40. H. C. Urey, Primitive planetary atmospheres and the origin of life, in *the Origin of Life on Earth*, ed. by A. I. Oparin. London, Pergamon Press, 1959.
41. G. Wald, The origins of life. *Proc. Nat. Acad. Sci.* **52,** 595–611, 1964.

ON THE SYNTHESIS OF PEPTIDES UNDER PRIMITIVE EARTH CONDITIONS

by
J. H. REUTER
Institut für Biologie der Kernforschungsanlage Jülich, Germany

IN recent years it has been clearly established through a series of experiments that essentially all monomers necessary for the origin of living systems could have been formed under conditions assumed to have prevailed on the primitive earth (Oró, 1965). Using gas mixtures according to Urey's model of a reducing primitive atmosphere with electric discharge and electron bombardment or ultraviolet irradiation as energy sources, a series of aminoacids has been produced: glycine, alanine, serine, proline, leucine, isoleucine, aspartic acid, glutamic acid, cysteic acid, lysine and others. Criticism of this model of the nature of the primaeval atmosphere led to experiments with gas mixtures of less reducing character. Although the reaction mechanisms are different, the final steps still lead to the formation of some amino acids: glycine, serine, alanine, aspartic acid, and glutamic acid have been observed (cf. Abelson, 1966).

The formation of the peptide bond is of special interest, since peptides and proteins possess catalytic functions that are basic to the chemistry of the most primitive living system. Peptide formation requires a dehydrating mechanism. This can be provided by working under anhydrous conditions at temperatures above 100° C (Fox and Harada, 1960). A geochemically more reasonable approach is the use of certain dehydrating agents in dilute aqueous solutions. Recently, Schimpl *et al.* (1965) have demonstrated the formation of cyanamide, and its dimer dicyandiamide, upon ultraviolet irradiation of dilute cyanide solutions. Steinmann *et al.* (1964, 1965a, b) and Ponnamperuma and Peterson (1965) were able to show that dicyandiamide promotes certain dehydration condensations under possible primitive Earth conditions, notably the formation of peptide bonds in dilute aqueous solutions on exposure to ultraviolet light. In 1955 Akabori proposed a somewhat different approach to the problem of peptide formation by the polymerization of aminoacetonitrile on a solid surface, followed by the hydrolysis of the polymer to polyglycine and ammonia. Hanafusa and Akabori (1959) succeeded in demonstrating that under anhydrous conditions and at temperatures considerably above 100° C di- and tri-glycyl peptides were formed from aminoacetonitrile adsorbed on clay. In his work on the mechanism of synthesis of amino acids by electric discharges, Miller (1957) showed that the reaction in the gaseous phase was essentially a Strecker synthesis and that the resulting amino-nitriles were hydrolysed to the amino acids in the boiling, alkaline aqueous phase. Under less hydrolysing conditions, i.e. in a primitive ocean where the buffering system of weathered silicates would maintain the pH at about 8–9 (Sillén, 1961; Holland, 1965), the nitriles would have a considerable lifetime. The experiments reported here study possible condensation reactions of aminonitriles in dilute aqueous solutions on exposure to ultraviolet light.

In a typical experiment, about 230 ml of a solution containing aminoacetonitrile hydrochloride (supplied by Th. Schuchardt GmbH., München, Germany) and dicyandiamide were used. Their concentrations were of the order of 10^{-2} mole/litre. The dicyandiamide was prepared from cyanamide (m.p. 44° C), and had a melting point of 209–210° C. Dissolved air was removed by bubbling a stream of methane through the solution for 20 min. The sample was then transferred to the cooling jacket of a photochemical reactor† and an all-glass centrifugal pump (KPG-Umlaufpumpe UP 60, Schott + Gen., Mainz, Germany), which was connected with inlet and outlet of the lamp cooling jacket, thus yielding a closed pumping circuit. (This arrangement had the advantage that pumping through the cooling jacket during the irradiation assured that in experiments with clay mineral suspensions all parts of the optically very dense solution were forced through the narrow space around the ultraviolet light source.) The sample was then irradiated for 48 hours. The temperature during the irradiation was slightly above room temperature. The pH of the solution was about 5. The reaction mixture, which had turned dark yellow during the irradiation, was then concentrated to 50 ml under reduced pressure.

† Hanovia 1 l. photochemical reactor, with a low pressure mercury resonance lamp. The cooling jacket around the lamp has windows made of high purity synthetic quartz, which transmits over 95% of the ultraviolet light above 1800 Å.

The dicyandiamide was precipitated with silver acetate, any residual silver was then removed with N/10 hydrochloric acid. After adjusting to pH 1 with more HCl, the solution was refluxed for 1 hour in order to hydrolyze the imino group of the polymerized product. The sample was then neutralized with sodium hydroxide and further concentrated to a final volume of 4 ml. During and after the concentration procedures, dark colloidal matter precipitated from the solution.

In the experiments with clay minerals, 2·5 g of montmorillonite was added to the solution of aminoacetonitrile and dicyandiamide. The montmorillonite had been prepared in the following way: raw clay was fractionated by sedimentation. The fraction of particle size < 6 micron was brought to the sodium form by equilibration with sodium chloride. Then it was carefully washed 3 times with 30% hydrogen peroxide at 90° C in order to destroy any residual organic matter, and finally dried at 60° C in an oven. The irradiation procedure was the same as in the experiment without montmorillonite. After the irradiation the clay minerals were removed by centrifugation. However, with montmorillonite the reaction solution was not, or was only faintly, discoloured. Instead, the clay fraction had turned dark brown from an initial almost white appearance. The supernatant liquid was treated in the same fashion as described above. The dark colloidal precipitate, however, was not observed. Two irradiation experiments in the dry state were performed. 500 mg aminoacetonitrile was dissolved in an aqueous suspension containing 2·5 g montmorillonite in a 2-l. flask. The suspension was evaporated to complete dryness under reduced pressure in a rotating evaporator, thus spreading a thin clay mineral film over the inner surface of the flask. This film was irradiated in a nitrogen atmosphere with a low pressure mercury resonance lamp for 85 hours. The originally white montmorillonite film turned brown. It was resuspended and succesively extracted with water, N/10 hydrochloric acid and N/10 sodium hydroxide. The combined extracts were refluxed at pH 1 for 1 hour. They were further treated and analysed in the same way as described above. In another experiment the dry montmorillonite film was irradiated for 15 hours with X-rays from the outside, while the flask was being rotated below the X-ray source. No discolouring appeared. The extraction and analysis procedure was the same as in the other experiments.

An aliquot of each concentrate was chromatographed on a Sephadex G-10 column ($0{\cdot}9 \times 130$ cm) with 0·1 M pyridinacetate buffer. This gave a crude prefractionation and desalting. The Sephadex fractions 0–37 ml,

38–39 ml, and 40–41 ml were concentrated to 0·25 ml and further chromatographed in parallel with true glycine, glycylglycine and diglycylglycine standard solution (supplied by Th. Schuchardt GmbH, München, Germany) by high voltage electrophoresis on paper at 75 V/cm with an acetic acid-formic acid-water buffer at pH 2·1, which gave complete resolution of the three compounds. The spots were detected with a modified ninhydrin-collidine reagent that yielded differential colour development for the three compounds under investigation. This, as well as the position of the spots relative to the authentic standards, was used for identification. For further confirmation of the identity of the reaction products, their DNP (2,4 dinitrobenzene) and Dansyl (1-dimethylamino naphthalene-5-sulphonyl) derivatives were formed. For the preparation of the DNP derivatives the following microtechnique was applied (Waley, 1956, modified): 100 microlitre aliquots of the concentrated Sephadex fractions 0–37 ml were lyophilized and the residue dissolved in 10 μl water. Then 0·1 ml triethylamine carbonate buffer (pH 8·3) and 10 mg 1-fluoro-2,4-dinitrobenzene in 0·25 ml ethanol were added. The solution was kept for about 3 hours in the dark, and the ethanol distilled under reduced pressure. More buffer (0·2 ml) was added, and the excess reagent extracted with ether. The solution was then evaporated to dryness under reduced pressure. The residue was dissolved in 0·2 ml N/10 HCl and extracted with ethylacetate. The combined extracts were evaporated to dryness *in vacuo* and finally taken up in 0·1 ml acetone. Standard solutions of DNP derivatives of glycine, glycylglycine, and glycylglycylglycine were prepared in the same fashion in parallel. Two-dimensional polyamide thin layer chromatography (MN Polyamide for TLC; Macherey, Nagel u. Co Düren, Germany) was applied, the solvents being in the first dimension dimethylformamide-acetic acid-water-ethanol (1:2:6:4), in the second dimension 2-butanone-chloroform-formic acid (3:7:1) (Wang and Huang, 1965). The samples under investigation were run together with the DNP standard solutions on two plates in the same tank. Ultraviolet contact photographs were made and the spots identified by coincidence of the spots on sample and standard chromatograms. Co-chromatography of samples and standards from the same origin was also used for identification. The Dansyl derivatives were prepared in the following way (Gray and Hartley, 1963): 100 μl aliquots of Sephadex fractions 0–37 ml were lyophilized and the residue dissolved in 0·1 ml 0·1 M $NaHCO_3$. Then 0·1 ml Dansyl reagent (6 mg of 1-dimethyl-aminonaphthalen-5-sulphonyl chloride in 1 ml acetone) was added. The mixture was allowed to react overnight. Then more acetone (0·8 ml) was added and the sample was centrifuged. Standard solutions of Dansyl derivatives of glycine, glycylglycine, and glycylglycylglycine were prepared in parallel in a similar way. Thin-layer chromatography (Deyl and Rosmus, 1965) was performed on activated Silicagel G layers (Macherey, Nagel u. Co. Düren, Germany). For development, the following solvent systems were used: chloroform-ethanol-acetic acid (38:4:3), n-butanol-pyridine-acetic acid-water (15:10:3:12), benzene-pyridine-acetic acid (16:4:1). The spots were detected by their yellow fluorescence in ultraviolet light. Identification of the compounds was made by comparison with standards run in parallel on the same chromatogram, as well as by co-chromatography of sample and standard from the same origin.

From the electropherogram it was found that the Sephadex fraction 0–37 ml contained glycine, diglycine and triglycine. These results were confirmed from the thin-layer chromatograms of both the DNP and Dansyl derivatives. The presence of glycine would be expected from the hydrolysis of aminoacetonitrile during the irradiation and subsequent refluxing at pH 1; the successive Sephadex fractions contained less peptide and more glycine. No peptide could be found if the hydrolysis after irradiation was omitted. This indicated that the bulk of glycyl peptide was not formed via dehydration condensation of glycine resulting from the hydrolysis of aminoacetonitrile in the course of the irradiation, but rather suggested a condensation of the aminoacetonitrile itself to an amidine type oligomer:

$$n(H_2H\text{—}CH_2\text{—}CN) \rightarrow (\text{—}NH\text{—}CH_2\text{—}\underset{\underset{NH}{\|}}{C}\text{—})_n$$

No glycyl peptides could be detected when dicyandiamide was absent during the irradiation. The role of dicyandiamide in the condensation of aminoacetonitrile is not understood, since dicyandiamide has only been shown to promote certain dehydration condensations in aqueous solutions (Steinman *et al.*, 1964, 1965a, b).

The yield of di- and tri-peptides from the starting material aminoacetonitrile was only about 0·1%. But since only less than 2% of glycine was formed by the hydrolysis of aminoacetonitrile during the irradiation, the amounts of both diglycine and triglycine relative to glycine are approximately more than 5% for each peptide. With higher pH values (up to 11) the yields of peptides decreased. The presence of the clay mineral montmorillonite did not appear to enhance the yield of glycyl peptides. However, one interesting observation was made: In the experiments without montmorillonite, the formation of black colloidal precipitate continued in the final concentrates. Analysis for peptides on subsequent days after irradiation showed a considerable decrease in the amount of peptide with increasing amount of black polymer in the vial. After several days virtually no peptide could be detected. In the experiment with montmorillonite present during the irradiation, the clay minerals adsorbed dark material, and consequently no precipitate formed in the concentrates. The concentration of peptide did not decrease. This observation could be of interest when considering the role of clay minerals in a primitive earth environment. It could mean that substances incompatible with each other would be separated by selective adsorption to clay minerals. Thus, removal of compounds from the primitive oceans by adsorption does not in all cases imply a limitation on the production and preservation of organic compounds in any primitive "soup" (Abelson, 1966).

Extraction of the montmorillonite with dilute acid and alkali after the irradiation experiment yielded only traces of ninhydrin-positive substances. No peptide was found in this "sediment". The irradiation of aminoacetonitrile in the dry state adsorbed on clay yielded small amounts of of glycylglycine only. In these cases the condensation of aminoacetonitrile occurred without the presence of dicyandiamide. These findings compare well with the results of Hanafusa and Akabori (1959), who assumed an activation of the starting material by the adsorption on the clay. The results of this study agree with recent findings of other workers (Steinman *et al.*, 1965 a, b; Ponnamperuma and Peterson, 1965), that peptide synthesis under conditions possible on the primitive earth is not a rare event. Yields of several percent relative to the starting material have been reported. Of special interest is the finding of Steinman *et al.* (1965 b) that dicyanamide promotes the condensation of triglycine and glycine to tetraglycine. Since in dilute aqueous solutions glycil peptides seem to be more readily formed than peptides with different amino acid residues, there still exists the problem of the formation of the latter.

Akabori (1959) has demonstrated some possible mechanisms for the introduction of side chains into polyglycine, giving rise to aminoacid residues of the same stereochemical configuration throughout one single peptide chain. From here the step to a primitive enzyme appears to be relatively short. Eck and Dayhoff (1966) proposed a detailed reconstruction of the evolutionary history of present-day ferredoxin, which occurs in primitive anaerobic organisms and has functions basic to cell chemistry. According to their analysis, ferredoxin had a shorter ancestor which was based on a repeating sequence of the amino acids alanine, aspartic acid, serine, and glycine. Thus the synthesis of peptides under primitive earth conditions supports the hypothesis of chemical evolution.

References

Abelson, P. H. (1966) Chemical events on the primitive earth. *Proc. Nat. Acad. Sci. U.S.* **55,** 1365–1372.

Akabori, S. (1959) On the origin of the fore-protein, in *The Origin of Life on the Earth*, I.U.B. Symp. Ser., vol. I, pp. 189–196. Pergamon Press London.

Deyl, Z. and Rosmus, J. (1965) Thin layer chromatography of Dansyl amino acid derivatives. *J. Chromatog.* **20,** 514–520.

Eck, R. V. and Dayhoff, M. O. (1966) Evolution of the structure of ferredoxin based on living relics of primitive amino acid sequences. *Science*, **152,** 363–365.

Fox, S. W. and Harada, K. (1960) The thermal copolymerization of amino acids common to protein. *J. Amer. Chem. Soc.* **82,** 3745–3751.

Gray, W. R. and Hartley, B. S. (1963) A fluorescent end-group reagent for proteins and peptides. *Biochem. J.* **89,** 59 P.

Hanafusa, H. and Akabori, S. (1959) Polymerization of amino-acetonitrile. *Bull. Chem. Soc. of Japan*, **32,** 626–630.

Holland, H. D. (1965) The history of ocean water and its effect on the chemistry of the atmosphere. *Proc. Nat. Acad. Sci. U.S.* **53,** 1169–1183.

Miller, S. L. (1957) The mechanism of synthesis of amino acids by electric discharges. *Biochem. Biophys. Acta*, **23,** 480–489.

Oró, J. (1965) Stages and mechanisms of prebiological organic synthesis, in *The Origins of Prebiological Systems and of Their Molecular Matrices* pp. 137–162. ed. S.W. Fox. Academic Press New York.

Ponnamperuma, C. and Peterson, E. (1965) Peptide synthesis from amino acids in aqueous solution. *Science*, **147,** 1572–1574.

Schimpl, A., Lemmon, R. M. and Calvin, M. (1965) Cyanamide formation under primitive earth conditions. *Science*, **147,** 149–150.

Sillén, L. G. (1961) The physical chemistry of sea water, in *Oceanography*, pp. 549–581. ed. Mary Sears. Washington D.C.: A.A.A.S. Public. No. 67.

Steinman, G., Lemmon, R. M. and Calvin, M. (1964) Cyanamide: a possible key compound in chemical evolution. *Proc. Nat. Acad. Sci. U.S.* **52,** 27–30.

Steinman, G., Lemmon, R. M. and Calvin, M. (1965a) Dicyandiamide: possible role in peptide synthesis during chemical evolution. *Science*, **147,** 1574–1575.

Steinman, G., Kenyon, D. H. and Calvin, M. (1965b) Dehydration condensation in aqueous solution. *Nature*, **206,** 707–708.

Waley, S. G. (1956) Acidic peptides of the lens. *Biochem. J.* **64,** 715–726.

Wang, K. and Huang, J. M. (1965) Polyamide layer chromatography of DNP amino acids. *Nature*, **208,** 281–282.

AUTHOR INDEX

SUBJECT INDEX

OTHER TITLES IN THE SERIES IN EARTH SCIENCES

Vol. 1. BENIOFF *et al.—Contributions in Geophysics*
Vol. 2. SWINEFORD—*Clays and Clay Minerals (Sixth Conference)*
Vol. 3. GINZBURG—*Principles of Geochemical Prospecting*
Vol. 4. WAIT—*Overvoltage Research in Geophysical Applications*
Vol. 5. SWINEFORD—*Clays and Clay Minerals (Seventh Conference)*
Vol. 6. TYUTYUNOV—*An Introduction to the Theory of the Formation of Frozen Rocks*
Vol. 7. KRINOV—*Principles of Meteoritics*
Vol. 8. NALIVKIN—*The Geology of the U.S.S.R.*
Vol. 9. SWINEFORD—*Clays and Clay Minerals (Eighth Conference)*
Vol. 10. POKORNY—*Principles of Zoological Micropalaeontology, Volume I*
Vol. 11. SWINEFORD—*Clays and Clay Minerals (Ninth Conference)*
Vol. 12. SWINEFORD—*Clays and Clay Minerals (Tenth Conference)*
Vol. 13. BRADLEY—*Clays and Clay Minerals (Eleventh Conference)*
Vol. 14. ROSENQVIST and GRAFF-PETERSEN—*International Clay Conference 1963, Volume 11*
Vol. 15. COLOMBO and HOBSON—*Advances in Organic Geochemistry*
Vol. 16. BREGER—*Organic Geochemistry*
Vol. 17. HELGESON—*Complexing and Hydrothermal Ore Deposition*
Vol. 18. BATTEY and TOMKEIEFF—*Aspects of Theoretical Mineralogy in the U.S.S.R.*
Vol. 19. BRADLEY—*Clays and Clay Minerals (Twelfth Conference)*
Vol. 20. POKORNY—*Principles of Zoological Micropalaeontology, Volume II*
Vol. 21. ROSENQVIST and GRAFF-PETERSEN—*International Clay Conference 1963, Volume 11*
Vol. 22. YERMAKOV *et al.—Research on the Nature of Mineral-Forming Solutions*
Vol. 23. BRIDGE and BRIDGE—*Clays and Clay Minerals (Index of First to Tenth Conferences)*
Vol. 24. HOBSON and LOVIS—Advances in *Organic Geochemistry, 1964*
Vol. 25. BRADLEY—*Clays and Clay Minerals (Thirteenth Conference)*
Vol. 26. BAILEY—*Clays and Clay Minerals (Fourteenth Conference)*
Vol. 27. BAILEY—*Clays and Clay Minerals (Fifteenth Conference)*
Vol. 28. MANSKAYA and DROZDOVA—*Geochemistry of Organic Substances*
Vol. 29. ANDREEV *et al.—Transformation of Petroleum in Nature*
Vol. 30. AHRENS—*Origin and Distribution of the Elements*
Vol. 31. SCHENK and HAVENAAR—*Advances in Organic Geochemistry, 1968.*

"Franklin" Printing House, Budapest, Hungary